CAMBRIDGE STUDIES IN
ADVANCED MATHEMATICS : 46

INTRODUCTION TO ANALYTIC AND PROBABILISTIC NUMBER THEORY

Already published

1 W.M.L. Holcombe *Algebraic automata theory*
2 K. Petersen *Ergodic theory*
3 P.T. Johnstone *Stone spaces*
4 W.H. Schikhof *Ultrametric calculus*
5 J.-P. Kahane *Some random series of functions, 2nd edition*
6 H. Cohn *Introduction to the construction of class fields*
7 J. Lambek & P.J. Scott *Introduction to higher-order categorical logic*
8 H. Matsumura *Commutative ring theory*
9 C.B. Thomas *Characteristic classes and the cohomology of finite groups*
10 M. Aschbacher *Finite group theory*
11 J.L. Alperin *Local representation theory*
12 P. Koosis *The logarithmic integral I*
13 A. Pietsch *Eigenvalues and s-numbers*
14 S.J. Patterson *An introduction to the theory of the Riemann zeta-function*
15 H.J. Baues *Algebraic homotopy*
16 V.S. Varadarajan *Introduction to harmonic analysis on semisimple Lie groups*
17 W. Dicks & M. Dunwoody *Groups acting on graphs*
18 L.J. Corwin & F.P. Greenleaf *Representations of nilpotent Lie groups and their applications*
19 R. Fritsch & R. Piccinini *Cellular structures in topology*
20 H Klingen *Introductory lectures on Siegel modular forms*
21 P. Koosis *The logarithmic integral II*
22 M.J. Collins *Representations and characters of finite groups*
24 H. Kunita *Stochastic flows and stochastic differential equations*
25 P. Wojtaszczyk *Banach spaces for analysts*
26 J.E. Gilbert & M.A.M. Murray *Clifford algebras and Dirac operators in harmonic analysis*
27 A. Frohlich & M.J. Taylor *Algebraic number theory*
28 K. Goebel & W.A. Kirk *Topics in metric fixed point theory*
29 J.F. Humphreys *Reflection groups and Coxeter groups*
30 D.J. Benson *Representations and cohomology I*
31 D.J. Benson *Representations and cohomology II*
32 C. Allday & V. Puppe *Cohomological methods in transformation groups*
33 C. Soulé et al *Lectures on Arakelov geometry*
34 A. Ambrosetti & G. Prodi *A primer of nonlinear analysis*
35 J. Palis & F. Takens *Hyperbolicity and sensitive chaotic dynamics at homoclinic bifurcations*
36 M. Auslander, I. Reiten & S. Smalo *Representation theory of Artin algebras*
37 Y. Meyer *Wavelets and operators*
38 C. Weibel *An introduction to homological algebra*
39 W. Bruns & J. Herzog *Cohen-Macaulay rings*
40 V. Snaith *Explicit Brauer induction*
42 E.B. Davies *Spectral theory and differential operators*
43 J. Diestel, H. Jarchow & A. Tonge *Absolutely summing operators*
44 P. Mattila *Geometry of sets and measures in Euclidean spaces*
45 R. Pinsky *Positive harmonic functions and diffusion*

Introduction to Analytic and Probabilistic Number Theory

Gérald Tenenbaum
Professor at Université Henri Poincaré–Nancy I

CAMBRIDGE
UNIVERSITY PRESS

Published by the Press Syndicate of the University of Cambridge
The Pitt Building, Trumpington Street, Cambridge CB2 1RP
40 West 20th Street, New York, NY 10011-4211, USA
10 Stamford Road, Oakleigh, Victoria 3166, Australia

Translated by C.B. Thomas, University of Cambridge

First published in English 1995

Printed in Great Britain at the University Press, Cambridge

Library of Congress cataloguing in publication data available
A catalogue record for this book is available from the British Library

ISBN 0 521 41261 7 hardback

TAG

À Catherine Jablon,

pour la douceur du jour,
ce bouquet de symboles
dont ta conversation
éclaire les secrets.

Contents

Preface

Arising, as it does, from graduate courses given in Bordeaux, Paris and Nancy during the past fifteen years, this book is a revised, updated, and slightly expanded version of the text which appeared (in French) as issue number 13 of the series *Publications de l'Institut Élie Cartan* in the autumn of 1990. It was written with the double purpose of providing younger researchers with a self-contained account of analytic methods in number theory, and their elders with a source of references for certain basic questions. Such an undertaking entails making some choices. In general, these have been a matter of rather arbitrary personal aesthetic considerations but, of course, the available choices also have been subject to the undeniable constraints of ignorance.

These twin motivations have led us to employ a slight variation of the traditional subdivision into text, notes and exercises. Thus, the main text, although generally restricted to statements proved in full detail, may also contain comments on additional references when they provide a useful background for a first reading. Conversely, the notes often give way to statements, and even proofs, of related results which may be omitted on first contact. In a similar way, the exercises serve a double purpose. Some are traditionally designed to facilitate the mastery of concepts introduced in the text itself. Others, mainly in Part III, lead to genuine research results which are sometimes new. In this context we have tried to break away from an unfortunate modern tendency by only proposing exercises which are soluble without excessive ingenuity or exceptional technical skill. We usually avoid questions to which the answers are not provided; the results aimed at are systematically stated at the outset and the main steps are indicated. This part of the book may therefore be used, even without making the effort of solving the problems, as an informal source of references. Complete solutions will appear shortly as a joint volume with my colleague Jie Wu, published by the *Société Mathématique de France.*

We have been guided by the constant concern of emphasising the methods rather more than the results, a strategy which we believe to be specifically heuristic. This has led to the somewhat artificial subdivision into three parts, respectively devoted to elementary, complex-analytical, and probabilistic methods. It will be easy to criticise this taxonomy: what makes the van der Corput method, which rests upon the Poisson summation formula, more elementary than that of Selberg–Delange, which appeals to contour integration? Why qualify as probabilistic the saddle-point method, whose initial step amounts to an inverse Laplace integral? and so on... One could multiply the examples of inconsistency with respect to this or that criterion, and it is obvious that the choices that have been made rest on rather questionable grounds. Thus, we regard as elementary a method which exclusively employs the *real* variable, and we choose to view the saddle-point method as 'probabilistic' as much because it is an ever-present tool in probability theory, as for being a specific method implemented to solve problems in probabilistic number theory... One might as well say at the very outset that the classification made in this book is anything but a Bourbakist choice. Our ambition is limited to the mere wish that it might, at least for a while, light the way for the neophyte.

Without aiming at complete originality, the text tries to avoid well-trodden paths. We have, when this seemed desirable and indeed possible, rethought the presentation of classical results: either by employing new approaches (such as Nair's method for Chebyshev's estimates), or by occasionally introducing technical simplifications that are invisible in the table of contents but will hopefully be useful to the active reader.

Some developments are, however, new or unpublished in book form. This concerns in particular the following: the uniform results derived from the Selberg–Delange method for the asymptotic study of the coefficients of Dirichlet series that are 'close' to a complex power of the Riemann zeta function (Chapter II.5); the version with explicit remainder term of the Ikehara–Ingham Tauberian theorem (§ II.7.4); and the study of the sieve function $\Phi(x, y)$ via the saddle-point method (Chapter III.6). The effective form of Ikehara's theorem turns out to be closely connected to the Berry–Esseen inequality. This connection can actually be viewed almost as a conceptual identity, which we continue to find fascinating. Besides, a concern of complementarity with respect to the existing literature (and especially the fine book of Elliott) has influenced some of our decisions, such as the choice of the method of proof for the theorems of Erdős–Wintner, Erdős–Kac or Halász—see Chapter III.4. As for this last result, the innovative method of Halász has been reappraised in the light of Montgomery's improvements.

This book owes a great deal to a number of friends and colleagues who provided scientific and linguistic assistance at various stages of the preparation of the manuscript. In the first rank of these I would like to warmly thank Mohan Nair who steadily reread and corrected the translation, and answered a non-enumerable set of silly questions. This English version would never have appeared without his invaluable help. It is also a special pleasure to thank here Michel Balazard, Gautami Bhomwik, Régis de la Bretèche, Paul Erdős, Misha Katz, Michel Mendès France, Olivier Ramaré, Jean-Luc Rémy, Imre Ruzsa, Patrick Sargos, András Sárközy, Marijke Wijsmuller and Jie Wu: as long as the list of errata might turn out to be, it would have been much longer without their aid. Finally, I want to express my deep gratitude to David Tranah of Cambridge University Press for his unfailing, effective help and infinite patience in the process of editing this book.

Nancy, G.T.

Notation

The following notation and conventions are used systematically in the text.

We use $\mathbb{Z}$, $\mathbb{Q}$, $\mathbb{R}$, $\mathbb{C}$ in their usual meaning, and $\mathbb{N}$ to denote the set of non-negative integers.

The letter p, with or without subscript, denotes a prime number.

$a|b$ means: a divides b; $p^\nu \| a$ means: $p^\nu | a$ but $p^{\nu+1} \nmid a$; $a|b^\infty$ means: $p|a \Rightarrow p|b$.

$P^+(n)$ (resp. $P^-(n)$) denotes the largest (resp. smallest) prime factor of the integer $n > 1$. By convention $P^+(1) = 1$, $P^-(1) = \infty$.

The integer and fractional parts of the real number x are respectively written $[x]$ and $\{x\}$. We write $\|x\| := \min_{n \in \mathbb{Z}} |x - n|$.

When the letter s denotes a complex number, we implicitly define the real numbers σ and τ by the relation $s = \sigma + i\tau$.

We write $x^+ := \max(x, 0)$, and we put

$$e(x) := \exp\{2\pi i x\} \quad (x \in \mathbb{R}), \qquad \log^+ x := \max\{0, \log x\} \quad (x > 0).$$

Furthermore, we denote by $\log_k$ the k-fold iterated logarithm.

We use interchangeably Landau's notation $f = O(g)$ and Vinogradov's $f \ll g$ to both mean that $|f| \leq C|g|$ for some positive constant C, which may be absolute or depend upon various parameters, in which case these may be indicated in subscript. Moreover, we write $f \asymp g$ to indicate that $f \ll g$ and $g \ll f$ hold simultaneously. We draw the reader's attention to the fact that we have therefore extended the common use of these symbols to complex-valued functions.

We denote the number of elements of a finite set $\mathcal{A}$ either by $\operatorname{card} \mathcal{A}$ or by $|\mathcal{A}|$.

We list below page numbers where various notations in the body of the text are introduced.

Part I

Elementary methods

I.0

Some tools from real analysis

§ 0.1 Abel summation

Classically one calls Abel summation the process whereby one transforms a finite sum of products of two terms by means of the partial sums of one of them. Thus, by letting $A_0 = 0$, we have $A_n = \sum_{m=1}^{n} a_m \quad (n \geq 1)$,

$$\begin{aligned}\sum_{n=1}^{N} a_n b_n &= \sum_{n=1}^{N} (A_n - A_{n-1})\, b_n = \sum_{n=1}^{N} A_n b_n - \sum_{n=1}^{N-1} A_n b_{n+1} \\ &= \sum_{n=1}^{N-1} A_n (b_n - b_{n+1}) + A_N b_N.\end{aligned}$$

In the setting of the Stieltjes integral, Abel summation takes the innocuous form of partial integration, and therefore it is sometimes refered to as partial summation. It constitutes a simple but effective tool for handling arithmetical sums. The reader can find the essential notions concerning the Stieltjes integral in Chapter 1 of Widder's book (1946).

Theorem 1. *Let $\{a_n\}_{n=1}^{\infty}$ be a sequence of complex numbers. Set*

$$A(t) = \sum_{n \leq t} a_n \qquad (t > 0).$$

Let $b(t)$ be a continuously differentiable function on the interval $[1, x]$. Then we have

$$\sum_{1 \leq n \leq x} a_n b(n) = A(x) b(x) - \int_1^x A(t)\, b'(t) \mathrm{d}t.$$

Proof. The quantity to calculate is

$$\int_{1-}^{x} b(t)\, \mathrm{d}A(t) = \Big[A(t)\, b(t)\Big]_{1-}^{x} - \int_1^x b'(t) A(t)\, \mathrm{d}t,$$

by partial integration. This gives the required result.

The same formalism allows us to easily recover other classical results. Let us begin with two important examples.

Theorem 2 (Comparison of a sum and an integral). *Let f be a real monotonic function on the interval $[a, b]$, with $a, b \in \mathbb{Z}$. Then there exists some real number $\theta = \theta(a, b)$, $0 \le \theta \le 1$, such that*

$$\sum_{a<n\le b} f(n) = \int_a^b f(t)\,\mathrm{d}t + \theta\big(f(b) - f(a)\big).$$

Proof. Introduce the Stieltjes integral of f with respect to the measure d[t]. Then

$$\sum_{a<n\le b} f(n) - \int_a^b f(t)\,\mathrm{d}t = \int_a^b f(t)\,\mathrm{d}[t] - \int_a^b f(t)\,\mathrm{d}t = -\int_a^b f(t)\,\mathrm{d}\{t\}.$$

Integrating this expression by parts gives

$$\Big[-f(t)\{t\}\Big]_a^b + \int_a^b \{t\}\,\mathrm{d}f(t) = \int_a^b \{t\}\,\mathrm{d}f(t).$$

Let us suppose for example that f is increasing, so that the measure $\mathrm{d}f$ is positive. The last integral then has the value $\theta(f(b) - f(a))$ with $0 \le \theta \le 1$, from which the result follows.

Corollary 2.1. *For $n \ge 1$, we have*

$$\log n! = n \log n - n + 1 + \theta \log n$$

with $\theta = \theta_n \in [0, 1]$.

Theorem 3 (Second mean value formula). *Let $f(t)$ be monotonic and $g(t)$ be integrable on the real interval $[a, b]$. Then there exists a real number ξ, with $a \le \xi \le b$, such that*

$$\int_a^b f(t)\,g(t)\,\mathrm{d}t = f(a)\int_a^\xi g(t)\,\mathrm{d}t + f(b)\int_\xi^b g(t)\,\mathrm{d}t.$$

Proof. Write $G(t) = \int_a^t g(v)\,\mathrm{d}v$. The left-hand side equals

$$\int_a^b f(t)\,\mathrm{d}G(t) = G(b)\,f(b) - \int_a^b G(t)\,\mathrm{d}f(t),$$

by partial integration. Let us suppose for example that f is increasing. Then $\mathrm{d}f(t)$ is a positive Stieltjes measure, and since $G(t)$ is continuous the last integral equals $G(\xi)\,(f(b) - f(a))$ for some ξ, $a \le \xi \le b$. The desired result follows by rearranging the terms.

§ 0.2 The Euler–Maclaurin summation formula

Consider the sequence $\{b_r(x)\}_{r=0}^{\infty}$ of polynomials defined on $[0,1]$ by the conditions

$$b_0(x) \equiv 1, \tag{1}$$

$$b_r'(x) \equiv r\, b_{r-1}(x) \qquad (r \geq 1), \tag{2}$$

$$\int_0^1 b_r(x)\mathrm{d}x = 0 \qquad (r \geq 1). \tag{3}$$

One easily verifies that these assumptions imply the identity

$$\sum_{r=0}^{\infty} b_r(x)\,\frac{y^r}{r!} = \frac{y\,e^{xy}}{e^y - 1}$$

allowing us to calculate the b_r. We have

$$\begin{aligned} b_0(x) &= 1 & b_3(x) &= x^3 - \tfrac{3}{2}x^2 + \tfrac{1}{2}x \\ b_1(x) &= x - \tfrac{1}{2} & b_4(x) &= x^4 - 2x^3 + x^2 - \tfrac{1}{30} \\ b_2(x) &= x^2 - x + \tfrac{1}{6} & b_5(x) &= x^5 - \tfrac{5}{2}x^4 + \tfrac{5}{3}x^3 - \tfrac{1}{6}x \end{aligned} \quad .$$

One then defines the rth Bernoulli function $B_r(x)$ as the periodic function of period 1 which coincides with b_r on $[0,1[$. Set

$$B_r := B_r(0).$$

B_r is the rth Bernoulli number. It is easy to see that $B_{2r+1} = 0$ for $r \geq 1$. One has the numerical values

r	0	1	2	4	6	8	10
B_r	1	$-\frac{1}{2}$	$\frac{1}{6}$	$-\frac{1}{30}$	$\frac{1}{42}$	$-\frac{1}{30}$	$\frac{5}{66}$

Let f be a numerical function of class C^{k+1} on the interval $[a,b]$, with $a, b \in \mathbb{Z}$. Since $B_1(x) = \{x\} - \frac{1}{2}$, we can write

$$\sum_{a<n\leq b} f(n) = \int_a^b f(t)\,\mathrm{d}[t] = \int_a^b f(t)\,\mathrm{d}t - \int_a^b f(t)\,\mathrm{d}B_1(t).$$

Let us calculate the last integral by parts:

$$\int_a^b f(t)\mathrm{d}B_1(t) = B_1\cdot(f(b) - f(a)) - \int_a^b B_1(t)f'(t)\,\mathrm{d}t$$
$$= B_1\cdot(f(b) - f(a)) - \frac{1}{2!}\int_a^b f'(t)\,\mathrm{d}B_2(t).$$

Indeed, it can be easily verified that $B_2(t)$ is continuous on $\mathbb{R}$ and differentiable on $\mathbb{R}\smallsetminus\mathbb{Z}$, where it satisfies $B_2'(t) = 2B_1(t)$. Furthermore, for $r \geq 3$, $B_r(t)$ is differentiable on the entire real line and satisfies

$$B_r'(t) = rB_{r-1}(t).$$

We can then calculate the integral with respect to $B_2(t)$ by a new partial integration, involving $B_3(t)$. Iterating the process, we obtain in this way the following famous theorem.

Theorem 4 (Euler–Maclaurin summation formula). *For any integer $k \geq 0$ and for any function f of class C^{k+1} on $[a,b]$, $a, b \in \mathbb{Z}$, we have*

$$\sum_{a<n\leq b} f(n) = \int_a^b f(t)\,\mathrm{d}t + \sum_{r=0}^{k} \frac{(-1)^{r+1}B_{r+1}}{(r+1)!}\left(f^{(r)}(b) - f^{(r)}(a)\right)$$
$$+ \frac{(-1)^k}{(k+1)!}\int_a^b B_{k+1}(t)\, f^{(k+1)}(t)\,\mathrm{d}t.$$

By way of application, we give an estimate of the partial sums of the harmonic series.

Theorem 5. *For $n \geq 1$, we have*

$$\sum_{m\leq n} \frac{1}{m} = \log n + \gamma + \frac{1}{2n} - \frac{1}{12n^2} + \frac{\theta}{60n^4}$$

where γ is Euler's constant and $\theta = \theta_n \in [0,1]$.

Proof. Apply Theorem 4 with $f(t) = 1/t$, $a = 1$, $b = n$, $k = 3$. We have

$$\sum_{2\leq m\leq n} \frac{1}{m} = \log n + \tfrac{1}{2}\Big(\frac{1}{n} - 1\Big) - \tfrac{1}{12}\Big(\frac{1}{n^2} - 1\Big) + \tfrac{1}{120}\Big(\frac{1}{n^4} - 1\Big) - \int_1^n t^{-5}B_4(t)\,\mathrm{d}t.$$

Adding the term corresponding to $m = 1$ and letting n tend to infinity, we obtain

$$\gamma = \tfrac{1}{2} + \tfrac{1}{12} - \tfrac{1}{120} - \int_1^\infty t^{-5}B_4(t)\,\mathrm{d}t.$$

The result is then implied by the inequality

$$\left|\int_n^\infty t^{-5} B_4(t)\,\mathrm{d}t\right| \leq \frac{1}{120n^4}$$

which may be immediately deduced from the fact that $|B_4(t)| \leq \frac{1}{30}$ for all t.

Remark. An immediate generalisation of the preceding calculations provides the formula

$$\gamma = \tfrac{1}{2} + \sum_{r=2}^{k} \frac{B_r}{r} - \int_1^\infty t^{-k-1} B_k(t)\,\mathrm{d}t \qquad (k \geq 1)$$

which can be used to calculate γ numerically: one subtracts the above expression from the expansion for $\sum_{m=1}^{n} 1/m$ and optimises k as a function of n. We have $\gamma \approx 0.577215663$.

Exercises

1. For $k \in \mathbb{Z}^+$, calculate

$$\frac{1}{2\pi i} \oint_{|z|=\pi(2k+1)} \frac{z e^{xz}}{(e^z - 1)} \frac{\mathrm{d}z}{z^{2r+1}}$$

and thus establish the Fourier expansion for Bernoulli functions of even order:

$$B_{2r}(x) = (-1)^{r-1} 2(2r)!\,(2\pi)^{-2r} \sum_{m=1}^{\infty} \frac{\cos(2\pi m x)}{m^{2r}} \qquad (r \geq 1).$$

Deduce a general formula for $\zeta(2r)$.

2. Establish by the method described in the previous exercise the Fourier expansion of $B_{2r+1}(x)$, i.e.

$$B_{2r+1}(x) = (-1)^{r-1}(2r+1)!\,2(2\pi)^{-2r-1} \sum_{m=1}^{\infty} \frac{\sin(2\pi m x)}{m^{2r+1}} \qquad (r \geq 0),$$

where equality is restricted to $x \in \mathbb{R}\smallsetminus\mathbb{Z}$ when $r = 0$.

Explain why this does not yield a formula for $\zeta(2r+1)$.

3. *Stirling's formula.*

(a) By applying the Euler-Maclaurin formula of order zero to $f(t) = \log t$, show that one has

$$n! = n^n e^{-n}\sqrt{(An)}\,\{1 + O(1/n)\} \qquad (n \geq 1)$$

with $\log A = 2 + 2\int_1^\infty B_1(t)t^{-1}\mathrm{d}t$.

(b) Show using integration by parts that the sequence

$$W_n = \int_0^{\pi/2} (\cos t)^n \mathrm{d}t$$

of Wallis integrals satisfies the recurrence relation

$$n\,W_n = (n-1)W_{n-2} \quad (n \geq 2).$$

Deduce that as $n \to \infty$, $W_{2n} \sim \pi/\sqrt{(2An)}$ and $W_{2n+1} \sim \sqrt{(A/8n)}$.

(c) Show that $W_n \sim W_{n+1}$ $(n \to \infty)$ and deduce that $A = 2\pi$.

I.1

Prime numbers

§ 1.1 Introduction

Addition and multiplication equip the set of positive natural numbers $\{1, 2, 3, \ldots\}$ with a double structure of Abelian semi-group. The first is associated with a total order relation, and is generated by the single number 1. The second, reflecting the partial order of divisibility, has an infinite number of generators: the prime numbers. Defined since antiquity, this key concept has yet to deliver up all its secrets—and there are plenty of them. The central position of prime number theory in arithmetic is amply justified by the following result, the proof of which we sketch, using Euclid's first theorem, in Exercises 1 to 4.

Theorem 1 (Fundamental theorem of arithmetic). *Each natural number > 1 can be decomposed uniquely, up to the order of the factors, as a product of prime numbers.*

Euclid's second theorem asserts the infinity of the set of prime numbers. It is an immediate consequence of the fundamental theorem of arithmetic: if $p_1 = 2$, $p_2 = 3, \ldots, p_n$ are the n first primes, then the integer

$$N = 1 + \prod_{j=1}^{n} p_j$$

is divisible by none of the numbers $p_1, p_2, \ldots, p_n$. Its smallest prime factor is thus a prime $> p_n$.

One usually denotes by $\pi(x)$ the number of primes not exceeding x, so that for each integer n we have $\pi(p_n) = n$. Euclid's second theorem expresses the fact that

$$\pi(x) \to \infty \qquad (x \to \infty).$$

For more than 23 centuries, mathematicians have been concerned with providing quantitative versions of this qualitative relation. One of the aims of this work is to describe the various methods which they have invented and implemented to achieve this.

The proof given above of Euclid's second theorem is too simple to be ineffective. Indeed we have

$$p_{n+1} \le 1 + \prod_{j=1}^{n} p_j$$

from which we deduce by an immediate induction that

$$p_n \le 2^{2^n} \qquad (n \ge 1).$$

We obtain in this way the following lower bound.

Theorem 2. *We have*

$$\pi(x) > \frac{\log_2 x}{\log 2} - \frac{1}{2} \qquad (x \ge 2).$$

Proof. Given the upper bound for p_n established above, we can write

$$\pi(x) \ge \max\{m : 2^{2^m} \le x\} = \left[\frac{\log(\log x/\log 2)}{\log 2}\right]$$
$$\ge \frac{\log_2 x}{\log 2} - \left(1 + \frac{\log_2 2}{\log 2}\right),$$

which implies the stated result.

The lower bound of Theorem 2 is far from being optimal. After having been conjectured for more than a century (notably by Legendre and Gauss) the prime number theorem, viz.

$$\pi(x) \sim \frac{x}{\log x} \qquad (x \to \infty),$$

was established independently in 1896 by Hadamard (1865–1963) and de La Vallée–Poussin (1866–1962). Their methods rest on techniques of complex analysis, which will be described in Part II. One had to wait until 1949 for the appearance of the first elementary proofs of the prime number theorem, due to Erdős and Selberg. An elegant alternative elementary argument was recently discovered by Daboussi (1984).

§ 1.2 Chebyshev's estimates

The first serious work on the function $\pi(x)$ is due to the Russian mathematician Chebyshev. In 1852, he proved Bertrand's postulate according to which each interval $]n, 2n]$, $n \ge 1$, contains at least one prime. He obtained this result by establishing an effective form of the estimate

$$\{c_1 + o(1)\}\frac{x}{\log x} \le \pi(x) \le \{c_2 + o(1)\}\frac{x}{\log x} \qquad (x \to \infty),$$

with $c_1 = \log(2^{1/2}\,3^{1/3}\,5^{1/5}\,30^{-1/30}) \approx 0.92129$, and $c_2 = \frac{6}{5}\,c_1 \approx 1.10555$.

We shall prove by a simple method the following result, which implies a slightly weaker version of Bertrand's postulate: for each $\varepsilon > 0$, there exists some $n_0 = n_0(\varepsilon)$ such that each interval $]n, (2+\varepsilon)n]$, $n \ge n_0$, contains at least one prime—for a complete proof of Bertrand's postulate, see Exercise 10.

Theorem 3. *For $n \geq 4$, we have*

$$(\log 2)\frac{n}{\log n} \leq \pi(n) \leq \left\{\log 4 + \frac{8\log_2 n}{\log n}\right\}\frac{n}{\log n}.$$

Proof. The upper bound is an easy consequence of the following classical result.

Theorem 4. *For $n \geq 1$, we have*

$$\prod_{p \leq n} p \leq 4^n.$$

Indeed, accepting this result for the moment, we have for all t, $1 < t \leq n$,

$$t^{\pi(n)-\pi(t)} \leq \prod_{t<p\leq n} p \leq 4^n$$

from which we deduce, by taking logarithms, that

$$\pi(n) \leq \frac{n\log 4}{\log t} + t.$$

The stated result follows by choosing $t = n/(\log n)^2$—the numerical details being left to the reader.

Proof of Theorem 4. We proceed by induction on n, which we can clearly take to be at least 3. If n is even, n is not prime, and hence

$$\prod_{p\leq n} p = \prod_{p\leq n-1} p \leq 4^{n-1} < 4^n.$$

If n is odd, let us write $n = 2m+1$. The argument rests on the integrality of the binomial coefficients of order n. Since $\binom{2m+1}{m} = (2m+1)!/m!(m+1)!$ we have

$$\Big(\prod_{m+1<p\leq 2m+1} p\Big)\Big|\binom{2m+1}{m} \leq \tfrac{1}{2}2^{2m+1} = 4^m,$$

where the last inequality arises from the fact that the coefficient $\binom{2m+1}{m}$ equals $\binom{2m+1}{m+1}$ and thus appears twice in the binomial expansion of $(1+1)^{2m+1}$. By the inductive assumption applied to $m+1 < n$ we then have

$$\prod_{p\leq n} p = \prod_{p\leq m+1} p \prod_{m+1<p\leq 2m+1} p \leq 4^{m+1}\,4^m = 4^n,$$

which completes the proof.

The lower bound of Theorem 3 will be obtained by a new method, which is remarkably simple and effective, due to Nair (1982a, b). It rests on the inequality

$$\pi(n) \geq (\log d_n)/\log n \qquad (n \geq 2),$$

where d_n denotes the least common multiple of the numbers $1, 2, \ldots, n$. Indeed, if $p^\nu \| d_n$ there exists some $m \leq n$ such that $p^\nu \| m$, so $p^\nu \leq n$ and

$$d_n = \prod_{p \leq n,\, p^\nu \| d_n} p^\nu \leq \prod_{p \leq n} n = n^{\pi(n)},$$

which is equivalent to the stated inequality. The desired result is now implied by the following theorem.

Theorem 5 (Nair). *For $n \geq 7$, we have $d_n \geq 2^n$.*

Proof. The essential idea introduced by Nair consists in considering the integral

$$I(m,n) = \int_0^1 x^{m-1}(1-x)^{n-m}\,dx \qquad (1 \leq m \leq n).$$

On the one hand the binomial expansion of $(1-x)^{n-m}$ shows that $I(m,n)$ is a rational number with denominator dividing d_n: we have

$$I(m,n) = \sum_{j=0}^{n-m} (-1)^j \binom{n-m}{j} \frac{1}{m+j} \in \frac{1}{d_n}\mathbb{Z}.$$

On the other hand $I(m,n)$ is "small". Its actual value may easily be calculated by noting that for all y, $0 \leq y \leq 1$,

$$\sum_{m=1}^{n} \binom{n-1}{m-1} y^{m-1} I(m,n) = \int_0^1 (1-x+xy)^{n-1}\mathrm{d}x = \frac{1}{n}\sum_{m=1}^{n} y^{m-1},$$

so that

$$I(m,n) = 1\Big/n\binom{n-1}{m-1} = 1\Big/m\binom{n}{m}, \qquad (1 \leq m \leq n).$$

This shows that

$$m\binom{n}{m} \,\Big|\, d_n, \qquad (1 \leq m \leq n),$$

from which we get

$$n\binom{2n}{n} \,\Big|\, d_{2n} \,\Big|\, d_{2n+1}, \qquad \text{and} \qquad (n+1)\binom{2n+1}{n} = (2n+1)\binom{2n}{n} \,\Big|\, d_{2n+1}.$$

Since n and $2n+1$ are coprime, it follows that

$$n(2n+1)\binom{2n}{n} \mid d_{2n+1},$$

and finally, since $\binom{2n}{n}$ is the largest of the $(2n+1)$ binomial coefficients occurring in the expansion of $(1+1)^{2n}$, that

$$d_{2n+1} \geq n\,4^n \qquad (n \geq 1).$$

We deduce that

$$d_{2n+1} \geq 2{\cdot}4^n = 2^{2n+1} \qquad (n \geq 2),$$

and

$$d_{2n+2} \geq d_{2n+1} \geq 4^{n+1} \qquad (n \geq 4),$$

which proves the stated inequality $\mathrm{d}_n \geq 2^n$ for all $n \geq 9$. It can be easily checked that it also holds for $n = 7$ and $n = 8$: $d_7 = 420$, $d_8 = 840$.

§ 1.3 p-adic valuation of $n!$

For each prime number p, the p-adic valuation, denoted by v_p, is defined as the arithmetic function which associates to each integer n the exponent of p in its canonical factorisation. The following simple theorem will be useful in what follows.

Theorem 6. *For each prime p, we have*

$$v_p(n!) = \sum_{k=1}^{\infty} [n/p^k] \qquad (n \geq 1).$$

Remark. The sum over k is genuinely finite since the general term vanishes for $k > (\log n)/\log p$.

Proof. We have

$$v_p(n!) = \sum_{m \leq n} v_p(m) = \sum_{m \leq n} \sum_{1 \leq k \leq v_p(m)} 1 = \sum_{k=1}^{\infty} \sum_{m \leq n,\, v_p(m) \geq k} 1.$$

The inner sum equals the number of integers $m \leq n$ which are divisible by p^k. It therefore has value $[n/p^k]$, which provides the desired expression.

Corollary 6.1. *For each prime p, we have*

$$\frac{n}{p} - 1 < v_p(n!) \leq \frac{n}{p} + \frac{n}{p(p-1)} \qquad (n \geq 1).$$

This readily follows from Theorem 6 and the bounds $x - 1 < [x] \leq x$, valid for all real x.

§ 1.4 Mertens' first theorem

Certain quantities depending on the set of primes not exceeding x have an asymptotic behaviour which is more easily accessible than that of the function $\pi(x)$. This is the case for the expression evaluated in the following theorem.

Theorem 7 (Mertens' first theorem). *For $x \geq 2$, we have*

$$\sum_{p \leq x} \frac{\log p}{p} = \log x + O(1).$$

Moreover, the term $O(1)$ arising in this formula lies in the open interval $]-1-\log 4, \log 4[$.

NB. $\log 4 \approx 1.38629$.

Proof. We evaluate the quantity $\log(n!)$ in two different ways for $n = [x]$.

On the one hand we have from Corollary 0.2.1 that

$$\log(n!) = n \log n - n + 1 + \theta_n \log n$$

with $0 \leq \theta_n \leq 1$.

On the other hand, by expanding $n!$ as a product of prime factors, viz.

$$\log n! = \sum_{p \leq n} v_p(n!) \log p,$$

we deduce from the corollary to Theorem 6 that

$$\log n! < n \sum_{p \leq n} \frac{\log p}{p} + n \sum_{p \leq n} \frac{\log p}{p(p-1)}$$

and

$$\log n! > n \sum_{p \leq n} \frac{\log p}{p} - \sum_{p \leq n} \log p.$$

By Theorem 4, the last p-sum does not exceed $n \log 4$. We thus infer

$$n \sum_{p \leq n} \frac{\log p}{p} - n \log 4 < n \log n - n + (1 + \log n) < n \log n$$

from which it follows that

$$\sum_{p \leq x} \frac{\log p}{p} = \sum_{p \leq n} \frac{\log p}{p} < \log n + \log 4 \leq \log x + \log 4.$$

Next, we also have

$$\sum_{p\le n} \frac{\log p}{p(p-1)} < \sum_{m=2}^{\infty} \frac{\log m}{m(m-1)}$$
$$\le \sum_{r=1}^{\infty} \sum_{2^{r-1}<m\le 2^r} \frac{r\log 2}{m(m-1)} = \sum_{r=1}^{\infty} \frac{r\log 2}{2^r} = \log 4,$$

from which we get

$$n\sum_{p\le n} \frac{\log p}{p} + n\log 4 > n\log n - n + 1$$

and

$$\sum_{p\le x} \frac{\log p}{p} > \log n + \frac{1}{n} - (1+\log 4) \ge \log x - (1+\log 4).$$

This concludes the proof.

§ 1.5 Two new asymptotic formulae

Mertens' first theorem is conceptually different from Chebyshev's estimates inasmuch as it provides, for a weighted sum over primes, a genuine asymptotic formula. In this sense it is the prototype of a family of results culminating in the prime number theorem—which corresponds to the case when the weighting coefficient equals the constant 1.

We shall see that Theorem 7 actually contains other results of the same kind. In particular, it allows one to evaluate the expressions

$$\sum_{p\le x} \frac{1}{p} \quad \text{and} \quad \prod_{p\le x} \Big(1 - \frac{1}{p}\Big).$$

Let us begin by exhibiting a close link between these two quantities.

Theorem 8. *Let* $c_0 := \sum_p \Big\{ \log\Big(\frac{1}{1-1/p}\Big) - \frac{1}{p}\Big\} \approx 0.315718$. *Then we have, for* $x \ge 2$,

$$\sum_{p\le x} \frac{1}{p} = \log\Big\{1 \Big/ \prod_{p\le x} \Big(1 - \frac{1}{p}\Big)\Big\} - c_0 + \frac{\theta}{2(x-1)}$$

where $\theta = \theta(x) \in]0, 1[$.

Proof. With the given expression for c_0, we see that the stated formula holds with

$$\begin{aligned} 0 < \theta(x) &= 2(x-1)\sum_{p>x}\Big\{\log\Big(\frac{1}{1-1/p}\Big)-\frac{1}{p}\Big\} \\ &= 2(x-1)\sum_{p>x}\sum_{k=2}^{\infty}\frac{1}{k}p^{-k} < \sum_{p>x}\frac{2(x-1)}{2p(p-1)} \\ &< \sum_{n>x}\frac{(x-1)}{n(n-1)} = \frac{x-1}{N-1} \end{aligned}$$

where N is the smallest integer $> x$. This gives the required estimate.

Theorem 9. *There exists a constant c_1 such that, for $x \geq 2$, one has that*

$$\sum_{p\leq x}\frac{1}{p} = \log_2 x + c_1 + O\Big(\frac{1}{\log x}\Big).$$

Furthermore, the implicit constant in Landau's symbol does not exceed $2(1+\log 4) < 5$.

Remark. Theorem 11 below easily yields a numerical approximation for c_1. We have $c_1 = \gamma - c_0 \approx 0.261497$.
Proof. By Mertens' first theorem we have for $t \geq 2$

$$R(t) := \sum_{p\leq t}\frac{\log p}{p} - \log t = O(1).$$

Now we may also write

$$\begin{aligned} \sum_{p\leq x}\frac{1}{p} &= \int_{2-}^{x}\frac{1}{\log t}\,\mathrm{d}\Big\{\sum_{p\leq t}\frac{\log p}{p}\Big\} = \int_{2}^{x}\frac{\mathrm{d}t}{t\log t} + \int_{2-}^{x}\frac{\mathrm{d}R(t)}{\log t} \\ &= \log_2 x - \log_2 2 + \frac{R(x)}{\log x} - \frac{R(2-)}{\log 2} + \int_2^x \frac{R(t)}{t(\log t)^2}\mathrm{d}t, \end{aligned}$$

where we have applied Abel summation to the integral involving $R(t)$.

Let $R = \sup_{t\geq 2-}|R(t)|$. From the upper bound in Theorem 7 we have

$$\left|\frac{R(x)}{\log x} - \int_x^{\infty}\frac{R(t)}{t(\log t)^2}\,\mathrm{d}t\right| \leq \frac{2R}{\log x} < \frac{2(1+\log 4)}{\log x}.$$

From this we deduce the stated formula with

$$c_1 = -\log_2 2 + 1 + \int_2^{\infty}\frac{R(t)}{t(\log t)^2}\,\mathrm{d}t.$$

Theorem 10. *If c_0 and c_1 are given the values introduced in Theorems 8 and 9, then we have for $x \geq 2$*

$$\prod_{p \leq x} \left(1 - \frac{1}{p}\right) = \frac{e^{-(c_0+c_1)}}{\log x} \left\{1 + O\left(\frac{1}{\log x}\right)\right\}.$$

This is an immediate corollary of Theorems 8 and 9.

§ 1.6 Mertens' formula

Mertens' second theorem, known as "Mertens' formula", provides an explicit value for the constant appearing in Theorem 10.

Theorem 11 (Mertens' formula). *With the notation of the previous section, we have $c_0 + c_1 = \gamma$, where γ denotes Euler's constant. Thus*

$$\prod_{p \leq x} \left(1 - \frac{1}{p}\right) = \frac{e^{-\gamma}}{\log x} \left\{1 + O\left(\frac{1}{\log x}\right)\right\} \qquad (x \geq 2).$$

Proof. For $\sigma > 0$ let us write

$$\zeta(1+\sigma) := \sum_{n=1}^{\infty} n^{-1-\sigma}.$$

Comparing the sum to an integral, we easily check that

$$\zeta(1+\sigma) = \frac{1}{\sigma} + O(1) \qquad (\sigma > 0).$$

Moreover, we have

$$\sum_{n \leq x} n^{-1-\sigma} \leq \prod_{p \leq x} \left(1 - p^{-1-\sigma}\right)^{-1} \leq \zeta(1+\sigma)$$

since the product over p equals the sum $\sum_{n=1}^{\infty} \varepsilon_n n^{-1-\sigma}$ where ε_n equals 1 if all the prime factors of n are $\leq x$ and is 0 otherwise. Letting x tend to infinity, we obtain the famous formula of Euler

$$\zeta(1+\sigma) = \prod_{p} \left(1 - p^{-1-\sigma}\right)^{-1}.$$

Now let us consider the function

$$f(\sigma) = \log \zeta(1+\sigma) - \sum_{p} p^{-1-\sigma} = \sum_{p} \left\{\log\left(\frac{1}{1 - p^{-1-\sigma}}\right) - p^{-1-\sigma}\right\}.$$

Since the general term is positive and bounded above by $1/p(p-1)$, the series $f(\sigma)$ converges uniformly for $\sigma \geq 0$ and in particular the sum is continuous at $\sigma = 0$. Hence

$$\lim_{\sigma \to 0} f(\sigma) = f(0) = c_0.$$

We shall transform the two terms of the sum $f(\sigma)$. On the one hand

$$\log \zeta(1+\sigma) = \log\big(1/\sigma + O(1)\big) = \log\big(1/\sigma\big) + O(\sigma) = \log\Big(\frac{1}{1-e^{-\sigma}}\Big) + O(\sigma)$$
$$= \sum_{n=1}^{\infty} e^{-\sigma n} n^{-1} + O(\sigma) = \int_0^{\infty} e^{-\sigma t} \mathrm{d}H(t) + O(\sigma)$$

where we have set

$$H(t) := \sum_{1 \leq n \leq t} 1/n.$$

Integrating the Stieltjes integral by parts then gives

$$\log \zeta(1+\sigma) = \sigma \int_1^{\infty} e^{-\sigma t} H(t)\,\mathrm{d}t + O(\sigma).$$

On the other hand, writing $P(u) := \sum_{p \leq u} 1/p$, we have

$$\sum_p p^{-1-\sigma} = \int_1^{\infty} u^{-\sigma} \mathrm{d}P(u) = \sigma \int_1^{\infty} u^{-1-\sigma} P(u)\,\mathrm{d}u$$
$$= \sigma \int_0^{\infty} e^{-\sigma t} P(e^t)\,\mathrm{d}t.$$

We hence obtain

$$f(\sigma) = \sigma \int_0^{\infty} e^{-\sigma t}\big(H(t) - P(e^t)\big)\mathrm{d}t + O(\sigma).$$

Now, we have seen in Theorem 0.5 that

$$H(t) = \log t + \gamma + O(1/t) \qquad (t \geq 1).$$

Moreover, it follows from Theorem 9 that

$$P(e^t) = \log t + c_1 + O(1/t) \qquad (t \geq 1).$$

Hence

$$f(\sigma) = \sigma \int_0^{\infty} \Big\{\gamma - c_1 + O\Big(\frac{1}{t+1}\Big)\Big\} e^{-\sigma t}\mathrm{d}t + O(\sigma)$$
$$= \gamma - c_1 + O\Big(\sigma + \sigma \int_0^{\infty} e^{-\sigma t} \frac{\mathrm{d}t}{t+1}\Big)$$
$$= \gamma - c_1 + O\big(\sigma \log(1/\sigma)\big) \qquad (\sigma \leq \tfrac{1}{2})$$

and finally $c_0 = f(0) = \gamma - c_1$. This is all we need.

§ 1.7 Another theorem of Chebyshev

Chebyshev has shown that, if some asymptotic formula of the type

$$\pi(x) \sim c\frac{x}{\log x}$$

holds, then the constant c must equal 1.

Theorem 12. *We have*

$$\liminf_{x\to\infty} \frac{\pi(x)}{x/\log x} \leq 1 \leq \limsup_{x\to\infty} \frac{\pi(x)}{x/\log x}.$$

Proof. The two inequalities are handled analogously; we confine ourselves to establishing that on the left. Let

$$\ell := \liminf_{x\to\infty} \frac{\pi(x)}{x/\log x}.$$

For each $\varepsilon > 0$, there exists some $x_0 = x_0(\varepsilon) \geq 2$ such that

$$\pi(t) \geq (\ell - \varepsilon)\frac{t}{\log t}, \qquad (t \geq x_0(\varepsilon)).$$

For $x > x_0$, this implies that

$$\begin{aligned}\sum_{p\leq x} \frac{1}{p} &\geq \int_{x_0}^{x} \frac{\mathrm{d}\pi(t)}{t} = \frac{\pi(x)}{x} - \frac{\pi(x_0)}{x_0} + \int_{x_0}^{x} \pi(t)t^{-2}\,\mathrm{d}t \\ &\geq -1 + (\ell - \varepsilon)\int_{x_0}^{x} \frac{\mathrm{d}t}{t\log t} \geq (\ell - \varepsilon)\log_2 x + O_\varepsilon(1).\end{aligned}$$

By Theorem 9, it follows that $\ell - \varepsilon \leq 1$ and hence $\ell \leq 1$, since ε can be chosen arbitrarily small.

Notes

§ **1.2.** The proof given here of Theorem 4 was found independently by Erdős and Kalmár in 1939.

The prime number theorem implies that $\prod_{p\le n} p \le e^{(1+\varepsilon)n}$ for all $\varepsilon > 0$ and $n \ge n_0(\varepsilon)$. It is however useful to have at one's disposal uniform upper bounds, like those of Theorem 4, valid without any restriction on n. In this spirit, Hanson showed in 1972 that one has

$$\prod_{p\le n} p \le 3^n \qquad (n \ge 2).$$

For numerical bounds on $\pi(x)$ and Chebyshev functions (cf. § 3.6) see Rosser & Schoenfeld (1962, 1975) and Schoenfeld (1976). For example, we have

$$\frac{x}{\log x}\Big(1+\frac{1}{2\log x}\Big) < \pi(x) < \frac{x}{\log x}\Big(1+\frac{3}{2\log x}\Big) \qquad (x \ge 52).$$

The idea exploited by Nair (1982a) for the proof of Theorem 5 is not new: see Gelfond (1946). Gorshkov showed that the prime number theorem cannot be obtained by this method with polynomials in one variable and gave numerical bounds for the best estimates that can be achieved in this way (cf. Gorshov, 1956). These bounds were refined by Aparicio Bernardo (1981). The generalisation given by Nair (1982b) is genuinely new. It provides very precise numerical bounds and might in principle lead to a proof of the prime number theorem.

For more classical proofs of Chebyshev-type bounds for $\pi(n)$, see Exercises 6 to 10.

Exercises

1. Show by induction that each integer > 1 can be written as a product of primes.

2. (a) Let I be an *ideal* of $\mathbb{Z}$ (i.e. an additive subgroup stable under multiplication by elements of $\mathbb{Z}$). Show that $I = k\mathbb{Z}$, where k is the least positive element of I.

(b) Let $m, n \in \mathbb{Z}$. Show that $m\mathbb{Z} + n\mathbb{Z}$ is an ideal of $\mathbb{Z}$.

(c) Let $d = (m, n)$ be the greatest positive common divisor of m and n. Prove *Bachet's theorem* (1624): $m\mathbb{Z} + n\mathbb{Z} = d\mathbb{Z}$.

3. *Euclid's first theorem.* Let $a, b \in \mathbb{Z}$ and p be a prime number. Suppose that $p \mid ab$ and $p \nmid a$. Use Bachet's theorem to prove that there exist $u, v \in \mathbb{Z}$ such that $up + va = 1$. Deduce that $p \mid b$.

4. Using Euclid's first theorem, give an inductive proof of the *fundamental theorem of arithmetic*: the decomposition of each integer > 1 as a product of primes is unique up to the order of the factors.

5. Let p_n be the nth prime, and $d_n := p_{n+1} - p_n$. Assuming the minimal form of the prime number theorem, namely $\pi(x) \sim x/\log x$ $(x \to \infty)$, show that

(a) $p_n \sim n \log n \quad (n \to \infty)$

(b) $\displaystyle\sum_{1<n\leq x} d_n/\log n \sim x \quad (x \to \infty)$

(c) $\liminf_{n\to\infty} (d_n/\log n) \leq 1 \leq \limsup_{n\to\infty} (d_n/\log n)$.

(d) For each $\alpha > 0$ there exists a sequence of integers $\{n_1, n_2, \ldots\}$, increasing in the weak sense, such that $p_{n_j} \sim \alpha j$ $(j \to \infty)$.

(e) The set of rational numbers of the form p'/p, where p and p' are prime, is dense in $[0, +\infty[$.

6. Let $\{a_n\}_{n=1}^{\infty}$ be a sequence of non-negative real numbers. Assume that

$$B(x) := \sum_{n\leq x} a_n [x/n] = x \log x + Cx + o(x).$$

Prove *Shapiro's Tauberian theorem (1950): there exist two positive constants α and β such that $\alpha x \leq A(x) := \sum_{n\leq x} a_n \leq \beta x$ $(x > x_0)$.*

Give precise values for α and β. Consider the case when one only assumes that $B(x) = x \log x + O(x)$. [*Hint: Consider $A(x) - A(x/2)$ and use the properties of the function $u \mapsto [u] - 2[u/2]$.*]

7. Define $\Lambda(n) := \log p$, if $n = p^\nu$, $\Lambda(n) := 0$ where n is not a prime power. Show that

(a) $\sum_{d|n} \Lambda(d) = \log n$

(b) $\sum_{d\leq x} \Lambda(d) [x/d] = x \log x - x + O(\log x) \qquad (x \geq 2)$.

8. Deduce from the two previous exercises that the function $\psi(x) := \sum_{n\leq x} \Lambda(n)$ satisfies

$$\alpha x \leq \psi(x) \leq \beta x \qquad (x > x_0)$$

for two positive constants α and β which may be computed.

9. From Exercise 8, deduce the existence of two positive constants a and b, which may be determined, such that

$$ax/\log x \leq \pi(x) \leq bx/\log x \qquad (x \geq 2).$$

[*For a refinement of Shapiro's Tauberian theorem leading to an elementary proof of the prime number theorem, see Smith (1980).*]

10. *Chebyshev's estimates.* Set $M > 1$ and let h be an arithmetical function such that

$$\text{(i)} \quad h(1) = 1, \quad \textstyle\sum_{m=1}^{M} h(m)/m = 0.$$

$$\text{(ii)} \quad \chi(x) := \textstyle\sum_{m=1}^{M} h(m)[x/m] \in [0,1] \quad (x \in \mathbb{R}).$$

Show that, with the hypotheses of Exercise 6, we have for large x

$$(H - \varepsilon)x < A(x) < \Big(\frac{a}{a-1}H + \varepsilon\Big)x$$

where we have set $H := -\sum_{m=1}^{M} h(m)\log m/m$ and a is defined as the largest integer such that $\chi(x) = 1$ for $1 \leq x < a$. To what choice for h does Shapiro's Tauberian theorem correspond? Show that selecting

$$M = 6,\ h(1) = h(6) = 1,\ h(2) = -1,\ h(3) = -2,\ h(4) = h(5) = 0$$

leads to a proof of Bertrand's postulate. Recover Chebyshev's estimates for $\psi(x)$ with his original choice

$$\begin{aligned} &M = 30,\ h(1) = h(30) = 1,\ h(2) = h(3) = h(5) = -1,\\ &h(m) = 0 \quad (m = 4 \quad \text{or} \quad 6 \leq m \leq 29). \end{aligned}$$

11. Show that $\lim_{x\to\infty} \sum_{\sqrt{x}<p\leq x} 1/p = \log 2$. Deduce that, as $x \to \infty$, a positive proportion of integers $n \leq x$ have their greatest common prime factor $> \sqrt{n}$. Can this statement be made more precise?

12. Evaluate $\sum_{pq\leq x} 1/pq$, where p and q denote primes. Generalise.

13. Show that from Bertrand's postulate it is possible to deduce simply that the number $H_n = \sum_{1\leq m\leq n} 1/m$ is not integral for any $n > 1$. Give another proof of this result by considering powers of 2.

14. Show that there exist infinitely many primes of the form $4n + 3$. [*Hint: consider the integer* $N = 4{\cdot}n! - 1$.]

I.2

Arithmetic functions

§ 2.1 Definitions

An *arithmetic function* is a complex-valued function which is defined on $\mathbb{N}^* = \{1, 2, 3, \ldots\}$. Two classes of arithmetic functions play a particularly important role: additive functions and multiplicative functions. A function is called *additive* if it satisfies

$$f(mn) = f(m) + f(n) \quad (\text{whenever } (m,n) = 1); \tag{1}$$

it is said to be *multiplicative* if $f(1) = 1$ and

$$f(mn) = f(m)f(n) \quad (\text{whenever } (m,n) = 1). \tag{2}$$

The condition $f(1) = 1$ is a convention designed to exclude the zero function from the class of multiplicative functions.

The main interest in these notions is that an additive or multiplicative function respects the multiplicative structure of $\mathbb{N}^*$ in the sense that the image of an integer is the sum or product of the images of the prime powers arising in its canonical factorisation. We can thus write

$$f(n) = \sum_{p^\nu \| n} f(p^\nu), \quad f(n) = \prod_{p^\nu \| n} f(p^\nu),$$

when f is respectively additive or multiplicative.

A function f is called *completely additive* or *completely multiplicative* if the conditions (1) or (2) hold even when $(m,n) \neq 1$, that is if $f(p^\nu) = \nu f(p)$ or $f(p^\nu) = f(p)^\nu$ respectively. One says that f is *strongly additive* or *strongly multiplicative* if, besides (1) or (2), we have $f(p^\nu) = f(p)$ for all $\nu \geq 1$.

§ 2.2 Examples

The following arithmetic functions are classical and define fundamental concepts attached to the multiplicative structure of n:

• The counting functions of the total number of prime factors of n, taken with or without multiplicity, thus

$$\Omega(n) := \sum_{p^\nu \| n} \nu, \qquad \omega(n) := \sum_{p^\nu \| n} 1 = \sum_{p|n} 1.$$

• The "number of divisors" function and the function "sum of kth powers of divisors", traditionally denoted by

$$\tau(n) := \sum_{d|n} 1, \qquad \sigma_k(n) := \sum_{d|n} d^k \qquad (k \in \mathbb{C}).$$

Usually $\tau(n)$ is referred to as the divisor function and one writes $\sigma_1(n) = \sigma(n)$.

• Euler's totient function, counting the number of invertible residues modulo n, that is

$$\varphi(n) := \sum_{1 \le h \le n,\, (h,n)=1} 1.$$

• The Möbius function, defined by

$$\mu(n) := \begin{cases} (-1)^{\omega(n)}, & \text{if } n \text{ is squarefree,} \\ 0, & \text{otherwise.} \end{cases}$$

• Von Mangoldt's function

$$\Lambda(n) := \begin{cases} \log p, & \text{if } n = p^\nu \text{ for some } \nu \ge 1, \\ 0, & \text{if } n \text{ is not a prime power.} \end{cases}$$

(In §2.6, we shall give a different definition of Λ, which can easily be seen to be equivalent to the above.)

It is immediate from their definitions that Ω and ω are additive, the former completely, the latter strongly. The case of the divisor function $\tau(n)$ is less obvious. However, representing the divisors of n as all integers of the form

$$d = \prod_{p|n} p^{\alpha_p}$$

with $0 \le \alpha_p \le v_p(n)$ for each prime p, we deduce that

$$\tau(n) = \prod_{p|n} (v_p(n) + 1).$$

Thus we can state the following result.

Theorem 1. *The divisor function is multiplicative. We have*

$$\tau(n) = \prod_{p^\nu \| n} (\nu + 1) \qquad (n \ge 1).$$

We shall investigate later Euler's function and the σ_k-functions. Let us now consider the case of the Möbius function. We have

$$\mu(p^\nu) = \begin{cases} -1 & (\nu = 1) \\ 0 & (\nu > 1) \end{cases}$$

and hence immediately obtain that $\mu(n) = \prod_{p^\nu \| n} \mu(p^\nu)$.

Theorem 2. *The Möbius function is multiplicative.*

§ 2.3 Formal Dirichlet series

An essential tool for the study of arithmetic functions is the concept of a Dirichlet series, which we shall study more completely in Part II. Here we restrict ourselves to extracting certain algebraic properties of arithmetic functions, which can be conveniently displayed by employing the concept of a formal Dirichlet series.

Definition. *Let f be an arithmetic function. The formal Dirichlet series associated with f is the formal series*

$$D(f;s) := \sum_{n=1}^{\infty} f(n)\, n^{-s}.$$

The sum and product of two formal Dirichlet series are defined in a natural way by

$$D(f;s) + D(g;s) = \sum_{n=1}^{\infty} \big(f(n)+g(n)\big)n^{-s}, \tag{3}$$

$$D(f;s)\, D(g;s) = \sum_{n=1}^{\infty} h(n)\, n^{-s}, \tag{4}$$

with

$$h(n) = \sum_{dd'=n} f(d)g(d'). \tag{5}$$

This second definition agrees with the formal computation

$$\begin{aligned}\sum_{m=1}^{\infty} f(m)m^{-s} \sum_{k=1}^{\infty} g(k)k^{-s} &= \sum_{m,k=1}^{\infty} f(m)g(k)(mk)^{-s} \\ &= \sum_{n=1}^{\infty} n^{-s} \sum_{km=n} f(m)g(k).\end{aligned}$$

It may be easily checked that the set of formal Dirichlet series equipped with these two operations has the structure of a commutative ring with unity given by the series

$$D(\delta;s) = 1,$$

associated with the arithmetic function

$$\delta(n) := \begin{cases} 1 & (n=1) \\ 0 & (n>1). \end{cases}$$

§ 2.4 The ring of arithmetic functions

The correspondence between arithmetic functions and formal Dirichlet series induces on the set of arithmetic functions an addition $+$ and a product $*$ for which the fundamental properties are respectively

$$D(f+g;s) = D(f;s) + D(g;s)$$
$$D(f*g;s) = D(f;s)\,D(g;s).$$

We thus have $(f+g)(n) = f(n)+g(n)$, and $(f*g)(n) = h(n)$ where h is defined by (5).

The product $*$ is called *Dirichlet convolution.* These operations give the set $\mathcal{A}$ of arithmetic functions the structure of a commutative ring with unity isomorphic to that of formal Dirichlet series. Cashwell & Everett (1959) showed that this ring is *factorial*, that is to say an integral domain whose quotient by the group of units satisfies the fundamental theorem of arithmetic.

A necessary and sufficient condition for $f \in \mathcal{A}$ to be invertible is that $f(1) \neq 0$. Indeed under this assumption the family of equations

$$\sum_{d|n} f(n/d)g(d) = \delta(n) \qquad (n \geq 1) \tag{6}$$

allows us to calculate the inverse function $g(n)$ recursively: we have

$$\begin{cases} g(1) = f(1)^{-1} \\ g(n) = -f(1)^{-1} \displaystyle\sum_{d|n,\, d<n} f(n/d)g(d) \qquad (n > 1). \end{cases} \tag{7}$$

Conversely, if $f(1) = 0$, then (6) is not soluble for $n = 1$ and f is not invertible.

Theorem 3. *The group $\mathcal{G}$ of units in the ring $\mathcal{A}$ consists of arithmetic functions f such that $f(1) \neq 0$.*

An element π in $\mathcal{A}$ is prime if it is not a unit and if the relation $\pi = u * v$ implies that either u or v is a unit. One checks easily that the set of prime elements in $\mathcal{A}$ strictly contains the set of functions f such that $f(1) = 0$ and $f(p) \neq 0$ for some prime number p—cf. Exercises 8 and 9.

Multiplicative functions are units, since they satisfy $f(1) = 1$ by definition. The following result will help us to show that they actually form a subgroup of $\mathcal{G}$.

Theorem 4. *A necessary and sufficient condition for a function f in $\mathcal{A}$ to be multiplicative is that its associated formal series $D(f;s)$ be expandable as an infinite formal product of Eulerian type, that is*

$$D(f;s) = \prod_p \Big(1 + \sum_{\nu=1}^{\infty} f(p^\nu)p^{-\nu s}\Big). \tag{8}$$

This is immediate since relation (8) is algebraically equivalent to the conditions

$$f(1) = 1, \quad f(n) = \prod_{p^\nu \| n} f(p^\nu) \qquad (n > 1).$$

Theorem 5. *The set $\mathcal{M}$ of multiplicative arithmetic functions is a subgroup of the group $\mathcal{G}$ of units of $\mathcal{A}$.*

Proof. If f and g are in $\mathcal{M}$, formal calculation immediately yields the relation

$$\begin{aligned} D(f;s)D(g;s) &= \prod_p \Big(1 + \sum_{\nu=1}^{\infty} f(p^\nu)p^{-\nu s}\Big)\Big(1 + \sum_{\nu=1}^{\infty} g(p^\nu)p^{-\nu s}\Big) \\ &= \prod_p \Big(1 + \sum_{\nu=1}^{\infty} h(p^\nu)p^{-\nu s}\Big) \end{aligned}$$

where $h(p^\nu)$ is defined by the formula

$$h(p^\nu) = \sum_{j=0}^{\nu} f(p^j)g(p^{\nu-j}). \tag{9}$$

Since, by definition, we have $D(f;s)D(g;s) = D(f*g;s)$, it follows that $f*g$ coincides with the multiplicative function determined by (9) on prime powers.

It remains to verify that the inverse $\tilde{f}$ of a function f in $\mathcal{M}$ is also in $\mathcal{M}$. Relation (6), applied with $g = \tilde{f}$, $n = 1$ and then $n = p^\nu$, immediately implies that $\tilde{f}(1) = 1$ and that

$$\Big(1 + \sum_{\nu=1}^{\infty} f(p^\nu)p^{-\nu s}\Big)\Big(1 + \sum_{\nu=1}^{\infty} \tilde{f}(p^\nu)p^{-\nu s}\Big) = 1$$

for all primes p. Hence

$$D(f;s)\prod_p \Big(1 + \sum_{\nu=1}^{\infty} \tilde{f}(p^\nu)p^{-\nu s}\Big) = 1 = D(f;s)D(\tilde{f};s)$$

and relation (8) is certainly satisfied by $\tilde{f}$. By Theorem 4 this implies that $\tilde{f} \in \mathcal{M}$.

Let $\mathbf{1}$ be the arithmetic function defined by

$$\mathbf{1}(n) = 1 \qquad (n \geq 1). \tag{10}$$

Then **1** is trivially multiplicative and for all $n \geq 1$

$$\tau(n) = \sum_{d|n} 1 = \sum_{d|n} \mathbf{1}(d)\mathbf{1}(n/d),$$

from which

$$\tau = \mathbf{1} * \mathbf{1}. \tag{11}$$

By Theorem 5 this provides a new proof of the multiplicativity of the divisor function τ.

Let j denote the identity function, viz.

$$j(n) = n \qquad (n \geq 1). \tag{12}$$

We clearly have

$$\sigma = \mathbf{1} * j \tag{13}$$

and hence:

Theorem 6. *The function "sum of divisors" $\sigma(n)$ is multiplicative.*

Of course the same holds for the functions

$$\sigma_k(n) = \sum_{d|n} d^k = (\mathbf{1} * j^k)(n)$$

for all real or complex values of the parameter k.

§ 2.5 The Möbius inversion formulae

For any prime number p and any non-negative integer ν, we have

$$(\mathbf{1} * \mu)(p^\nu) = \sum_{j=0}^{\nu} \mu(p^j) = \begin{cases} 1 & (\nu = 0) \\ 0 & (\nu \geq 1) \end{cases}$$
$$= \delta(p^\nu).$$

Since $\mathbf{1} * \mu$ and δ are multiplicative these two functions must be equal.

Theorem 7. *The Möbius function is the convolution inverse of the function* $\mathbf{1}$, *that is*

$$\mathbf{1} * \mu = \delta. \tag{14}$$

In other words,

$$\sum_{d|n} \mu(d) = \begin{cases} 1 & (n = 1) \\ 0 & (n > 1). \end{cases} \tag{15}$$

However trivial it may appear, formula (15) is rich in applications. In particular it is the starting point of combinatorial sieve theory—cf. §§4.1–4.3.

Before we consider actual applications of (15), we note that we could have calculated effectively the convolution inverse of μ by the method described in the course of the proof of Theorem 5. In general, the inverse $\tilde{f}$ of a multiplicative function f is the multiplicative function determined on prime powers by the formal identity

$$\Big(1 + \sum_{\nu=1}^{\infty} \tilde{f}(p^\nu)\xi^\nu\Big)\Big(1 + \sum_{\nu=1}^{\infty} f(p^\nu)\xi^\nu\Big) = 1. \tag{16}$$

This remark can be very useful when bounds for $|\tilde{f}(p^\nu)|$ are needed. One applies (16) to complex values of ξ in a common disc of convergence of the two series and gets the desired estimate via Cauchy's integral formula.

Theorem 8 (First Möbius inversion formula). *Let f and g be arithmetic functions. The two following properties are equivalent*

$$\text{(i)} \qquad g(n) = \sum_{d|n} f(d) \qquad (n \geq 1)$$

$$\text{(ii)} \qquad f(n) = \sum_{d|n} g(d)\,\mu(n/d) \qquad (n \geq 1).$$

Proof. Condition (i) is equivalent to $g = f * \mathbf{1}$, and condition (ii) to $f = g * \mu$. The result thus follows from (14).

Let us now consider generalisations to functions of a real variable.

Theorem 9 (Second Möbius inversion formula). *Let F and G be functions defined on* $[1, \infty[$. *The two following conditions are equivalent:*

$$\text{(i)} \qquad F(x) = \sum_{n \leq x} G(x/n) \qquad (x \geq 1)$$

$$\text{(ii)} \qquad G(x) = \sum_{n \leq x} \mu(n) F(x/n) \qquad (x \geq 1).$$

Proof. Let us for instance prove the implication (i) $\Rightarrow$ (ii), the converse being similar. For $x \geq 1$ we have

$$\begin{aligned}\sum_{n\leq x}\mu(n)F(x/n) &= \sum_{n\leq x}\mu(n)\sum_{m\leq x/n}G(x/mn)\\ &= \sum_{k\leq x}G(x/k)\sum_{mn=k}\mu(m).\end{aligned}$$

By (15), the inner sum has value $\delta(k)$. This yields (ii).

Applying Theorem 9 to the case $G(x) \equiv 1$, we obtain

$$\sum_{n\leq x}\mu(n)[x/n] = 1 \qquad (x \geq 1). \tag{17}$$

This suggests that

$$\lim_{x\to\infty}\sum_{n\leq x}\frac{\mu(n)}{n} = 0. \tag{18}$$

We shall see in Section 3.6 that (18) is equivalent to the prime number theorem.

§ 2.6 Von Mangoldt's function

Traditionally denoted by Λ, *von Mangoldt's function* is defined as the arithmetic function

$$\Lambda := \mu * \log. \tag{19}$$

For each $n \geq 1$, we can hence write

$$\begin{aligned}\Lambda(n) &= \sum_{d|n}\mu(d)\log(n/d) = \delta(n)\log n - \sum_{d|n}\mu(d)\log d\\ &= -\sum_{d|n}\mu(d)\log d\end{aligned}$$

that is

$$\Lambda = -\mu\log * \mathbf{1}. \tag{20}$$

For all m, n such that $(m,n) = 1$ it follows from (20) that

$$\begin{aligned}\Lambda(mn) &= -\sum_{d|m}\sum_{t|n}\mu(dt)\log(dt) = -\sum_{d|m}\mu(d)\sum_{t|n}\mu(t)\{\log d + \log t\}\\ &= \sum_{d|m}\mu(d)\{-\delta(n)\log d + \Lambda(n)\} = \delta(n)\Lambda(m) + \delta(m)\Lambda(n).\end{aligned}$$

Thus $\Lambda(n)$ is zero whenever n is not a positive power of a prime number. Moreover it follows from (20) that

$$\Lambda(n) = \begin{cases} \log p & (n = p^\nu,\, \nu \geq 1) \\ 0 & (n \neq p^\nu). \end{cases} \tag{21}$$

Chebyshev's summatory functions

$$\psi(x) := \sum_{n \leq x} \Lambda(n) = \log \operatorname{lcm}\{n : n \leq x\}, \tag{22}$$

$$\theta(x) := \sum_{p \leq x} \log p, \tag{23}$$

are important in the analytic theory of prime numbers. From Theorems 1.5 and 1.4 respectively, we deduce the inequalities

$$\psi(x) \geq [x] \log 2 \qquad (x \geq 7), \tag{24}$$

$$\theta(x) \leq x \log 4 \qquad (x \geq 1). \tag{25}$$

An immediate consequence of (21) is that

$$\psi(x) = \sum_{k=1}^{\infty} \theta(x^{1/k}) \qquad (x \geq 1). \tag{26}$$

For each x, summation over k is finite: the general term vanishes for $2^k > x$.

Theorem 10. *For $x \geq 2$, we have*

$$\psi(x) = \theta(x) + O(\sqrt{x}), \tag{27}$$

$$\pi(x) = \frac{\theta(x)}{\log x} + O\Big(\frac{x}{(\log x)^2}\Big). \tag{28}$$

Proof. Formula (27) follows immediately from (25) and (26). In order to establish (28) we evaluate $\theta(x)$ by partial summation:

$$\theta(x) = \int_1^x \log t \,\mathrm{d}\pi(t) = \pi(x) \log x - \int_1^x \frac{\pi(t)}{t} \,\mathrm{d}t.$$

By Chebyshev's upper bound (Theorem 1.3) the integral is $O(x/\log x)$, giving the stated formula.

Corollary 10.1. *Let α, β be constants such that $0 < \alpha < \log 2$, $\beta > \log 4$. For all sufficiently large x, we have*

$$\alpha x \leq \theta(x) \leq \psi(x) \leq \beta x. \tag{29}$$

This follows immediately from Theorem 10 and Theorem 1.3.

§ 2.7 Euler's totient function.

For each $n \geq 1$ we have defined $\varphi(n)$ as the number of invertible residues modulo n. We can hence write

$$\varphi(n) = \sum_{1 \leq m \leq n} \delta\big((m,n)\big),$$

from which we infer by applying (14) that

$$\varphi(n) = \sum_{m \leq n} \sum_{d|(m,n)} \mu(d) = \sum_{d|n} \mu(d) \sum_{\substack{m \leq n \\ m \equiv 0 \,(\mathrm{mod}\, d)}} 1 = \sum_{d|n} \mu(d) \frac{n}{d}.$$

This relation is equivalent to

(30) $$\varphi = \mu * j.$$

In particular for each prime p we have

$$\varphi(p^\nu) = \mu(1)p^\nu + \mu(p)p^{\nu-1} = p^\nu \Big(1 - \frac{1}{p}\Big).$$

We have therefore proved the following theorem.

Theorem 11. *Euler's totient function φ is multiplicative. For any integer $n \geq 1$, we have*

(31) $$\varphi(n) = n \prod_{p|n} \Big(1 - \frac{1}{p}\Big).$$

Another proof of this result consists in noting that each rational fraction with denominator n can be written in the form $h/n = a/d$, with $d|n$, $(a,d) = 1$. For any function F of a real variable we can hence write

(32) $$\sum_{1 \leq h \leq n} F(h/n) = \sum_{d|n} \sum_{\substack{1 \leq a \leq d \\ (a,d)=1}} F(a/d).$$

Applying this identity to the special case $F(x) \equiv 1$, we obtain that

(33) $$n = \sum_{d|n} \varphi(d).$$

In other words, $j = \mathbf{1} * \varphi$, and the desired result follows from (14).

The *inclusion–exclusion principle* (cf. Notes) provides an easy third proof of Theorem 11.

Notes

§ **2.4.** For other properties of the ring of arithmetic functions, see Shapiro (1972).

§ **2.5.** *The inclusion–exclusion principle.* A purely combinatorial version of the Möbius inversion formula in its basic form $\mathbf{1} * \mu = \delta$ is known as the inclusion–exclusion principle. It can be stated thus:

Let $\mathcal{A}$ be a finite set of cardinality N and $\mathcal{P} = \{(1), \ldots, (k)\}$ be a family of numbered properties. For each subset I of $\mathcal{P}$ let $A(I)$ be the number of elements of $\mathcal{A}$ which satisfy all of the properties in I. Writing $S(\mathcal{A}, \mathcal{P})$ for the number of elements of $\mathcal{A}$ which do not satisfy any property in $\mathcal{P}$, we have

$$S(\mathcal{A}, \mathcal{P}) = N + \sum_{s=1}^{k} (-1)^s \sum_{\substack{I \subseteq \mathcal{P} \\ \operatorname{card} I = s}} A(I). \tag{34}$$

It is easy to prove the identity (34) directly by introducing, for each element a of $\mathcal{A}$, the number $m(a)$, $0 \le m(a) \le k$, of properties satisfied by a. It is also possible to argue indirectly as follows. Let $2 = p_1 < p_2 < \cdots < p_k$ be the k smallest prime numbers, write

$$P = \prod_{i=1}^{k} p_i$$

and to each a in $\mathcal{A}$ associate the integer $F(a) = \prod_{a \in (i)} p_i$ where "$a \in (i)$" means that a satisfies property (i). Then we have

$$S(\mathcal{A}, \mathcal{P}) = \sum_{a \in \mathcal{A}} \delta\big(F(a)\big) = \sum_{a \in \mathcal{A}} \sum_{d | F(a)} \mu(d) = \sum_{d | P} \mu(d) \sum_{\substack{a \in \mathcal{A} \\ F(a) \equiv 0 (\operatorname{mod} d)}} 1.$$

If $d = \prod_{(i) \in I} p_i$, we certainly have

$$\mu(d) = (-1)^{\operatorname{card} I}, \qquad \text{and} \qquad \sum_{\substack{a \in \mathcal{A} \\ F(a) \equiv 0 (\operatorname{mod} d)}} 1 = A(I).$$

This implies (34).

Exercises

1. Let $\zeta(s)$ be the Riemann zeta function, which we identify with the formal Dirichlet series associated with the arithmetic function $\mathbf{1}(n) = 1$ $(n \geq 1)$. Express in terms of $\zeta(s)$ the Dirichlet series associated with the following arithmetic functions: $\mu(n)$, $\mu(n)^2$, $\varphi(n)$, $\sigma(n)$, $\tau(n)$, $2^{\omega(n)}$, $\Lambda(n)$. Find the convolution relations which correspond to these identities.

2. Show, for all $n \geq 1$, that $2^{\omega(n)} \leq \tau(n) \leq 2^{\Omega(n)}$.

3. Show, for all $n > 1$, that $6n^2/\pi^2 < \sigma(n)\varphi(n) < n^2$.

4. Show that each integer $n \geq 1$ can be uniquely decomposed as $n = qm^2$, where q is squarefree. Let $Q(x)$ denote the number of squarefree integers q not exceeding x. Show that:

(a) $$[x] = \sum_{m \leq \sqrt{x}} Q(x/m^2) \qquad (x \geq 1)$$

(b) $$Q(x) = \sum_{d \leq \sqrt{x}} \mu(d)[x/d^2] \qquad (x \geq 1)$$

(c) $$Q(x) = 6x/\pi^2 + O(\sqrt{x}) \qquad (x \geq 1).$$

Generalise these considerations to "k-free" integers, that is, integers such that $p|n \Rightarrow p^k \nmid n$.

5. *Ramanujan sums.* Define the *Ramanujan sums* $c_n(m)$ by

$$c_n(m) := \sum_{1 \leq h \leq n,\, (h,n)=1} e(hm/n) \qquad (m, n \geq 1)$$

with the notation $e(u) := \exp\{2\pi i u\}, u \in \mathbb{R}$. Show that:

(a) $$c_n(m) = \sum_{d|(m,n)} \mu(n/d)d$$

(b) $c_n(m) = \mu(t)\varphi(n)/\varphi(t)$, with $t = n/(m, n)$.

Deduce that $$\mu(n) = \sum_{1 \leq h \leq n,\, (h,n)=1} e(h/n).$$

Show that, for each fixed m, $c_n(m)$ is a multiplicative function of n. Give a direct proof of this result.

6. Let f be a function defined on $\mathbb{R}^+$ and such that

$$\sum_{m=1}^{\infty} |f(mx)|\tau(m) < \infty.$$

Write $g(x) = \sum_{m=1}^{\infty} f(mx)$. Then $f(x) = \sum_{n=1}^{\infty} \mu(n)g(nx)$. Find a converse to this property.

7. Define an arithmetic function $R(n)$ by

$$R(n) := \operatorname{card}\{(d, d') : d \geq 1,\, d' \geq 1,\, n = [d, d']\}.$$

(a) Show that $R(n)$ is a multiplicative function of n.

(b) Calculate $\sum_{d|n} R(d)$ and deduce an alternative proof of (a).

(c) Show that the formal Dirichlet series $F(s) := \sum_{d,d' \geq 1} [d, d']^{-s}$ can be simply expressed as a function of $\zeta(s)$.

(d) Study the generalisation $\sum_{d_1,\ldots,d_k \geq 1} [d_1, \ldots, d_k]^{-s}$.

8. Let f be a real-valued arithmetic function. Suppose that there exist prime numbers p and q such that $f(pq)^2 < 4f(p^2)f(q^2)$. Show that f is either a unit or a prime element in the ring $\mathcal{A}_r$ of real arithmetic functions.

9. Let f be a complex-valued arithmetic function such that $f(1) = 0$.

(a) Show that if $f = u * v$, with u and v not units, then for each prime pair (p, q) the solutions of the equation $z^2 - f(pq)z + f(p^2)f(q^2) = 0$ are $u(p)v(q)$ and $u(q)v(p)$.

(b) Define $g_i(p, q)$ $(i = 1, 2)$ to be the two complex numbers of the form

$$g_i(p, q) = f(pq) \pm \sqrt{f(pq)^2 - 4f(p^2)f(q^2)}$$

where a determination of a complex square root has been chosen. Show that if there exist four prime numbers p, q, r, s such that the 16 determinants of the form

$$\begin{vmatrix} g_i(p, q) & g_k(q, s) \\ g_j(p, r) & g_\ell(r, s) \end{vmatrix} \qquad (i, j, k, \ell = 1, 2)$$

are all non-zero, then f is a prime element in $\mathcal{A}$.

10. *On monotone multiplicative functions.* Let $f : \mathbb{Z}^+ \to \mathbb{R}$ be a non-decreasing multiplicative arithmetic function. For integers $a \geq 3$, $t \geq 1$, put $R_t := \sum_{j=0}^{t} a^j$, $S_t := a^t - \sum_{j=0}^{t-1} a^j$. Show that $f(R_t) \geq f(a)^t \geq f(S_t)$. Deduce that for all integers a, b, n exceeding 2 we have $f(b)^{s-1} \leq f(n) \leq f(a)^{r+2}$ with $r := [\log n / \log a]$, $s := [\log n / \log b]$. Show that $f(a)^{1/\log a} = f(b)^{1/\log b}$ and hence that there exists some non-negative constant k such that $f(n) = n^k$ for all $n \geq 1$. [*This result was first proved by Erdős (1946). The simple proof given above is due to Moser & Lambek (1953).*]

I.3

Average orders

§ 3.1 Introduction

We define an average order of an arithmetic function f to be any elementary function g of a real variable such that

$$\sum_{n \le x} f(n) \sim \sum_{n \le x} g(n). \tag{1}$$

A given function f may possess several average orders. In general one seeks a function g which is simply expressible in terms of elementary functions, and whose asymptotic behaviour is easy to determine. Of course, given the choice between several candidates, it is preferable to use the one which minimises the error term in the asymptotic relation (1).

Arithmetic functions often have very erratic behaviour, which is difficult to analyse, even with the aid of very extensive numerical tables. For example this is the case for the divisor function $\tau(n)$, which oscillates between the value 2 (attained at each prime number) and occasional very large values—see Chapter 5. The averaging procedure often has a substantial regularising effect by masking atypical values which occur only rarely. Normally, and in particular when f takes only positive values, the average order is the easiest non-trivial piece of information to determine for an arithmetic function.

§ 3.2 Dirichlet's problem and the hyperbola method

We investigate here the average order of the divisor function $\tau(n)$. A simple summation interchange gives a first estimate. Indeed we have

$$\begin{aligned}\sum_{n \le x} \tau(n) &= \sum_{n \le x} \sum_{d \mid n} 1 = \sum_{d \le x} \sum_{\substack{n \le x \\ n \equiv 0 (\mathrm{mod}\, d)}} 1 = \sum_{d \le x} \left[\frac{x}{d}\right] \\ &= \sum_{d \le x} \left(\frac{x}{d} + O(1)\right) = x \log x + O(x).\end{aligned}$$

It is possible to improve this estimate using a general principle, due to Dirichlet in its primitive form, concerning the average of a convolution product.

Theorem 1 (Dirichlet's hyperbola method). *Let f, g be two arithmetic functions, with respective summatory functions F, G. For $1 \le y \le x$, we have*

$$\sum_{n \le x} f * g(n) = \sum_{n \le y} g(n) F(x/n) + \sum_{m \le x/y} f(m) G(x/m) - F(x/y) G(y). \tag{2}$$

Proof. The left-hand side of (2) can be written

$$\begin{aligned}\sum_{md \le x} f(m) g(d) &= \sum_{md \le x,\, d \le y} f(m) g(d) + \sum_{md \le x,\, d > y} f(m) g(d) \\ &= \sum_{d \le y} g(d) F(x/d) + \sum_{m \le x/y} f(m) \{G(x/m) - G(y)\}.\end{aligned}$$

Expanding the last term, this immediately implies the stated result.

As noted above this result yields more precise information on the average behaviour of $\tau(n)$.

Theorem 2. *As x tends to infinity, we have*

$$\sum_{n \le x} \tau(n) = x(\log x + 2\gamma - 1) + O(\sqrt{x}) \tag{3}$$

where γ denotes Euler's constant.

Proof. Apply (2) with $f = g = \mathbf{1}$, $F(x) = G(x) = [x]$ and $y = \sqrt{x}$. We find

$$\begin{aligned}\sum_{n \le x} \tau(n) &= 2 \sum_{m \le \sqrt{x}} [x/m] - [\sqrt{x}]^2 \\ &= 2x \sum_{m \le \sqrt{x}} \frac{1}{m} - x + O(\sqrt{x}).\end{aligned} \tag{4}$$

The sum over m may be evaluated by Theorem 0.5 which yields

$$\tfrac{1}{2} \log x + \gamma + O(1/\sqrt{x}).$$

Inserting in (4), this yields (3).

The name *hyperbola method* comes from the following geometric interpretation of Dirichlet's problem.

In the (u, v)-plane, let us consider the rectangular hyperbola $uv = x$.

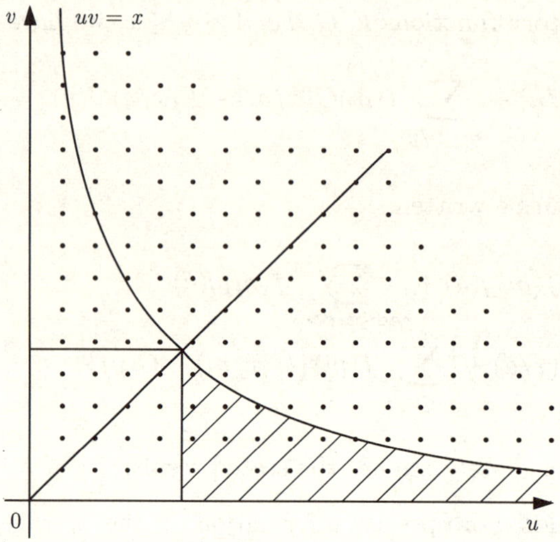

Each lattice point (a, b) lying between the curve and the coordinate axes (excluded) corresponds to a decomposition of the integer $n = ab \leq x$ as a product of two factors. Hence the left-hand side of (3) is exactly equal to the number of lattice points lying under the curve. The calculation carried out in the proof of Theorem 2 consists simply in exploiting the symmetry of the figure: the sought-after quantity equals the number of points contained in the square $[1, \sqrt{x}]^2$ augmented by twice the number of points contained in the hatched zone.

Let us write

$$\Delta(x) := \sum_{n \leq x} \tau(n) - x(\log x + 2\gamma - 1).$$

We have seen that $\Delta(x) = O(\sqrt{x})$. In 1903 Voronoï showed that

$$\Delta(x) \ll x^{1/3} \log x$$

and in 1915 Hardy and Landau proved independently that $\Delta(x)$ is not $o(x^{1/4})$. Let α be the infimum of the set of exponents ξ such that

$$\Delta(x) \ll x^{\xi}.$$

The exact value of α remains unknown. It is generally conjectured that $\alpha = 1/4$, and the best upper bound known to date is that of Huxley (1993a)

$$\alpha \leq 23/73 \approx 0.315068,$$

improving on the estimate of Iwaniec & Mozzochi (1988) $\alpha \leq 7/22$. In Chapter 6 we shall show how van der Corput's method (1922) can be applied in order to prove Voronoï's result in a simple way.

§ 3.3 The sum of divisors function

The following theorem shows that the average order of the function

$$\sigma(n) = \sum_{d|n} d$$

is $(\pi^2/6)n$. Actually we obtain a more precise result.

Theorem 3. *As x tends to infinity, we have*

$$\sum_{n \le x} \sigma(n) = \frac{\pi^2}{12} x^2 + O(x \log x). \tag{5}$$

Proof. The idea of the proof is very simple: we write the function under consideration as a sum taken over divisors of n, and we interchange summations. We already met this approach in the previous section and we shall meet it again in other contexts in this chapter. Thus

$$\sum_{n \le x} \sigma(n) = \sum_{md \le x} d = \tfrac{1}{2} \sum_{m \le x} \left[\frac{x}{m}\right] \left(\left[\frac{x}{m}\right] + 1\right)$$

$$= \tfrac{1}{2} \sum_{m \le x} \frac{x^2}{m^2} + O\Big(x \sum_{m \le x} \frac{1}{m}\Big).$$

The stated result follows, given the classical formula

$$\sum_{m=1}^{\infty} \frac{1}{m^2} = \frac{\pi^2}{6}. \tag{6}$$

In Exercise 1 we suggest two methods for obtaining (6).

The best remainder term known to date for the asymptotic formula (5) is $O\big(x(\log x)^{2/3}\big)$. It is due to Walfisz (1963), p. 99.

§ 3.4 Euler's totient function

Starting from the convolution formula

$$\varphi(n) = \sum_{md=n} \mu(d) m, \tag{7}$$

the method of Theorem 3 applies without change.

Theorem 4. *As x tends to infinity, we have*

$$\sum_{n\le x}\varphi(n)=\frac{3}{\pi^2}x^2+O(x\log x). \tag{8}$$

Proof. By (7), the sum to be evaluated is

$$\sum_{d\le x}\mu(d)\sum_{m\le x/d} m=\tfrac12\sum_{d\le x}\mu(d)\left[\frac{x}{d}\right]\left(\left[\frac{x}{d}\right]+1\right)$$
$$=\tfrac12 x^2\sum_{d\le x}\frac{\mu(d)}{d^2}+O\Big(x\sum_{d\le x}\frac{1}{d}\Big).$$

Since the Möbius function is the convolution inverse of **1**, we have formally

$$\Big(\sum_{d=1}^{\infty}\frac{\mu(d)}{d^2}\Big)\Big(\sum_{d=1}^{\infty}\frac{1}{d^2}\Big)=1.$$

Each of the two series being absolutely convergent, this equality is in fact valid in the usual sense, and it follows from (6) that we have

$$\sum_{d=1}^{\infty}\frac{\mu(d)}{d^2}=\frac{6}{\pi^2}. \tag{9}$$

This implies the stated formula.

The best error term for (8) known at present is $O\big(x(\log x)^{2/3}(\log_2 x)^{4/3}\big)$— cf. Walfisz (1963), p. 144.

For each $N\ge 1$, the *Farey series of order* N, denoted $\mathcal{F}_N$, is the ordered sequence of rational fractions in lowest terms having denominator not greater than N. Thus we have

$$\mathcal{F}_1=\{\tfrac01,\tfrac11\},\quad \mathcal{F}_2=\{\tfrac01,\tfrac12,\tfrac11\},\qquad \mathcal{F}_3=\{\tfrac01,\tfrac13,\tfrac12,\tfrac23,\tfrac11\},\quad \text{etc}\ldots$$

From the definition of $\varphi(n)$, we immediately note that

$$F(N):=\operatorname{card}\mathcal{F}_N=1+\sum_{n\le N}\varphi(n).$$

Hence it follows from Theorem 4 that

$$F(N)\sim\frac{3}{\pi^2}N^2\qquad (N\to\infty).$$

This estimate may be interpreted in the following way: $F(N)-1$ equals the number of rational fractions m/n, $1\le m\le n\le N$, which are in lowest terms. Since the total number of pairs $\{m,n\}$ is $\frac12 N(N+1)$, it follows that *the probability that a rational fraction in* $]0,1]$ *having denominator not exceeding* N *is in lowest terms tends to* $6/\pi^2$ *as* N *tends to infinity.*

We now generalise this statement.

Theorem 5. *Let* $G(x,y)$ *denote the number of pairs of integers* $\{m,n\}$ *such that* $1 \le m \le x$, $1 \le n \le y$, $(m,n) = 1$. *Then* $G(x,y) \sim (6/\pi^2)xy$ *as* x *and* y *tend to infinity. More precisely, setting* $z := \min(x,y)$ *we have*

$$G(x,y) = xy\Big\{\frac{6}{\pi^2} + O\Big(\frac{\log z}{z}\Big)\Big\} \qquad (x,y \ge 2).$$

Proof. From the Möbius inversion formula (2.14), we have

$$\begin{aligned} G(x,y) &= \sum_{m\le x,\, n\le y} \delta\big((m,n)\big) = \sum_{d\le z} \mu(d) \left[\frac{x}{d}\right] \left[\frac{y}{d}\right] \\ &= xy \sum_{d\le z} \frac{\mu(d)}{d^2} + O\Big((x+y)\sum_{d\le z}\frac{1}{d}\Big) \\ &= xy\Big\{\frac{6}{\pi^2} + O\Big(\frac{1}{z} + \Big(\frac{1}{x}+\frac{1}{y}\Big)\log z\Big)\Big\}, \end{aligned}$$

as required.

§ 3.5 The functions ω and Ω

Theorem 6. *As* x *tends to infinity, we have*

$$\sum_{n\le x} \omega(n) = x\log_2 x + c_1 x + O\Big(\frac{x}{\log x}\Big) \tag{10}$$

where $c_1 \approx 0.261497$ *is the constant appearing in Theorem 1.9.*

Proof. We have

$$\sum_{n\le x} \omega(n) = \sum_{n\le x}\sum_{p\mid n} 1 = \sum_{p\le x} [x/p] = x\sum_{p\le x} 1/p + O(\pi(x)).$$

Formula (10) now follows from Theorem 1.9 and the upper estimate of Chebyshev (Theorem 1.3).

The case of the function $\Omega(n) = \sum_{p^\nu \| n} \nu$ is similar, but a little more delicate.

Theorem 7. *As* x *tends to infinity, we have*

$$\sum_{n\le x} \Omega(n) = x\log_2 x + c_2 x + O\left(\frac{x}{\log x}\right) \tag{11}$$

with

$$c_2 = c_1 + \sum_p \frac{1}{p(p-1)} \approx 1.034653. \tag{12}$$

Proof. We have

$$A(x) := \sum_{n\le x}\{\Omega(n)-\omega(n)\} = \sum_p\sum_{\nu\ge 2}[x/p^\nu].$$

On the one hand

$$A(x) \le x\sum_p\sum_{\nu\ge 2}p^{-\nu} = x\sum_p\frac{1}{p(p-1)},$$

and on the other

$$\begin{aligned}A(x) \ge \sum_{p\le\sqrt{x}}\ \sum_{2\le\nu\le\log x/\log p}(xp^{-\nu}-1) &= \sum_{p\le\sqrt{x}}\Big\{\frac{x}{p(p-1)}+O\Big(\frac{\log x}{\log p}\Big)\Big\}\\ &= x\sum_p\frac{1}{p(p-1)}+O(\sqrt{x}),\end{aligned}$$

by Theorem 1.3. Hence

$$A(x) = x(c_2-c_1)+O(\sqrt{x})$$

and formula (11) follows immediately from (10).

§ 3.6 Mean value of the Möbius function and the summatory functions of Chebyshev

We saw in Section 2.6 that von Mangoldt's function

$$\Lambda = \mu * \log$$

is closely linked to the characteristic function of the prime numbers and has a summatory function

$$\psi(x) = \sum_{n\le x}\Lambda(n)$$

satisfying

$$\psi(x)\sim\pi(x)\log x \qquad (x\to\infty).$$

Therefore it is natural to ask if the average value of $\mu(n)$ has a simple interpretation in terms of the asymptotic behaviour of Chebyshev's functions $\pi(x)$, $\theta(x)$, $\psi(x)$. The following theorem, due to Landau (1909), provides a complete answer to this question.

Theorem 8. *The following three statements are elementarily equivalent:*

(i) $\psi(x) \sim x \qquad (x \to \infty),$

(ii) $M(x) := \sum_{n \le x} \mu(n) = o(x) \qquad (x \to \infty),$

(iii) $\sum_{n=1}^{\infty} \mu(n)/n = 0.$

Remark. Statement (iii) means of course that the series appearing on the left-hand side converges and has sum zero.

Proof. The implication (iii) ⇒ (ii) results from a simple partial summation. Indeed, assuming that

$$m(x) := \sum_{n \le x} \mu(n)/n = o(1) \quad (x \to \infty),$$

it follows that

$$M(x) = \int_{1-}^{x} t \, \mathrm{d}m(t) = xm(x) - \int_{1}^{x} m(t) \, \mathrm{d}t = o(x).$$

In order to establish the implication (ii) ⇒ (i), we first set up a convolution identity for the function $\Lambda - \mathbf{1}$, the summatory function of which we have to show to be $o(x)$. We have

$$\begin{aligned} \Lambda - \mathbf{1} = (\log - \tau) * \mu &= (\log - \tau + 2\gamma \mathbf{1}) * \mu - 2\gamma\delta \\ &= f * \mu - 2\gamma\delta, \end{aligned}$$

say, where the function f satisfies

(13) $$F(x) := \sum_{n \le x} f(n) = O(\sqrt{x}).$$

This follows from Theorem 2 and from the estimate of Corollary 0.2.1 for $\sum_{n \le x} \log n$.

We shall show that

(14) $$H(x) := \sum_{n \le x} f * \mu(n) = o(x)$$

using (13) and the hyperbola method. By Theorem 1 we can write for each fixed $y > 2$

$$H(x) = \sum_{n \le x/y} \mu(n) F(x/n) + \sum_{m \le y} f(m) M(x/m) - F(y) M(x/y).$$

Under assumption (ii) it then follows that

$$\begin{aligned} \limsup_{x \to \infty} |x^{-1} H(x)| &\le \limsup_{x \to \infty} \Big| x^{-1} \sum_{n \le x/y} F(x/n) \Big| \\ &\ll \limsup_{x \to \infty} x^{-1} \sum_{n \le x/y} \sqrt{(x/n)} \ll y^{-1/2}. \end{aligned}$$

Since y is arbitrarily large, this implies (14).

It remains to show that (i) $\Rightarrow$ (iii). Note first of all that the convolution relation $\mu * \mathbf{1} = \delta$ yields by summation that

$$1 = \sum_{n \leq x} \mu(n)\,[x/n] = xm(x) + O(x)$$

from which we deduce that

$$m(x) = O(1). \tag{15}$$

Again the same relation implies

$$\sum_{md=n} \frac{\mu(m)}{md} = \delta(n)$$

from which we obtain

$$\sum_{m \leq x} \frac{\mu(m)}{m} \sum_{d \leq x/m} \frac{1}{d} = 1 \qquad (x \geq 1).$$

Working out the inner sum by means of Theorem 0.5, we deduce that

$$\begin{aligned} 1 &= \sum_{m \leq x} \frac{\mu(m)}{m} \Big\{ \log\Big(\frac{x}{m}\Big) + \gamma + O\Big(\frac{m}{x}\Big) \Big\} \\ &= m(x)(\log x + \gamma) - G(x) + O(1), \end{aligned}$$

where we have written

$$G(x) := \sum_{m \leq x} \frac{\mu(m) \log m}{m}.$$

It hence suffices to establish that

$$G(x) = o(\log x) \qquad (x \to \infty). \tag{16}$$

To this end we use the convolution formula (2.20):

$$\mu \log = -\Lambda * \mu = (\mathbf{1} - \Lambda) * \mu - \delta.$$

Thus

$$G(x) = \sum_{jk \leq x} \Big(\frac{1 - \Lambda(j)}{j} \Big) \frac{\mu(k)}{k} - 1 = -1 + \int_{1-}^{x} \frac{m(x/t)}{t} \, \mathrm{d}R(t)$$

with

$$R(t) := [t] - \psi(t) \qquad (t \geq 1),$$

so that by hypothesis

$$R(t) = o(t) \qquad (t \to \infty).$$

By partial summation, it follows that

$$\begin{aligned} G(x) &= -1 + \int_1^x t^{-2} m(x/t) R(t)\,\mathrm{d}t - \int_{1-}^x t^{-1} R(t)\,\mathrm{d}m(x/t) \\ &= O(1) + \int_1^x o(1/t)\,\mathrm{d}t - \int_{1-}^x o(1)|\,\mathrm{d}m(x/t)|. \end{aligned}$$

The first integral is $o(\log x)$. In view of the inequality between Stieltjes measures

$$|\,\mathrm{d}m(y)| \leq \mathrm{d}\Big\{\sum_{n\leq y} 1/n\Big\},$$

we note that the same holds for the second, from which we deduce (16) and finally (iii).

Corollary 8.1. *The following assertions are elementarily equivalent to those of Theorem 8:*

(iv) $\quad \pi(x) \sim x/\log x \qquad (x \to \infty),$

(v) $\quad \theta(x) \sim x \qquad (x \to \infty),$

(vi) $\quad \sum_{n\leq x} \Lambda(n)/n = \log x - \gamma + o(1) \qquad (x \to \infty).$

Proof. The cases of (iv) and (v) follow immediately from Theorem 2.10. The implication (vi) ⇒ (i) is proved by mere partial summation, the details of which we leave to the reader. For the converse, we again use the function $f = \log - \tau + 2\gamma\mathbf{1}$ introduced in the proof of Theorem 8. We have

$$\begin{aligned} \sum_{n\leq x} \frac{\Lambda(n)-1}{n} &= \sum_{kd\leq x} \frac{f(k)}{k}\frac{\mu(d)}{d} - 2\gamma \\ &= \sum_{d\leq x/y} \frac{\mu(d)}{d} E\Big(\frac{x}{d}\Big) + \sum_{k\leq y} \frac{f(k)}{k} m\Big(\frac{x}{k}\Big) - E(y)\, m\Big(\frac{x}{y}\Big) - 2\gamma \end{aligned} \tag{17}$$

with

$$E(z) := \sum_{k\leq z} f(k)/k = C + O(1/\sqrt{z}) \qquad (z \geq 1).$$

This last relation follows by partial summation from the estimate (13) for a suitable constant C. Letting x and then y tend to infinity, we obtain as before, using (iii), that (17) has value $-2\gamma + o(1)$, giving the required conclusion.

§ 3.7 Squarefree integers

When an arithmetic function takes only the values 0 or 1, it can be regarded as the characteristic function of an integer sequence $\mathcal{A}$. The study of the average order then merges with that of the counting function

$$A(x) := \big|\mathcal{A} \cap [1, x]\big|.$$

The multiplicative function $\mu(n)^2$ is a good example of such a situation. Its summatory function

$$Q(x) := \sum_{n \le x} \mu(n)^2$$

equals the number of squarefree integers not exceeding x.

Theorem 9. *As x tends to infinity, we have*

$$Q(x) = \frac{6}{\pi^2}x + O(\sqrt{x}). \tag{18}$$

Proof. As in the cases of $\tau(n)$, $\sigma(n)$ or $\varphi(n)$, we want to write $\mu(n)^2$ as a convolution product. In order to achieve this, let us canonically decompose each integer n in the form

$$n = qm^2, \quad \mu(q)^2 = 1. \tag{19}$$

This factorisation is unique: indeed q may be defined as the product of those prime factors of n with an odd exponent. Hence we have

$$\mu(n)^2 = \delta(m) = \sum_{d|m} \mu(d). \tag{20}$$

Noting that $d \mid m$ is equivalent to $d^2 \mid n$, we obtain that

$$Q(x) = \sum_{n \le x} \sum_{d^2|n} \mu(d) = \sum_{d \le \sqrt{x}} \mu(d) \left[\frac{x}{d^2}\right] = \frac{6}{\pi^2}x + O\Big(x \sum_{d > \sqrt{x}} d^{-2} + \sqrt{x}\Big),$$

where the last step follows from (9). This yields (18).

Remark. We could have adopted the following slightly different argument. Denote, for each $m \le \sqrt{x}$, by $\mathcal{A}_m$ the set of integers $n \le x$ satisfying (19). Then $\{n : n \le x\}$ is the disjoint union of the $\mathcal{A}_m$, and we have

$$|\mathcal{A}_m| = Q(x/m^2),$$

from which

$$\sum_{m\le\sqrt{x}} Q(x/m^2) = [x] \qquad (x \ge 1).$$

Write $x = y^2$, and apply the second Möbius inversion formula (Theorem 2.9) to the functions $F(y) = [y^2]$ and $G(y) = Q(y^2)$. Then it follows, as before, that

$$Q(x) = \sum_{d\le\sqrt{x}} \mu(d)[x/d^2] \qquad (x \ge 1).$$

The following result shows that the prime number theorem furnishes an improvement on the error term in (18).

Theorem 10. *Under the assumption*

(21) $$M(x) = \sum_{n\le x} \mu(n) = o(x),$$

we have

(22) $$Q(x) = \frac{6}{\pi^2}x + o(\sqrt{x}) \qquad (x \to \infty).$$

Proof. Relation (20) can be rewritten as

$$\mu^2 = \lambda * \mathbf{1}$$

where λ is the arithmetic function defined by

$$\lambda(n) = \begin{cases} \mu(d) & \text{if } n = d^2, \\ 0 & \text{if } n \text{ is not a square.} \end{cases}$$

By the hyperbola method, we obtain, for $1 \le y \le x$, that

$$\begin{aligned} Q(x) &= \sum_{d\le\sqrt{(x/y)}} \mu(d)\,[x/d^2] + \sum_{m\le y} M\big(\sqrt{(x/m)}\big) - [y]M\big(\sqrt{(x/y)}\big) \\ &= \sum_{d\le\sqrt{(x/y)}} \mu(d)\,\{x/d^2 + O(1)\} + o_y(\sqrt{x}) \\ &= (6/\pi^2)x - x\int_{\sqrt{(x/y)}}^{\infty} t^{-2}\,\mathrm{d}M(t) + O\big(\sqrt{(x/y)}\big) + o_y(\sqrt{x}). \end{aligned}$$

Now for all $z \ge 1$ we have

$$\begin{aligned} \int_z^{\infty} t^{-2}\,\mathrm{d}M(t) &= -z^{-2}M(z) + 2\int_z^{\infty} t^{-3}M(t)\,\mathrm{d}t \\ &= o(1/z) + \int_z^{\infty} o(1/t^2)\,\mathrm{d}t = o(1/z). \end{aligned}$$

We can therefore write

$$Q(x) = \frac{6x}{\pi^2} + O\big(\sqrt{(x/y)}\big) + o_y(\sqrt{x}),$$

as a consequence of which we have for each fixed $y \geq 1$

$$\limsup_{x\to\infty} |Q(x) - 6x/\pi^2|/\sqrt{x} \ll 1/\sqrt{y}.$$

This implies the required conclusion (22) by letting y tend to infinity.

§ 3.8 Mean value of a multiplicative function with values in $[0,1]$

Theorem 9 states that the average order of the function $\mu(n)^2$ is the constant

$$\frac{6}{\pi^2} = \prod_p \left(1 - 1/p^2\right) = \prod_p (1 - 1/p)\left(1 + \mu(p)^2/p\right).$$

One says that $\mu(n)^2$ possesses a *mean value*, equal to $6/\pi^2$. It is easy to construct arithmetic functions, even with values in $[0,1]$, which fail to have a mean value. For example, with

$$f(n) := \begin{cases} 1 & \text{if } 2^{2k} < n \leq 2^{2k+1} \quad (k = 0, 1, \ldots), \\ 0 & \text{if } 2^{2k+1} < n \leq 2^{2k+2} \quad (k = 0, 1, \ldots), \end{cases}$$

it can easily be shown that

$$\liminf_{x\to\infty} x^{-1} \sum_{n \leq x} f(n) = \tfrac{1}{3}, \qquad \limsup_{x\to\infty} x^{-1} \sum_{n\leq x} f(n) = \tfrac{2}{3}.$$

The following theorem settles the question in the case of multiplicative functions.

Theorem 11. *Let f be a multiplicative function with values in $[0,1]$. Set*

$$M(f) := \prod_p (1 - p^{-1}) \sum_{\nu=0}^{\infty} f(p^\nu) p^{-\nu}$$

where the infinite product is considered to be 0 when it diverges. Then we have

$$\sum_{n\leq x} f(n) = x\{M(f) + o(1)\} \qquad (x \to \infty). \tag{23}$$

Proof. For each $y > 2$, we introduce the auxiliary completely multiplicative functions α_y and β_y defined by

$$\alpha_y(p) = \begin{cases} 1 & \text{if } p \le y, \\ 0 & \text{if } p > y, \end{cases} \qquad \beta_y(p) = \begin{cases} 0 & \text{if } p \le y, \\ 1 & \text{if } p > y. \end{cases}$$

It is easy to check that for any multiplicative function f one has

$$f = f\alpha_y * f\beta_y.$$

When f takes values in $[0,1]$, this implies that

$$f \le f_y := f\alpha_y * \beta_y.$$

Moreover $\beta_y = \mathbf{1} * h_y$ where $h_y := \beta_y * \mu$ is defined by

$$h_y(p^\nu) = \begin{cases} -1 & \text{if } p \le y \text{ and } \nu = 1, \\ 0 & \text{if } p > y \text{ or } \nu \ge 2. \end{cases}$$

In particular $h_y(n) = 0$ whenever $n > N(y) := \prod_{p \le y} p$. This implies that, as $x \to \infty$,

$$x^{-1} \sum_{n \le x} \beta_y(n) = x^{-1} \sum_{m \le N(y)} h_y(m) \left[\frac{x}{m}\right] \to \sum_{m=1}^{\infty} \frac{h_y(m)}{m} = \prod_{p \le y} (1 - p^{-1}).$$

Thus, for each fixed y, β_y has a mean value equal to $M(\beta_y)$. Now we can write

$$x^{-1} \sum_{n \le x} f_y(n) = \sum_{d \le x} \frac{f(d)\alpha_y(d)}{d} \cdot \frac{d}{x} \sum_{m \le x/d} \beta_y(m). \tag{24}$$

Since the series of non-negative terms

$$\sum_{d=1}^{\infty} \frac{f(d)\alpha_y(d)}{d} = \prod_{p \le y} \sum_{\nu=0}^{\infty} f(p^\nu) p^{-\nu}$$

converges, we deduce from (24), by the theorem of dominated convergence, that f_y possesses a mean value, equal to

$$\sum_{d=1}^{\infty} \frac{f(d)\alpha_y(d)}{d} M(\beta_y) = \prod_{p \le y} (1 - p^{-1}) \sum_{\nu=0}^{\infty} f(p^\nu) p^{-\nu} = M(f_y).$$

Hence, for each fixed $y > 2$,

$$x^{-1} \sum_{n \le x} f(n) \le M(f_y) + o(1) \qquad (x \to \infty). \tag{25}$$

Now $M(f_y)$ is a decreasing function of y. Thus if the infinite product $M(f)$ diverges, $M(f_y) \to 0$ as $y \to \infty$. In this case we conclude that f has zero mean value. If the product $M(f)$ converges, the same holds for the series

$$\sum_p \frac{1 - f(p)}{p}.$$

For all $n \ge 1$, we then have

$$\begin{aligned} f_y(n) - f(n) = \prod_{p^\nu \| n,\, p \le y} f(p^\nu) - \prod_{p^\nu \| n} f(p^\nu) &\le 1 - \prod_{p^\nu \| n,\, p > y} f(p^\nu) \\ &\le \sum_{p^\nu \| n,\, p > y} (1 - f(p^\nu)) \end{aligned}$$

from which we infer that

$$\begin{aligned} \sum_{n \le x} (f_y(n) - f(n)) &\le \sum_{p > y} \sum_{\nu = 1}^{\infty} (1 - f(p^\nu))[x/p^\nu] \\ &\le x \left\{ \sum_{p > y} \frac{1 - f(p)}{p} + \frac{1}{p(p-1)} \right\} =: x\varepsilon(y), \end{aligned}$$

say, where $\varepsilon(y)$ tends to 0 as y tends to infinity. The desired result follows from this estimate, in view of (25), by making x and then y tend to infinity.

Notes

§ **3.2.** This presentation of the hyperbola method is due to Diamond (1982).

The original proof of Voronoï is elementary and is based on the Euler–Maclaurin summation formula. There are two other classical elementary proofs of the inequality $\alpha \leq 1/3$: that of Landau (1912) using the theory of Bessel functions and that of I.M. Vinogradov (1917), explained in the book of Gelfond & Linnik (1965), which also depends on the Euler–Maclaurin formula. The upper bound $\alpha \leq 23/73$ of Huxley (1993a) stems from a refinement of the method of Iwaniec & Mozzochi (1988) which yielded $\alpha \leq 7/22$, thus improving on a result of Kolesnik (1985), $\alpha \leq 139/429$. The numerical gain is relatively slight ($139/429 \approx 0.324$, $23/73 \approx 0.315$), as in the whole history of this problem, but the ideas involved are important. Kolesnik's method is an extension to many variables of that of van der Corput (1922) sketched in Chapter 6. That of Iwaniec & Mozzochi essentially amounts, by a sophisticated process, to bounding an average of double exponential sums. This treatment is a bidimensional version of that of Bombieri & Iwaniec (1986), which led to a new upper bound for the Riemann zeta function on the critical line, $|\zeta(\frac{1}{2}+it)| \ll_\varepsilon t^{9/56+\varepsilon}$. In this direction, the best result known to date is due to Huxley (1993b), who obtains, by elaborating upon the same method, the bound

$$|\zeta(\tfrac{1}{2}+it)| \ll_\varepsilon t^{89/570+\varepsilon}.$$

The fact that $\Delta(x)$ is not $o(x^{1/4})$ follows directly from the quadratic mean evaluation of Tong (1956):

$$\int_0^x \Delta(y)^2\,\mathrm{d}y = \frac{\zeta(3/2)^4}{6\pi^2\zeta(3)}x^{3/2} + O\big(x(\log x)^5\big).$$

§ **3.4.** Sharpening a result of Erdős & Shapiro (1951), Montgomery (1987) showed that

$$\lim {\sup \atop \inf} \Big\{ \sum_{n\leq x} \varphi(n) - \frac{3}{\pi^2}x^2 \Big\} \Big/ x\sqrt{(\log_2 x)} \gtrless 0.$$

§ **3.6.** The proof of Theorem 8 is essentially that proposed by Diamond (1982). The implication (ii) $\Rightarrow$ (i) in Theorem 8 is a special case of a general result concerning Möbius inverses of functions of bounded variation—cf. Ellison & Mendès France (1975), Theorem 3.1.

§ **3.7.** Theorem 10 is due to Landau (1909). The best error term known at present for the formula (22) is

$$O\Big(\sqrt{x}\,\exp\Big\{-c\frac{(\log x)^{3/5}}{(\log_2 x)^{1/5}}\Big\}\Big),$$

due to Walfisz (1963). Improving a result of Montgomery & Vaughan (1981), S. Graham (1981a) obtained under the Riemann hypothesis

$$O_\varepsilon(x^{8/25+\varepsilon}).$$

§ **3.8.** The method of the functions α_y, β_y goes back at least to Erdős in the thirties. It has been used in a remarkable manner by Daboussi—see in particular 1979, 1984. The case of convergence of the product $M(f)$ can be generalised as follows: *Let f be a complex-valued multiplicative function such that*

$$\sum_{n=1}^{\infty} \frac{|f * \mu(n)|}{n} < \infty. \tag{26}$$

Then we have

$$\sum_{n \le x} f(n) = x\{M(f) + o(1)\}$$

where the product $M(f)$, defined as in Theorem 11, is absolutely convergent.

Indeed, putting $h = f * \mu$, we have for $x \ge 1$

$$\frac{1}{x}\sum_{n \le x} f(n) = \frac{1}{x}\sum_{d \le x} h(d)\left[\frac{x}{d}\right] \to \sum_{d=1}^{\infty} \frac{h(d)}{d} \qquad (x \to \infty)$$

by the theorem of dominated convergence. It is then easy to verify that the hypothesis of absolute convergence implies

$$\sum_{d=1}^{\infty} \frac{h(d)}{d} = M(f).$$

We shall see in Theorem II.1.2 that assumption (26) is equivalent to

$$\sum_p \sum_{\nu \ge 1} |f(p^\nu) - f(p^{\nu-1})| p^{-\nu} < \infty. \tag{27}$$

The case of zero mean value in Theorem 11 can be handled by Theorem III.3.5. Such an approach actually provides an effective estimate.

Exercises

1. Show that $\sum_{n=1}^{\infty} n^{-2} = \pi^2/6$:

(a) by integrating the Fourier series for $\{t\}$;

(b) by integrating the Taylor series for $(\operatorname{arctanh} x)/x$. [*Calculate the integral by the method of residues—see for example Cartan (1961), pp. 107–109.*]

2. Show that $\sum_{n \le x} 2^{\omega(n)} = (6/\pi^2)x \log x + O(x)$.

3. Determine an arithmetic function h such that $h * 2^\omega = \mu^2 \, 2^\omega$.

(a) Show that $\sum_{n>x} |h(n)|n^{-1} \ll_\varepsilon x^{-1/2+\varepsilon}$ for all $\varepsilon > 0$.

(b) Deduce that $\sum_{n \le x} \mu(n)^2 2^{\omega(n)} = Cx \log x + O(x)$ where

$$C := \prod_p (1 - 1/p)^2 (1 + 2/p).$$

4. Show that

$$\sum_{n \le x} 3^{\omega(n)} = \tfrac{1}{2} Cx(\log x)^2 + O(x \log x),$$

where C is the constant defined in Exercise 3. Generalise.

5. (a) Show that $\sum_{n \le x} 2^{\Omega(n)} = \sum_{k \le \log x/\log 2} 2^k \sum_{m \le x/2^k} f(m)$ where f is the multiplicative function defined by $f(p^\nu) := 0$ if $p = 2$, $:= 2^\nu$ if $p \ge 3$.

(b) By writing $f = \tau * h$, where h is some function to be determined, show that $\sum_{n \le x} f(n) = \frac{1}{4} C_0 x \log x + O(x)$ for $C_0 = \prod_{p>2} \big(1 + 1/p(p-2)\big)$.

(c) Show that $\sum_{n \le x} 2^{\Omega(n)} = (C_0/8 \log 2) x (\log x)^2 + O(x \log x)$.

6. Assuming the prime number theorem, show that

$$\sum_{p \le x} \frac{\log p}{p-1} = \log x - \gamma + o(1).$$

7. Let $A(n) := \sum_{p^\nu \| n} \nu p$ be the function of Alladi & Erdős (1977, 1979). Assuming the prime number theorem, show that

$$\sum_{n \le x} A(n) \sim \frac{\pi^2}{12} \frac{x^2}{\log x} \qquad (x \to +\infty).$$

Make the error term in this relation more precise by appealing to a strong form of the prime number theorem—for example, Theorem II.4.1.

8. Let f be a multiplicative arithmetic function such that $f*\mu \geq 0$. Show that, for $x \geq 1$,

$$\sum_{n\leq x} f(n) \leq x \prod_{p\leq x} (1-p^{-1}) \sum_{\nu\geq 0} f(p^\nu)p^{-\nu}.$$

9. For $\alpha > 0$, $t \geq 1$, write $f_\alpha(n) := \big(n/\varphi(n)\big)^\alpha$ and

$$F_x(t) := \big|\{n \leq x : n > t\varphi(n)\}\big|.$$

(a) Applying the result of the previous exercise to f_α with suitable $\alpha = \alpha(t)$, show that there exists a positive absolute constant c such that

$$F_x(t) \ll x \exp\{-e^{ct}\} \qquad (x \geq 1,\, t \geq 1).$$

(b) Show that for fixed $\alpha > 0$, $\varepsilon > 0$, one has

$$\sum_{n\leq x} \Big(\frac{n}{\varphi(n)}\Big)^\alpha = x \prod_p \Big(1 + \frac{f_\alpha(p)-1}{p}\Big) + O(x^\varepsilon).$$

10. *Squarefull integers.*
Let $S := \{n \geq 1 : p|n \Rightarrow p^2|n\}$, $S(x) := \big|S \cap [1,x]\big|$.
(a) By estimating $\sum_{n\leq x,\, n\in S} \sqrt{(x/n)}$, show that

$$S(x) \ll \sqrt{(x)} \log x.$$

(b) Show that each n in S can be written in a unique way as $n = m^3d^2$ where m is squarefree. Deduce from this that

$$S(x) \sim \frac{\zeta(3/2)}{\zeta(3)} \sqrt{x}.$$

[*See Suryanarayana & Sita Rama Chandra (1973) for a more precise result.*]

11. Let $k(n) := \prod_{p|n} p$ be the *squarefree kernel* of an integer n.
(a) Deduce from Theorem 11 that $\sum_{n\leq x} k(n)/n \sim Cx$ with

$$C := \prod_p \big(1 - 1/p(p+1)\big).$$

(b) Using the result of Exercise 10(b), show that

$$\sum_{n\leq x} \frac{k(n)}{n} = Cx + O(\sqrt{x}).$$

(c) Show that $\sum_{n\leq x} k(n) = \frac{1}{2}Cx^2 + O(x^{3/2})$. [*See also Cohen (1960, 1964).*]

12. Prove the existence of a positive constant A such that

$$\sum_{n\leq x}\omega(\tau(n)) = Ax + O\Big(\frac{x}{\log x}\Big)$$

and show that

$$\sum_{n\leq x}\Omega(\tau(n)) = x\log_2 x + O(x).$$

[*See Rieger (1972); Heppner (1974).*]

13. Show that the prime number number implies that

$$\int_1^{+\infty}\frac{\psi(t)-t}{t^2}\mathrm{d}t = -1-\gamma.$$

14. Let $\lambda(n)$ be Liouville's function defined by $\lambda(n) := (-1)^{\Omega(n)}$.

(a) Show that $\lambda = \mu * \kappa$, where κ denotes the characteristic function of the squares.

(b) Show that the prime number theorem is equivalent to

$$\sum_{n\leq x}\lambda(n) = o(x).$$

15. *Selberg's identity (1949).*

Define the arithmetic function $\Lambda_2 := \log^2 *\mu$.

(a) Show that, for all primes p, q and all integers $\alpha \geq 1$, $\beta \geq 1$, one has $\Lambda_2(p^\alpha) = (2\alpha-1)\log^2 p$, $\Lambda_2(p^\alpha q^\beta) = 2\log p\log q$ and $\Lambda_2(n) = 0$ if $\omega(n) > 2$.

(b) Show that $\sum_{n\leq x}\Lambda_2(n) = \sum_{p\leq x}\log^2 p + \sum_{pq\leq x}\log p\log q + O(x)$.

(c) Show that $\sum_{n\leq x}\tau * \mathbf{1}(n) = \frac{1}{2}x\log^2 x + ax\log x + bx + O\big(x^{2/3}\log(2x)\big)$, where a and b are real constants. Deduce that for suitable constants A and B, the function $h := 2\tau*\mathbf{1}+A\tau+B\mathbf{1}-\log^2$ has a summatory function $H(x) \ll x^{3/4}$.

(d) Prove that $\sum_{n\leq x}\Lambda_2(n) = 2x\log x + O(x)$. [*Selberg deduces the prime number theorem elementarily from this formula. This derivation is actually valid in a general setting: see Shapiro (1959).*]

16. Let g be a complex multiplicative function with values in the unit disc. Set $S(n) := \sum_{d|n} g(d)$. Show the existence of the mean value $\lim_{x\to\infty} x^{-1}\sum_{n\leq x}|S(n)/\tau(n)|^2$ and express it as an Eulerian product. Deduce that the condition

$$\sum_p \big(1-\Re e\, g(p)\big)/p = \infty$$

is necessary and sufficient so that $S(n) = o(\tau(n))$ for almost all integers n, i.e. that there exists a function $\varepsilon(x)\to 0$ as $x\to\infty$ with

$$\big|\{n\leq x : |S(n)| > \varepsilon(x)\tau(n)\}\big| = o(x) \qquad (x\to\infty).$$

I.4

Sieve methods

§ 4.1 The sieve of Eratosthenes

The inclusion–exclusion principle or the Möbius inversion formula can theoretically be used to calculate $\pi(x)$. For sufficiently large x, let us write

$$P = \prod_{p \le \sqrt{x}} p.$$

Then an integer n with $\sqrt{x} < n \le x$ is a prime number if and only if $(n, P) = 1$. Thus, we can write

$$\pi(x) - \pi(\sqrt{x}) + 1 = \sum_{n \le x} \delta((n,P)) = \sum_{d \mid P} \mu(d) \left[\frac{x}{d}\right]. \tag{1}$$

At this stage, if we insert the simple estimate $[x/d] = x/d + O(1)$, we obtain

$$\pi(x) - \pi(\sqrt{x}) + 1 = x \prod_{p \le \sqrt{x}} \left(1 - \frac{1}{p}\right) + O(2^{\pi(\sqrt{x})}).$$

By Mertens' theorem the main term of this formula is

$$(1 + o(1)) 2e^{-\gamma} x / \log x,$$

but Chebyshev's estimates show that the error term is greater than any power of x.

This calls for two comments. On the one hand the exact formula (1)—called the sieve formula of Eratosthenes—involves too many terms for reasonable practical validity. On the other hand, the estimate of the main term, taking the prime number theorem into account, shows *a posteriori* that the "error terms" created by replacing $[x/d]$ by x/d have made a global contribution of the same order of magnitude as the "main terms". This suggests that, even suitably adapted, this method will never allow a proof of the prime number theorem. However we will see that it can provide Chebyshev-type estimates in a very general context.

In order to obtain a non-trivial result starting from formula (1) one may introduce a parameter y, $1 \le y \le x$, and bound $\pi(x) - \pi(y) + 1$ by the number of integers n not exceeding x and having no prime factor $\le y$. With the same calculations we obtain

$$\begin{aligned}\pi(x) &\le x \prod_{p \le y}\Big(1 - \frac{1}{p}\Big) + O(2^y)\\ &= x\frac{\{e^{-\gamma} + o(1)\}}{\log y} + O(2^y)\\ &\le \{e^{-\gamma} + o(1)\}\frac{x}{\log_2 x}\end{aligned}$$

for the essentially optimal choice $y = \log x$.

It was with the aim of improving the efficiency of this method that the Norwegian mathematician Viggo Brun invented the theory of the combinatorial sieve between 1917 and 1924.

§ 4.2 Brun's combinatorial sieve

Eratosthenes' sieve rests on the identity

$$\mu * \mathbf{1} = \delta.$$

Brun's idea consists in introducing two auxiliary functions satisfying

$$\mu_1 * \mathbf{1} \le \delta \le \mu_2 * \mathbf{1}, \tag{2}$$

and vanishing sufficiently often so that the number of non-zero terms in the resulting formula analogous to (1) is not prohibitive.

Brun's initial choice, leading to what in the literature is called the " Brun pure sieve", is the following.

Theorem 1 (Brun). *Denote by χ_t the characteristic function of the set of integers n such that $\omega(n) \le t$. Then for each integer $h \ge 0$ the functions defined by*

$$\mu_i(n) := \mu(n)\chi_{2h+2-i}(n) \qquad (i = 1, 2) \tag{3}$$

satisfy the inequalities (2).

Proof. Since $\mu_i * \mathbf{1}(n)$ depends only on the squarefree kernel of n, it suffices to consider the case $\mu(n)^2 = 1$. If $\omega(n) = k$, then, for each r with $0 \le r \le k$, it is clear that n has exactly $\binom{k}{r}$ divisors d such that $\omega(d) = r$. For any given $t \ge 0$ we can thus write

$$\mu\chi_t * \mathbf{1}(n) = \sum_{d|n,\, \omega(d) \le t} \mu(d) = \sum_{r \le t}(-1)^r\binom{k}{r} = (-1)^t\binom{k-1}{t},$$

where the last equality is easily obtained by induction on t. This is all we need.

Corollary 1.1. *Let $\mathcal{A}$ be a finite set of integers and let $\mathcal{P}$ be a set of prime numbers. Write*

$$A_d := \operatorname{card}\{a \in \mathcal{A} : a \equiv 0 \,(\operatorname{mod} d)\},$$
$$P(y) := \prod_{p \in \mathcal{P},\, p \le y} p,$$
$$S(\mathcal{A}, \mathcal{P}, y) := \operatorname{card}\{a \in \mathcal{A} : (a, P(y)) = 1\}.$$

Then for each integer $h \geq 0$ we have

$$\sum_{d|P(y),\, \omega(d) \le 2h+1} \mu(d) A_d \le S(\mathcal{A}, \mathcal{P}, y) \le \sum_{d|P(y),\, \omega(d) \le 2h} \mu(d) A_d. \tag{4}$$

Let us see how this result enables us to improve considerably the upper bound for $\pi(x)$ provided by the sieve of Eratosthenes.

We select in the above corollary $\mathcal{A} = \{n : n \leq x\}$ and $P = P(y) := \prod_{p \leq y} p$. Then $S(\mathcal{A}, \mathcal{P}, y) \geq \pi(x) - \pi(y) + 1$, from which we obtain

$$\begin{aligned} \pi(x) &\le \sum_{\substack{d|P(y)\\ \omega(d)\le 2h}} \mu(d) \left[\frac{x}{d}\right] + y \\ &= x \sum_{\substack{d|P(y)\\ \omega(d)\le 2h}} \frac{\mu(d)}{d} + O\Big(y + \sum_{\substack{d|P(y)\\ \omega(d)\le 2h}} 1\Big) \\ &= x \prod_{p \le y}\Big(1 - \frac{1}{p}\Big) + O\Big(y + \sum_{\substack{d|P(y)\\ \omega(d)\le 2h}} 1 + x \sum_{\substack{d|P(y)\\ \omega(d)> 2h}} \frac{1}{d}\Big). \end{aligned} \tag{5}$$

The second of the three error terms does not exceed y^{2h} since this is an upper bound for all integers d such that $d \mid P(y)$, $\omega(d) \leq 2h$. The d-sum arising in the third remainder term is bounded, for each value of the parameter $u \geq 1$, by

$$\sum_{d|P(y)} u^{\omega(d)-2h}/d = u^{-2h} \prod_{p \le y} (1 + u/p) \le \exp\Big\{-2h \log u + u \sum_{p \le y} 1/p\Big\}.$$

For the optimal choice $u = 2h \big/ \sum_{p \leq y} p^{-1}$, we obtain that this quantity is

$$\ll_u (\log y)^{-v}$$

where $v = u \log u - u$. This follows from Theorem 1.9. We note that when $u > 5$, then $v > 3$. It is easy to show that for sufficiently large y there exists some $u = u(y)$, $5 < u < 6$, such that

$$h := \tfrac{1}{2} u \sum_{p \le y} \frac{1}{p}$$

is an integer. With this choice of the parameters, we have for sufficiently large x

$$y^{2h} \leq \exp\{6 \log y(\log_2 y + O(1))\} < x^{2/3}$$

from which we deduce that

$$y \leq \exp\{\log x/10 \log_2 x\} =: Y(x). \tag{6}$$

Collecting all previous estimates, and selecting $y = Y(x)$, we see that

$$\pi(x) \ll \frac{x \log_2 x}{\log x}.$$

Although inferior to that of Chebyshev, this result is remarkable because of the great generality of the argument. The corresponding lower bound for $S(\mathcal{A}, \mathcal{P}, y)$ can be obtained in the same way. This quantity has actually an intrinsic arithmetic interest: it equals the number of integers not exceeding x all prime factors of which are greater than y. Hence we can state the following result.

Theorem 2. *For each integer $n \geq 1$, let $P^-(n)$ denote the smallest prime factor of n, with the convention that $P^-(1) = \infty$. Write*

$$\Phi(x, y) := \operatorname{card}\{n \leq x : P^-(n) > y\}.$$

Then we have under condition (6)

$$\Phi(x, y) = x \prod_{p \leq y}\Big(1 - \frac{1}{p}\Big)\Big\{1 + O\Big(\frac{1}{(\log y)^2}\Big)\Big\}. \tag{7}$$

The choice of the functions μ_1, μ_2 defined by (3) is certainly not optimal. The method can be refined, at the price of certain technical complications, by introducing a partition of $]1, y]$ into subintervals $]y_j, y_{j+1}]$, $0 \leq j \leq k$ and selecting, for $i = 1, 2$,

$$\mu_i(d) = \mu(d)\chi_i^*(d)$$

where χ_i^* is the characteristic function of those integers d having at most $2h_j + 2 - i$ prime factors in $\mathcal{P} \cap]y_j, y]$ for each j, $0 \leq j \leq k$—see Exercise 7. There are thus two families of parameters to optimise, the y_j and the h_j.

It would take too long to develop here the theory of the combinatorial sieve, which is still in progress at this very moment. The interested reader will find an expository account of this subject in the book of Halberstam & Richert, *Sieve Methods* (1974)—see also the notes for this chapter.

We confine ourselves to quoting the basic result of the theory, obtained with the choice of the $\mu_i(d)$ indicated above. It is this theorem which is intended, when in the literature one invokes "Brun's method" without further qualification.

Theorem 3 (Fundamental lemma of the combinatorial sieve). *With the notation of Corollary 1.1, assume there exist a non-negative multiplicative function w, some real number X, and positive constants κ, A such that*

$$\text{(a)}\qquad A_d =: Xw(d)/d + R_d \qquad (d \mid P(y))$$

$$\text{(b)}\qquad \prod_{\eta\le p\le\xi}\Big(1-\frac{w(p)}{p}\Big)^{-1} < \Big(\frac{\log\xi}{\log\eta}\Big)^{\kappa}\Big(1+\frac{A}{\log\eta}\Big) \qquad (2\le\eta\le\xi).$$

Then we have, uniformly for $\mathcal{A}$, X, y and $u \ge 1$,

$$(8)\quad S(\mathcal{A},\mathcal{P},y) = X\prod_{p\le y,\,p\in\mathcal{P}}\Big(1-\frac{w(p)}{p}\Big)\{1+O(u^{-u/2})\}+O\Big(\sum_{d\le y^u,\,d\mid P(y)}|R_d|\Big).$$

From this result one easily deduces a Chebyshev-type upper bound for $\pi(x)$, as well as an evaluation of the order of magnitude of $\Phi(x,y)$ for $y \le x^{\delta}$, where δ is some fixed positive real number. We leave the details to the reader.

§ 4.3 Application to prime twins

Here we illustrate the results of the previous section by an application to Brun's theorem on prime twins.

It is obvious that the difference between two odd prime numbers must be at least 2. When it is exactly 2 one says that the pair $\{p, p+2\}$ is composed of *prime twins*, thus $\{3,5\}$, $\{5,7\}$, $\{11,13\}$, $\{17,19\}$, $\{29,31\}, \ldots$.

A famous conjecture states that there are infinitely many prime twins. Let us write

$$\mathcal{J} := \{p : p+2 \text{ is prime}\} \qquad\text{and}\qquad J(x) := \big|\mathcal{J}\cap[1,x]\big|.$$

On the basis of an analytic approach, and in agreement with a heuristic probabilistic calculation, Hardy & Littlewood (1922) conjectured that

$$(9)\qquad J(x) \sim 2\prod_{p>2}\Big(1-\frac{1}{(p-1)^2}\Big)\frac{x}{(\log x)^2} \qquad (x\to\infty).$$

By Brun's pure method, we establish the following result.

Theorem 4 (Brun). *As x tends to infinity, we have*

$$(10)\qquad J(x) \ll x\Big(\frac{\log_2 x}{\log x}\Big)^2.$$

Corollary 4.1. *We have*

$$(11)\qquad \sum_{p\in\mathcal{J}}\frac{1}{p} < \infty.$$

Proof. Apply Corollary 1.1 with $\mathcal{A} = \{m(m+2) : m \le x\}$, choosing $\mathcal{P}$ to be the set of all prime numbers. For all y, $1 \le y \le x$, we have

$$J(x) \le S(\mathcal{A}, \mathcal{P}, y) + y \le \sum_{d|P(y),\, \omega(d)\le 2h} \mu(d) A_d + y, \tag{12}$$

where A_d is the number of integral solutions $m \le x$ to the congruence

$$m(m+2) \equiv 0 \,(\mathrm{mod}\, d). \tag{13}$$

This relation is equivalent to

$$m \equiv 0 \,(\mathrm{mod}\, 2^\nu), \quad m \equiv 0 \text{ or } -2 \,(\mathrm{mod}\, p) \quad (p|m,\, p \ne 2), \tag{14}$$

where ν equals 1 or 0 according as to whether d is even or odd. By the Chinese remainder theorem there are therefore $\rho(d)$ solutions modulo d, where ρ is the strongly multiplicative function defined by

$$\rho(2) = 1, \quad \rho(p) = 2 \qquad (p \ge 3). \tag{15}$$

Each interval of length d contains $\rho(d)$ integers m counted in A_d. We can therefore write

$$A_d = x\frac{\rho(d)}{d} + O(\rho(d)) \qquad (\mu(d)^2 = 1). \tag{16}$$

Inserting this back into (12) and performing a calculation parallel to (5), it follows that

$$J(x) \le x \sum_{d|P(y)} \frac{\mu(d)\rho(d)}{d} + O\Big(y + \sum_{\substack{d|P(y)\\ \omega(d)\le 2h}} \rho(d) + x \sum_{\substack{d|P(y)\\ \omega(d)>2h}} \frac{\rho(d)}{d}\Big). \tag{17}$$

The main term has value

$$\tfrac{1}{2}x \prod_{2<p\le y} \left(1 - \frac{2}{p}\right) \le 2x \prod_{p \le y} \left(1 - \frac{1}{p}\right)^2 \sim 2e^{-2\gamma} x (\log y)^{-2}.$$

Selecting, as in the application of §4.2, $h = c \log_2 y + O(1)$ for some suitable constant c, and $\log y \sim c' \log x / \log_2 x$ with c' sufficiently small, we check that the error term is of smaller order than $x/(\log y)^2$. This implies (10), and thereby completes the proof of the theorem. The corollary follows at once by partial summation.

§ 4.4 The large sieve—analytic form

The large sieve is one of the most powerful tools in analytic number theory. Invented by Linnik in 1941, it has since been developed systematically by several mathematicians, as much from the point of view of its fundamental principle as for its arithmetic applications.

The reader will find complete expositions and an exhaustive bibliography in the monograph of Bombieri (1974) and the survey by Montgomery (1978a). Here we restrict ourselves to mentioning that the crucial steps in the elaboration of the current theory of the large sieve are due to Rényi (cf. his article of 1950), Roth (1965) and Bombieri (1965). The optimal version has been obtained by Montgomery & Vaughan (1973, 1974).

Davenport & Halberstam (1966) were the first to isolate the analytic foundations of the large sieve. Let $\{a_n\}_{n=0}^{\infty}$ be a sequence of complex numbers and, for given arbitrary integers M, $N \geq 0$, let

$$S(\alpha) := \sum_{M<n\leq M+N} a_n e(\alpha n) \tag{18}$$

be a trigonometric polynomial, where $e(u) := \exp\{2\pi i u\}$ $(u \in \mathbb{R})$. The analytic form of the large sieve is an inequality of type

$$\sum_{i=1}^{R} |S(\alpha_i)|^2 \leq \Delta(N, \delta) \sum_{M<n\leq M+N} |a_n|^2 \tag{19}$$

valid for all R-tuples $\{\alpha_1, \ldots, \alpha_R\}$ of real numbers, which are δ-*well spaced* in the sense that

$$\min_{1\leq i<j\leq R} \|\alpha_j - \alpha_i\| \geq \delta > 0, \tag{20}$$

where $\|u\|$ denotes the distance of the real number u to the set of integers. Our aim in this section is to prove the following optimal result.

Theorem 5 (Montgomery & Vaughan; Selberg). *Under the above conditions the large sieve inequality (19) holds for the choice*

$$\Delta(N, \delta) = N + \delta^{-1} - 1. \tag{21}$$

The proof which we give here, due to Selberg, is presented in Montgomery's article (1978a). The value for $\Delta(N, \delta)$ obtained by Montgomery & Vaughan in 1974 was only marginally weaker: $\Delta = N+\delta^{-1}$. Selberg's improvement is unimportant for the applications—where an upper bound of the type $\Delta \ll N + \delta^{-1}$ is often sufficient—and it is mainly the nature of the argument which has motivated our choice.

However let us observe that the value given in (21) is attained for certain values of the α_i, N and δ. Indeed for each integer $R \geq 1$, set $\alpha_j = j/R$ $(1 \leq j \leq R)$, so that $\delta = 1/R$, and when $N \equiv 1 \pmod{R}$ let us consider the case $a_n := 1$ if $R \mid n$, $a_n := 0$ if $R \nmid n$. One then has

$$\sum_{j=1}^{R} |S(\alpha_j)|^2 = \sum_{j=1}^{R} \Big| \sum_{\substack{0 \leq n \leq N-1 \\ n \equiv 0 (\mathrm{mod}\, R)}} 1 \Big|^2 = R\Big(\frac{N-1}{R} + 1\Big)^2$$

$$= (N - 1 + R)\Big(1 + \frac{N-1}{R}\Big) = (N - 1 + \delta^{-1}) \sum_{0 \leq n \leq N-1} |a_n|^2.$$

The proof of Theorem 5 rests on a duality principle asserting equality of norms for a Banach spaces operator and its adjoint. We will only use the case of endomorphisms of $\ell_2(\mathbb{C})$, when the duality principle takes the following simple form.

Lemma 5.1. *Let (c_{nr}) be an $N \times R$ matrix with complex coefficients. The three following assertions concerning the real positive number D are equivalent:*

(i) $$\sum_r \Big| \sum_n c_{nr} x_n \Big|^2 \leq D \sum_n |x_n|^2 \qquad (\forall x_n \in \mathbb{C}),$$

(ii) $$\Big| \sum_{n,r} c_{nr} y_r x_n \Big|^2 \leq D \sum_n |x_n|^2 \sum_r |y_r|^2 \qquad (\forall x_n, y_r \in \mathbb{C}),$$

(iii) $$\sum_n \Big| \sum_r c_{nr} y_r \Big|^2 \leq D \sum_r |y_r|^2 \qquad (\forall y_r \in \mathbb{C}).$$

Proof. Let us show the equivalence of (i) and (ii). That of (ii) and (iii) follows by interchanging the roles of the indices r and n.

(i) $\Rightarrow$ (ii): We have

$$\Big| \sum_{n,r} c_{nr} y_r x_n \Big|^2 = \Big| \sum_r y_r \sum_n c_{nr} x_n \Big|^2$$

$$\leq \sum_r |y_r|^2 \sum_r \Big| \sum_n c_{nr} x_n \Big|^2 \leq D \sum_n |x_n|^2 \sum_r |y_r|^2$$

where the first upper bound stems from the Cauchy–Schwarz inequality.

(ii) $\Rightarrow$ (i): For each r, set $L_r := \sum_n c_{nr} x_n$ and apply (ii) with $y_r = \overline{L_r}$. Then

$$\Big(\sum_r |L_r|^2 \Big)^2 \leq D \sum_n |x_n|^2 \sum_r |L_r|^2,$$

that is to say (i) holds.

In the sequel we will systematically use notation (18). The following result is an immediate application of Lemma 5.1.

Lemma 5.2. *Let α_r $(1 \le r \le R)$ be fixed real numbers. The two following assertions, concerning the sequence of real numbers $b_n \ge 0$ $(n \in \mathbb{Z})$ such that $b_n > 0$ $(M < n \le M+N)$ and the real positive number B, are equivalent:*

$$\text{(i)} \qquad \sum_{1\le r\le R} |S(\alpha_r)|^2 \le B \sum_{M<n\le M+N} |a_n|^2/b_n \qquad (\forall a_n \in \mathbb{C})$$

$$\text{(ii)} \qquad \sum_{M<n\le M+N} b_n \Big| \sum_{1\le r\le R} y_r\, e(n\alpha_r)\Big|^2 \le B \sum_{1\le r\le R} |y_r|^2 \qquad (\forall y_r \in \mathbb{C}).$$

Proof. We use Lemma 5.1 with $c_{nr} = e(n\alpha_r)\sqrt{b_n}$. Up to replacing a_n by $a_n\sqrt{b_n}$, expression (i) is equivalent to

$$\sum_{1\le r\le R} \Big| \sum_{M<n\le M+N} a_n \sqrt{b_n}\, e(\alpha_r n)\Big|^2 \le B \sum_n |a_n|^2 \quad (\forall a_n \in \mathbb{C}).$$

Using the equivalence of statements (i) and (iii) in Lemma 5.1, we see that the above condition can be written

$$\sum_{M<n\le M+N} \Big| \sum_{1\le r\le R} y_r e(n\alpha_r)\sqrt{b_n}\Big|^2 \le B \sum_{1\le r\le R} |y_r|^2 \quad (\forall y_r \in \mathbb{C}),$$

which is the required inequality.

Corollary 5.2.1. *Let $B(\alpha) := \sum_{n\in\mathbb{Z}} b_n\, e(n\alpha)$ be a convergent Fourier series such that $b_n \ge 0$ $(n \in \mathbb{Z})$, and $b_n > 0$ for $M < n \le M+N$. Then for each positive real number B, the inequality*

$$\text{(i)} \qquad \sum_{1\le r\le R} |S(\alpha_r)|^2 \le B \sum_{M<n\le M+N} |a_n|^2/b_n \qquad (\forall a_n \in \mathbb{C})$$

is satisfied provided that

$$\text{(ii)} \qquad \sum_{1\le r,s\le R} y_r \overline{y_s}\, B(\alpha_r - \alpha_s) \le B \sum_{1\le r\le R} |y_r|^2 \qquad (\forall y_r \in \mathbb{C}).$$

Proof. By expanding $B(\alpha_r - \alpha_s)$ as a series and interchanging summations, we see that (ii) is equivalent to the second inequality of Lemma 5.2, in which the sum over n is extended to $\mathbb{Z}$. Since $b_n \ge 0$ for all n, the conclusion is immediate.

An obvious procedure for constructing functions satisfying the conditions of the Corollary is to insist that

(a) $b_n \geq 0 \ (n \in \mathbb{Z}), \quad b_n \geq 1 \ (M < n \leq M + N)$,

(b) $B(\alpha) = 0 \quad (\|\alpha\| \geq \delta)$,

where δ is defined by (20). Here it is convenient to suppose that $0 < \delta < \frac{1}{2}$, the case $\delta = \frac{1}{2}$ (possible only if $R = 2$) following then by a straightforward limit procedure. When (a) and (b) are realised, assertion (i) of Corollary 5.2.1 shows that the inequality (19) of the large sieve is satisfied with

$$\Delta(N, \delta) = B(0). \tag{22}$$

The remainder of this section is devoted to making explicit a suitable choice for the sequence $\{b_n\}_{n\in\mathbb{Z}}$ of Fourier coefficients for $B(\alpha)$.

It is natural to try and write b_n as the value at n of a function $b \in L^1(\mathbb{R})$ for which the Fourier transform

$$\widehat{b}(\theta) := \int_{-\infty}^{+\infty} b(t)e(-\theta t)\,\mathrm{d}t$$

has support contained in $[-\delta, \delta]$. The Poisson formula

$$B(\alpha) := \sum_{n\in\mathbb{Z}} b(n)e(\alpha n) = \sum_{k\in\mathbb{Z}} \widehat{b}(k - \alpha) \tag{23}$$

then guarantees the validity of condition (b).

In order to verify that the Poisson summation formula is effectively applicable in this context, we first observe that $\widehat{b} \in L^1(\mathbb{R})$, from which we deduce (cf. for example Katznelson (1968), p. 126) the integral representation

$$b(t) = \int_{-\delta}^{\delta} \widehat{b}(\theta)e(\theta t)\,\mathrm{d}\theta. \tag{24}$$

In particular $b(n)$ is the Fourier coefficient of the periodic continuous function $\beta(\alpha) := \sum_{k\in\mathbb{Z}} \widehat{b}(k - \alpha)$. Now by (24), we have for $N \geq 1$, $|\alpha| \leq \frac{1}{2}$,

$$\begin{aligned} &\sum_{|n|\leq N} b(n)e(\alpha n) - \int_{-N-\frac{1}{2}}^{N+\frac{1}{2}} b(t)e(\alpha t)\,\mathrm{d}t \\ &\qquad\qquad = \int_{-\delta+\alpha}^{\delta+\alpha} \lambda_\alpha(\theta)\sin\{(2N + 1)\pi\theta\}\,\mathrm{d}\theta, \end{aligned} \tag{25}$$

with

$$\lambda_\alpha(\theta) := \widehat{b}(\theta - \alpha)\Big\{\frac{1}{\sin(\pi\theta)} - \frac{1}{\pi\theta}\Big\}.$$

Since $|\pm\delta+\alpha|<1$ (it is here that we make use of the assumption $\delta<\frac{1}{2}$), we have

$$\lambda_\alpha\in L^1[-\delta+\alpha,\,\delta+\alpha].$$

The Riemann–Lebesgue lemma then allows us to conclude that the last θ-integral tends to 0 as $N\to+\infty$. Since $b\in L^1(\mathbb{R})$, this implies the (symmetric) convergence of the series

$$\sum_{n\in\mathbb{Z}} b(n)e(\alpha n)$$

to $\widehat{b}(-\alpha)=\beta(\alpha)$. This establishes (23) for $|\alpha|\le\frac{1}{2}$, and thus for all α, by periodicity.

We are thus led to look for an integrable function b such that the quantity

$$B(0)=\widehat{b}(0)=\int_{-\infty}^{+\infty} b(t)\,\mathrm{d}t \tag{26}$$

is as small as possible, subject to the constraints

$$\begin{cases} b(t)\ge 0 & (t\in\mathbb{R}),\\ b(t)\ge 1 & (M+1\le t\le M+N),\\ \widehat{b}(\theta)=0 & (|\theta|\ge\delta). \end{cases} \tag{27}$$

The Fejér kernel easily allows us to exhibit a first possibility. For

$$b(t):=C\sum_{\delta(M+1)\le n\le\delta(M+N)}\left(\frac{\sin\left(\frac{1}{2}\pi(n-\delta t)\right)}{\frac{1}{2}\pi(n-\delta t)}\right)^2$$

we have

$$\widehat{b}(\theta)=2C\delta^{-1}(1-|\theta\delta^{-1}|)^+\sum_{\delta(M+1)\le n\le\delta(M+N)} e(-n\theta\delta^{-1})$$

with the consequence that conditions (27) are certainly realised for the choice $C=\frac{1}{4}\pi^2$. This proves the inequality

$$\widehat{b}(0)\le\tfrac{1}{2}\pi^2(N-1+\delta^{-1})$$

which suffices for most applications. Selberg remarked that the following lemma allows a better choice for $b(t)$.

Lemma 5.3. *Let*

$$F(z) := \Big(\frac{\sin \pi z}{\pi}\Big)^2 \Big\{ \sum_{n=0}^{\infty} \frac{1}{(z-n)^2} - \sum_{n=1}^{\infty} \frac{1}{(z+n)^2} + \frac{2}{z} \Big\}.$$

Then F defines an entire function of z, such that

$$F(z) = O(\exp\{2\pi |\Im m\, z|\}), \qquad F(x) \geq \operatorname{sgn}(x) \quad (x \in \mathbb{R}), \qquad F(0) = 1,$$

and

$$\int_{-\infty}^{+\infty} \big(F(x) - \operatorname{sgn}(x)\big)\, \mathrm{d}x = 1. \tag{28}$$

Remark. It plainly follows from (28) that $F \notin L^1(\mathbb{R})$. One can however interpret the upper bound for $F(z)$ as meaning that, in a certain sense, $\widehat{F}(\theta) = 0$ for $|\theta| \geq 1$.

Proof. The first two assertions are clear: writing $z = x + iy$, then, assuming for instance that $|y| \geq 1$, we have $|z \pm n|^2 \geq 1 + (|x| - n)^2$ $(n \geq 0)$. In order to prove the third assertion we first recall Euler's formula

$$\Big(\frac{\sin \pi z}{\pi}\Big)^2 \sum_{n \in \mathbb{Z}} \frac{1}{(z-n)^2} = 1 \qquad (z \in \mathbb{C})$$

and note that for $x > 0$

$$\sum_{n=1}^{\infty} \frac{1}{(x+n)^2} \leq \sum_{n=1}^{\infty} \int_{x+n-1}^{x+n} \frac{\mathrm{d}u}{u^2} = \frac{1}{x} = \sum_{n=0}^{\infty} \int_{x+n}^{x+n+1} \frac{\mathrm{d}u}{u^2} \leq \sum_{n=0}^{\infty} \frac{1}{(x+n)^2}.$$

This implies, still for $x > 0$,

$$F(x) = \Big(\frac{\sin \pi x}{\pi}\Big)^2 \Big\{ \sum_{n \in \mathbb{Z}} \frac{1}{(x-n)^2} - 2 \sum_{n=1}^{\infty} \frac{1}{(x+n)^2} + \frac{2}{x} \Big\} \geq 1,$$

and for $x < 0$ (setting $y = -x$)

$$F(x) = \Big(\frac{\sin \pi x}{\pi}\Big)^2 \Big\{ - \sum_{n \in \mathbb{Z}} \frac{1}{(x-n)^2} + 2 \sum_{n=0}^{\infty} \frac{1}{(y+n)^2} - \frac{2}{y} \Big\} \geq -1.$$

Finally

$$F(0) = \lim_{z \to 0} \Big(\frac{\sin \pi z}{\pi z}\Big)^2 = 1 \geq 0 = \operatorname{sgn}(0).$$

Let us prove (28). We have

$$\int_{-\infty}^{+\infty} (F(x) - \operatorname{sgn}(x))\,\mathrm{d}x = \int_0^{\infty} (F(x) - 1)\,\mathrm{d}x + \int_0^{\infty} (F(-y) + 1)\,\mathrm{d}y$$
$$= \int_0^{\infty} (F(x) + F(-x))\,\mathrm{d}x = 2\int_0^{\infty} \Big(\frac{\sin \pi x}{\pi x}\Big)^2\,\mathrm{d}x = 1.$$

Conclusion of the proof of Theorem 5.

Set

$$b(t) := \tfrac{1}{2}\big\{F\big(\delta(t - M - 1)\big) + F\big(\delta(M + N - t)\big)\big\}.$$

Then $b(t)$ satisfies the first of conditions (27), and relation (28) shows that b is integrable over $\mathbb{R}$ together with

$$\int_{-\infty}^{+\infty} b(t)\,\mathrm{d}t = N - 1 + \delta^{-1}.$$

This results immediately from the following identity, valid for $t \neq M+1, M+N$,

$$b(t) = \chi(t) + \tfrac{1}{2}\big\{F\big(\delta(t - M - 1)\big) - \operatorname{sgn}\big(\delta(t - M - 1)\big)\big\}$$
$$+ \tfrac{1}{2}\big\{F\big(\delta(M + N - t)\big) - \operatorname{sgn}\big(\delta(M + N - t)\big)\big\},$$

where χ is the characteristic function of $[M + 1, M + N]$. Moreover, as a consequence of Lemma 5.3, we have

$$b(z) = O(\exp\{2\pi\delta|\Im m\, z|\}) \qquad (z \in \mathbb{C}). \tag{29}$$

In particular b is bounded on $\mathbb{R}$. Since $b \in L^1(\mathbb{R})$, we hence also have $b \in L^2(\mathbb{R})$. The estimate (29) hence implies, by the Paley–Wiener theorem (cf. Katznelson (1968), theorem 7.4, p. 176) that

$$\widehat{b}(\theta) = 0 \qquad (|\theta| \geq \delta).$$

This completes the proof of Theorem 5.

§ 4.5 The large sieve—arithmetic form

Let $\{a_n\}_{n=M+1}^{M+N}$ be a finite sequence of complex numbers and set

$$S(\alpha) := \sum_{M<n\leq M+N} a_n e(n\alpha).$$

We apply Theorem 5 to the case when the α_r are all rational numbers of the form $\alpha_r = a/q$ with $(a, q) = 1$, $q \leq Q$. For $r \neq s$ one clearly has

$$\|\alpha_r - \alpha_s\| = \|a/q - a'/q'\| = \|(aq' - a'q)/qq'\| \geq 1/Q^2,$$

which shows that the α_r are Q^{-2}-well spaced. We can thus write

$$\sum_{q\le Q}\ \sum_{1\le a\le q,\,(a,q)=1} |S(a/q)|^2 \le (N-1+Q^2) \sum_{M<n\le M+N} |a_n|^2. \tag{30}$$

The usefulness of this inequality rests in the observation that one can bound the inner sum from below by an explicit function of q linked to the number $w(p)$ of classes modulo p (p dividing q) in which a_n vanishes identically. More precisely, let us write, for each prime p,

$$w(p) := \operatorname{card}\{h : 0 \le h < p,\ n \equiv h \ (\operatorname{mod} p) \Rightarrow a_n = 0\}, \tag{31}$$

and put

$$g(q) := \mu(q)^2 \prod_{p|q} \frac{w(p)}{p-w(p)}. \tag{32}$$

(We may assume that $w(p) < p$ for all p, since otherwise $a_n \equiv 0$.) The foundations of the arithmetic form of the large sieve are set out in the following result.

Theorem 6. *With the previous notation, we have for all $q \ge 1$*

$$\Big| \sum_{M<n\le M+N} a_n \Big|^2 g(q) \le \sum_{1\le a\le q,\,(a,q)=1} |S(a/q)|^2. \tag{33}$$

Corollary 6.1 (Arithmetic large sieve). *For any finite sequence of complex numbers $\{a_n : M < n \le M+N\}$, we have*

$$\Big| \sum_{M<n\le M+N} a_n \Big|^2 \le \frac{N-1+Q^2}{L} \sum_{M<n\le M+N} |a_n|^2 \tag{34}$$

with

$$L := \sum_{q\le Q} g(q), \tag{35}$$

where $g(q)$ is defined by (31) and (32).

Proof of Theorem 6. We have to show that, for any given sequence $\{a_n\}$,

$$|S(0)|^2 g(q) \le \sum_{1\le a\le q,\,(a,q)=1} |S(a/q)|^2. \tag{36}$$

Since the definition of $w(p)$ is unchanged when one replaces a_n by $a_n e(n\beta)$, relation (36) is equivalent to

$$|S(\beta)|^2 g(q) \le \sum_{1\le a\le q,\,(a,q)=1} |S(a/q+\beta)|^2 \quad (\beta\in\mathbb{R}). \tag{37}$$

Assume (37) is satisfied for q and q' with $(q,q')=1$. By the Chinese remainder theorem we can then write

$$\begin{aligned}\sum_{\substack{1\le c\le qq'\\(c,qq')=1}} |S(c/qq')|^2 &= \sum_{\substack{1\le a\le q\\(a,q)=1}} \sum_{\substack{1\le b\le q'\\(b,q')=1}} |S(a/q+b/q')|^2\\ &\ge \sum_{1\le a\le q,\,(a,q)=1} |S(a/q)|^2 g(q') \ge |S(0)|^2 g(q)g(q').\end{aligned}$$

Since g is multiplicative, it follows that (36) (and consequently (37)) is true for qq'. Moreover $g(q)=0$ when q is not squarefree, so we may confine ourselves to establishing (36) when q is prime.

For any prime number p we have

$$\begin{aligned}|S(0)|^2+\sum_{a=1}^{p-1}|S(a/p)|^2 &= \sum_{a,a'=0}^{p-1} S(a/p)\overline{S(a'/p)}\,\frac{1}{p}\sum_{h=0}^{p-1} e((a-a')h/p)\\ &= \frac{1}{p}\sum_{h=0}^{p-1}\Big|\sum_{a=0}^{p-1} e(-ah/p)S(a/p)\Big|^2 = \frac{1}{p}\sum_{h=0}^{p-1}\Big|\sum_n a_n\sum_{a=0}^{p-1} e\big(a(n-h)/p\big)\Big|^2\\ &= p\sum_{h=0}^{p-1}|S(p,h)|^2, \text{ say,}\end{aligned} \tag{38}$$

where we have set

$$S(p,h) := \sum_{\substack{M<n\le M+N\\ n\equiv h\,(\mathrm{mod}\,p)}} a_n.$$

Note that by assumption $S(p,h)$ is zero for at least $w(p)$ values of h modulo p. So, by the Cauchy–Schwarz inequality, we have

$$|S(0)|^2 = \Big|\sum_{h=0}^{p-1} S(p,h)\Big|^2 \le (p-w(p))\sum_{h=0}^{p-1}|S(p,h)|^2,$$

from which it follows by (38) that

$$\begin{aligned}\sum_{a=1}^{p-1}\big|S(a/p)\big|^2 &= p\sum_{h=0}^{p-1}|S(p,h)|^2 - |S(0)|^2\\ &\ge \Big(\frac{p}{p-w(p)}-1\Big)|S(0)|^2 = g(p)|S(0)|^2.\end{aligned}$$

This establishes (36) for $q=p$ and thereby completes the proof.

From the identity $S(0) = \sum_{h=0}^{p-1} S(p,h)$, we obtain that

$$p\sum_{h=0}^{p-1}\Big|S(p,h) - \frac{1}{p}S(0)\Big|^2 = p\sum_{h=0}^{p-1}|S(p,h)|^2 - |S(0)|^2,$$

from which we deduce by (38) that

$$p\sum_{h=0}^{p-1}\Big|S(p,h) - \frac{1}{p}S(0)\Big|^2 = \sum_{a=1}^{p-1}|S(a/p)|^2.$$

Inserting this in (30), we obtain the following result.

Theorem 7. *With the previous notation, we have*

$$\sum_{p\le Q} p\sum_{h=0}^{p-1}\Big|S(p,h) - \frac{1}{p}S(0)\Big|^2 \le (N-1+Q^2)\sum_{M<n\le M+N}|a_n|^2. \tag{39}$$

Relation (39) is a weakened form of the large sieve inequality, since only the contribution of those q that are primes is estimated. It is however a very useful result for applications—cf. Notes. Moreover, it may be extended to congruence classes for composite moduli. Montgomery (1968) showed that for all squarefree q one has

$$q\sum_{h=0}^{q-1}\Big|\sum_{d\mid q}\frac{\mu(d)}{d}\,S(q/d,h)\Big|^2 = \sum_{1\le a\le q,\,(a,q)=1}\big|S(a/q)\big|^2.$$

Inserting this into (30) we see that $S(q/d,h)$ is, on average over d dividing q and h in $[0,q-1]$, close to $(d/q)S(0)$.

§ 4.6 Applications

By comparison with Brun's method, the large sieve provides a remarkably effective upper bound for the number $J(x)$ of prime twins not exceeding x.

Theorem 8. *As x tends to infinity, we have*

$$J(x) \le (8C + o(1))x/(\log x)^2 \tag{40}$$

with

$$C := 2\prod_{p\ge 3}(1-(p-1)^{-2}). \tag{41}$$

This upper bound is thus asymptotically equal to eight times the conjectured value for $J(x)$—cf. the Notes for references on improvements.

Proof. We use (34) with $N = [x]$, $Q = x^{1/2-\varepsilon}$, $M = 0$ and $a_n := 1$ if $P^-\big(n(n+2)\big) > Q$, $a_n := 0$ otherwise. Then

$$J(x) - J(\sqrt{x}) \le (1 + o(1))x/L \tag{42}$$

where L is defined by (35) with

$$w(2) = 1, \quad w(p) = 2 \qquad (p \ge 3).$$

We have $g(q) = 2^\omega * h(q)/q$ where h is the multiplicative function defined by

$$h(2) = 0, \quad h(2^\nu) = 2(-1)^{\nu-1} \qquad (\nu \ge 2)$$

$$h(p) = \frac{4}{p-2}, \quad h(p^\nu) = 2(-1)^{\nu-1}\Big(\frac{p+2}{p-2}\Big) \qquad (\nu \ge 2).$$

It is easy to check that the series $\sum_{d=1}^\infty h(d)d^{-\sigma}$ is absolutely convergent for $\sigma > \frac{1}{2}$, from which we deduce that

$$\sum_{q\le y} g(q) = \sum_{md\le y} \frac{h(d)}{d}\frac{2^{\omega(m)}}{m} \sim \frac{3}{\pi^2}(\log y)^2 \sum_{d=1}^\infty \frac{h(d)}{d} \qquad (y \to \infty)$$

where the sum over m has been evaluated by partial summation from the estimate

$$\sum_{m\le y} 2^{\omega(m)} = \sum_{m\le y} \mathbf{1} * \mu^2(m) \sim \frac{6}{\pi^2} y \log y.$$

From (42) we then deduce that

$$J(x) \le (2C + o(1))x(\log\sqrt{x})^{-2}$$

with

$$\begin{aligned} C &= \frac{\pi^2}{6}\Big(\sum_{d=1}^\infty \frac{h(d)}{d}\Big)^{-1} \\ &= \prod_p (1-p^{-2})^{-1}\, \tfrac{3}{2} \prod_{p\ge 3}\Big(1 + \frac{4}{p(p-2)} - \frac{2(p+2)}{p^2(p-2)(1+p^{-1})}\Big)^{-1} \\ &= 2\prod_{p\ge 3}\Big(1 - \frac{1}{(p-1)^2}\Big). \end{aligned}$$

This yields the required conclusion.

Our second application is concerned with primes in arithmetic progressions. Let us write

$$\pi(x; \ell, k) := \operatorname{card}\{p \leq x : p \equiv \ell \pmod k\}$$

so that $\pi(x; \ell, k)$ is possibly unbounded only when ℓ takes one of the $\varphi(k)$ values of the invertible residues modulo k. It is thus natural to conjecture that in a suitable range for k and x, we have

$$\pi(x; \ell, k) \sim \frac{\pi(x)}{\varphi(k)} \sim \frac{x}{\varphi(k) \log x}.$$

We shall show this in Part II for fixed k (Dirichlet's theorem) and even for $k \leq (\log x)^c$ (Siegel–Walfisz theorem). The large sieve allows us to obtain an upper bound of the same type, equally valid for "short intervals".

Theorem 9 (Brun–Titchmarsh). *Let x, y be positive numbers and k, ℓ be integers. If $y/k \to \infty$, we have*

$$\pi(x + y; \ell, k) - \pi(x; \ell, k) \leq (2 + o(1)) \frac{y}{\varphi(k) \log(y/k)}. \tag{43}$$

Proof. The left-hand side of (43) is at most equal to

$$\sum_n a_n + \pi\big(\sqrt{(y/k)}\big), \tag{44}$$

where $a_n := 1$ if $x < \ell + nk \leq x + y$ and $P^-(\ell + kn) > \sqrt{(y/k)}$, and $a_n := 0$ otherwise. The second term of (44) is plainly absorbed by the remainder term of (43). With the notation of the large sieve, we have $N \leq y/k + 1$ and $w(p) \geq 1$ for any prime number $p \leq \sqrt{(y/k)}$ with $p \nmid k$. We therefore obtain that, for all $Q \leq \sqrt{(y/k)}$,

$$\sum_n a_n \leq \frac{y/k + Q^2}{L} \tag{45}$$

with

$$L := \sum_{q \leq Q,\, (q,k)=1} \mu(q)^2 \prod_{p|q} \frac{1}{p-1} = \sum_{q \leq Q,\, (q,k)=1} \frac{\mu(q)^2}{\varphi(q)}.$$

Noting that each integer $m \leq Q$ factorises uniquely in the form $m = qdt$ with $(q, k) = 1$, $\mu(q)^2 = 1$, $d|q^\infty$, $t|k^\infty$, we can write

$$\sum_{m \leq Q} \frac{1}{m} \leq \sum_{q \leq Q,\, (q,k)=1} \frac{\mu(q)^2}{q} \sum_{d|q^\infty} \frac{1}{d} \sum_{t|k^\infty} \frac{1}{t} = \sum_{q \leq Q,\, (q,k)=1} \frac{\mu(q)^2}{q} \frac{q}{\varphi(q)} \frac{k}{\varphi(k)},$$

from which we get

$$L \geq \frac{\varphi(k)}{k} \log Q.$$

Inserting this estimate in (45) and selecting $Q = \sqrt{(y/k)}/\log(y/k)$, we obtain the stated result.

Notes

§ **4.2.** In order to estimate the third error term of (5) we have appealed to the *parametric method*, presented in detail in Chapter 0 of the book of Hall & Tenenbaum (1988). The famous *Rankin's method* (cf. § III.5.1) is another example of this fruitful computational trick, which consists in bounding the characteristic function of a set of integers by a constant multiple of a multiplicative function depending on a parameter to optimise.

The application of Brun's method to an upper bound for $\pi(x)$ is of course a mere illustration of the basic ideas, and should not be considered as a genuine result. Not only, as we already observed, does it give a weaker estimate than Chebyshev's bound, but also one could object, *stricto sensu*, to a loss of information. Indeed, we have appealed to Theorem 1.9, which itself rests on Mertens' first theorem. But this last result immediately yields a Chebyshev-type upper bound for $\pi(x)$ since

$$\pi(x) - \pi(\sqrt{x}) \ll \frac{x}{\log x} \sum_{\sqrt{x}<p\leq x} \frac{\log p}{p} \ll \frac{x}{\log x}.$$

However it can be easily checked that what is really needed in Brun's treatment is an asymptotic formula for $\sum_{p\leq x} 1/p$. This is a much weaker assertion than Theorems 1.9 or 1.7, and reveals precisely the reason why the scope of the method is so large.

For other results concerning $\Phi(x,y)$, see Chapter III.6.

Alladi (1988) explains recent developments of Brun's method relating to sums of multiplicative functions on certain subsets of $\mathbb{N}$.

The reader interested in the up-to-date theory of the combinatorial sieve can look at the survey of Diamond & Halberstam (1985), and at the deep articles of Iwaniec (1980, 1981).

§ **4.4.** Although the proof that we give here for the analytic version of the large sieve appeals (at least for the optimal form of the result) to the theory of analytic functions of a complex variable and to a deep result of harmonic analysis, it is possible to consider the large sieve as an essentially elementary tool. We have seen that the proof of the inequality

$$\Delta(N,\delta) \leq \tfrac{1}{2}\pi^2(N-1+\delta^{-1})$$

is elementary. One can also look at the proofs, all different, of Montgomery (1971) [$\Delta \leq N + 2\delta^{-1}$], Bombieri (1974) [*idem*], and Elliott (1980) [$\Delta \leq N + \delta^{-1}$].

The function $F(z)$ of Lemma 5.3 was studied by Beurling at the end of the thirties. For an exhaustive study of the optimisation problem (26)–(27) and of the numerous applications of various generalisations of this question, see Graham & Vaaler (1981, 1984), Vaaler (1985).

The large sieve is equally useful to estimate mean values of weighted averages of Dirichlet characters—cf. § II.8.1. Let us put

$$T(\chi) := \sum_{M<n\leq M+N} a_n \chi(n).$$

Gallagher (1967) gave a simple proof of the inequality

$$\sum_{\chi}{}^{*} |T(\chi)|^2 \leq \frac{\varphi(q)}{q} \sum_{1\leq a\leq q,\,(a,q)=1} \left|S\left(\frac{a}{q}\right)\right|^2$$

where the sum is extended to primitive characters modulo q, that is to characters which are not induced by a character modulo d with $d|q$, $d<q$. Then (30) implies

$$\sum_{q\leq Q} \frac{q}{\varphi(q)} \sum_{\chi}{}^{*} |T(\chi)|^2 \leq (N-1+Q^2) \sum_{M<n\leq M+N} |a_n|^2.$$

This inequality plays an essential role in the study of L-functions (cf. § II.8.2) and the distribution of prime numbers in arithmetic progressions. It is, in particular, one of the fundamental ingredients of the proof of the Bombieri–Vinogradov theorem (1965, 1966)—cf. Chapter II.8, Notes.

§ **4.5.** Theorem 7 has been used in numerous elegant solutions of arithmetic problems. In particular it underpins the original proof of Daboussi's theorem (Daboussi & Delange, 1974). It is also the starting point of a new method of Hildebrand (1986c, d, e, 1987a) for studying the mean value of multiplicative functions of modulus at most 1. Hildebrand (1986b) also applies this inequality to manufacture a new elementary proof of the prime number theorem. A variant of Theorem 7 has been established by Elliott (1979, lemma 4.7) where only the congruence classes modulo p are taken into account. Elliott proves the inequality (see also Theorem III.3.2)

$$\sum_{p\leq N} p\left|S(0,p) - \frac{1}{p}S(0)\right|^2 \leq 16N \sum_{1\leq n\leq N} |a_n|^2.$$

This result is not directly comparable to (39): summation over p is longer (since (39) implies an upper bound of the same order only when $Q \ll \sqrt{N}$), but here only a single congruence class for each p is considered.

§ **4.6.** The best upper bound at present for $J(x)$ is due to Wu (1990), improving on a result of Fouvry & Grupp (1986). The factor 8 in Theorem 8 is replaced by 3.418. The basic result in this problem provides a factor 4. It is due to Bombieri & Davenport (1966) — cf. Halberstam & Richert (1974). Besides the sieve (Selberg's, but the large sieve is also applicable) the proof requires the Bombieri–Vinogradov theorem.

The inequality (43) is in fact valid uniformly *without* any error term, i.e.

$$\pi(x+y;\ell,k)-\pi(x;\ell,k)<\frac{2y}{\varphi(k)\log(y/k)}\qquad(1\le k<y\le x).$$

This handy result is due to Montgomery & Vaughan (1973).

Exercises

1. *Integers with only small or large prime factors.*

For $x \ge z \ge y \ge 1$, let $\Psi_0(x,y,z)$ denote the number of integers $n \le x$ which have no prime factor in $]y,z]$.

(a) Use Eratosthenes' sieve to show that

$$\lim_{x\to\infty} x^{-1}\Psi_0(x,y,z)=\prod_{y<p\le z}(1-1/p).$$

(b) Use Brun's "pure" method to show that there is a positive absolute constant c, such that one has, uniformly for $1\le y\le z\le \exp\{c\log x/\log_2 x\}$,

$$\Psi_0(x,y,z)\sim x\prod_{y<p\le z}(1-1/p)\qquad(x\to\infty).\tag{46}$$

(c) Extend this result, using the fundamental lemma of combinatorial sieve theory.

(d) Show, assuming the prime number theorem, that (46) is not valid uniformly for $1\le y<z\le\sqrt{x}$.

2. *Primes and quasi-primes of the form* n^2+1.

(a) Show that the number $\rho(p)$ of solutions of the equation $\xi^2+1 \equiv 0 \pmod{p}$ equals 1 if $p=2$, 2 if $p \equiv 1 \pmod 4$, and 0 if $p \equiv 3 \pmod 4$.

(b) Deduce that the equation $\xi^2+1 \equiv 0 \pmod{d}$ has $\rho(d) := \prod_{p|d} \rho(p)$ solutions for each squarefree integer d.

(c) Using the fundamental lemma of combinatorial sieve theory, show that the number $S(x)$ of primes $p \le x$ of the form $p = n^2+1$ satisfies

$$S(x) \ll \sqrt{x} \prod_{\substack{p \le x \\ p \equiv 1 \pmod 4}} \left(1 - \frac{2}{p}\right).$$

(d) Make the preceding calculation more precise by using Dirichlet's theorem

$$\sum_{\substack{p \le x \\ p \equiv 1 \pmod 4}} \frac{1}{p} = \tfrac{1}{2} \log_2 x + O(1).$$

(e) Under the same assumptions show that there exists some absolute positive constant B such that

$$\operatorname{card}\{n \le \sqrt{x} : n^2+1 \in Q(B,x)\} \asymp (\sqrt{x})/\log x$$

where $Q(B,x) := \{m \le x : p|m \Rightarrow p > x^{1/B}\}$. [*The elements of this set are usually referred to as "quasi-primes"—see Halberstam & Richert (1974), §2.8. In particular a quasi-prime has only a bounded number of prime factors. Moreover* $|Q(B,x)| \asymp \pi(x)$.]

3. *Almost squares.*

(a) Let p be an odd prime. Show that the kernel of the endomorphism of $(\mathbb{Z}/p\mathbb{Z})^* : x \mapsto x^2$ is $\{\pm 1\}$ and deduce that the number of quadratic non-residues modulo p is $(p-1)/2$.

(b) By sieving the set $\mathcal{A} = \{n : n \le x\}$ by the prime numbers $\le \sqrt{x}$ for suitable classes, show that the number S of integers $n \le x$ such that n is a quadratic residue modulo p for all $p \le \sqrt{x}$ satisfies $[\sqrt{x}] \le S \le C\sqrt{x}$ where C is an absolute constant to be calculated.

4. *Majorising the number of representations in Goldbach's problem.* Let N be an even integer and let $r(N)$ denote the number of representations of N in the form $N = p+q$ where p and q are primes.

(a) Show that, for any multiplicative function $f \ge 0$ with support included in the set of squarefree numbers, one has

$$\sum_{n \le x} f(n) \le \sum_{d|N} f(d) \sum_{m \le x,\, (m,N)=1} f(m) \qquad (1 \le x \le N).$$

(b) Prove that there exists an absolute constant C such that

$$r(N) \le C \prod_{p|N} \left(1 + \frac{2}{p}\right) \frac{N}{(\log N)^2}.$$

(c) Let h be a multiplicative function satisfying the conditions:

$$|h(p)| \le 1 \ (p|N), \quad |h(p)| \le p^{-\delta} \ (p \nmid N), \quad |h(p^\nu)| \ll 1 \ (\nu \ge 2),$$

where δ is a positive constant. Show that, uniformly for $0 \le \alpha \le 1/\log_2 N$, one has

$$\sum_{d=1}^{\infty} |h(d)| d^{\alpha - 1} \ll \log_2 N.$$

(d) Applying the above result to a suitable function h, establish the following inequality, which sharpens the result obtained in (b)

$$r(N) \le (16 + o(1)) C_N \frac{N}{(\log N)^2},$$

with $C_N := \prod_{p>2}(1-1/(p-1)^2) \prod_{p|N,\, p>2}((p-1)/(p-2))$. [*This upper bound is asymptotically eight times the conjectured value for* $r(N)$. *Halberstam & Richert (1974, Theorem 3.11) obtain an estimate where the factor 16 is replaced by 8.*]

5. *Poisson summation formula.*

Let $f \in L^1(\mathbb{R})$.

(a) Show that the series

$$\varphi(t) := \sum_{n \in \mathbb{Z}} f(n+t)$$

converges for almost all t.

(b) Assume in addition that f is continuous and of bounded variation on $\mathbb{R}$. Show that the Poisson formula

$$\sum_{n \in \mathbb{Z}} f(n+t) = \lim_{N \to \infty} \sum_{|n| \le N} \widehat{f}(n) e(tn)$$

holds for all $t \in [0, 1[$.

6. *Integers coprime to q.*

For each integer $q \geq 1$, set $N_q(x) := |\{n \leq x : (n,q) = 1\}|$.

(a) Show, for each fixed $q \geq 1$, that

$$\lim_{x\to\infty} N_q(x)/x = \varphi(q)/q.$$

(b) Given q, show that each integer $n \geq 1$ can be written uniquely in the form $n = hdt$, with $(h,q) = 1$, $d \mid q$, $\mu(d)^2 = 1$, $p \mid t \Rightarrow p \mid d$.

(c) Calculate $\sum_{p|t \Rightarrow p|d} 1/t$.

(d) For $q \geq 1$, $Q \geq 1$ set $L(q,Q) := \sum_{d|q,\, d\leq Q} \mu(d)^2/\varphi(d)$. Show that

$$\log(Q+1) \leq L(q,Q) \prod_{p\leq Q,\, p\nmid q} (1 - 1/p)^{-1}.$$

Deduce that, for $x \geq Q \geq 1$, $q \geq 1$, $P^+(q) \leq x$, one has

$$\frac{1}{L(q,Q)} \leq \frac{e^\gamma \varphi(q)\log x}{q\log(Q+1)}\Big\{1 + O\Big(\frac{1}{\log x}\Big)\Big\}.$$

(e) Show that, uniformly for $x \geq 3$, $q \geq 1$, $P^+(q) \leq x$, one has

$$N_q(x) \leq \Big\{1 + O\Big(\frac{\log_2 x}{\log x}\Big)\Big\} 2e^\gamma \frac{\varphi(q)}{q} x.$$

[*This upper bound can be halved: see Exercise III.3.13.*]

7. *Functions of the combinatorial sieve.*

(a) Show that for each arithmetic function χ one has

$$\mu\chi * \mathbf{1}(n) = \sum_{d|m/q} \mu(d)\{\chi(d) - \chi(qd)\} \qquad (n \geq 1)$$

with $q := P^-(n)$, $m := \prod_{p|n} p$.

(b) Let $\mathcal{P}$ be a set of prime numbers, y a real number. Set $P(y) := \prod_{p\leq y,\, p\in\mathcal{P}} p$. Show that if χ_1, χ_2 satisfy the three following properties

(α) $\chi_i(d) = 0$ or 1 $\qquad (d \mid P(y))$

(β) $\chi_i(d) = 1 \Rightarrow \chi(t) = 1 \qquad (t \mid d)$

(γ) $\chi_i(d) = 1$, $\mu(d) = (-1)^{i-1} \Rightarrow \chi_i(pd) = 1 \quad (pd \mid P(y),\, p < P^-(d))$,

then for n dividing $P(y)$, $q = P^-(n)$, one has for $i = 1, 2$

$$(-1)^i \mu(d)\{\chi_i(d) - \chi_i(qd)\} \geq 0 \qquad (d \mid m/q).$$

Deduce that

(47) $$\mu\chi_1 * \mathbf{1}(n) \leq \delta(n) \leq \mu\chi_2 * \mathbf{1}(n) \qquad (n \mid P(y)).$$

(c) Show that the functions μ_1, μ_2 described after Theorem 2 satisfy relation (47).

I.5

Extremal orders

§ 5.1 Introduction and definitions

Although the average order of an arithmetic function gives an intuitive idea of its overall behaviour, it can only reflect very crudely the variations of its values. In this chapter we describe, by means of several significant examples, the standard methods leading to individual bounds that are *optimal* in a sense to be made precise.

Let f be an arithmetic function. Let g, h denote, generally, "elementary" monotone functions such as logarithms or powers. In accord with the perspective described above the information concerning f may be of one the following types:

$$\begin{array}{lll} \text{(a)} & f(n) = O(g(n)) & (n \geq n_0) \\ \text{(b)} & f(n) = o(g(n)) & (n \to \infty) \\ \text{(c)} & f(n) = (1 + o(1))g(n) & (n \to \infty) \\ \text{(d)} & f(n) = g(n) + O(h(n)) & (n \geq n_0). \end{array}$$

In the last case it is naturally desirable that $h(n)$ be $o(g(n))$. A measure of optimality of these asymptotic relations can be expressed by estimates of the following form (where $g(n)$ is assumed to be ultimately positive):

$$\begin{array}{lll} (\alpha) & f(n) = \Omega_+(g(n)) & [\text{i.e. } \limsup f(n)/g(n) > 0] \\ (\beta) & f(n) = \Omega_-(g(n)) & [\text{i.e. } \liminf f(n)/g(n) < 0] \\ (\gamma) & f(n) = \Omega(g(n)) & [\text{i.e. } \limsup |f(n)|/g(n) > 0]. \end{array}$$

The symbol $f(n) = \Omega_\pm(g(n))$ indicates that the relations (α) and (β) hold simultaneously. The use of the letter Ω in this context is traditional; in practice, no confusion arises with the function "number of prime factors". The following definition introduces an appreciably finer concept than the previous ones for determining the individual order of an arithmetic function.

Definition. *Let f be an arithmetic function and let g be a non-decreasing function which is ultimately positive. We say that g is a maximal (resp. minimal) order for f if*

$$\limsup_{n\to\infty} f(n)/g(n) = 1 \qquad (\text{resp. } \liminf_{n\to\infty} f(n)/g(n) = 1).$$

Thus knowledge of the two extremal orders for an arithmetic function f provides, up to a multiplicative factor tending to 1, *the best possible individual bounds for* $f(n)$. In what follows we investigate such information for the main additive and multiplicative functions presented in Chapter 2.

§ 5.2 The function $\tau(n)$

Theorem 1. *Let f be a multiplicative function. If* $\lim_{p^\nu\to\infty} f(p^\nu) = 0$, *then* $\lim_{n\to\infty} f(n) = 0$.

Corollary 1.1. *For all $\varepsilon > 0$ we have $\tau(n) = O_\varepsilon(n^\varepsilon)$.*

We obtain the Corollary immediately by applying Theorem 1 to $f(n) = \tau(n)n^{-\varepsilon}$. For $q = p^\nu$ we have

$$f(q) = (\nu+1)p^{-\nu\varepsilon} \le 2(1+\log q)q^{-\varepsilon} \to 0 \quad (q \to \infty).$$

Proof of Theorem 1. Let q denote, in general, a prime power p^ν. By assumption, for each $\varepsilon > 0$, there exists some integer $Q = Q(\varepsilon)$ such that

$$q > Q \Rightarrow |f(q)| \le \varepsilon.$$

Next consider the following partition of the set of all possible q.

$$\begin{aligned} Q_1 &:= \{q : q \le Q, |f(q)| \le 1\}, \\ Q_2 &:= \{q : q \le Q, |f(q)| > 1\}, \\ Q_3 &:= \{q : q > Q\}. \end{aligned}$$

Each integer n decomposes uniquely in the form

$$n = n_1 n_2 n_3 \quad \text{with} \quad n_i := \prod_{q\|n,\, q\in Q_i} q \qquad (i = 1, 2, 3).$$

It is clear that the n_i are mutually coprime, so

$$f(n) = f(n_1)f(n_2)f(n_3). \tag{1}$$

By definition of Q_1, we have $|f(n_1)| \le 1$. Since Q_2 is contained in the finite set of q such that $|f(q)| > 1$, there exists some constant A independent of ε such that $|f(n_2)| \le A$. Moreover, except for a finite number of integers n, $|f(n_3)| \le \varepsilon$. It then follows from (1) that

$$\limsup_{n\to\infty} |f(n)| \le A\varepsilon,$$

and hence the desired conclusion.

The following result shows that Corollary 1.1 is not far from being optimal: ε can only tend to 0 as $c/\log_2 n$.

Theorem 2. *A maximal order for the function* $\log \tau(n)$ *is*

$$\log 2 \cdot \log n / \log_2 n.$$

Proof. For each $\varepsilon > 0$ we must show that

$$\tau(n) < \exp\{(1+\varepsilon)\log 2 \cdot \log n / \log_2 n\} \tag{2}$$

for all $n \geq n_0(\varepsilon)$, and

$$\tau(n) > \exp\{(1-\varepsilon)\log 2 \cdot \log n / \log_2 n\} \tag{3}$$

for infinitely many integers n.

The upper bound (2) is easy to prove. For each value of the parameter t with $2 \leq t \leq n$, we can write

$$\begin{aligned}\tau(n) = \prod_{p^\nu \| n} (\nu+1) &\leq \prod_{p^\nu \| n,\, p \leq t} (\nu+1) \prod_{p^\nu \| n,\, p>t} 2^\nu \\ &\leq \Big(1 + \frac{\log n}{\log 2}\Big)^t \Big(\prod_{p^\nu \| n} p^\nu\Big)^{\log 2/\log t} \\ &\leq \exp\Big\{t(2+\log_2 n) + \frac{\log 2 \cdot \log n}{\log t}\Big\}.\end{aligned}$$

Choosing $t = \log n / (\log_2 n)^3$ it follows that

$$\tau(n) \leq \exp\Big\{\frac{\log 2 \cdot \log n}{\log_2 n}\Big(1 + O\Big(\frac{\log_3 n}{\log_2 n}\Big)\Big)\Big\} \tag{4}$$

from which we immediately deduce (2).

In order to establish the lower bound (3), it is necessary to find integers having many divisors. Natural candidates are the products

$$n_k := \prod_{j=1}^{k} p_j \qquad (k = 1, 2, \ldots) \tag{5}$$

where p_j denotes the jth prime number. We have

$$\tau(n_k) = 2^k$$

and

$$\log n_k = \theta(p_k) = \sum_{p \leq p_k} \log p \leq \pi(p_k) \log p_k = k \log p_k.$$

This implies that

$$\log \tau(n_k) \geq (\log 2 \cdot \log n_k)/\log p_k. \tag{6}$$

From Chebyshev's estimate (2.29) in the form

$$\log n_k = \theta(p_k) \geq A p_k \tag{7}$$

with a suitable positive constant A, we deduce, substituting (7) in (6), that

$$\log \tau(n_k) \geq \frac{\log 2 \cdot \log n_k}{\log_2 n_k}\Big\{1 + O\Big(\frac{1}{\log_2 n_k}\Big)\Big\}. \tag{8}$$

This implies (3).

Remark. The problem of the minimal order of $\tau(n)$ is trivial: we have $\tau(n) \geq 2$ with equality whenever n is prime.

§ 5.3 The functions $\omega(n)$ and $\Omega(n)$

By multiplicativity one easily sees that for all $n \geq 1$

$$2^{\omega(n)} \leq \tau(n) \leq 2^{\Omega(n)} \leq n. \tag{9}$$

Each of these inequalities is optimal: the first two are equalities if and only if n is squarefree, and the third whenever n is a power of 2. This allows a rapid evaluation of the maximal orders for $\omega(n)$ and $\Omega(n)$—the minimal orders being trivial.

First, (4) and (9) imply

$$\omega(n) \leq (1 + o(1)) \log n/\log_2 n \quad (n \to \infty)$$

and inequality (8) shows that this upper bound is optimal up to the factor $(1 + o(1))$, since the n_k occurring in (8) are squarefree.

The situation is even simpler for $\Omega(n)$ since the last inequality in (9) is attained for an infinitely many integers n.

We can therefore state the following result.

Theorem 3. (i) *A maximal order for the function* $\omega(n)$ *is* $\log n/\log_2 n$.
(ii) *A maximal order for the function* $\Omega(n)$ *is* $\log n/\log 2$.

§ 5.4 Euler's function $\varphi(n)$

Euler's phi-function is multiplicative and its value at prime powers is given by

$$\varphi(p^\nu) = p^\nu(1-p^{-1}) \tag{10}$$

—cf. § 2.7. We thus have

$$\varphi(n) \le n \qquad (n \ge 1) \tag{11}$$

and, for all $\varepsilon > 0$,

$$\varphi(n) \ge n^{1-\varepsilon} \qquad (n \ge n_0(\varepsilon)). \tag{12}$$

The upper bound (11) is trivial, the lower bound (12) follows from Theorem 1 applied to $f(n) = n^{1-\varepsilon}\varphi(n)^{-1}$.

We can make these estimates more precise in the following result.

Theorem 4. *A maximal order for $\varphi(n)$ is n. A minimal order is*

$$e^{-\gamma} n / \log_2 n$$

where γ denotes Euler's constant.

Proof. The first assertion follows immediately from (10). In order to establish the second, we note that

$$\varphi(n) = n \prod_{p|n}(1-p^{-1}) \ge n \prod_{p \le q}(1-p^{-1}) \tag{13}$$

for all integers q with $\pi(q) \ge \omega(n)$. By Chebyshev's lower estimate (Theorem 1.3) this condition is realised as soon as

$$q/\log q \ge (\omega(n)/\log 2) + 4$$

and therefore, in particular, for

$$q = [A\omega(n) \log \omega(n)]$$

where A is some suitable positive constant. By Theorem 3(i) we have with this choice

$$q \ll \log n$$

from which, by (13), applying Mertens' formula,

$$\varphi(n) \geq n \frac{e^{-\gamma}}{\log q}\Big\{1 + O\Big(\frac{1}{\log q}\Big)\Big\} \geq n \frac{e^{-\gamma}}{\log_2 n}\Big\{1 + O\Big(\frac{1}{\log_2 n}\Big)\Big\}. \tag{14}$$

In order to complete the proof of Theorem 4, it remains to show that this lower bound is asymptotically attained on a suitable subsequence. Now with n_k defined by (5) we have

$$\begin{aligned}\frac{\varphi(n_k)}{n_k} &= \prod_{p \leq p_k} (1 - p^{-1}) = \frac{e^{-\gamma}}{\log p_k}\Big\{1 + O\Big(\frac{1}{\log p_k}\Big)\Big\} \\ &= \frac{e^{-\gamma}}{\log_2 n_k}\Big\{1 + O\Big(\frac{1}{\log_2 n_k}\Big)\Big\},\end{aligned}$$

by Chebyshev's estimate (2.29) in the form

$$\log_2 n_k = \log \theta(p_k) = \log p_k + O(1).$$

This concludes the proof.

§ 5.5 The functions $\sigma_\kappa(n)$, $\kappa > 0$

In § 2.2 we defined the arithmetic function

$$\sigma_\kappa(n) = \sum_{d|n} d^\kappa \qquad (\kappa \in \mathbb{R}).$$

It is a multiplicative function ($\sigma_\kappa = j^\kappa * \mathbf{1}$) and the identity

$$\sigma_\kappa(n) = n^\kappa \sigma_{-\kappa}(n)$$

shows that it suffices to study the extremal orders for positive values of the parameter—the case $\kappa = 0$, corresponding to $\sigma_0 = \tau$, having already been considered.

We have

$$\sigma_\kappa(p^\nu) = \frac{p^{(\nu+1)\kappa} - 1}{p^\kappa - 1} = p^{\nu\kappa} \frac{1 - p^{-(\nu+1)\kappa}}{1 - p^{-\kappa}}, \tag{15}$$

so that

$$\sigma_\kappa(n) \geq n^\kappa \qquad (n \geq 1) \tag{16}$$

and, for all $\varepsilon > 0$,

$$\sigma_\kappa(n) \leq n^{\kappa(1+\varepsilon)} \qquad (\kappa > 0,\ n \geq n_0(\varepsilon)), \tag{17}$$

by Theorem 1.

When $\kappa \geq 1$, the study of $\sigma_\kappa(n)$ can be undertaken in very much the same way as that of $\varphi(n)$ carried out above. The lower bound (16) trivially provides a minimal order for $\sigma_\kappa(n)$, and the upper bound (17) can easily be made more precise. Indeed choosing q as in (13), we have

$$\sigma_\kappa(n)n^{-\kappa} \leq \prod_{p \leq q}(1 - p^{-\kappa})^{-1} =: \zeta(\kappa, q), \tag{18}$$

say. The product converges for $\kappa > 1$, and equals $e^\gamma \log q + O(1)$ when $\kappa = 1$. Moreover $\log q \leq \log_2 n + O(1)$. In particular, we can then deduce that

$$\sigma_\kappa(n)n^{-\kappa} \leq \begin{cases} e^\gamma \log_2 n + O(1) & \text{if } \kappa = 1, \\ \zeta(\kappa)(1 + O_\kappa(\log^{-\kappa} n)) & \text{if } \kappa > 1. \end{cases} \tag{19}$$

These upper bounds are achieved asymptotically for the sequence $\{m_k\}$ defined by

$$m_k := \Big(\prod_{j=1}^{k} p_j\Big)^{\ell(k)} \qquad (k = 1, 2, \ldots)$$

where $p_1 = 2 < p_2 = 3 < \ldots$ denotes the increasing sequence of prime numbers and we have written $\ell(k) = [\log k]$. Checking this last point is left to the reader.

When $0 < \kappa < 1$, the above study still holds, but the asymptotic behaviour of the factor $\zeta(\kappa, q)$ in (18) is more complicated. By partial summation the prime number theorem implies that

$$\log \zeta(\kappa, q) \sim q^{1-\kappa}/(1-\kappa)\log q \tag{20}$$

but even the form with the best known remainder term does not allow us to infer an elementary asymptotic formula for $\zeta(\kappa, q)$. Another intrinsic difficulty resides in the fact that the value of q in (18) is not determined with any great precision. In the case when $\kappa \geq 1$, this does not have any important consequence since $\zeta(\kappa, q)$ is a slowly increasing function of $\log q$. The situation is radically different for $0 < \kappa < 1$.

We can summarise the results obtained so far in the following theorem.

Theorem 5. *For $\kappa > 0$ a minimal order for $\sigma_\kappa(n)$ is n^κ. For $\kappa > 1$, a maximal order is $\zeta(\kappa)n^\kappa$. For $\kappa = 1$, a maximal order is $e^\gamma n \log_2 n$. For $0 < \kappa < 1$, we have*

$$\sigma_\kappa(n) \leq n^\kappa \exp\left\{(1 + o(1))\frac{(\log n)^{1-\kappa}}{(1-\kappa)\log_2 n}\right\}$$

and the opposite inequality is satisfied when n runs through a suitable infinite sequence of integers.

Notes

§§ **5.1–5.5.** The first systematic study of the maximal order of an arithmetic function is due to Ramanujan (1915), who was concerned not only with the determination of large values of $\tau(n)$, but also with making explicit the structure of the numbers for which these large values are attained. Today the study of maximal orders for arithmetic functions has become an active area in the theory of numbers, with powerful methods and original techniques. An excellent survey, including an exhaustive bibliography, has recently been written by Nicolas (1988). The reader who is interested in this subject will find other examples of problems and conjectures in Nicolas (1974/5, 1978, 1983a), Erdős & Nicolas (1981a and b, 1989), Erdős & Tenenbaum (1989b) and Erdős & Sárközy (1994).

Questions linked to maximal orders are often extremely non-trivial, particularly in the extent to which they are concerned with the fine structure of integers. In general the problem of "large values" induces in a natural way a study of the distribution of those integers at which they are attained. This is often very difficult. Even in the well known case of the divisor function $\tau(n)$ and highly composite numbers of Ramanujan, the results obtained are incomplete—cf. Nicolas (1988).

§ **5.5.** For the asymptotic behaviour of $\zeta(\kappa, q)$, see Lemma III.5.9.1.

Exercises

1. What are the extremal orders for the function "sum of digits in base r"?

2. Write $\delta_r(n) := \operatorname{card}\{m_1, \ldots, m_r : n = [m_1, \ldots, m_r]\}$. Calculate $\delta_r * \mathbf{1}$ and show that δ_r is a multiplicative function. Determine the maximal order of $\log \delta_r$.

3. Our aim here is to recover a result of Erdős & Nicolas (1981b) concerning the extremal orders of the function $F(n) = \varphi(n) + \varphi(n+1)$.

(a) Show that $F(n) \leq 3n/2 \quad (n > 1)$.

(b) Let p_j denote the jth prime and put $N_k := 1 + \prod_{1 \le j \le k} p_j$. Show that $p | N_k \Rightarrow p > p_k$ and that $p | (N_k + 1) \Rightarrow p = 2$ or $p > p_k$. Show that $\log N_k \asymp p_k \asymp k \log k$. Deduce that, as $k \to \infty$, one has $\varphi(N_k) \sim N_k$ and $\varphi(N_k + 1) \sim \frac{1}{2} N_k$. What is the maximal order of $F(n)$?

(c) Show that, as $n \to \infty$, one has $\varphi(n) \sim n \prod_{p|n,\, p \le \log n} (1 - p^{-1})$.

(d) Deduce from the previous result that

$$F(n) \ge (1 + o(1)) 2 e^{-\gamma/2} n / \surd(\log_2 n) \quad (n \to \infty). \tag{21}$$

[*Use the inequality* $a + b \ge 2\surd(ab)$.]

(e) Show that there exists some integer $r = r(k)$ such that

$$\prod_{1 \le j \le r} (1 - p_j^{-1}) \sim \prod_{r < j \le k} (1 - p_j^{-1}) \quad (k \to \infty).$$

(f) Let M_k be the smallest solution of the congruence

$$M \equiv 0 \,(\operatorname{mod} p_1 \dots p_r), \quad M \equiv -1 \,(\operatorname{mod} p_{r+1} \dots p_k).$$

Show that

$$\max\left(\frac{\varphi(M_k)}{M_k}, \frac{\varphi(M_k + 1)}{(M_k + 1)} \right) \le (1 + o(1)) \prod_{1 \le j \le r} (1 - 1/p_j) \quad (k \to \infty).$$

Deduce that the right-hand side of (21) is a minimal order for $F(n)$.

4. (a) Determine the maximal order of the function $n \mapsto \log \tau(n^2)$.

(b) Let $S := \{s : p | s \Rightarrow p^2 | s\}$ be the set of "squarefull" integers. What is the maximal order of the function $s \mapsto \log \tau(s)$ $(s \in S)$?

(c) Let $\omega_1(n)$ be the number of prime factors p of n such that $p^2 \nmid n$. Show that

$$\tau(n) \le D(n)^{\lambda + o(1)} 2^{(1-\lambda)\omega_1(n)} \qquad (n \to \infty)$$

with $\lambda := (\log 3)/\log 4$ and $D(n) := \max_{m \le n} \tau(m)$.

5. Let $\alpha > 1$. Set $F_\alpha(n) := \sum_{1 \le i < \tau(n)} \big((d_{i+1}/d_i) - 1\big)^\alpha$, where $\{d_j : 1 \le j \le \tau(n)\}$ denotes the increasing sequence of divisors of n.

(a) Writing $\chi(n; u, v) := \begin{cases} 1 & (\text{if } d \mid n \Rightarrow d \notin]u, v]) \\ 0 & (\text{if } \exists d \mid n,\ u < d \le v) \end{cases}$, show that

$$F_\alpha(n) = \alpha \int_1^n \int_1^v \chi(n; u, v)(v - u)^{\alpha - 2} (\alpha v - u) u^{-1-\alpha} \, du \, dv.$$

(b) We define a *decomposition* of the integer n to be any finite sequence $\{n_j : 1 \le j \le k\}$ such that $n_1 = 1$, $n_j | n_{j+1}$ $(1 \le j < k)$, $n_k = n$. We then say that a sequence of real numbers $\{\varepsilon_j : 1 < j \leq k\}$ is *admissible* for $\{n_j\}$ if for any $j \in [2, k]$ and each $z \in \left]\sqrt{(n_{j-1})}, \sqrt{n_j}\right]$ the interval $]z, (1+\varepsilon_j)z]$ contains at least one divisor of n_j. Show that

$$F_\alpha(n) \leq \alpha \sum_{j=2}^{k} (1+\varepsilon_j)\varepsilon_j^{\alpha-1} \log(n_j/n_{j-1})$$

for any decomposition $\{n_j : 1 \leq j \leq k\}$ of n and each associated admissible sequence $\{\varepsilon_j : 1 < j \leq k\}$. [*This remark is the first step in the proof of Erdős' conjecture that* $\liminf F_\alpha(n) < \infty$ *for each* $\alpha > 1$. *See Vose (1984). Using this and a crude version of the saddle-point method, Tenenbaum (1987) showed that* $F_\alpha(n!)$, $F_\alpha([1, 2, \ldots, n])$, $F_\alpha(\prod_{p \leq n} p)$ *are all uniformly bounded for each* $\alpha > 1$.]

6. We consider the arithmetic function $f(n) := \omega(n)\sigma(n)/n$.

(a) For $k \geq 1$, let $r = r(k) := [\log k]$ and $n_k := \prod_{j=1}^{r} p_j^r \prod_{r<j\leq k} p_j$, where p_j denotes the jth prime number. Assuming the prime number theorem, show that $f(n_k) \sim e^\gamma \log n_k \quad (k \to \infty)$.

(b) Show that a maximal order for $f(n)$ is $e^\gamma \log n$.

(c) Application. Let $A(n) := \sum_{p^\nu \| n} \nu p$ denote the Alladi–Erdős function (1977, 1979). Put $A^*(n) := \sum_{d|n} A(d)/d$. Determine an additive function $g(n)$ such that $A^*(n) = \sigma(n)\{\omega(n) + g(n)\}/n$. Show that $g(p^\nu) \ll 1/p$ and that $g(n) \ll \log_3 n$. Deduce that $e^\gamma \log n$ is a maximal order for $A^*(n)$. [*This result sharpens estimates of Sitaramaiah & Subbarao (1993).*]

I.6

The method of van der Corput

§ 6.1 Introduction

In the nineteen twenties van der Corput developed a method of estimating trigonometric sums which has numerous applications in number theory. This theory is typically relevant to the Dirichlet divisor problem and to the circle problem, i.e. the evaluation of the number of pairs $(m, n) \in (\mathbb{Z}^+)^2$ such that $m^2 + n^2 \leq x$. It is more generally applicable to counting the number of points with integral coordinates within a given contour: see Exercise 5.

As early as 1922, van der Corput had obtained that the remainder term in the Dirichlet problem has order $\ll_\varepsilon x^{33/100+\varepsilon}$. This problem, and others of a similar nature, have subsequently given rise to a prolific literature. A large part of it rests on the ideas of van der Corput — see also the notes on section 3.2.

In this chapter our aim consists essentially of describing the principle of the method in its simplest form, which leads to the theorem of Voronoï, to the estimate $O(x^{1/3})$ for the remainder in the circle problem and to the exponent 1/6 in the bound for the zeta function on the critical line (cf. Section II.3.4). The reader interested in a deeper study of the method is referred to the work of Titchmarsh (1951), to the original articles of van der Corput (1922–1937), to the works of Kolesnik (1981, 1985), as well as to the more recent developments of Bombieri & Iwaniec (1986), Iwaniec & Mozzochi (1988), Huxley & Watt (1988), Watt (1989), Huxley (1990, 1993a, b), and Huxley & Kolesnik (1991). The monograph of Graham & Kolesnik (1991) furnishes a thorough introduction to the modern state of the subject.

We shall make use of the Poisson summation formula, which undoubtedly constitutes the most natural approach to the study of trigonometric sums. Define the Fourier transform by

$$\widehat{f}(\theta) := \int_{-\infty}^{+\infty} f(t)e(-\theta t)\,\mathrm{d}t \qquad (f \in L^1(\mathbb{R})). \tag{1}$$

We apply the Poisson formula with the hypotheses appearing in the following statement.

Theorem 1. *Let $f \in L^1(\mathbb{R})$. Assume that the series*

$$\varphi(t) = \sum_{n\in\mathbb{Z}} f(n+t) \tag{2}$$

converges for all t and that its sum defines a function of bounded variation on $[0,1]$ which is continuous at 0. Then we have

$$\lim_{N\to\infty} \sum_{|\nu|\le N} \widehat{f}(\nu) = \sum_{n\in\mathbb{Z}} f(n). \tag{3}$$

Proof. Jordan's theorem on Fourier series (see, for example, Titchmarsh (1939), p. 406) states that each periodic function of bounded variation over a period is the sum of its Fourier series at each point of continuity. This implies (3) since the Fourier coefficient of order ν of φ is $\widehat{f}(\nu)$.

§ 6.2 Trigonometric integrals

We collect here two results for bounding trigonometric integrals which will be useful in what follows.

Theorem 2. *Let $f \in C^1]a,b[$ be such that $f'(t)$ is monotone and of constant sign on $]a,b[$. Write*

$$m := \inf_{a<t<b} |f'(t)|.$$

Then we have

$$\left|\int_a^b e(f(t))\,\mathrm{d}t\right| \le 2/\pi m. \tag{4}$$

Proof. Without loss of generality we may assume that f' is non-increasing on $]a,b[$. We can then write

$$\begin{aligned}\left|2\pi \int_a^b e(f(t))\,\mathrm{d}t\right| &= \left|\int_a^b (1/f'(t))\,\mathrm{d}e(f(t))\right| \\ &= \left|\big[e(f(t))/f'(t)\big]_a^b - \int_a^b e(f(t))\,\mathrm{d}\{1/f'(t)\}\right| \\ &\le 2/m + \int_a^b \mathrm{d}\{1/f'(t)\} \le 4/m.\end{aligned}$$

Theorem 3. *Let $f \in C^2]a,b[$ such that $f''(t)$ has constant sign on $]a,b[$. Set*

$$r := \inf_{a<t<b} |f''(t)|.$$

Then we have

$$\left|\int_a^b e(f(t))\,\mathrm{d}t\right| \le 4\sqrt{(2/\pi r)}. \tag{5}$$

Proof. Let us suppose, for example, that $f''(t) \le -r < 0$ for $a < t < b$. Then $f'(t)$ vanishes at most once on $]a, b[$, say at $t = c$. (If $f'(t)$ does not vanish, the argument is similar, but simpler.) Now we can write

$$I := \int_a^b e(f(t))\,dt = \int_a^{c-\delta} + \int_{c-\delta}^{c+\delta} + \int_{c+\delta}^b = I_1 + I_2 + I_3,$$

say, where the positive parameter δ satisfies $a + \delta \le c \le b - \delta$. We have

$$|f'(t)| = \left|\int_c^t f''(v)\,dv\right| \ge r|t - c| \ge r\delta$$

for t in $[a, c - \delta] \cup [c + \delta, b]$. By Theorem 2, it follows that

$$|I_1| + |I_3| \le 4/\pi r\delta.$$

Since trivially $|I_2| \le 2\delta$, it follows that

$$|I| \le 2\delta + 4/\pi r\delta$$

and hence we have the stated result, by choosing $\delta = \sqrt{(2/\pi r)}$. It is clear that this upper bound remains valid if, with this choice of δ, one has either $c < a + \delta$ or $c > b - \delta$.

§ 6.3 Trigonometric sums

We now come to the heart of the matter, i.e. the connection between trigonometric sums and integrals, and bounding the former using results from the previous section on the latter.

Theorem 4. *Let $f \in C^1[a, b]$ such that $f'(t)$ is monotone on $]a, b[$. Set*

$$\alpha := \inf_{a<t<b} f'(t), \quad \beta := \sup_{a<t<b} f'(t).$$

Then, for each $\varepsilon > 0$, we have

$$\sum_{a<n\le b} e(f(n)) = \sum_{\alpha-\varepsilon<\nu<\beta+\varepsilon} \int_a^b e(f(t) - \nu t)\,dt + O_\varepsilon\big(\log(\beta - \alpha + 2)\big). \tag{6}$$

Proof. The number ε being fixed we can assume that

$$-1 \le \alpha - \varepsilon < 0. \tag{7}$$

Indeed, formula (6) is invariant on replacing $f(t)$ by $f(t) + kt$, for any $k \in \mathbb{Z}$. We can equally well restrict ourselves to the case when a and b are of the form $m + \frac{1}{2}$, with $m \in \mathbb{Z}$. Indeed, the error involved is $O(1)$ on the left-hand side, and $O\big(\log(\beta - \alpha + 2)\big)$ on the right. This last estimate follows easily from a summation interchange, using the trivial upper bound $\sum_\nu e(-\nu t) \ll \min(\beta - \alpha + 2, 1/\|t\|)$, where $\|t\|$ denotes the distance from t to the set of integers. Finally we suppose that f' is decreasing on $]a, b[$. Let us set

$$F(t) = \begin{cases} e(f(t)) & \text{if } a < t \leq b, \\ 0 & \text{otherwise.} \end{cases} \tag{8}$$

It is immediate that $\varphi(t) := \sum_{n\in\mathbb{Z}} F(n+t)$ is continuous at 0 (since $a, b \notin \mathbb{Z}$) and has bounded variation on $[0, 1]$. The Poisson formula (3) then implies

$$\sum_{a<n\leq b} e(f(n)) = \sum_{|\nu|\leq N} \widehat{F}(\nu) + o(1) \quad (N \to \infty)$$

with

$$\widehat{F}(\nu) = \int_a^b e(f(t) - \nu t)\,\mathrm{d}t.$$

Taking account of (7), it remains to show that we have

$$\sum_{\substack{|\nu|\leq N \\ \nu\notin[0,\beta+\varepsilon[}} \widehat{F}(\nu) = O_\varepsilon(\log(\beta + 2)). \tag{9}$$

Now

$$2\pi i\widehat{F}(\nu) = \int_a^b \frac{\mathrm{d}\{e(f(t)-\nu t)\}}{(f'(t)-\nu)} = \left[\frac{e\big(f(t)-\nu t\big)}{(f'(t)-\nu)}\right]_a^b - \int_a^b e\big(f(t)-\nu t\big)\,\mathrm{d}\Big\{\frac{1}{f'(t)-\nu}\Big\}$$
$$= (-1)^\nu \frac{e\big(f(b)\big)}{\alpha - \nu} + (-1)^{\nu+1} \frac{e\big(f(a)\big)}{\beta - \nu} + O\Big(\frac{1}{\alpha - \nu} - \frac{1}{\beta - \nu}\Big).$$

It is plain that the contribution to (9) of the main terms is $O_\varepsilon(1)$. Moreover, that of the error term is

$$\ll \sum_{\nu\notin[0,\beta+\varepsilon[} \frac{\beta+1}{\nu(\nu-\beta)} = (\beta+1)\sum_{\nu=1}^{\infty} \frac{1}{\nu(\nu+\beta)} + (\beta+1) \sum_{\nu>\beta+\varepsilon} \frac{1}{\nu(\nu-\beta)}$$
$$\ll_\varepsilon \sum_{1\leq\nu\leq\beta+1} \frac{1}{\nu} + \sum_{\beta+\varepsilon<\nu\leq 2\beta} \frac{1}{\nu-\beta} + 2(\beta+1)\sum_{\nu>2\beta} \frac{1}{\nu^2} \ll_\varepsilon \log(\beta+2).$$

This completes the proof.

The following result is an easy consequence of Theorems 3 and 4.

Theorem 5 (van der Corput). *Let $f \in C^2]a,b]$, such that*

$$|f''(t)| \asymp \lambda > 0 \qquad (a < t < b).$$

Then we have

$$\sum_{a<n\leq b} e\big(f(n)\big) \ll (b-a+1)\lambda^{1/2} + \lambda^{-1/2}. \tag{10}$$

Proof. We can assume that $\lambda \leq 1$, since otherwise (10) is trivially satisfied. With the notation of Theorem 4, the left-hand side of (10) is

$$\ll (\beta-\alpha+1)\max_{\nu}\Big|\int_a^b e\big(f(t)-\nu t\big)\,\mathrm{d}t\Big| + \log(\beta-\alpha+2)$$
$$\ll (\beta-\alpha+1)\lambda^{-1/2} + \log(\beta-\alpha+2),$$

where the second upper bound follows from Theorem 3. Now we have

$$\beta-\alpha = \Big|\int_a^b f''(t)\,\mathrm{d}t\Big| \asymp \lambda(b-a).$$

The previous bound is thus

$$\ll \lambda^{1/2}(b-a) + \lambda^{-1/2} + 1 + \lambda(b-a) \ll \lambda^{1/2}(b-a) + \lambda^{-1/2}.$$

For of a function of class C^3, we establish the following variant of Theorem 5—cf. Titchmarsh (1951), § 5.9.

Theorem 6. *Let $f \in C^3[a,b]$, with $b-a \geq 1$. Suppose that*

$$|f'''(t)| \asymp \lambda > 0 \qquad (a < t < b).$$

Then

$$\sum_{a<n\leq b} e\big(f(n)\big) \ll (b-a)\lambda^{1/6} + (b-a)^{1/2}\lambda^{-1/6}. \tag{11}$$

Lemma 6.1. *Let f be a real function defined on $]a,b]$. For any integer q with $1 \leq q \leq b-a$, we have*

$$\Big|\sum_{a<n\leq b} e\big(f(n)\big)\Big| \leq \frac{2(b-a)}{\sqrt{q}} + 2\Big\{\frac{(b-a)}{q}\sum_{r=1}^{q-1}\Big|\sum_{a<n\leq b-r} e\big(f(n+r)-f(n)\big)\Big|\Big\}^{1/2}. \tag{12}$$

Proof of Lemma. Let F be the function defined by (8) and $S := \sum_{n\in\mathbb{Z}} F(n)$ be the sum to be estimated. We have trivially

$$S = q^{-1}\sum_{m=1}^{q}\sum_{n\in\mathbb{Z}} F(n+m).$$

Interchanging the summations and applying the Cauchy–Schwarz inequality, we obtain

$$|S|^2 \le q^{-2}\sideset{}{'}\sum_{n\in\mathbb{Z}} 1\cdot\sideset{}{'}\sum_{n\in\mathbb{Z}}\sum_{m,m'=1}^{q} F(n+m)\overline{F(n+m')} \tag{13}$$

where the dash indicates that the summation is restricted to integers n with $a < n+m \le b$ for at least one m such that $1 \le m \le q$. Thus the first sum over n does not exceed $b-a+q \le 2(b-a)$. The inner sum equals

$$q + 2\Re e\sum_{1\le m'<m\le q} F(n+m)\overline{F(n+m')},$$

so the second sum over n is at most

$$2(b-a)q + 2\Big|\sum_{1\le m'<m\le q}\sum_{n\in\mathbb{Z}} F(n+m)\overline{F(n+m')}\Big|.$$

Let us perform the change of variables $m+n=\nu$, $m-m'=r$. Then ν runs through $\mathbb{Z}$ and r can take the values $1, 2, \dots, q-1$. Moreover, when ν and r are fixed, there are exactly $q-r$ solutions for $\{n, m, m'\}$, namely $\{\nu-j-r,\ j+r, j\}$ for $1 \le j \le q-r$. The above expression is thus

$$\le 2(b-a)q + 2\Big|\sum_{r=1}^{q-1}(q-r)\sum_{\nu\in\mathbb{Z}} F(\nu)\overline{F(\nu-r)}\Big|$$

$$\le 2q\Big\{(b-a) + \sum_{r=1}^{q-1}\Big|\sum_{\nu\in\mathbb{Z}} F(\nu+r)\overline{F(\nu)}\Big|\Big\}.$$

Inserting this upper bound in (13) we obtain the stated result.

Proof of Theorem 6. We appeal to the previous lemma, noting that, if we set $g(x) := f(x+r) - f(x)$, then

$$|g''(x)| \asymp r\lambda \qquad (a < x < b-r).$$

This follows immediately from Taylor's theorem (of order 1) for f''. Using Theorem 5 to bound the sum over n in the right-hand side of (12), we have that

$$\sum_{a<n\le b} e(f(n)) \ll Lq^{-1/2} + \Big\{Lq^{-1}\sum_{r=1}^{q-1}\Big(L(r\lambda)^{1/2} + (r\lambda)^{-1/2}\Big)\Big\}^{1/2}$$

where we have written $L := b - a$. We observe that this bound is

$$\ll Lq^{-1/2} + L(q\lambda)^{1/4} + L^{1/2}(q\lambda)^{-1/4}.$$

If λ satisfies $1 \le \lambda^{-1/3} \le L$, we can choose $q = [\lambda^{-1/3}]$. This makes the first two terms of the same order of magnitude $L\lambda^{1/6}$, and hence establishes (11) in this case. It is easy to check that the estimate (11) is trivially valid when $\lambda > 1$ or $\lambda < L^{-3}$.

§ 6.4 Application to the theorem of Voronoï

Here we show that the error term in the Dirichlet problem is $O_\varepsilon(x^{1/3+\varepsilon})$. The same technique provides an identical bound in the circle problem (cf. Exercise 6) and also the estimate $\zeta(\frac{1}{2} + it) \ll_\varepsilon t^{1/6+\varepsilon}$ (cf. Exercise 2 and § II.3.4).

Theorem 7 (Voronoï, 1903). *For $x \ge 2$, we have*

$$\sum_{n\le x} \tau(n) = x\big(\log x + 2\gamma - 1\big) + O\big(x^{1/3}\log x\big). \tag{14}$$

Proof. Write $N := \big[\sqrt{x}\big]$. The hyperbola method—cf. formula (3.4)—shows that the left-hand side of (14) can be written as

$$2\sum_{d\le N}\Big[\frac{x}{d}\Big] - N^2 = 2\sum_{d\le N}\Big(\frac{x}{d} - B_1\Big(\frac{x}{d}\Big)\Big) - N - N^2$$

where $B_1(t) = \{t\} - \frac{1}{2}$ denotes the first Bernoulli function. (Recall that $\{t\}$ is the fractional part of t.) Using Theorem 0.5 to estimate $\sum_{d\le N} 1/d$, we can write the above expression in the form

$$P(x) - 2R(x)$$

with

$$P(x) := 2x\Big(\log N + \gamma + \frac{1}{2N} + O\Big(\frac{1}{x}\Big)\Big) - N - N^2$$

and

$$R(x) := \sum_{d\le N} B_1(x/d).$$

Writing $N = \sqrt{x} - \theta$, with $0 \le \theta < 1$, and expanding $P(x)$ (up to order 1) as a function of $\theta/\sqrt{x}$, we obtain

$$P(x) = x(\log x + 2\gamma - 1) + O(1).$$

Thus formula (14) is equivalent to

$$R(x) \ll x^{1/3} \log x. \tag{15}$$

In order to establish (15), van der Corput's technique consists essentially in expanding $B_1(x/d)$ as a Fourier series, interchanging summations, and estimating the sums in $\sin(2\pi jx/d)$ for fixed j and x, by appealing to the results of the previous section. The fact that the Fourier series for B_1 is not absolutely convergent introduces a technical difficulty which we overcome by replacing $B_1(t)$ with

$$B(t) := \tfrac{1}{2} J \int_{-1/J}^{1/J} B_1(t+u)\,du$$

for large J. Since B is Lipschitz of order 1, its Fourier series is absolutely convergent. Explicitly

$$B(t) = \sum_{j=1}^{\infty} a_j \sin(2\pi jt) \tag{16}$$

with

$$a_j = -\frac{J}{2\pi^2 j^2} \sin\left(\frac{2\pi j}{J}\right) \qquad (j = 1, 2, \ldots). \tag{17}$$

In addition, the function

$$h(t) := |B(t) - B_1(t)|$$

is also Lipschitz of order 1. Actually we have

$$h(t) = \tfrac{1}{2}(1 - J\|t\|)^{+} \qquad (J > 1).$$

An easy calculation shows that

$$h(t) = \frac{1}{2J} + \sum_{j=1}^{\infty} b_j \cos(2\pi jt) \tag{18}$$

with

$$b_j = \frac{J}{\pi^2 j^2} \sin^2\Big(\frac{\pi j}{J}\Big). \tag{19}$$

From (17) and (19) it follows, in particular, that

$$|a_j| + |b_j| \ll \min(j, J)/j^2 \qquad (j \geq 1). \tag{20}$$

We are now in a position to embark on the last stage of the proof. For $M < T \leq 2M$, we set

$$R(x; M, T) := \sum_{M<d\leq T} B_1(x/d).$$

Clearly we have

$$\Big|R(x; M, T) - \sum_{M<d\leq T} B(x/d)\Big| \leq \sum_{M<d\leq T} h(x/d),$$

and hence, by (16) and (18), we can write

$$\begin{aligned} R(x; M, T) = &\sum_{j=1}^{\infty} a_j \sum_{M<d\leq T} \sin(2\pi jx/d) \\ &+ O\Big(\frac{M}{J} + \sum_{j=1}^{\infty} b_j \Big| \sum_{M<d\leq T} \cos(2\pi jx/d)\Big|\Big). \end{aligned} \tag{21}$$

Now it follows from Theorem 5 that for each real number y we have

$$\sum_{M<d\leq T} e(y/d) \ll \Big(\frac{y}{M}\Big)^{1/2} + \Big(\frac{M^3}{y}\Big)^{1/2}. \tag{22}$$

Applying this estimate with $y = jx$ and substituting in (21) we have

$$\begin{aligned} R(x; M, T) &\ll \frac{M}{J} + \sum_{j=1}^{\infty} (|a_j| + |b_j|)\Big(\Big(\frac{jx}{M}\Big)^{1/2} + \Big(\frac{M^3}{jx}\Big)^{1/2}\Big) \\ &\ll \frac{M}{J} + \Big(\frac{Jx}{M}\Big)^{1/2} + \Big(\frac{M^3}{x}\Big)^{1/2}, \end{aligned}$$

where the sum over j has been estimated taking account of (20). For the optimal choice $J := Mx^{-1/3}$, we obtain

$$R(x; M, T) \ll x^{1/3} + \Big(\frac{M^3}{x}\Big)^{1/2}. \tag{23}$$

This bound is *a priori* valid only if $J > 1$, i.e. if $M > x^{1/3}$. It is clear that it still holds, trivially, in the complementary case. Set $r_0 := [\log N/\log 2] - 1$. We have

$$\begin{aligned} R(x) &= \sum_{r=-1}^{r_0} R(x; 2^r, 2^{r+1}) + R(x; 2^{r_0+1}, N) \\ &\ll \sum_{r=-1}^{r_0} \left(x^{1/3} + 2^{3r/2}x^{-1/2}\right) + x^{1/3} \\ &\ll r_0 x^{1/3} + N^{3/2}x^{-1/2} \ll x^{1/3}\log x. \end{aligned}$$

This establishes (15) and completes the proof of Theorem 7.

Notes

§§ **6.2–6.3.** The content of these sections is essentially included in chapters 4 and 5 of Titchmarsh (1951).

§ **6.3.** Theorem 5 represents a relatively crude application of van der Corput's method, inasmuch as formula (6), i.e.

$$\sum_n F(n) = \sum_{\alpha-\varepsilon<\nu<\beta+\varepsilon} \widehat{F}(\nu) + \text{remainder}, \tag{24}$$

is used in the form

$$\sum_n F(n) \ll \sum_\nu |\widehat{F}(\nu)|.$$

Indeed the integrals

$$\widehat{F}(\nu) = \int_a^b e(f(t) - \nu t)\,\mathrm{d}t$$

have an oscillating asymptotic behaviour, which may be revealed by applying the stationary phase method, due to Laplace. With the hypotheses of Theorem 4, let then x_ν be the unique solution, if it exists, of the equation

$$f'(x_\nu) = \nu.$$

We obtain (cf. Titchmarsh (1951), chapter 4) that

$$\widehat{F}(\nu) = e^{\pm i\pi/4}|f''(x_\nu)|^{-1/2}e(f(x_\nu) - \nu x_\nu) + \text{remainder}$$

where the error term can be bounded explicitly as a function of parameters λ, μ such that

$$|f'(t)| \asymp \lambda, \quad |f'''(t)| \ll \mu \qquad (a < t \leq b).$$

With a suitable degree of approximation, the second term of (24) can hence be considered as a new trigonometric sum, weighted by the factors $|f''(x_\nu)|^{-1/2}$. A non-trivial treatment of this sum allows an improvement of the final estimate. Of course, an immediate further application of the Poisson formula would only lead back to the initial sum, but one can, nevertheless, hope for a worthwhile gain by inserting a simple transformation, based on the Cauchy–Schwarz inequality, such as that which appears in Lemma 6.1, which is known under the name of the Weyl–van der Corput transformation—see Weyl (1916, 1921) and van der Corput (1922). One can thus alternate repeated applications of this transformation (let us say process A) with that of the Poisson formula (let us say, process B) according to the algorithm

$$A^{r_1}BA^{r_2}\ldots BA^{r_n} \qquad (r_i \geq 0,\ 1 \leq i \leq n).$$

The choice of the r_i then poses a difficult optimisation problem. This subject was formalised by van der Corput in the twenties and then simplified by Phillips (1933). This is known today as the *theory of exponent pairs*—cf. Ivić (1985), chapter 2, and Graham & Kolesnik (1991).

Exercises

1. Set $\tau(n,\theta) := \sum_{d|n} d^{i\theta}$ $(\theta \subset \mathbb{R})$. Using Theorem 4, show that

$$x^{-1}\Big|\sum_{n\leq x}\tau(n,\theta)\Big| \ll \log(2+|\theta|) + |\theta|^{-1} \qquad (\theta \neq 0,\ x > 1+|\theta|).$$

[*See also Corollary II.3.5.1 and Theorem II.3.7.*]

2. Show that

$$\sum_{n\leq t} n^{-\frac{1}{2}-it} \ll t^{1/6}\log t \qquad (t \geq 2).$$

[*Use Theorems 5 and 6 in suitable regions.*]

3. Give an upper bound for $|\sum_{n\le x}\exp\{in^\alpha\}|$ for $0<\alpha<2$.

4. Let a, $b\in\mathbb{Z}$, and $f\in C^2[a,b]$ such that $|f''(t)|\asymp\lambda$ for $a\le t\le b$. Show that

$$\sum_{a<n\le b}\{f(n)\}=\tfrac12(b-a)+O(\lambda^{1/3}(b-a)+\lambda^{-1/2}).$$

5. Use the Euler–Maclaurin formula to show, with the assumptions of Exercise 4, that, for $\lambda\le 1$, we have

$$\sum_{a<n\le b}[f(n)]=\int_a^b f(t)\,\mathrm{d}t+\tfrac12\big(f(b)-f(a)+a-b\big)+O\big(\lambda^{1/3}(b-a)+\lambda^{-1/2}\big).$$

6. *On the circle problem.*

Deduce from the previous exercise that, as $x\to\infty$,

$$\operatorname{card}\{(m,n)\in\mathbb{Z}^2: m^2+n^2\le x\}=\pi x+O(x^{1/3}).$$

[*The best estimate known to date for the remainder term is* $\ll x^{23/73+\varepsilon}$, *due to Huxley (1993a).*]

7. Show that for all $\lambda\in\mathbb{R}\smallsetminus\{0\}$ we have

$$\sum_{n\le x}\exp\{i\lambda n\log n\}=o(x)\qquad(x\to\infty).$$

8. Let $f:[1,+\infty[\to\mathbb{R}^+$ be differentiable, such that $f'(x)$ is non-increasing, $f'(x)\to 0\quad(x\to\infty)$, and $xf'(x)\to\infty\quad(x\to\infty)$.

(a) Using either Theorem 4, or the Euler–Maclaurin formula at order 0, show that

$$\sum_{1\le n\le N}e\big(f(n)\big)=o(N)\qquad(N\to\infty).\tag{25}$$

(b) Integrating by parts $\int_1^N e\big(f(t)\big)\,\mathrm{d}t$, show that (25) does not hold if $xf'(x)$ tends to a finite limit.

Part II

Methods of complex analysis

II.1

Generating functions: Dirichlet series

§ 1.1 Convergent Dirichlet series

Let f be an arithmetic function. If the power series

$$S(z) := \sum_{n=1}^{\infty} f(n) z^n$$

converges in some neighbourhood of 0, several classical theorems allow us to connect the analytic behaviour of the sum with the values of the coefficients—for example, the relation $S^{(n)}(0) = n! f(n)$ or Cauchy's integral formulae.

Similarly, the analytic behaviour of a convergent Dirichlet series

$$F(s) := \sum_{n=1}^{\infty} f(n) n^{-s} \tag{1}$$

is closely linked with the asymptotic nature of the sequence $f(n)$. One of the principal aims of Part II is to develop methods which make this link explicit.

The letter s denotes a complex variable. The real numbers σ and τ are implicitly defined by the relation

$$s = \sigma + i\tau.$$

Definition. *Let f be an arithmetic function. The Dirichlet series associated with f is the function of the complex variable $F(s)$ defined by (1) for those points s where the series is convergent.*

The properties of formal Dirichlet series studied in Chapter I.2 suggest that the concept of a convergent Dirichlet series might have interesting arithmetical consequences. We begin with a simple but fundamental result, concerning Dirichlet convolution.

Theorem 1. *Let f, g and h be arithmetic functions, with respective Dirichlet series F, G and H. Let us suppose that*

$$h = f * g. \tag{2}$$

If F and G converge absolutely at any given point s, then so does H and, further,

$$H(s) = F(s)G(s). \tag{3}$$

Proof. If F and G are absolutely convergent at s, then for each $x \geq 1$ we have

$$\begin{aligned}\sum_{n\leq x}|h(n)n^{-s}| &= \sum_{md\leq x}|f(m)g(d)m^{-s}d^{-s}| \\ &\leq \sum_{m\leq x}|f(m)m^{-s}|\cdot\sum_{d\leq x}|g(d)d^{-s}|.\end{aligned}$$

This implies the absolute convergence of $H(s)$. Relation (3) follows by the associated formal identity—cf. § I.2.4.

Remark. The conclusion of Theorem 1 fails, if we consider conditional convergence instead of absolute convergence. It is relatively easy to construct functions f, g and h satisfying (2), and such that, at certain points s, the series $F(s)$ and $G(s)$ converge, while $H(s)$ diverges—cf. Exercise 1; see also the Notes.

§ 1.2 Dirichlet series of multiplicative functions

We have seen in Theorem I.2.4 that formal Dirichlet series of multiplicative functions have the characteristic property of being representable as formal Euler products. The following theorem provides a sufficient condition for such an algebraic identity to have an analytic meaning.

Theorem 2. *Let f be a multiplicative function and let s be a complex number. If the condition*

$$\sum_p\sum_{\nu\geq 1}|f(p^\nu)p^{-\nu s}| < \infty \tag{4}$$

is realised then the Dirichlet series (1) is absolutely convergent, and we have

$$F(s) = \prod_p\sum_{\nu=0}^{\infty}f(p^\nu)p^{-\nu s}. \tag{5}$$

Remarks. (a) If the series (1) is absolutely convergent, the same holds for (4). So the absolute convergence of (1) is actually equivalent to that of (4).

(b) When F is completely (resp. strongly) multiplicative, the right-hand side of (5) takes the simpler form

$$\prod_p\Bigl(1-\frac{f(p)}{p^s}\Bigr)^{-1}, \qquad \Bigl(\text{resp. } \prod_p\Bigl(1+\frac{f(p)}{p^s-1}\Bigr)\Bigr).$$

Proof. Observe first of all that condition (4) implies the convergence of the infinite product $M := \prod_p \left(1 + \sum_{\nu=1}^{\infty} |f(p^\nu)p^{-\nu s}|\right)$. Then for all $x \geq 1$ we can write

$$\sum_{n \leq x} |f(n)n^{-s}| \leq \sum_{P^+(n) \leq x} |f(n)n^{-s}| = \prod_{p \leq x} \left(1 + \sum_{\nu=1}^{\infty} |f(p^\nu)p^{-\nu s}|\right) \leq M.$$

This shows that the series $F(s)$ is absolutely convergent. The inequality

$$\left|\sum_{n=1}^{\infty} f(n)n^{-s} - \prod_{p \leq x} \sum_{\nu=0}^{\infty} f(p^\nu)p^{-s}\right| = \left|\sum_{P^+(n) > x} f(n)n^{-s}\right| \leq \sum_{n > x} |f(n)n^{-s}|$$

then implies (5), by letting x tend to infinity.

Corollary 2.1 (Euler's formula). *For $\sigma > 1$ we have*

$$\zeta(s) = \prod_p (1 - p^{-s})^{-1}. \tag{6}$$

This simple formula highlights the explicit link between Riemann's zeta function and the sequence of primes. This is one of the most important tools in analytic number theory.

§ 1.3 Fundamental analytic properties of Dirichlet series

Let $n \mapsto a_n$ be an arithmetic function. Set

$$A(t) := \sum_{n \leq e^t} a_n.$$

Then the Dirichlet series associated with (a_n) can be written in the form

$$F(s) := \sum_{n=1}^{\infty} a_n n^{-s} = \int_{0-}^{\infty} e^{-ts}\, \mathrm{d}A(t).$$

This integral is called the *Laplace–Stieltjes transform* of the function $A(t)$. Most of the fundamental theorems concerning Dirichlet series can be generalised in the context of this transform, where the Stieltjes integral reveals itself to be a handy technical tool.

Let $\mathcal{V}$ be the class of functions defined on $\mathbb{R}$ and with bounded variation on every finite interval. Although we are primarily concerned with Dirichlet series, we consider, whenever possible, the Laplace–Stieltjes transforms of functions in $\mathcal{V}$—with more concern for clarity than generality.

Theorem 3. *Let A be a function in $\mathcal{V}$ and let*

$$F(s) := \int_{0-}^{+\infty} e^{-st}\,\mathrm{d}A(t) \tag{7}$$

be its Laplace–Stieltjes transform.

(i) If the integral (7) converges for $s = s_0 = \sigma_0 + i\tau_0$, then it converges for all s such that $\sigma > \sigma_0$, and the convergence is uniform in any sector

$$S(\theta) := \{s \in \mathbb{C} : |\arg(s - s_0)| \le \theta\} \qquad (0 \le \theta < \frac{\pi}{2}).$$

(ii) If the integral (7) converges absolutely for $s = s_0$ then it converges absolutely and uniformly in the closed half-plane $\sigma \ge \sigma_0$.

(iii) The function $F(s)$ is holomorphic in any open domain of convergence of (7), and we have

$$F^{(k)}(s) = \int_{0-}^{\infty} (-t)^k e^{-st}\,\mathrm{d}A(t) \qquad (k = 0, 1, 2, \ldots). \tag{8}$$

Proof. (i). Let $\theta \in [0, \pi/2[$. The sector $S(\theta)$ is the set of complex numbers s such that

$$|s - s_0| \le \frac{\sigma - \sigma_0}{\cos\theta}.$$

For any $\varepsilon > 0$, we show there exists a real number $x_0 = x_0(\varepsilon, \theta)$ such that, whenever $y \ge x \ge x_0$, we have

$$\Big|\int_x^y e^{-ts}\,\mathrm{d}A(t)\Big| \le \varepsilon \qquad (s \in S(\theta)).$$

Let us write

$$g(u) := \int_{0-}^{u} e^{-ts_0}\,\mathrm{d}A(t) \qquad (u \ge 0).$$

By hypothesis, there exists some $x_0 = x_0(\varepsilon, \theta)$ such that, for $v \ge u \ge x_0$,

$$|g(v) - g(u)| \le \tfrac{1}{2}\,\varepsilon\cos\theta.$$

Now, for $y \ge x \ge x_0$ and $s \in S(\theta)\smallsetminus\{s_0\}$ we can write

$$\begin{aligned}\Big|\int_x^y e^{-ts}\,\mathrm{d}A(t)\Big| &= \Big|\int_x^y e^{-u(s-s_0)}\,\mathrm{d}\{g(u) - g(x)\}\Big| \\ &= \Big|e^{-y(s-s_0)}\{g(y) - g(x)\} + (s - s_0)\int_x^y e^{-u(s-s_0)}\{g(u) - g(x)\}\,\mathrm{d}u\Big| \\ &\le \tfrac{1}{2}\,\varepsilon\cos\theta + |s - s_0|\tfrac{1}{2}\,\varepsilon\cos\theta \int_x^y e^{-u(\sigma-\sigma_0)}\,\mathrm{d}u \\ &\le \tfrac{1}{2}\,\varepsilon\cos\theta\Big(1 + \frac{|s - s_0|}{\sigma - \sigma_0}\Big) \le \tfrac{1}{2}\,\varepsilon(\cos\theta + 1) \le \varepsilon.\end{aligned}$$

This completes the proof of (i).

Assertion (ii) is merely a reformulation of the trivial upper bound

$$\int_x^y |e^{-ts}|\,|\mathrm{d}A(t)| \leq \int_x^y |e^{-ts_0}|\,|\mathrm{d}A(t)|,$$

valid for $y \geq x$ and $\sigma \geq \sigma_0$.

Let us show (iii). The uniform convergence of the exponential series allows us to write, for all $x \geq 0$,

$$\int_{0-}^x e^{-ts}\,\mathrm{d}A(t) = \sum_{n=0}^{\infty} \frac{s^n}{n!} \int_{0-}^x (-t)^n\,\mathrm{d}A(t).$$

The left-hand side of this equality is thus an entire function of s, for which the kth derivative takes the value

$$\sum_{n=k}^{\infty} \frac{s^{n-k}}{(n-k)!} \int_{0-}^x (-t)^n\,\mathrm{d}A(t) = \int_{0-}^x (-t)^k e^{-ts}\,\mathrm{d}A(t).$$

Assertion (iii) follows from the above, by appealing to Weierstrass' theorem on uniform limits of sequences of analytic functions on compact sets —see for example Cartan (1961), chapter V, theorem 1.

Let us consider an integral $F(s)$ of type (7). By Theorem 3, (i) and (ii), the set of real parts σ of those points s where the integral converges (resp. converges absolutely) is a half-line with origin σ_c (resp. σ_a). One defines σ_c and σ_a to be respectively the *abscissa of convergence* and the *abscissa of absolute convergence* of the Stieltjes integral $F(s)$. By convention we allow these numbers to be $\pm\infty$.

It is easy to construct examples of Dirichlet series converging everywhere or diverging everywhere on the *line of convergence* $\sigma = \sigma_c$. The following result shows that the sum of the series may be evaluated on the line of convergence by analytic continuation, if this exists.

Theorem 4. *Suppose that the integral $F(s)$ defined by (7) for $\sigma > \sigma_c$ has an analytic continuation $\widetilde{F}(s)$ for certain points s with real part $\sigma = \sigma_c$. Then we have*

$$\widetilde{F}(s) = \int_{0-}^{+\infty} e^{-ts}\,\mathrm{d}A(t)$$

at each point of convergence of the integral.

Proof. We use uniformity, as established in Theorem 3(i), in the form

$$F(s) = \lim_{\delta\to 0+} F(s+\delta) \qquad (\sigma = \sigma_c).$$

Since $\widetilde{F}$ is analytic, this implies that

$$\widetilde{F}(s) = \lim_{\delta\to 0+} \widetilde{F}(s+\delta) = \lim_{\delta\to 0+} F(s+\delta) = F(s),$$

as required.

As an application of Theorem 4, let us consider the Riemann zeta function. For $\sigma > 1$

$$(9)\qquad \zeta(s) = \int_{1-}^{\infty} t^{-s}\,\mathrm{d}[t] = s\int_1^{\infty} [t]t^{-s-1}\,\mathrm{d}t = \frac{s}{s-1} - s\int_1^{\infty} \{t\}t^{-s-1}\,\mathrm{d}t.$$

Since the last integral is absolutely convergent for $\sigma > 0$, this shows that $\zeta(s)$ can be analytically continued to a meromorphic function for $\sigma > 0$, with just one singularity, a simple pole, at $s = 1$. From Theorem 4, assuming that the series converges, one must necessarily have

$$(10)\qquad \sum_{n=1}^{\infty} \mu(n)n^{-1} = \zeta(1)^{-1} = 0.$$

Thus, taking account of Theorem I.3.8, it follows that the prime number theorem is equivalent to the *convergence* of the series (10).

The following theorem shows that σ_c and σ_a cannot be chosen independently when $F(s)$ is a Dirichlet series.

Theorem 5. *Let $F(s)$ be a Dirichlet series, defined by (1). Then we have*

$$(11)\qquad \sigma_c \le \sigma_a \le \sigma_c + 1.$$

Proof. Let $\varepsilon > 0$. The convergence of the series $\sum_{n=1}^{\infty} f(n)/n^{\sigma_c+\varepsilon}$ implies the upper bound

$$f(n) \ll_\varepsilon n^{\sigma_c+\varepsilon}$$

and, consequently, the absolute convergence of the series $F(s)$ at $s = \sigma_c+1+2\varepsilon$. Thus we have $\sigma_a \le \sigma_c + 1 + 2\varepsilon$, from which the result follows by letting ε tend to 0.

It is easy to see that the bounds (11) are optimal. For the series

$$(12)\qquad G(s) := \sum_{n=1}^{\infty} (-1)^n n^{-s} = (2^{1-s} - 1)\zeta(s)$$

we have $\sigma_c = 0$ (by the theorem for alternating series), and, of course, $\sigma_a = 1$.

By analogy with power series one might be tempted to believe that a Dirichlet series necessarily possesses a singularity on the line of convergence. This is not the case. The series (12) constitutes in this respect an excellent counter-example. We shall indeed see in Chapter 3 that the Riemann zeta function can be continued as a meromorphic function on all of $\mathbb{C}$ with only a single, simple, pole at $s = 1$. This implies that $G(s)$ may be continued as an entire function. One can also prove directly that $G(s)$ has a regular analytic continuation for $\sigma > -1$—cf. Exercise 2.

The following theorem, usually referred to as *Landau's theorem* (cf. Notes), describes a situation in which the line of convergence always contains a singularity.

Theorem 6 (Phragmén–Landau). *Let A be a function in $\mathcal{V}$ and let $F(s)$ be its Laplace–Stieltjes transform defined by (7). If A is non-decreasing, then the point $s = \sigma_c$ is a singularity of $F(s)$.*

Corollary 6.1. *The abscissa of convergence is a singularity of any Dirichlet series with non-negative coefficients.*

Proof. Let us argue by contradiction and suppose that F has an analytic continuation in some suitable neighbourhood of $s = \sigma_c$. There then exist two numbers $\sigma > \sigma_c$ and $r > \sigma - \sigma_c$ such that the Taylor series of F at σ, namely

$$F(s) = \sum_{k=0}^{\infty} \frac{1}{k!} F^{(k)}(\sigma)(s-\sigma)^k,$$

converges in the disc $|s - \sigma| < r$. Using (8), this property implies that

$$\begin{aligned} F(s) &= \sum_{k=0}^{\infty} \frac{1}{k!}(s-\sigma)^k \int_{0-}^{+\infty} (-t)^k e^{-\sigma t}\,\mathrm{d}A(t) \\ &= \sum_{k=0}^{\infty} \frac{1}{k!} \int_{0-}^{\infty} t^k(\sigma - s)^k e^{-\sigma t}\,\mathrm{d}A(t). \end{aligned}$$

When s is real with $\sigma - r < s < \sigma$, we may interchange summations since the integrand and the measure $\mathrm{d}A(t)$ are both non-negative. It follows that

$$\begin{aligned} F(s) &= \int_{0-}^{\infty} \sum_{k=0}^{\infty} \frac{1}{k!} t^k(\sigma - s)^k e^{-\sigma t}\,\mathrm{d}A(t) \\ &= \int_{0-}^{\infty} e^{t(\sigma - s)} e^{-\sigma t}\,\mathrm{d}A(t) = \int_{0-}^{\infty} e^{-st}\,\mathrm{d}A(t). \end{aligned}$$

Thus the Laplace–Stieltjes integral converges at s, which leads to a contradiction by selecting $s < \sigma_c$. This concludes the proof.

The Phragmén–Landau theorem assumes a particular importance in practice because it forms the basis of the proofs of most *oscillation theorems*. Here we confine ourselves to the following two results which are typical of its use.

Theorem 7. *Let $A : [1, \infty[\to \mathbb{R}$ be measurable and locally bounded. If the integral*

$$H(s) := \int_1^{\infty} A(t) t^{-s-1}\,\mathrm{d}t \tag{13}$$

has a finite abscissa of convergence σ_c and if $H(s)$ has a regular analytic continuation at the point $s = \sigma_c$, then for each $\varepsilon > 0$ we have

$$A(x) = \Omega_{\pm}(x^{\sigma_c - \varepsilon}). \tag{14}$$

Proof. Assume, for example, that there exists some constant $K = K(\varepsilon)$ such that
$$A(t) \le Kt^{\sigma_c - \varepsilon}$$
for all sufficiently large t. By modifying the value of K we can suppose that this inequality is actually satisfied for all $t \ge 1$. So we can write
$$H(s) - \frac{K}{s - \sigma_c + \varepsilon} = \int_1^\infty (A(t) - Kt^{\sigma_c - \varepsilon})t^{-s-1}\,\mathrm{d}t = -\int_0^\infty B(u)e^{-su}\,\mathrm{d}u$$
with
$$B(u) := Ke^{(\sigma_c - \varepsilon)u} - A(e^u) \ge 0.$$
The abscissa of convergence of the last integral is still σ_c, since
$$\int_1^\infty t^{\sigma_c - \varepsilon - s - 1}\,\mathrm{d}t$$
converges absolutely for $\sigma > \sigma_c - \varepsilon$. By the Phragmén–Landau theorem, the point $s = \sigma_c$ must be a singularity of $H(s) - K/(s - \sigma_c + \varepsilon)$ and hence $H(s)$ is not holomorphic at the point σ_c. This yields the required contradiction.

Theorem 8. *Let $F(s) = \sum_{n=1}^\infty a_n n^{-s}$ be a Dirichlet series with real coefficients having a finite abscissa of convergence. Suppose there exists a real number $\sigma_0 > 0$ such that $F(s)$ has an analytic continuation which is regular at all points of the half-line $[\sigma_0, \infty[$ and has a pole on the vertical line $\sigma = \sigma_0$. Then the associated summatory function satisfies*
$$\sum_{n \le x} a_n = \Omega_\pm(x^{\sigma_0}).$$

Proof. Write $A(t) := \sum_{n \le t} a_n$ and let σ_c denote the abscissa of convergence of the integral $H(s)$ defined by (13). By partial summation, we have, for sufficiently large σ,
$$F(s) = sH(s)$$
so that $H(s)$ has a continuation which is regular on the half-line $[\sigma_0, \infty[$. We assume that there exists some constant K such that $A(t) + Kt^{\sigma_0}$ has constant sign, say positive, for sufficiently large t. Then there exists some constant C such that
$$B(t) := A(t) + Kt^{\sigma_0} + C \ge 0 \qquad (t \ge 1).$$
The integral
$$L(s) := \int_1^\infty B(t)t^{-s-1}\,\mathrm{d}t = H(s) + \frac{K}{s - \sigma_0} + \frac{C}{s},$$

being holomorphic on $]\sigma_0, \infty[$, is convergent for $\sigma > \sigma_0$ by the Phragmén–Landau theorem. If $s_0 = \sigma_0 + i\tau_0$ is a pole of $F(s)$ with principal part $\lambda(s - s_0)^{-m}$, then, on the one hand,

$$|L(s_0 + \delta)| \leq L(\sigma_0 + \delta) \qquad (\delta > 0)$$

and, on the other,

$$L(\sigma_0 + \delta) \sim \frac{K}{\delta}, \qquad L(s_0 + \delta) \sim \frac{\lambda}{s_0 \delta^m} \quad (\delta \to 0+).$$

This implies that $|K| \geq |\lambda / s_0|$. In particular, $A(t) + Kt^{\sigma_0}$ does not have constant sign for all sufficiently large t if

$$|K| < |\lambda / s_0|.$$

In Exercise 3 we meet three classical examples of oscillation theorems.

We end this section by establishing the uniqueness of the Dirichlet series expansion.

Theorem 9. *Let* $F(s) = \sum_{n=1}^{\infty} f(n) n^{-s}$ *be a Dirichlet series that vanishes identically for sufficiently large* σ. *Then* $f(n) = 0$ *for* $n \geq 1$.

Proof. Let m be the smallest integer such that $f(m) \neq 0$. For sufficiently large σ we have

$$0 = F(s) = f(m) m^{-s} \{1 + G(s)\} \tag{15}$$

with

$$G(s) := \sum_{k=1}^{\infty} \frac{f(m+k)}{f(m)} \Big(\frac{m}{m+k}\Big)^s.$$

By Theorem 5, this series converges absolutely for $\sigma > \sigma_c + 1$. For $\sigma > \sigma_1 > \sigma_c + 1$, we then have

$$\begin{aligned} |G(s)| &\leq |f(m)|^{-1} (1 + m^{-1})^{\sigma_1 - \sigma} m^{\sigma_1} \sum_{k=1}^{\infty} |f(m+k)| (m+k)^{-\sigma_1} \\ &\ll_{\sigma_1} (1 + m^{-1})^{-\sigma}. \end{aligned}$$

Thus as $\sigma \to \infty$, $G(s)$ tends to 0, contradicting (15) and hence concluding the proof.

§ 1.4 Abscissa of convergence and mean value

Let A be a function in $\mathcal{V}$ and let

$$F(s) = \int_{0-}^{\infty} e^{-ts}\,\mathrm{d}A(t) \tag{16}$$

be its Laplace–Stieltjes transform. In this section our aim is to give an explicit formula for the abscissa of convergence of (16) in terms of the asymptotic behaviour of A. It is clear that the value of A in any bounded neighbourhood of the origin does not affect the value of σ_c. Hence, without loss of generality, we can assume that

$$A(0\pm) = 0. \tag{17}$$

The purpose of Theorem 11 is to provide an explicit expression for σ_c which is analogous to Cauchy's formula for the radius of convergence of a power series. The proof rests on the following result.

Theorem 10. *Let σ_c be the abscissa of convergence of the integral (16).*

(i) If we have $A(x) \ll e^{\delta x}$ for some real δ, then $\sigma_c \leq \delta$.

(ii) If the integral (16) converges for $s = s_0$ with $\sigma_0 > 0$, then

$$A(x) = o(e^{\sigma_0 x}) \qquad (x \to \infty).$$

(iii) If the integral (16) converges for $s = s_0$ with $\sigma_0 < 0$, then there exists some real number α such that

$$A(x) = \alpha + o(e^{\sigma_0 x}) \qquad (x \to \infty).$$

Proof. (i). For all $x > 0$, we have

$$\int_0^x e^{-st}\,\mathrm{d}A(t) = A(x)e^{-sx} + s\int_0^x e^{-st}A(t)\,\mathrm{d}t.$$

The assumption concerning the rate of increase of $A(t)$ then implies the convergence of the integral (16) for all s such that $\sigma > \delta$. Hence $\sigma_c \leq \delta$.

(ii). By hypothesis,

$$B(x) := \int_0^x e^{-s_0 t}\,\mathrm{d}A(t) = F(s_0) + o(1) \qquad (x \to \infty). \tag{18}$$

We can then write

$$\begin{aligned} A(x) = \int_0^x e^{s_0 t}\,\mathrm{d}B(t) &= e^{s_0 x}B(x) - s_0\int_0^x e^{s_0 t}B(t)\,\mathrm{d}t \\ &= s_0\int_0^x \{B(x) - B(t)\}e^{s_0 t}\,\mathrm{d}t + B(x). \end{aligned}$$

The expected result then follows from (18).

(iii). By Theorem 3(i), we can assert that $F(s)$ converges for $s = 0$. Setting $\alpha := F(0)$, we have, with the notation (18),

$$\begin{aligned}\alpha - A(x) &= \int_x^\infty e^{s_0 t}\,\mathrm{d}B(t) = -e^{s_0 x}B(x) - s_0 \int_x^\infty e^{s_0 t} B(t)\,\mathrm{d}t \\ &= s_0 \int_x^\infty \{B(x) - B(t)\} e^{s_0 t}\,\mathrm{d}t = s_0 \int_x^\infty o(e^{\sigma_0 t})\,\mathrm{d}t = o(e^{\sigma_0 x}).\end{aligned}$$

Remark. Assertions (ii) and (iii) of Theorem 10 are false for $\sigma_0 = 0$. A counter-example to the first is provided by the function

$$A(x) = \begin{cases} 0 & (0 \le x \le 1) \\ 1 & (x > 1). \end{cases}$$

The integral (16) then converges for $s_0 = 0$, but we do not have $A(x) = o(1)$ as x tends to infinity. For the second statement, consider

$$A(x) = \begin{cases} 0 & (0 \le x \le 1) \\ 2\sqrt{x} & (x > 1). \end{cases}$$

Clearly we have

$$F(i) = 2e^{-i} + \int_1^\infty t^{-1/2} e^{-it}\,\mathrm{d}t.$$

The integral is well known to be convergent; however $A(x)$ does not tend to a finite limit as $x \to \infty$.

Theorem 11. *Put* $\kappa := \limsup_{x\to\infty} x^{-1} \log |A(x)|$.

(i) If $\kappa \ne 0$, then we have $\sigma_c = \kappa$.

(ii) If $\kappa = 0$, then either $A(x)$ does not tend to a finite limit as $x \to \infty$ and we have $\sigma_c = 0$, or there exists some real number α such that $A(x) \to \alpha$ as $x \to \infty$, and we have

$$\sigma_c = \limsup_{x\to\infty} x^{-1} \log |A(x) - \alpha| \le 0.$$

Proof. For each fixed $\varepsilon > 0$, we have $A(x) \ll_\varepsilon e^{(\kappa+\varepsilon)x}$. Assertion (i) of Theorem 10 then implies that in all cases

$$\sigma_c \le \kappa. \tag{19}$$

Suppose, initially, that $\kappa > 0$. Then the integral $F(s)$ diverges when $0 < \sigma < \kappa$, since otherwise we would have, by Theorem 10(ii), that $A(x) = o(e^{\sigma x})$, contradicting the definition of κ. Thus $\sigma_c \ge \kappa$, and the required equality follows.

If $\kappa < 0$, then $A(x) \to 0$ as $x \to \infty$. Theorem 10(iii) then implies that

$$A(x) = o(e^{\sigma x}) \qquad (x \to \infty) \tag{20}$$

for all $\sigma > \sigma_c$. But, by definition of κ, no $\sigma < \kappa$ satisfies (20). Hence $\sigma_c \geq \kappa$, so once again we obtain $\sigma_c = \kappa$.

Let us now consider the case $\kappa = 0$. If $A(x)$ does not have a limit as $x \to \infty$, the integral $F(s)$ diverges at $s = 0$, so that $\sigma_c \geq 0$ and hence the required conclusion still follows from (19). On the other hand, if $A(x) = \alpha + o(1)$, we need to show that $\sigma_c = \xi$, where ξ is the infimum of real numbers σ_1 such that

$$A(x) = \alpha + o(e^{\sigma_1 x}). \tag{21}$$

From Theorem 10(iii) we have that $\sigma_c \geq \xi$. Conversely, a trivial integration by parts shows, by (21), that $F(s)$ converges for $\sigma > \xi$, from which we deduce that $\sigma_c \leq \xi$. This completes the proof.

§ 1.5 An arithmetic application: the kernel of an integer

The abscissa of convergence of a Dirichlet series with an Euler product is often relatively easy to calculate. By using Theorems 10 and 11 one can sometimes deduce non-trivial information about the associated summatory function.

Consider, for example, the case of the *kernel* of an integer n, i.e. the largest squarefree divisor of n, which we denote by

$$k(n) := \prod_{p|n} p. \tag{22}$$

Theorem 10 readily provides some information concerning the distribution function

$$N(x, y) := \text{card}\,\{n \leq x : k(n) \leq y\}.$$

Theorem 12. *For any $\varepsilon > 0$, and uniformly for $1 \leq y \leq x$, we have*

$$N(x, y) \ll_\varepsilon yx^\varepsilon. \tag{23}$$

Proof. The function $k(n)$ is multiplicative. Since the series of positive terms

$$\sum_p \sum_{\nu=1}^{\infty} k(p^\nu)^{-1} p^{-\nu\varepsilon} = \sum_p p^{-1}(p^\varepsilon - 1)^{-1} \qquad (\varepsilon > 0),$$

converges, we deduce from Theorem 2 that the abscissa of convergence of

$$F(s) := \sum_{n=1}^{\infty} k(n)^{-1} n^{-s}$$

is $\sigma_c = 0$. By Theorem 10, it follows that, for any $\varepsilon > 0$, we have

$$\sum_{n \le x} 1/k(n) \ll_\varepsilon x^\varepsilon. \tag{24}$$

The estimate (23) is now immediate from this bound, since

$$N(x,y) \le \sum_{n \le x} y/k(n).$$

A lower bound of the form

$$N(x,y) \gg_m y\big(\log(2x/y)\big)^m,$$

valid for all $m > 0$ and $x \ge y \ge y_0(m)$, is outlined in Exercise 5. (See the Notes for the best results to date concerning $N(x,y)$.)

The idea of comparing the summatory function of an arithmetic function with an Euler product underpins *Rankin's method* (cf. III.5.1). We show below how to apply this simple method in order to refine (23).

Theorem 13. *Uniformly for $x \ge y \ge 2$, we have*

$$N(x,y) \ll y(\log y)e^{\sqrt{(8\log(x/y))}}.$$

Proof. If the integer n is counted in $N(x,y)$, then we have trivially, for each ε with $0 \le \varepsilon \le 1$, that

$$1 \le \left(\frac{x}{n}\right)^\varepsilon \left(\frac{y}{k(n)}\right)^{1-\varepsilon}.$$

Let us write $v := \log(x/y)$. We may assume that $v \ge 2$, since otherwise the conclusion is obvious. It follows that

$$\begin{aligned} N(x,y) &\le ye^{\varepsilon v} \sum_{\substack{n \le x \\ p|n \Rightarrow p \le y}} n^{-\varepsilon} k(n)^{-1+\varepsilon} \le ye^{\varepsilon v} \prod_{p \le y} \left(1 + \frac{1}{p} + \frac{1}{p(p^\varepsilon - 1)}\right) \\ &\le y \exp\left\{\varepsilon v + \frac{K}{\varepsilon} + \sum_{p \le y} \frac{1}{p}\right\} \end{aligned}$$

with $K := \sum_p 1/(p\log p) \le 2$. Choosing $\varepsilon = \sqrt{(2/v)}$, and estimating the last p-sum by Theorem I.1.9, we obtain the stated result.

§ 1.6 Order of magnitude in vertical strips

It is evident that a Dirichlet series $F(s)$ is bounded in any closed half-plane contained in its domain of absolute convergence. However $|F(s)|$ can take large values as $\sigma \to \sigma_a$ and $|\tau| \to \infty$. The following theorem provides an important example.

Theorem 14. *For all real $T > 0$, there exists some real number $\tau > T$ such that*

$$\sup_{\sigma>1} |\zeta(\sigma + i\tau)| \geq \tfrac{1}{10} \log_2(3 + \tau). \tag{25}$$

In order to establish this result we appeal to the classical theorem of Dirichlet on simultaneous approximation of N real numbers modulo 1. We recall this in the lemma below. For $x \in \mathbb{R}$, we write

$$\|x\| := \min_{n\in\mathbb{Z}} |x - n|.$$

Lemma 14.1 (Dirichlet). *Let $\alpha_1, \alpha_2, \ldots, \alpha_N$ be real numbers and let D be an integer ≥ 1. For any integer $Q \geq 2$, there exists somc integer q with $D \leq q \leq D{\cdot}Q^N$ such that*

$$\max_{1\leq j\leq N} \|q\alpha_j\| \leq 1/Q. \tag{26}$$

Proof of Lemma. Consider the partition

$$\bigcup_{0\leq j_1,\ldots,j_N<Q} \prod_{h=1}^{N} \Big[\frac{j_h}{Q}, \frac{j_h+1}{Q}\Big[$$

of the N-dimensional unit cube $[0,1[^N$ into Q^N subcubes. By the pigeonhole principle, at least two of the $Q^N + 1$ points

$$(mD\alpha_1, mD\alpha_2, \ldots, mD\alpha_N) \qquad (0 \leq m \leq Q^N)$$

belong, modulo 1, to the same sub-cube. If m and m' are the corresponding indices, (26) is satisfied with $q = |m' - m|D \in [D, D{\cdot}Q^N]$.

Proof of Theorem 14. For any $\sigma > 1$ and any integer $N \geq 1$, we can write

$$\Re e\, \zeta(s) \geq \sum_{n\leq N} \cos(\tau \log n) n^{-\sigma} - \sum_{n>N} n^{-\sigma}. \tag{27}$$

Apply the lemma with $Q = 6$, $D := Q^N$, and $\alpha_n := (1/2\pi)\log n$ $(1 \le n \le N)$. Then there exists some real number τ with $6^N \le \tau \le 6^{2N}$ such that

$$\min_{1\le n\le N} \cos(\tau \log n) \ge \cos(\pi/3) = \tfrac{1}{2}.$$

Hence we deduce from (27) that

$$\begin{aligned}\Re e\,\zeta(s) &\ge \tfrac{1}{2}\sum_{n\le N} n^{-\sigma} - \sum_{n>N} n^{-\sigma} = \tfrac{1}{2}\zeta(\sigma) - \tfrac{3}{2}\sum_{n>N} n^{-\sigma}\\ &\ge \tfrac{1}{2}(\sigma-1)^{-1}\{1 - 3N^{1-\sigma}\}\end{aligned}$$

where the last inequality follows from the usual comparison estimates between series and integrals. For $\sigma = 1 + 4/\log N$ we have

$$\Re e\,\zeta(s) > \tfrac{1}{9}\log N \ge \tfrac{1}{10}\log_2(3+\tau),$$

if $N \ge N_0$. The extra condition $\tau > T$ is satisfied by choosing $N > \log T/\log 6$. This completes the proof.

The following theorem shows that a Dirichlet series necessarily satisfies certain bounds in its domain of convergence.

Theorem 15. *Let $F(s) = \sum_{n=1}^{\infty} a_n n^{-s}$ be a Dirichlet series with abscissa of convergence σ_c. Let $\sigma_0 > \sigma_c$ and $\varepsilon > 0$. Then, uniformly for $\sigma_0 \le \sigma \le \sigma_c + 1$, we have that*

$$F(s) \ll |\tau|^{1-(\sigma-\sigma_c)+\varepsilon} \qquad (|\tau| \ge 1). \tag{28}$$

Proof. Clearly we may assume that $0 < \varepsilon < \sigma_0 - \sigma_c$. Put

$$A(t) := \sum_{n\le e^t} a_n n^{-\sigma_c-\varepsilon}$$

so that $A(t) = F(\sigma_c + \varepsilon) + o(1)$ as $t \to \infty$. Furthermore

$$\begin{aligned}|F(s)| &= \Big|\sum_{n\le N} a_n n^{-s} + \int_{\log N}^{\infty} e^{-t(s-\sigma_c-\varepsilon)}\,\mathrm{d}A(t)\Big|\\ &\le \sum_{n\le N} |a_n| n^{-\sigma} + |A(\log N)| N^{-(\sigma-\sigma_c-\varepsilon)}\\ &\qquad + |s|\int_{\log N}^{\infty} |A(t)| e^{-t(\sigma-\sigma_c-\varepsilon)}\,\mathrm{d}t.\end{aligned}$$

Since $|a_n| \ll_\varepsilon n^{\sigma_c+\varepsilon}$, we deduce that

$$|F(s)| \ll_\varepsilon N^{1-(\sigma-\sigma_c)+\varepsilon} + |s|N^{-(\sigma-\sigma_c)+\varepsilon}.$$

The required upper bound follows from this estimate by choosing $N = 1+[|\tau|]$.

The problem of determining whether a given function, analytic in a certain half-plane, is or is not representable in the form of a Dirichlet series is difficult in general. Theorem 15 shows that functions satisfying a relation of the type

$$F(s) \ll |\tau|^A \qquad (|\tau| \geq 1) \tag{29}$$

for some $A > 0$ play a special role. If F satisfies (29) in a domain $\mathcal{D}$, we say that F has *finite order* in $\mathcal{D}$. A Dirichlet series has finite order in each closed half-plane contained in the domain of convergence. If the sum of the series can be analytically continued, it can happen that the continuation still has finite order in a larger domain. For example, let us consider the series $G(s) := \sum_{n=1}^\infty (-1)^n n^{-s}$, which furnishes the analytic continuation of $\zeta(s)$ for the domain $\sigma > 0$, $s \neq 1$, in the form

$$\zeta(s) = G(s)/(2^{1-s} - 1).$$

By Theorem 15, we have, for $0 < \sigma < 1$, $|s-1| \gg 1$, that

$$\zeta(s) \ll G(s) \ll |\tau|^{1-\sigma+\varepsilon}$$

with the result that the continuation of $\zeta(s)$ has finite order for $0 < \sigma < 1$.

For any function F of finite order in a domain $\mathcal{D}$, we let $\mu(\sigma) = \mu_F(\sigma)$ denote the infimum of the set of real numbers ξ such that

$$F(s) \ll_{\sigma,\xi} |\tau|^\xi \qquad (s \in \mathcal{D},\ |\tau| \geq 1).$$

Theorem 16. *Let F be a function of finite order in a vertical strip $\sigma_1 \leq \sigma \leq \sigma_2$. Then the function $\mu(\sigma)$ is convex in this interval. In particular, it is continuous for $\sigma_1 < \sigma < \sigma_2$.*

Proof. It is an immediate consequence of the classical Phragmén–Lindelöf theorem (cf. for example Titchmarsh (1939), § 5.65, or Valiron (1955), § 242) that the upper bounds

$$(\forall \varepsilon > 0) \qquad F(s) \ll_\varepsilon e^{\varepsilon|\tau|} \qquad (\sigma_1 \leq \sigma \leq \sigma_2)$$

and

$$F(\sigma_1 + i\tau) \ll |\tau|^{k_1}, \quad F(\sigma_2 + i\tau) \ll |\tau|^{k_2} \qquad (|\tau| \geq 1)$$

imply the estimate

$$F(\sigma+i\tau) \ll |\tau|^{k(\sigma)} \qquad (\sigma_1 \le \sigma \le \sigma_2,\ |\tau| \ge 1)$$

where $k(\sigma)$ is the linear function taking the values k_1 and k_2 at σ_1 and σ_2 respectively. We hence deduce that

$$\text{(30)} \qquad \mu(\sigma) \le \frac{(\sigma_2-\sigma)\mu(\sigma_1)+(\sigma-\sigma_1)\mu(\sigma_2)}{\sigma_2-\sigma_1} \qquad (\sigma_1 \le \sigma \le \sigma_2).$$

Remark. The Phragmén–Lindelöf theorem in fact implies that, for any $\varepsilon > 0$, the estimate

$$F(s) \ll_{\varepsilon,\sigma_1,\sigma_2} |\tau|^{k(\sigma)+\varepsilon}$$

holds in the domain $\sigma_1 \le \sigma \le \sigma_2$, $|\tau| \ge 1$. We shall later have occasion to use this local uniformity in σ.

Theorem 17. *For any Dirichlet series $F(s)$, we have*

$$\text{(31)} \qquad \mu(\sigma) = 0 \qquad (\sigma > \sigma_a).$$

Moreover $\mu(\sigma)$ is a non-increasing function of σ in any region where F has finite order.

Proof. For $\sigma > \sigma_a$, $F(s)$ is bounded, so $\mu(\sigma) \le 0$. Furthermore, if $f(m)$ is the first non-zero coefficient, we have

$$|F(s)| \ge |f(m)|m^{-\sigma} - \sum_{n \ge m+1} |f(n)|n^{-\sigma} > 0$$

provided that σ is sufficiently large. Since this lower bound is independent of τ, for such σ, we have $\mu(\sigma) \ge 0$, and hence $\mu(\sigma) = 0$. Applying now (30) with σ and σ_2 sufficiently large (so that $\mu(\sigma) = \mu(\sigma_2) = 0$), we obtain that $\mu(\sigma_1) \ge 0$ for any $\sigma_1 > \sigma_a$, thereby establishing (31). The second assertion is also a consequence of (30). Selecting $\sigma_2 > \sigma_a$, so that $\mu(\sigma_2) = 0$, we infer

$$\mu(\sigma) \le \frac{\sigma_2-\sigma}{\sigma_2-\sigma_1}\mu(\sigma_1) \le \mu(\sigma_1) \qquad (\sigma_1 \le \sigma)$$

with strict inequality $\mu(\sigma) < \mu(\sigma_1)$ if $\mu(\sigma_1) \ne 0$ and $\sigma > \sigma_1$.

Corollary 17.1. *If $F(s)$ is a Dirichlet series of finite order for $\sigma > \sigma_0$ with $\sigma_0 < \sigma_a$, then $\mu(\sigma_a) = 0$.*

Proof. This follows immediately from (31) and the last assertion of Theorem 16.

Notes

§ **1.1**. It is easy to extend Theorem 1 in the following manner: *if the series $F(s)$ converges, and if the series $G(s)$ converges absolutely, then the series $H(s)$ converges and we have $H(s) = F(s)G(s)$.*

Indeed let us define, for fixed s, $A(x) := \sum_{m\le x} f(m)m^{-s}$. Then for any $\varepsilon > 0$ there exists some $y = y(\varepsilon)$ such that

$$|A(z) - F(s)| < \varepsilon \qquad (z > y).$$

Hence, for sufficiently large x, we can write

$$\sum_{n\le x} h(n)n^{-s} = \sum_{md\le x} f(m)g(d)(md)^{-s} = \sum_{d\le x} g(d)d^{-s}A(x/d)$$
$$= \sum_{d\le x/y} g(d)d^{-s}\{F(s) + O(\varepsilon)\} + O\Big(\sum_{x/y<d\le x} |g(d)d^{-s}|\Big).$$

We obtain the result stated above by letting x tend to infinity, and then making ε tend to 0.

Using Cesàro summability for Dirichlet series, it can also be shown that *the equality $H(s) = F(s)G(s)$ remains valid in any domain of convergence common to the three series F, G, H*—cf. Landau (1909), pp. 762, 904, or Hardy & Riesz (1915), p. 64.

As shown by the example in Exercise 1, it is not in general possible to deduce the convergence of $H(s) = \sum_{n=1}^{\infty}(f * g)(n)n^{-s}$ from that of $F(s)$ and $G(s)$. By Theorem 5, if $F(s_0)$ and $G(s_0)$ both converge, then, for any $\varepsilon > 0$, $F(s_0 + 1 + \varepsilon)$ and $G(s_0 + 1 + \varepsilon)$ converge absolutely, so $H(s)$ is (absolutely) convergent for $\sigma > \sigma_0 + 1$. With slightly more effort, this result can be made much more precise (cf. Landau (1909) pp. 759 et seq., or Hardy & Riesz (1915), p. 67).

Theorem 18 (Stieltjes, 1887). *If $F(s)$ and $G(s)$ converge for $s = s_0$, then $H(s) = F(s)G(s)$ converges at $s = s_0 + \frac{1}{2}$.*

Proof. It can be assumed, without loss of generality, that $s_0 = 0$. Moreover partial summation readily shows that the hypotheses imply

$$(32) \qquad \sum_{n>y} f(n)n^{-1/2} = o(y^{-1/2}), \quad \sum_{n>y} g(n)n^{-1/2} = o(y^{-1/2}) \quad (y \to \infty).$$

Now we can write

$$\sum_{n\le x} h(n)n^{-1/2} = \sum_{m\le\sqrt{x}} f(m)m^{-1/2} \sum_{d\le\sqrt{x}} g(d)d^{-1/2} + R_1 + R_2 \tag{33}$$

where we have defined

$$R_1 := \sum_{d\le\sqrt{x}} g(d)d^{-1/2} \sum_{\sqrt{x}<m\le x/d} f(m)m^{-1/2}$$

and where R_2 is the analogous quantity obtained by interchanging the roles of f and g. By (32) we see that

$$R_1 = \sum_{d\le\sqrt{x}} g(d)d^{-1/2}\cdot o(x^{-1/4}) \ll \sum_{d\le\sqrt{x}} d^{-1/2}\cdot o(x^{-1/4}) = o(1).$$

We obtain similarly that $R_2 = o(1)$. By hypothesis, the first term of the right-hand side of (33) tends to $F(\frac{1}{2})G(\frac{1}{2})$ as $x\to\infty$. This completes the proof.

This theorem can have non-trivial applications. Consider for example the case of $F(s) = G(s) = \sum_{n=1}^{\infty}(-1)^n n^{-s}$. By writing each integer n in the form $n = 2^\nu m$ with $2\nmid m$, we see that

$$h(n) = \begin{cases} \tau(m) & (\nu = 0) \\ (\nu-3)\tau(m) & (\nu\ge 1). \end{cases}$$

Theorem 18 then immediately implies that the series associated to $h(n)$ converges for $\sigma > \frac{1}{2}$. Hence, by Theorem 10, we get

$$\sum_{n\le x} h(n) \ll_\varepsilon x^{1/2+\varepsilon}. \tag{34}$$

A deeper analysis allows us to elucidate this example further. By the identity

$$\sum_{m\le x,\, 2\nmid m} \tau(m) = T(x) - 2T(\tfrac{1}{2}x) + T(\tfrac{1}{4}x)$$

where

$$T(x) := \sum_{n\le x}\tau(n),$$

we can see, after an elementary calculation, that

$$\sum_{n\le x} h(n) = \Delta(x) - 4\Delta(\tfrac{1}{2}x) + 4\Delta(\tfrac{1}{4}x) \qquad (x>0)$$

where we have set

$$\Delta(x) := T(x) - x(\log x + 2\gamma - 1).$$

Voronoï's theorem I.6.7 thus allows us to replace the exponent 1/2 in (34) by 1/3, and it is natural to conjecture that the abscissa of convergence for $H(s) = F(s)^2$ is $\sigma_c = 1/4$.

In Exercise 1 we give a simple example of a Dirichlet series $F(s)$ converging everywhere on the line of convergence $\sigma = \sigma_c$ but such that the series $F(s)^2$ diverges everywhere on the same line. If $\sigma_c(F)$ and $\sigma_c(F^2)$ are the abscissae of convergence for $F(s)$ and $F(s)^2$ respectively, Theorem 18 implies that

$$\sigma_c(F^2) \leq \sigma_c(F) + \tfrac{1}{2}. \tag{35}$$

Disproving a conjecture of Cahen (1894), Landau (1909, p. 773) showed that it can happen that

$$\sigma_c(F^2) > \sigma_c(F). \tag{36}$$

His argument consists in making use of thc cstimate

$$\zeta(s) = \Omega(|\tau|^{(1/2)-\sigma}) \qquad (0 < \sigma < \tfrac{1}{2}) \tag{37}$$

which we shall see in § 3.4 to be a straightforward consequence of the functional equation for the function $\zeta(s)$. The by now familiar relation

$$G(s) := \sum_{n=1}^{\infty} (-1)^n n^{-s} = (2^{1-s} - 1)\zeta(s) \qquad (\sigma > 1)$$

implies that (37) is equally valid for $G(s)$, although $\sigma_c(G) = 0$. This implies that

$$G(s)^4 = \Omega(|\tau|^{2-4\sigma}) \qquad (0 < \sigma < \tfrac{1}{2}),$$

so Theorem 15 shows that $\sigma_c(G^4) \geq \frac{1}{4}$. Hence (36) holds either for $F = G$ or for $F = G^2$.

Actually one can even have equality in (35). Bohr (1910) gave an example of a Dirichlet series such that $\sigma_c = 0$, $\sigma_a = 1$, and

$$F(s) = \Omega_\varepsilon(|\tau|^{1-\sigma-\varepsilon}) \qquad (\varepsilon > 0,\ 0 < \sigma < 1). \tag{38}$$

Theorem 15 readily implies that, for this series, $\sigma_c(F^2) \geq \frac{1}{2}$, from which we have equality by (35).

Bohr's construction is ingenious but relatively easy. Let us consider a sequence $\{\tau_k\}_{k=1}^{\infty}$ of real numbers tending to infinity sufficiently fast ($\tau_k := \exp 2^k$ suffices), and let $\{\delta_k\}_{k=1}^{\infty}$ be any sequence such that $\delta_k \to 0$ and $\tau_k^{\delta_k} \to \infty$. Now let the sequence $\{a_n\}_{n=1}^{\infty}$ be defined by the formula

$$A_n := \sum_{m \le n} a_m = \begin{cases} 0 & (\tau_k^{1/2} < n \le \tau_k^{1+\delta_k}) \\ n^{i\tau_k} & (\tau_k^{1+\delta_k} < n \le \tau_k^2) \\ 1 & (\tau_k^2 < n \le \tau_{k+1}). \end{cases}$$

We shall show that the series $F(s) := \sum_{m=1}^{\infty} a_m m^{-s}$ satisfies (38). We first check that $\sigma_c = 0$. Indeed A_n does not have a limit at infinity (hence $\sigma_c \ge 0$) while Dirichlet's test shows that $F(\sigma)$ converges for all $\sigma > 0$ (hence $\sigma_c \le 0$). By Theorem 5, we have $\sigma_a \le 1$ and by Theorem 15, $\mu(\sigma) \le 1 - \sigma$ for $0 \le \sigma \le 1$. We now prove that for all σ with $0 < \sigma < 1$, we have

$$F(\sigma + i\tau_k) \sim (i/\sigma)\tau_k^{1-\sigma(1+\delta_k)} \qquad (k \to \infty),$$

which plainly establishes (38) and shows that

$$\text{(39)} \qquad \mu(\sigma) = 1 - \sigma \qquad (0 \le \sigma \le 1).$$

Setting $s_k = \sigma + i\tau_k$ we have

$$F(s_k) = \sum_{n=1}^{\infty} A_n\{n^{-s_k} - (n+1)^{-s_k}\} = \sum_{n \le \sqrt{\tau_k}} + \sum_{n > \tau_k^{1+\delta_k}} (\cdots)$$

since $A_n = 0$ when $\sqrt{\tau_k} < n \le \tau_k^{1+\delta_k}$. The first of the two sums above is clearly

$$\ll \sum_{n \le \sqrt{\tau_k}} n^{-\sigma} \ll_\sigma \tau_k^{(1-\sigma)/2}.$$

The second has value

$$\sum_{n > \tau_k^{1+\delta_k}} \left\{ s_k A_n n^{-1-s_k} + O(s_k^2 n^{-2-s_k}) \right\}$$

$$= s_k \sum_{\tau_k^{1+\delta_k} < n \le \tau_k^2} n^{-1-\sigma} + O\Bigg(\sum_{n > \tau_k^2} \tau_k n^{-1-\sigma} + \sum_{n > \tau_k^{1+\delta_k}} \tau_k^2 n^{-2-\sigma} \Bigg)$$

$$= s_k \Big\{ \sigma^{-1} \tau_k^{-\sigma(1+\delta_k)} + O_\sigma(\tau_k^{-2\sigma}) \Big\} + O_\sigma\big(\tau_k^{1-\delta_k-\sigma(1+\delta_k)}\big) \sim \frac{i}{\sigma} \tau_k^{1-\sigma(1+\delta_k)}.$$

§ **1.3.** For a historical clarification of the decisive contribution made by Phragmén to the so-called "Landau's theorem", the reader is invited to consult the detailed article of Dress (1983–84). This also contains an expository treatment and an up-to-date survey of oscillation theorems available in the literature. See also Kaczorowski & Pintz (1986–7).

The Phragmén–Landau theorem can be stated in the following form: *Let f be an arithmetic function ≥ 0. If the function $F(s)$ defined by $F(s) = \sum_{n=1}^{\infty} f(n)n^{-s}$ for $\sigma > \sigma_0$ has a regular analytic continuation to a domain containing the line $\sigma = \sigma_0$, then the series converges at $\sigma = \sigma_0$.*

Ingham (1935) showed that in this statement *one can replace the condition $f \geq 0$ by $|f| \leq M$ when $\sigma_0 = 1$.* This allows one to deduce the prime number theorem directly from the fact that $\zeta(s) \neq 0$ on $\sigma = 1$ (cf. § 3.7). Indeed we obtain the convergence of $\sum_{n=1}^{\infty} \mu(n)/n$, from which the desired conclusion follows by Theorem I.3.8.

Ingham's proof rests on considerations from Fourier analysis. D. J. Newman (1980) gave a much simpler proof of this theorem using an ingenious method of contour integration.

The proof given here of Theorem 8 provides an effective lower bound for the implicit constant in the $\Omega_{\pm}$ notation. For further information of this type see Dress (1983–84) or Grosswald (1972).

§ **1.5.** The best results known to date about $N(x, y)$ are due to Squalli (1985). Setting

$$v := \log(x/y) \qquad (1 \leq y \leq x),$$

he shows that for each $\varepsilon > 0$ we have

$$N(x, y) = yF(v)^{1+O(1/\log(v+2))} \qquad \left(\exp\{(\log x)^{(1/2)+\varepsilon}\} \leq y \leq x\right)$$

and

$$N(x, y) = yF(v)\{1 + O\left(\sqrt{(\log_2 x/\log x)}\right)\} \qquad \left(\exp\{(\log x)^{(3/4)+\varepsilon}\} \leq y \leq x\right)$$

where $F(v)$ is the differentiable function defined for $v \geq 0$ by

$$F(v) := e^v - \frac{6}{\pi^2} \sum_{m \leq e^v} \left(\frac{e^v}{m} - 1\right) \prod_{p|m} \left(\frac{1}{p+1}\right).$$

Squalli also establishes the existence of a sequence of polynomials Q_j $(j \geq 1)$, with $\deg Q_j \leq j$, such that for each $N \geq 1$ we have

$$F(v) = \exp\left\{\sqrt{\frac{8v}{\log v}}\left(1 + \sum_{j=1}^{N} \frac{Q_j(\log_2 v)}{(\log v)^j} + O_N\left(\left(\frac{\log_2 v}{\log v}\right)^{N+1}\right)\right)\right\} \qquad (v \geq 3).$$

In particular $Q_1(t) = 1 + \frac{1}{2}t$.

For an application of Theorem 13 see Exercise III.3.13.

§ **1.6.** Theorem 15, which states that

$$\mu(\sigma) \le 1 - (\sigma - \sigma_c) \qquad (\sigma_c \le \sigma \le \sigma_c + 1),$$

is optimal, as shown by Bohr's counter-example, quoted above—cf. Hardy & Riesz (1915) p. 19.

We shall see in § 2.2 (Theorem 2.4) that Theorem 17 has a partial converse: *if* $F(s)$ *is holomorphic and satisfies* $\mu(\sigma) = 0$ *for* $\sigma > \sigma_0$, *then* $\sigma_c \le \sigma_0$.

Thus, if $F(s) := 1 + \sum_{n=2}^{\infty} f(n)n^{-s}$ converges for $\sigma > \sigma_0$, is non-zero on this half-plane, and satisfies the lower bound $|F(s)| \gg_\varepsilon (1 + |\tau|)^{-\varepsilon}$ for any $\varepsilon > 0$, then the series $G(s) = F(s)^{-1}$ converges for all $\sigma > \sigma_0$. A theorem of Landau (1933) shows that the growth condition is actually superfluous.

Exercises

1. Show that the Dirichlet series $F(s) = \sum_{n=1}^{\infty} (-1)^n (\log 2n)^{-2} n^{-s}$ converges at each point of the line of convergence $\sigma = 0$. Define $H(s) = \sum_{n=1}^{\infty} h(n)n^{-s}$ by $H(s) := F(s)^2$. Show that $h(n)$ is unbounded, and deduce that $H(s)$ diverges everywhere on the line $\sigma = 0$.

2. Write $A(t) := \sum_{n \le t} (-1)^n n \qquad (t \ge 0)$.

(a) Show that $A(N) = (-1)^N \left[\frac{1}{2}(N+1)\right]$ for any integer $N \ge 1$.

(b) Calculate the abscissa of convergence of $G(s) := \int_{1-}^{\infty} t^{-s-1}\, \mathrm{d}A(t)$.

(c) Show that, for $\sigma > 0$, we have

$$G(s) = (s+1) \int_1^{\infty} t^{-s-2} A(t)\, \mathrm{d}t. \tag{40}$$

(d) Show that the series $\sum_{n=1}^{\infty} A(n) \int_n^{n+1} t^{-\varepsilon-1}\, \mathrm{d}t$ converges for each $\varepsilon > 0$.

(e) Deduce that (40) defines a regular analytic continuation of $G(s)$ to the half-plane $\sigma > -1$.

3. In this exercise assume the following weak consequence of Theorems 3.3 and 3.13: *The function* $\zeta(s)$ *has at least one zero in the closed half-plane* $\sigma \ge \frac{1}{2}$ *and the zeros in the strip* $0 < \sigma < 1$ *are placed symmetrically about the axis* $\sigma = \frac{1}{2}$.

(a) Using the properties of the series $G(s) := \sum_{n=1}^{\infty}(-1)^n n^{-s}$, show that $\zeta(s)$ is non-zero for real $s \in]0,\infty[$, and is continuable as a meromorphic function for $\sigma > 0$, having as sole singularity a simple pole at $s = 1$.

(b) By considering the series $\zeta(s)+\zeta'(s)/\zeta(s)$, show that $\psi(x) = x+\Omega_{\pm}(x^{\frac{1}{2}})$.

(c) By considering the series $1/\zeta(s)$, show that

$$M(x) := \sum_{n\leq x} \mu(n) = \Omega_{\pm}(x^{\frac{1}{2}}).$$

(d) By considering the series $\zeta(s)/\zeta(2s)$, show that

$$Q(x) := \sum_{n\leq x} \mu(n)^2 = \frac{6}{\pi^2}x + \Omega_{\pm}(x^{1/4}).$$

4. Let $P(n,m) = n^2 + m^3 + mn$. What is the convergence abscissa of $\sum_{n=1}^{\infty}\sum_{m=1}^{\infty} P(n,m)^{-s}$? What estimates can be deduced for

$$A(x) := \text{card}\,\{n, m \geq 1 : P(n,m) \leq x\} \ ?$$

5. (a) Show that the number of integral solutions $\nu_1 \geq 0, \nu_2 \geq 0, \ldots, \nu_m \geq 0$ of the inequality

$$\sum_{1\leq i\leq m} \nu_i \leq N$$

is $\binom{N+m}{m}$. [*Use the Taylor expansion for* $(1-x)^{-m-1}$.]

(b) Let $p_1 < p_2 < \ldots < p_m$ be distinct prime numbers. Show that each integer n of the form $n = p_1^{\nu_1}\ldots p_m^{\nu_m}r$, with $\mu(p_1\ldots p_m r)^2 = 1$, satisfies $k(n) \leq rp_1\ldots p_m$.

(c) Deduce from (a) and (b) that for any integer $m \geq 1$ we have

$$N(x,y) \gg_m \big(\log(2x/y)\big)^m \sum_{r\leq y/p_1\ldots p_m} \mu(p_1\ldots p_m r)^2.$$

(d) Establish the identity $\mu(n)^2 = \sum_{d^2|n}\mu(d)$ and deduce that, uniformly for $M \geq 1$, $\mu(M)^2 = 1$, $y \geq 1$, we have

$$\sum_{r\leq y} \mu(Mr)^2 = 6\pi^{-2}\prod_{p|M}(1+p^{-1})^{-1}y + O(2^{\omega(M)}\sqrt{y}).$$

(e) Show, for $m \geq 1$, $x \geq y \geq p_1\ldots p_m$, that

$$N(x,y) \gg_m y\big(\log(2x/y)\big)^m.$$

6. Let f be a multiplicative function. Show that the Dirichlet series associated with $n \mapsto f(kn)$ can, for each integer $k \geq 1$, be expanded as an infinite product of Eulerian type. Applications: for $\sigma > 1$, we have

$$\text{(a)} \qquad \sum_{n=1}^{\infty} \tau(kn)n^{-s} = \zeta(s)^2 \prod_{p^\nu \| k} (\nu + 1 - \nu p^{-s}),$$

$$\text{(b)} \quad \sum_{n=1}^{\infty} \tau(kn)^{-1}n^{-s} = \prod_{p \nmid k} \left\{ p^s \log\left(\frac{1}{1-p^{-s}}\right)\right\} \prod_{p^\nu \| k} \sum_{j=0}^{\infty} (j+\nu+1)^{-1} p^{-js}.$$

7. Show that, for any fixed real numbers $\alpha \in]0,1[$, $\theta \neq 0$, the abscissa of convergence of the Dirichlet series $F(s) := \sum_{n=1}^{\infty} e(\theta n^\alpha) n^{-s}$ equals $\sigma_c = 1 - \alpha$.

II.2

Summation formulae

§ 2.1 Perron formulae

Most of the theorems in the previous chapter were intended to obtain analytic properties of Dirichlet series, given the analytic behaviour of their coefficients, or of their summatory function. We are now going to proceed in the opposite direction. For the applications which we have in mind, the ultimate aim is to understand the behaviour of arithmetic functions, so that Dirichlet series are here to be regarded more as an indispensable tool than as an intrinsic subject of study.

In this context, the Perron formulae play the fundamental role of an inversion principle, analogous to that of the Cauchy formulae in the theory of power series.

Let

$$F(s) := \sum_{n=1}^{\infty} a_n n^{-s} \tag{1}$$

be a Dirichlet series with abscissa of convergence σ_c and abscissa of absolute convergence σ_a. Extending the definition of the function $n \mapsto a_n$ by setting $a_x = 0$ if $x \in \mathbb{R} \smallsetminus \mathbb{N}$, we introduce the *normalised summatory function*

$$A^*(x) := \sum_{n<x} a_n + \tfrac{1}{2} a_x \qquad (x \geq 0). \tag{2}$$

Theorem 1 (Perron's formula). *Let* $\kappa > \max(0, \sigma_c)$. *We have*

$$A^*(x) = \frac{1}{2\pi i} \int_{\kappa - i\infty}^{\kappa + i\infty} F(s) x^s s^{-1} \, \mathrm{d}s \qquad (x > 0), \tag{3}$$

where the integral is conditionally convergent for $x \in \mathbb{R} \smallsetminus \mathbb{N}$ *and converges in the sense of Cauchy's principal value when* $x \in \mathbb{N}$.

The proof rests on the following lemma, which amounts to an explicit calculation of the Laplace inversion formula for the function

$$h(x) = \begin{cases} 1 & (x > 1) \\ \frac{1}{2} & (x = 1) \\ 0 & (0 < x < 1). \end{cases} \tag{4}$$

Lemma 1.1. *For any positive* κ, T, T', *we have*

$$\text{(i)} \qquad \left|h(x) - \frac{1}{2\pi i}\int_{\kappa-iT'}^{\kappa+iT} x^s \frac{\mathrm{d}s}{s}\right| \leq \frac{x^\kappa}{2\pi|\log x|}\left(\frac{1}{T}+\frac{1}{T'}\right) \quad (x \neq 1),$$

$$\text{(ii)} \qquad \left|h(1) - \frac{1}{2\pi i}\int_{\kappa-iT}^{\kappa+iT} \frac{\mathrm{d}s}{s}\right| \leq \frac{\kappa}{T+\kappa}.$$

Provisionally assuming this lemma, let us see how we can deduce formula (3). First suppose that $\kappa > \sigma_a$. Then the series $F(s)$ is absolutely and uniformly convergent for $\sigma = \kappa$, hence

$$\frac{1}{2\pi i}\int_{\kappa-iT'}^{\kappa+iT} F(s)\frac{x^s}{s}\,\mathrm{d}s = \frac{1}{2\pi i}\sum_{n=1}^{\infty} a_n \int_{\kappa-iT'}^{\kappa+iT}\left(\frac{x}{n}\right)^s \frac{\mathrm{d}s}{s}.$$

Applying part (i) of the lemma for $x \in \mathbb{R}\smallsetminus\mathbb{N}$, it follows that

$$\text{(5)} \qquad \left|\frac{1}{2\pi i}\int_{\kappa-iT'}^{\kappa+iT} F(s)\frac{x^s}{s}\,\mathrm{d}s - A^*(x)\right| \leq \frac{x^\kappa}{2\pi}\left(\frac{1}{T}+\frac{1}{T'}\right)\sum_{n=1}^{\infty}\frac{|a_n|}{|n|^\kappa\,|\log(x/n)|}.$$

Since the factor $|\log(x/n)|$ is bounded from below independently of n, we obtain the first assertion of the theorem by allowing T and T' to tend independently to ∞. The second assertion is proved in much the same way: it suffices to take $T = T'$, and to replace on the right-hand side of (5) the (infinite) term corresponding to $n = x$ by $\kappa|a_x|/(T+\kappa)$.

Suppose now that $\sigma_c < \kappa \leq \sigma_a$. By Theorem 1.5, we have $\kappa + 1 > \sigma_a$. Consider the integral

$$I := \int_{\mathcal{R}} F(s)\frac{x^s}{s}\,\mathrm{d}s$$

where $\mathcal{R}$ is the rectangle formed by the lines $\sigma = \kappa$, $\tau = T$, $\sigma = \kappa + 1$ and $\tau = -T$. By Theorem 1.15, we have

$$F(s)x^s s^{-1} \ll \tau^{-(\sigma-\sigma_c)+\varepsilon}x^\sigma \quad (s \in \mathcal{R},\ |\tau| \geq 1)$$

so the contribution to I from the horizontal segments of $\mathcal{R}$ tends to 0 as T tends to infinity. Since $F(s)$ is analytic for $\sigma > \sigma_c$, the residue theorem shows that $I = 0$. Therefore, as $T \to \infty$, we have

$$\int_{\kappa-iT}^{\kappa+iT} F(s)x^s s^{-1}\,\mathrm{d}s = \int_{\kappa+1-iT}^{\kappa+1+iT} F(s)x^s s^{-1}\,\mathrm{d}s + o(1).$$

This completes the proof.

Proof of Lemma 1.1. Consider first the case when $x > 1$. Let k be a sufficiently large integer and let $\mathcal{R}_k$ denote the rectangle with vertices $\kappa - iT'$, $\kappa + iT$, $\kappa - k + iT$, $\kappa - k - iT'$. By the residue theorem, we may write

$$\frac{1}{2\pi i}\int_{\mathcal{R}_k} x^s \frac{\mathrm{d}s}{s} = 1 = h(x).$$

Now we have the following upper bounds

$$\Big|\int_{\kappa+iT}^{\kappa-k+iT} x^s s^{-1}\,\mathrm{d}s\Big| \le \frac{x^\kappa}{T|\log x|},$$
$$\Big|\int_{\kappa-k-iT'}^{\kappa-iT'} x^s s^{-1}\,\mathrm{d}s\Big| \le \frac{x^\kappa}{T'|\log x|},$$
$$\Big|\int_{\kappa-k+iT}^{\kappa-k-iT'} x^s s^{-1}\,\mathrm{d}s\Big| \le \frac{x^{\kappa-k}}{k-\kappa}(T+T').$$

We deduce the stated result by letting k tend to infinity.

The case $0 < x < 1$ can be dealt with in a symmetric way, applying the same argument with k replaced by $-k$. We omit the details. When $x = 1$, we simply note that

$$\frac{1}{2\pi i}\int_{\kappa-iT}^{\kappa+iT} s^{-1}\,\mathrm{d}s = \frac{1}{2\pi}\big(\arg(\kappa+iT) - \arg(\kappa - iT)\big) = \frac{1}{\pi}\arctan(T/\kappa).$$

The stated upper bound is now immediate from the following bounds, valid for all $y > 0$,

$$0 \le \frac{\pi}{2} - \arctan y = \int_y^\infty \frac{\mathrm{d}t}{1+t^2} \le \frac{2}{1+y}.$$

This concludes the proof of the lemma.

In practice, it is useful to have at hand effective versions of Perron's formula, i.e. explicit upper bounds for the contribution from the domain $|\tau| \ge T$ to the integral (3).

Theorem 2 (First effective Perron formula). *For $\kappa > \max(0, \sigma_a)$, $T \ge 1$ and $x \ge 1$, we have*

$$A(x) = \frac{1}{2\pi i}\int_{\kappa-iT}^{\kappa+iT} F(s)x^s\frac{\mathrm{d}s}{s} + O\Big(x^\kappa \sum_{n=1}^{\infty}\frac{|a_n|}{n^\kappa(1+T|\log(x/n)|)}\Big). \tag{6}$$

Proof. It suffices to show that, for any fixed $\kappa > 0$, we have, uniformly for $y > 0$, $T > 0$,

$$\Big|h(y) - \frac{1}{2\pi i}\int_{\kappa-iT}^{\kappa+iT} y^s s^{-1}\,\mathrm{d}s\Big| \ll y^\kappa/(1+T|\log y|). \tag{7}$$

Indeed, applying this estimate with $y = x/n$ and summing over $n \ge 1$ (after multiplication by a_n) we obtain precisely the stated formula.

When $T|\log y| > 1$, the estimate (7) follows from Lemma 1.1(i). Otherwise, we can write

$$\int_{\kappa-iT}^{\kappa+iT} y^s s^{-1}\,\mathrm{d}s = y^\kappa \int_{\kappa-iT}^{\kappa+iT} s^{-1}\,\mathrm{d}s + y^\kappa \int_{\kappa-iT}^{\kappa+iT} (y^{i\tau}-1)s^{-1}\,\mathrm{d}s.$$

The second integral is

$$\ll \int_0^T |(\tau\log y)/s|\,\mathrm{d}\tau \leq T|\log y| \leq 1$$

and consequently, by Lemma 1.1(ii), we see that the left-hand side of (7) is $\ll y^\kappa$. This concludes the proof.

Corollary 2.1 (Second effective Perron formula).

Let $F(s) := \sum_{n=1}^\infty a_n n^{-s}$ be a Dirichlet series with finite abscissa of absolute convergence σ_a. Suppose that there exists some real number $\alpha \geq 0$ such that

$$\text{(i)} \qquad \sum_{n=1}^\infty |a_n| n^{-\sigma} \ll (\sigma-\sigma_a)^{-\alpha} \qquad (\sigma > \sigma_a),$$

and that B is a non-decreasing function satisfying

$$\text{(ii)} \qquad |a_n| \leq B(n) \qquad (n \geq 1).$$

Then for $x \geq 2$, $T \geq 2$, $\sigma \leq \sigma_a$, $\kappa := \sigma_a - \sigma + 1/\log x$, we have

$$\sum_{n\leq x} \frac{a_n}{n^s} = \frac{1}{2\pi i}\int_{\kappa-iT}^{\kappa+iT} F(s+w)x^w \frac{\mathrm{d}w}{w} + O\Big(x^{\sigma_a-\sigma}\frac{(\log x)^\alpha}{T} + \frac{B(2x)}{x^\sigma}\Big(1 + x\frac{\log T}{T}\Big)\Big). \tag{8}$$

Proof. Apply formula (6) to the series $\sum_{n=1}^\infty b_n n^{-w}$ with $b_n := a_n n^{-s}$. The contribution of integers n which do not belong to $[\frac{1}{2}x, 2x]$ is

$$\ll x^\kappa T^{-1} \sum_{n=1}^\infty |a_n| n^{-\kappa-\sigma} \ll x^{\sigma_a-\sigma} T^{-1} (\log x)^\alpha.$$

When $\frac{1}{2}x \leq n \leq 2x$, we write $n = N + h$ where N is the nearest integer to x. We have $|\log(x/n)| \gg |h|/x$. This leads to the following estimate for the extra contribution:

$$x^{-\sigma} \sum_{x/2\leq n\leq 2x} \frac{|a_n|}{1+T|\log(x/n)|} \ll \frac{B(2x)}{x^\sigma} \sum_{0\leq h\leq x+1} \frac{1}{1+Th/x}$$
$$\ll \frac{B(2x)}{x^\sigma}\Big\{1 + \sum_{1\leq h\leq x/T} 1 + \sum_{x/T<h\leq x+1} \frac{x}{Th}\Big\} \ll \frac{B(2x)}{x^\sigma}\Big\{1 + x\frac{\log T}{T}\Big\}.$$

In certain circumstances, it is useful to work with an absolutely convergent Perron integral. This necessitates a suitable "smoothing" of the function $A(x)$. The following theorem provides two examples.

Theorem 3. *For $\kappa > \max(0, \sigma_c)$ and $x \geq 1$, we have*

$$\sum_{n \leq x} a_n \log(x/n) = \frac{1}{2\pi i} \int_{\kappa - i\infty}^{\kappa + i\infty} F(s) x^s \frac{ds}{s^2}, \tag{9}$$

and

$$\int_0^x A(t)\, dt = \frac{1}{2\pi i} \int_{\kappa - i\infty}^{\kappa + i\infty} F(s) x^{s+1} \frac{ds}{s(s+1)}. \tag{10}$$

Proof. For $w \geq 0$ and $x \in \mathbb{R}^+ \smallsetminus \mathbb{N}$, the Perron formula (3) implies

$$\sum_{n \leq x} a_n n^w = \frac{1}{2\pi i} \int_{\kappa - i\infty}^{\kappa + i\infty} F(s) x^{s+w} \frac{ds}{(s+w)}$$

when $\kappa > \max(0, \sigma_c)$. Thus

$$\sum_{n \leq x} a_n (x^w - n^w) = \frac{1}{2\pi i} \int_{\kappa - i\infty}^{\kappa + i\infty} F(s) x^{s+w} \frac{w}{s(s+w)}\, ds \tag{11}$$

and it is clear that this formula remains true also when x is an integer. By Theorem 1.15, for $\sigma = \kappa$, we have

$$F(s) \ll 1 + |\tau|^{1 - (\kappa - \sigma_c) + \varepsilon}, \tag{12}$$

so both the integral and its derivative with respect to w are absolutely and uniformly convergent. The conditions for differentiation under the integration sign are therefore satisfied, and (9) is obtained by taking the derivative at $w = 0$ of each side of (11). Formula (10) corresponds to the case $w = 1$ of (11), for which the left-hand side then equals

$$\int_0^x A(t)\, dt.$$

§ 2.2 Application: a convergence theorem

Perron's formula can be used to derive the following result, which is a particular case of an important theorem of Schnee–Landau—cf. Landau (1909, §238) and the Notes.

Theorem 4. *Let $F(s) := \sum_{n=1}^{\infty} a_n n^{-s}$ be a Dirichlet series with finite abscissa of convergence. If σ_0 is some real number such that $F(s)$ has a regular analytic continuation satisfying $\mu(\sigma) = 0$ for $\sigma > \sigma_0$, then we have*

$$\sigma_c \leq \sigma_0.$$

Proof. Without loss of generality, the value of σ_0 can be arbitrarily fixed. We let $\sigma_0 < 0$, and show that $F(s)$ is convergent at $s = 0$.

Let δ be some real number with $0 < \delta < -\sigma_0$. We apply Theorem 2 with fixed $\kappa > \sigma_a + 1$ and $x \in \frac{1}{2} + \mathbb{N}$. For each integer n, we have $|\log(x/n)| \geq \log\left((n+\frac{1}{2})/n\right) \gg 1/n$, so that the remainder term in (6) is

$$\ll x^\kappa \sum_{n=1}^{\infty} |a_n| n^{-\kappa}(1+T/n)^{-1} \ll x^\kappa T^{-1} \sum_{n=1}^{\infty} |a_n| n^{1-\kappa} \ll x^\kappa T^{-1}.$$

Next, we deform the contour of integration $[\kappa - iT, \kappa + iT]$ into the polygonal line passing through the points $-\delta - iT, -\delta + iT$. This alters the integral by the value of the residue at the pole $s = 0$, that is to say $F(0)$. We can then write

$$\sum_{n \leq x} a_n = F(0) + R_1 + R_2 + O(x^\kappa T^{-1}) \tag{13}$$

where R_1 and R_2 denote the respective contributions to the integral

$$\frac{1}{2\pi i} \int F(s) x^s s^{-1} \,\mathrm{d}s$$

from the horizontal segments and from the vertical segment of the new contour. Since, by assumption, $\mu(\sigma) = 0$ for $\sigma \geq -\delta$ for all $\varepsilon > 0$, we have

$$R_1 \ll_\varepsilon x^\kappa T^{\varepsilon - 1}, \qquad R_2 \ll_\varepsilon x^{-\delta} T^\varepsilon.$$

(Note that we have used here the fact that the bound

$$F(s) \ll T^\varepsilon \qquad (|\tau| \leq T)$$

is uniform for $-\delta \leq \sigma \leq \kappa$—cf. the remark following the proof of Theorem II.1.16.)

Substituting in (13) and choosing $T := x^{\kappa+\delta}$, $\varepsilon := \delta/2(\kappa+\delta)$, we obtain the convergence of the series $\sum a_n$ in the "explicit" form

$$\sum_{n \leq x} a_n = F(0) + O(x^{-\delta/2}).$$

This concludes the proof.

§ 2.3 The mean value formula

The following result is the analogue for Dirichlet series of Parseval's theorem concerning trigonometric series.

Theorem 5. *Let* $F(s) := \sum_{n=1}^{\infty} a_n n^{-s}$ *and* $G(s) := \sum_{n=1}^{\infty} b_n n^{-s}$ *be two Dirichlet series having respective abscissae of absolute convergence* σ_1 *and* σ_2*. For* $\alpha > \sigma_1$*,* $\beta > \sigma_2$*, we have*

$$\lim_{T\to\infty} \frac{1}{2T} \int_{-T}^{T} F(\alpha + i\tau)\, G(\beta - i\tau)\, \mathrm{d}\tau = \sum_{n=1}^{\infty} a_n b_n n^{-\alpha-\beta}. \tag{14}$$

Proof. We have

$$F(\alpha+i\tau)\,G(\beta-i\tau) = \sum_{n=1}^{\infty} a_n b_n n^{-\alpha-\beta} + \sum_{m\neq n} a_m b_n m^{-\alpha} n^{-\beta} (n/m)^{i\tau},$$

the series converging absolutely and uniformly in τ. We can therefore integrate term by term, and obtain

$$\frac{1}{2T}\int_{-T}^{T} F(\alpha+i\tau)\,G(\beta-i\tau)\,\mathrm{d}\tau = \sum_{n=1}^{\infty} \frac{a_n b_n}{n^{\alpha+\beta}} + \sum_{m\neq n} \frac{a_m b_n}{m^{\alpha} n^{\beta}} \left(\frac{\sin(T\log(n/m))}{T\log(n/m)}\right).$$

The factor involving T is uniformly bounded with respect to T, m, n and tends to 0 as $T \to \infty$. The stated result then follows from the theorem of dominated convergence.

Corollary 5.1. *For* $\sigma > \sigma_a$*, we have*

$$\lim_{T\to\infty} \frac{1}{2T}\int_{-T}^{T} |F(s)|^2\,\mathrm{d}\tau = \sum_{n=1}^{\infty} |a_n|^2 n^{-2\sigma}. \tag{15}$$

Corollary 5.2. *For* $\sigma > \sigma_a$*, we have*

$$\lim_{T\to\infty} \frac{1}{2T}\int_{-T}^{T} F(s) n^s\,\mathrm{d}\tau = a_n. \tag{16}$$

The formulae (15) and (16) provide two new instant proofs of the uniqueness theorem for the representation of a function as a Dirichlet series (Theorem 1.9). The range of validity of relation (15) is, in general, rather difficult to determine. It can be shown (see Titchmarsh (1939), § 9.5) that the set of points

s where $F(s)$ is holomorphic, of finite order and satisfies (15) is a half-plane with boundary the vertical line $\sigma = \sigma_m$, where

$$\sigma_m \geq \max\left(\sigma_c, \sigma_a - \tfrac{1}{2}\right).$$

It is, however, possible that, even when $F(s)$ is not holomorphic for $\sigma \geq \sigma_0$, the left-hand side of (15) converges for $\sigma = \sigma_0$. In the case of the zeta function, Titchmarsh (1951, § 7.2) showed that

$$\lim_{T\to\infty} \frac{1}{T} \int_1^T |\zeta(s)|^2 \, \mathrm{d}\tau = \zeta(2\sigma) \quad (\sigma > \tfrac{1}{2}) \tag{17}$$

and hence that (15) holds for $F = \zeta$ whenever $\sigma > \frac{1}{2}$, $\sigma \neq 1$.

Notes

§ **2.2.** The Schnee–Landau theorem can be stated thus: *If $a_n \ll_\varepsilon n^\varepsilon$ for any $\varepsilon > 0$ (so that $\sigma_a \leq 1$) and if $F(s) = \sum_{n=1}^{\infty} a_n n^{-s}$ has, for some σ_0, a regular continuation in $\sigma > \sigma_0$ which satisfies $\mu(\sigma) \leq \alpha$ in this half-plane, then*

$$\sigma_c \leq \min\left(\frac{\sigma_0 + \alpha}{1 + \alpha}, \ \sigma_0 + \alpha\right).$$

For a proof see Landau (1909), pp. 853 et seq.

When $\alpha = 0$, the Schnee–Landau theorem implies the convergence of $\sum_{n=1}^{\infty} a_n n^{-s-\kappa}$ for $\kappa > \sigma_a$, $\sigma > \sigma_0 - \kappa$. The condition $a_n \ll_\varepsilon n^\varepsilon$ hence becomes superfluous, and one obtains Theorem 4.

Theorem 4 is proved in Landau (1909), p. 848. Another proof appears in Hardy & Riesz (1915), theorem 50. See also Titchmarsh (1951), theorem 3.13.

Exercises

1. Show that, for $\kappa > \max(0, \sigma_c)$, one has

$$\sum_{n \leq x} a_n \left(\log \frac{x}{n}\right)^k = \frac{k!}{2\pi i} \int_{\kappa - i\infty}^{\kappa + i\infty} F(s) x^s s^{-k-1} \, \mathrm{d}s$$

where k is an arbitrary positive integer.

2. Let $w \in L^1(\mathbb{R})$ be a function such that the Fourier transform

$$\widehat{w}(\tau) := \int_{-\infty}^{+\infty} e^{-it\tau} w(t) \, \mathrm{d}t$$

is itself integrable on $\mathbb{R}$. Show that, for $\kappa > \sigma_a$, one has

$$\sum_{n=1}^{\infty} a_n \left(\frac{x}{n}\right)^{\kappa} w\left(\log \frac{x}{n}\right) = \frac{1}{2\pi i} \int_{\kappa - i\infty}^{\kappa + i\infty} F(s) x^s \widehat{w}(\tau) \, \mathrm{d}s.$$

3. By applying the conclusion of Exercise 2 to the function

$$w(t) := \left(\frac{\sin(t/2\varepsilon)}{t/2\varepsilon}\right)^2,$$

show that, if $a_n \geq 0$, then for any $\varepsilon > 0$, one has

$$\sum_{|\log(x/n)| \leq \varepsilon} a_n \leq C\varepsilon e^{\kappa\varepsilon} \int_{\kappa - iT}^{\kappa + iT} |F(s) x^s| \, |\mathrm{d}s| \qquad (\kappa > \sigma_a)$$

with $C = \frac{1}{4}\left(\sin \frac{1}{2}\right)^{-2}$, and $T = 1/\varepsilon$.

4. Let $F(s) := \sum a_n n^{-s}$ be a Dirichlet series such that $\lim_{\tau \to \infty} F(\sigma + i\tau)$ exists for at least one value of $\sigma > \sigma_a$. Let λ denote this limit.

(a) Show that $|\lambda|^2 = \sum_{n=1}^{\infty} |a_n|^2 n^{-2\sigma}$.

(b) Show that $\lambda = a_1$. [*Use Corollary 5.2.*]

(c) Deduce that $F(s)$ is constant.

II.3

The Riemann zeta function

§ 3.1 Introduction

The zeta function occupies a central role in arithmetic, particularly on account of its close correlation with prime numbers via Euler's formula (cf. §1.2). It also has the important property that a wide variety of Dirichlet series may be expressed in terms of it. In connection with the Perron formulae established in Chapter 2, the analytic study of the zeta function is hence of major interest.

The evaluation of an integral of the type

$$\int_{-T}^{T} F(s)x^s s^{-1} \, \mathrm{d}\tau,$$

where $F(s)$ is defined by a Dirichlet series, can, in general, be greatly simplified by the introduction of a closed contour and the use of the residue theorem—as we have seen in §2.2. Since $F(s)$ does not have a singularity in the half-plane $\sigma > \sigma_c$ (cf. Theorem 1.3), it is natural to use this approach in circumstances when the contour can be selected so as to intersect the complementary half-plane $\sigma \leq \sigma_c$. This requires the existence of an analytic continuation of $F(s)$. We shall see that this is realised in the case of $\zeta(s)$—and, consequently, for series expressed in terms of $\zeta(s)$.

From this point of view, the study of analytic properties of $\zeta(s)$ (identified with its continuation) constitutes for the number theorist an indispensable investment. In this chapter, we establish only the fundamental results of the theory—and invite the reader interested in more sophisticated developments to consult the classic work of Titchmarsh (1951), as updated by Heath-Brown (1986), or the comprehensive book of Ivić (1985).

§ 3.2 Analytic continuation

Theorem 1. *The function $\zeta(s)$ is analytically continuable to a meromorphic function on the entire complex plane having as sole singularity a simple pole at $s = 1$, with residue* 1.

Proof. For $\sigma > 1$, we have

$$G(s) := \sum_{n=1}^{\infty}(-1)^n n^{-s} = \sum_{m=1}^{\infty}(2m)^{-s} - \left(\sum_{n=1}^{\infty} n^{-s} - \sum_{m=1}^{\infty}(2m)^{-s}\right)$$
$$= (2^{1-s} - 1)\zeta(s).$$

Now $G(s)$ is convergent for $\sigma > 0$, and

$$G(1) = -\log 2 \neq 0, \qquad G(s_k) = 0 \quad (s_k := 1 + 2\pi ik/\log 2,\ k \neq 0).$$

The second formula follows from the estimate

$$\sum_{n\leq x}(-1)^n n^{-s_k} = 2^{1-s_k}\sum_{n\leq x/2} n^{-s_k} - \sum_{n\leq x} n^{-s_k} = -\sum_{x/2<n\leq x} n^{-s_k} \ll 1/x$$

obtained by partial summation. We thus see that the stated property holds for the half-plane $\sigma > 0$. The same result can also be obtained from the partial integration formula

$$\zeta(s) = \int_{1-}^{\infty} t^{-s}\,\mathrm{d}[t] = \int_{1}^{\infty} t^{-s}\,\mathrm{d}t - \int_{1-}^{\infty} t^{-s}\,\mathrm{d}\{t\}$$
$$= \frac{s}{s-1} - s\int_{1}^{\infty}\{t\}t^{-s-1}\,\mathrm{d}t,$$

initially valid for $\sigma > 1$, which provides an alternative explicit form of the continuation for $\sigma > 0$.

The stated result could be derived by successive integrations, starting from one or other of the preceding formulae—using, for example, in the second case, the Euler–Maclaurin formula (see Exercise 10). We employ another method, due to Riemann, which provides additional information useful in establishing the functional equation.

The starting point is the formula

(1) $$\Gamma(s)n^{-s} = \int_{0}^{\infty} t^{s-1}e^{-nt}\,\mathrm{d}t \qquad (\sigma > 0).$$

Summing over $n \geq 1$, we have, for $\sigma > 1$, that

$$\Gamma(s)\zeta(s) = \sum_{n=1}^{\infty}\int_{0}^{\infty} t^{s-1}e^{-nt}\,\mathrm{d}t = \int_{0}^{\infty} t^{s-1}(e^{t}-1)^{-1}\,\mathrm{d}t.$$

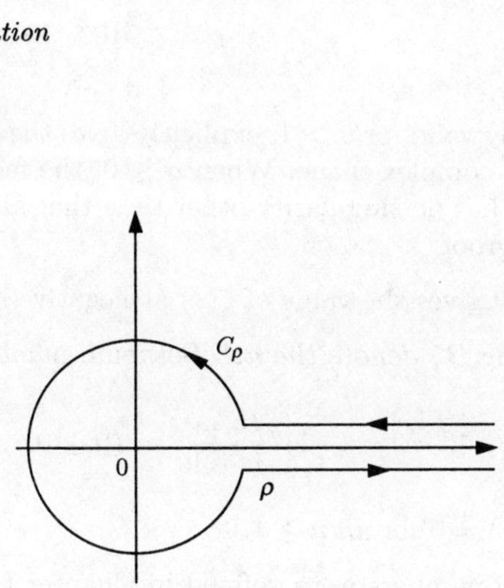

We obtain an analytic continuation of this integral by replacing the half-line of integration by the *Hankel contour* $\mathcal{C}_\rho$ where ρ is a real parameter, $0 < \rho < 2\pi$.

$\mathcal{C}_\rho$ is composed of three parts: the set of points with argument $0+$ on the real half-line $[\rho, +\infty[$ traced from right to left; the circle $|z| = \rho$, traced in the positive sense; and the set of points with argument $2\pi-$ on the half-line $[\rho, +\infty[$, traced from left to right.

Since the function $z \mapsto z^{s-1}(e^z - 1)^{-1}$ is holomorphic in the horizontal strip $|\Im m\, z| < 2\pi$ excluding the half-line $[0, +\infty[$, the integral

$$I(s) := \int_{\mathcal{C}_\rho} z^{s-1}(e^z - 1)^{-1}\,\mathrm{d}z$$

is independent of ρ in $]0, 2\pi[$. It is absolutely convergent for each $s \in \mathbb{C}$ and uniformly convergent on any compact subset. Thus it defines an entire function of s. We have

$$\text{(2)} \qquad I(s) = \oint_{|z|=\rho} z^{s-1}(e^z - 1)^{-1}\,\mathrm{d}z + (e^{2\pi i s} - 1)\int_\rho^\infty t^{s-1}(e^t - 1)^{-1}\,\mathrm{d}t.$$

Using the bound

$$|z^{s-1}(e^z - 1)^{-1}| \ll_s \rho^{\sigma-2} \qquad (|z| = \rho \leq \pi),$$

and letting ρ tend to 0, we obtain the formula

$$I(s) = (e^{2\pi i s} - 1)\Gamma(s)\zeta(s) \qquad (\sigma > 1).$$

The well-known functional equation $\Gamma(s)\Gamma(1-s) = \pi/\sin(\pi s)$ therefore implies

$$\text{(3)} \qquad \zeta(s) = \frac{1}{2\pi i} e^{-i\pi s}\Gamma(1-s)I(s).$$

This formula, initially valid for $\sigma > 1$, explicitly gives the analytic continuation of $\zeta(s)$ to the whole complex plane. When $\sigma \leq 0$, the factor $\Gamma(1-s)$ is holomorphic, hence $\zeta(s)$ has no singularity other than that already noted at $s = 1$. This concludes the proof.

Formula 2 readily gives the value of $\zeta(s)$ at negative integers:

Theorem 2. *Letting B_n denote the nth Bernoulli number, we have*

$$\zeta(-n) = (-1)^n \frac{B_{n+1}}{n+1} \qquad (n \geq 0). \tag{4}$$

In particular, $\zeta(-2n) = 0$ for all $n \geq 1$.

Proof. The Bernoulli numbers were defined in Chapter I.0 by the Laurent expansion

$$(e^z - 1)^{-1} = \sum_{m=0}^{\infty} \frac{1}{m!} B_m z^{m-1}.$$

Hence, by (2),

$$I(-n) = \oint_{|z|=\rho} (e^z - 1)^{-1} z^{-n-1} \, dz = 2\pi i \frac{B_{n+1}}{(n+1)!}.$$

We obtain the stated formula by substituting in (3).

§ 3.3 Functional equation

Theorem 3. *For each $s \neq 1$, we have*

$$\zeta(s) = 2^s \pi^{s-1} \sin\left(\tfrac{1}{2}\pi s\right) \Gamma(1-s) \zeta(1-s). \tag{5}$$

Remark. The functional equation takes the more symmetric form

$$\Phi(s) = \Phi(1-s) \qquad (s \neq 0, 1) \tag{6}$$

with the notation

$$\Phi(s) := \pi^{-s/2} \Gamma(s/2) \zeta(s). \tag{7}$$

This follows immediately from the previously stated functional equation for $\Gamma(s)$ and from the well-known duplication formula

$$\Gamma(s)\Gamma\left(s + \tfrac{1}{2}\right) = 2^{1-2s} \sqrt{\pi} \Gamma(2s). \tag{8}$$

Proof. For $k \geq 1$, let $\mathcal{H}_k$ be the Hankel contour with parameter $\rho_k := (2k+1)\pi$. Then $|e^z - 1|^{-1}$ is bounded for z on $\mathcal{H}_k$, and we have

$$\left|\int_{\mathcal{H}_k} z^{s-1}(e^z-1)^{-1}\,\mathrm{d}z\right| \ll_s k^\sigma. \tag{9}$$

Let ρ satisfy $0 < \rho < 2\pi$. The contour $\mathcal{C}_\rho - \mathcal{H}_k$ encircles, in the negative sense, the poles $z = 2n\pi i$ for $n = \pm 1, \pm 2, \ldots, \pm k$. The residue theorem then implies

$$I(s) = \int_{\mathcal{C}_\rho} z^{s-1}(e^z-1)^{-1}\,\mathrm{d}z = \int_{\mathcal{H}_k} z^{s-1}(e^z-1)^{-1}\,\mathrm{d}z - 2\pi i \sum_{1\leq|n|\leq k} (2n\pi i)^{s-1}$$
$$= \int_{\mathcal{H}_k} z^{s-1}(e^z-1)^{-1}\,\mathrm{d}z - (2\pi i)^s(1-e^{i\pi s}) \sum_{1\leq n\leq k} n^{s-1}.$$

Using (9), and letting k tend to infinity, we obtain, for each s in the half-plane $\sigma < 0$, that

$$I(s) = (2\pi i)^s(e^{i\pi s}-1)\zeta(1-s).$$

Substituting in (3), this gives equation (5) for $\sigma < 0$. By analytic continuation it remains valid for all s.

An immediate consequence of Theorems 2 and 3 is the following formula for $\zeta(2n)$, which can also be obtained as a special case of the Fourier series expansion for Bernoulli functions — cf. Exercise I.0.1.

Theorem 4. *We have*

$$\zeta(2n) = (-1)^{n-1}2^{2n-1}\frac{B_{2n}}{(2n)!}\pi^{2n} \qquad (n \geq 1). \tag{10}$$

§ 3.4 Approximations and bounds in the critical strip

Let us write

$$\chi(s) := 2(2\pi)^{s-1}\Gamma(1-s)\sin\left(\tfrac{1}{2}\pi s\right), \tag{11}$$

so that the functional equation (5) can be written as

$$\zeta(s) = \chi(s)\zeta(1-s). \tag{12}$$

The asymptotic behaviour of $\chi(s)$ when σ is fixed and $|\tau| \to \infty$ can be easily determined by means of the *complex Stirling formula* (cf. Titchmarsh (1939), p. 151), the proof of which is outlined in Exercise 1. We have

$$\log\Gamma(s) = \left(s-\tfrac{1}{2}\right)\log s - s + \tfrac{1}{2}\log(2\pi) - \int_0^\infty \frac{B_1(t)}{t+s}\,\mathrm{d}t, \tag{13}$$

where $B_1(t)$ is the first Bernoulli function, and where $\log s$ denotes the principal value of the complex logarithm. From (13), an easy calculation leads to the asymptotic formula

$$|\chi(s)| \sim \left(\frac{|\tau|}{2\pi}\right)^{\frac{1}{2}-\sigma} \qquad (|\tau| \to \infty). \tag{14}$$

Thus it follows from (12) that the order of magnitude of $\zeta(s)$ on vertical lines is completely determined when $\sigma < 0$. With the notation of §1.6, we deduce in particular that, for the zeta function,

$$\mu(\sigma) = \begin{cases} \frac{1}{2} - \sigma & (\sigma \leq 0) \\ 0 & (\sigma \geq 1). \end{cases} \tag{15}$$

The exact value of $\mu(\sigma)$ in the *critical strip* $0 < \sigma < 1$ is a deep problem which is still unsolved. The most simple hypothesis consistant with (15) is that the graph of $\mu(\sigma)$ is formed by two half-lines, viz

$$\mu(\sigma) = \max(\tfrac{1}{2} - \sigma, 0). \tag{16}$$

This formula is known as *Lindelöf's hypothesis.* Taking into account the convexity properties of $\mu(\sigma)$ (cf. Theorem 1.16), it is equivalent to $\mu(\frac{1}{2}) = 0$.

The relations $\mu(0) = \frac{1}{2}$ and $\mu(1) = 0$ immediately imply by convexity that $\mu(\sigma) \leq \frac{1}{2}(1-\sigma)$ for $0 \leq \sigma \leq 1$. The following result allows us to improve this estimate.

Theorem 5. *For* $\sigma > 0$ *and* $N \in \mathbb{Z}^+$, *we have*

$$\zeta(s) = \sum_{n \leq N} n^{-s} - \frac{N^{1-s}}{1-s} - s\int_N^\infty \{t\} t^{-s-1}\,\mathrm{d}t. \tag{17}$$

Proof. For $\sigma > 1$, we have

$$\begin{aligned} \zeta(s) - \sum_{n \leq N} n^{-s} &= \int_N^\infty t^{-s}\,\mathrm{d}[t] = -\frac{N^{1-s}}{1-s} - \int_N^\infty t^{-s}\,\mathrm{d}\{t\} \\ &= -\frac{N^{1-s}}{1-s} - s\int_N^\infty t^{-s-1}\{t\}\,\mathrm{d}t. \end{aligned} \tag{18}$$

By analytic continuation, this formula remains true for $\sigma > 0$.

Corollary 5.1. *Let $\sigma_0 > 0$, $0 < \delta < 1$. Uniformly for $\sigma \geq \sigma_0$, $x \geq 1$ and $0 < |\tau| \leq (1-\delta)2\pi x$, we have*

$$\zeta(s) = \sum_{n\leq x} n^{-s} - \frac{x^{1-s}}{1-s} + O(x^{-\sigma}). \tag{19}$$

Proof. For $|\tau| \leq (1-\delta)2\pi x$ and all $y \geq x$, we can write

$$\sum_{x<n\leq y} n^{-i\tau} = \int_x^y t^{-i\tau}\,\mathrm{d}t + O(1), \tag{20}$$

applying Theorem I.6.4 to the function $f(t) := -(\tau/2\pi)\log t$, which satisfies $|f'(t)| \leq 1-\delta$ for $x \leq t \leq y$. Then, for $N \geq x$, we obtain

$$\begin{aligned}\sum_{x<n\leq N} n^{-s} &= \int_x^N y^{-s}\,\mathrm{d}y + \int_x^N y^{-\sigma}\,\mathrm{d}\{O(1)\}\\ &= \frac{N^{1-s}-x^{1-s}}{1-s} + O(x^{-\sigma}).\end{aligned}$$

Substituting in (17) and letting $N \to \infty$, we obtain precisely (19).

Corollary 5.2. *We have*

$$\zeta(\tfrac{1}{2}+i\tau) \ll |\tau|^{1/6}\log|\tau| \qquad (|\tau|\to\infty). \tag{21}$$

Proof. We may assume that $\tau > 0$. Theorems I.6.5 and I.6.6 applied to $f(t) := -(\tau/2\pi)\log t$ readily give the estimate

$$\sum_{a<n\leq b} n^{-i\tau} \ll \min\left(\tau^{1/2} + a\tau^{-1/2}, a^{1/2}\tau^{1/6} + a\tau^{-1/6}\right)$$

valid uniformly for $\tau > 0$, $a < b \leq 2a$. Under the additional assumption $a \ll \tau$, we infer by partial integration that

$$\begin{aligned}\sum_{a<n\leq b} n^{-1/2-i\tau} &\ll \min\left((\tau/a)^{1/2} + (a/\tau)^{1/2},\ \tau^{1/6} + a^{1/2}\tau^{-1/6}\right)\\ &\ll \min\left(\tau^{1/6}, (\tau/a)^{1/2}\right).\end{aligned}$$

For $r \leq \log x/\log 2$, let us choose $a := 2^r$, $b := \min(2^{r+1}, x)$ and sum all the corresponding estimates. It follows that

$$\sum_{n\leq x} n^{-1/2-i\tau} \ll \tau^{1/6}\log\tau \qquad (x \ll \tau).$$

Substituting in (19) with $x = \tau$ we obtain the required estimate (21).

The convexity of the function $\mu(\sigma)$ allows us to deduce from the above result a new bound in the critical strip.

Theorem 6. *The zeta function satisfies the inequalities*

$$\mu(\sigma) \le \begin{cases} \frac{1}{3}(1-\sigma) & (\frac{1}{2} \le \sigma \le 1) \\ \frac{1}{6}(3-4\sigma) & (0 \le \sigma \le \frac{1}{2}). \end{cases} \tag{22}$$

Proof. This follows immediately from the estimates $\mu(0) = \frac{1}{2}$, $\mu(\frac{1}{2}) \le \frac{1}{6}$, $\mu(1) = 0$.

The bound in (22) does not provide any new information on the line $\sigma = 1$. However, using Corollary 5.1 directly, the estimate $\zeta(s) \ll |\tau|^{\varepsilon}$ $(\sigma = 1)$ can be considerably improved.

Theorem 7. *Let α be such that $0 < \alpha < 1$. We have*

$$|\zeta(s)| \le \frac{3|\tau|^{1-\alpha}}{2\alpha(1-\alpha)} \qquad (\sigma \ge \alpha,\ |\tau| \ge 1). \tag{23}$$

In particular, for each positive constant c, we have

$$\zeta(s) \ll \log|\tau| \qquad (|\tau| \ge 2,\ \sigma \ge 1 - c/\log|\tau|). \tag{24}$$

Remark. Clearly, the explicit bound (23) has no theoretical interest unless $\alpha = \alpha(\tau) \to 1$. Otherwise, relation (22) provides a superior result.

Proof. By (17), we can write

$$|\zeta(s)| \le \sum_{n \le N} n^{-\sigma} + N^{1-\sigma}|\tau|^{-1} + \tfrac{1}{2} N^{-\sigma} + \tfrac{1}{2}|s|\sigma^{-1}N^{-\sigma}.$$

Moreover we plainly have

$$\sum_{n \le N} n^{-\sigma} \le 1 + \int_1^N t^{-\sigma}\,\mathrm{d}t < \frac{N^{1-\sigma}}{1-\sigma} \qquad (0 < \sigma < 1).$$

Under the assumption $\sigma \ge \alpha$, it then follows that

$$|\zeta(s)| \le N^{1-\alpha}\left\{\frac{1}{1-\alpha} + \frac{1}{|\tau|} + \frac{2\alpha + |\tau|}{2\alpha N}\right\}.$$

Choosing $N = \big[|\tau|\big]$, we obtain

$$|\zeta(s)| \le \frac{N^{1-\alpha}}{\alpha(1-\alpha)}\{\alpha + \alpha(1-\alpha) + (1+\alpha)(1-\alpha)\}.$$

The maximum of the function in curly brackets is obtained for $\alpha = \frac{1}{2}$ and equals $\frac{3}{2}$.

Corollary 7.1. *For any positive constant c and any integer $k \geq 0$, we have*

$$\zeta^{(k)}(s) \ll (\log|\tau|)^{k+1} \quad (|\tau| \geq 2,\ \sigma \geq 1 - c/\log|\tau|). \tag{25}$$

Proof. Clearly we may suppose that $|\tau| \geq 3$. For some suitable constant c_0 and $r := c_0/\log|\tau|$, the disc $|z-s| \leq r$ is contained in a domain where (24) applies. The desired bound then follows from Cauchy's formula

$$\zeta^{(k)}(s) = \frac{k!}{2\pi i}\oint_{|z|=r} \zeta(s+z)\frac{\mathrm{d}z}{z^{k+1}}.$$

§ 3.5 Initial localisation of zeros

The convergence of the Euler product $\zeta(s) = \prod_p (1-p^{-s})^{-1}$ guarantees that $\zeta(s)$ does not vanish for $\sigma > 1$. We shall show in this section that this property extends to the closed half-plane $\sigma \geq 1$. The following result, simple but cunning, enables us to perform the limit process.

Theorem 8 (de La Vallée–Poussin). *Let $F(s) := \sum_{n=1}^{\infty} a_n n^{-s}$ be a Dirichlet series with non-negative coefficients and with abscissa of convergence σ_c. Then we have*

$$3F(\sigma) + 4\Re e\, F(\sigma + i\tau) + \Re e\, F(\sigma + 2i\tau) \geq 0 \quad (\sigma > \sigma_c). \tag{26}$$

Proof. Write $V(\vartheta) := 3 + 4\cos\vartheta + \cos 2\vartheta$ ($\vartheta \in \mathbb{R}$). A simple verification shows that $V(\vartheta) = 2(1+\cos\vartheta)^2 \geq 0$. Now the left-hand side of (26) equals

$$\sum_{n=1}^{\infty} a_n n^{-\sigma} V(\tau \log n).$$

This implies the stated conclusion.

Corollary 8.1. *We have*

$$\zeta(\sigma)^3 |\zeta(\sigma+i\tau)|^4 |\zeta(\sigma+2i\tau)| \geq 1 \quad (\sigma > 1). \tag{27}$$

Proof. It suffices to apply the theorem to the function

$$F(s) = \log\zeta(s) = -\sum_p \log(1-p^{-s}) = \sum_{n\geq 2} \frac{\Lambda(n)}{\log n} n^{-s}$$

which is convergent for $\sigma > 1$.

We are now in a position to prove the following important result.

Theorem 9. *The function $\zeta(s)$ has no zero in the half-plane $\sigma \geq 1$.*

Proof. Let us argue by contradiction and assume that $\zeta(1 + i\tau_0) = 0$. Then $\tau_0 \neq 0$ and $\zeta(s)$ is holomorphic in some neighbourhood of $1 + i\tau_0$. Therefore

$$\zeta(\sigma + i\tau_0) \ll \sigma - 1 \qquad (\sigma > 1).$$

On the other hand

$$\zeta(\sigma) \ll (\sigma - 1)^{-1}, \quad \zeta(\sigma + 2i\tau_0) \ll 1 \quad (\sigma > 1),$$

since $s = 1$ is a simple pole and $\zeta(s)$ is holomorphic everywhere except at $s = 1$. It follows that

$$\zeta(\sigma)^3 |\zeta(\sigma + i\tau_0)|^4 |\zeta(\sigma + 2i\tau_0)| \ll \sigma - 1 \quad (\sigma > 1),$$

which contradicts (27) as $\sigma \to 1+$.

Theorem 9 when used with the functional equation (5) clearly implies the following corollary.

Corollary 9.1. *In the half-plane $\sigma \leq 0$ the function $\zeta(s)$ has zeros only at the points $-2n$ ($n = 1, 2, \ldots$). These are simple zeros.*

The zeros at the negative even integers are called the *trivial zeros* of the zeta function. We will see in §3.7 that $\zeta(s)$ also has zeros in the critical strip $0 < \sigma < 1$. The equality

$$\sum_{n=1}^{\infty} (-1)^n n^{-\sigma} = (2^{1-\sigma} - 1)\zeta(\sigma) \quad (0 < \sigma < 1)$$

shows that *these non-trivial zeros are not real.* The functional equation and the fact that $\zeta(s)$ is real for real s imply that they are distributed symmetrically with respect to the line $\sigma = \frac{1}{2}$ and the real axis $\tau = 0$.

Another, particularly elegant proof of Theorem 9 is due to Ingham (1930). It rests on Ramanujan's identity

$$\frac{\zeta(s)^2 \zeta(s + i\theta)\zeta(s - i\theta)}{\zeta(2s)} = \sum_{n=1}^{\infty} |\tau(n, \theta)|^2 n^{-s} \tag{28}$$

valid for $\sigma > 1$, $\theta \in \mathbb{R}$, with

$$\tau(n, \theta) := \sum_{d|n} d^{i\theta}. \tag{29}$$

The formula (28) may be easily checked by identifying the factors of the Euler products on the two sides—see also the Notes. Let us consider the abscissa of convergence σ_c of (28). We have $\sigma_c \leq 1$ since $|\tau(n,\theta)| \leq \tau(n)$. By Theorem 1.3, the sum of the series is holomorphic for $\sigma > \sigma_c$. Since the coefficients are non-negative, the point $s = \sigma_c$ is a singularity. Now if $s = 1 + i\theta$ is a zero of $\zeta(s)$, so is $s = 1 - i\theta$ (since $\zeta(\bar{s}) = \overline{\zeta(s)}$) and these two zeros compensate for the double pole of $\zeta(s)^2$ at $s = 1$. But $\zeta(s)$ has no singularity other than at $s = 1$, hence it follows from the above that (28) is holomorphic *on the real axis* down to the first zero of $\zeta(2s)$, that is to say $s = -1$. We would then have $\sigma_c = -1$, which is absurd since $|\tau(n,\theta)|$ does not tend to 0.

§ 3.6 Lemmas from complex analysis

The following classical results will be useful in our further investigations concerning the zeta function.

Theorem 10 (Jensen's formula). *Let $F(s)$ be a holomorphic function for $|s| \leq R$, with $F(0) = 1$. For $0 < r \leq R$, let $n(r)$ denote the number of zeros (counted with multiplicities) of F in the disc $|s| \leq r$. Then we have*

$$\text{(30)} \qquad \int_0^R \frac{n(r)}{r}\,\mathrm{d}r = \frac{1}{2\pi}\int_0^{2\pi} \log|F(Re^{i\theta})|\,\mathrm{d}\theta.$$

Proof. If $F(s)$ is non-zero for $|s| = r$, we may write

$$\text{(31)} \qquad n(r) = \frac{1}{2\pi i}\oint_{|s|=r} \frac{F'(s)}{F(s)}\,\mathrm{d}s = \frac{1}{2\pi}\int_0^{2\pi} \frac{F'(re^{i\theta})}{F(re^{i\theta})} re^{i\theta}\,\mathrm{d}\theta.$$

Formally, equation (30) is then obtained by dividing by r, integrating over $[0, R]$, and taking the real parts of the two sides. It remains to show that the critical values of r corresponding to the moduli of the zeros do not interfere with this process.

Let $0 < r_1 < r_2 < \cdots < r_k \leq R$ denote the finite sequence of these exceptional moduli. Set $r_0 := 0$, $r_{k+1} := R$. We can integrate (31) over each interval

$$\big[(1+\varepsilon)r_j, (1-\varepsilon)r_{j+1}\big] \qquad (j = 0, \ldots, k),$$

and sum the equalities thus obtained. In order to establish (30), it therefore suffices to show that we can allow ε to tend to 0. In other words that, for each $j \geq 1$, we have

$$\text{(32)} \qquad \lim_{\varepsilon \to 0} \int_0^{2\pi} \log\left|\frac{F\big((1+\varepsilon)r_j e^{i\theta}\big)}{F\big((1-\varepsilon)r_j e^{i\theta}\big)}\right| \mathrm{d}\theta = 0.$$

Since F has only a finite number of zeros on the circle $|s| = r_j$, it is clear that it suffices to consider the case when $s = r_j$ is the only zero, say of multiplicity $m \geq 1$, of $F(s)$ on this circle. Then

$$\log|F(re^{i\theta})| = m\log\left|1 - \frac{r}{r_j}e^{i\theta}\right| + H(r,\theta)$$

for $|r - r_j| \leq \varepsilon r_j$, $0 \leq \theta \leq 2\pi$, where $H(r,\theta)$ is a continuous function of r and θ. Therefore (32) is equivalent to

$$\lim_{\varepsilon \to 0} \int_{-\pi}^{\pi} \log\left|\frac{1-(1+\varepsilon)e^{i\theta}}{1-(1-\varepsilon)e^{i\theta}}\right| \mathrm{d}\theta = 0. \tag{33}$$

By a straightforward calculation it can be checked that, uniformly for $|\theta| \leq \pi$, we have

$$\left|\frac{1-(1+\varepsilon)e^{i\theta}}{1-(1-\varepsilon)e^{i\theta}}\right|^2 = \frac{\varepsilon^2 + 4(1+\varepsilon)\sin^2\frac{1}{2}\theta}{\varepsilon^2 + 4(1-\varepsilon)\sin^2\frac{1}{2}\theta} = 1 + O(\varepsilon).$$

Hence the integral in (33) is $O(\varepsilon)$, which implies (33).

Theorem 11 (Borel–Carathéodory). *Let $F(s)$ be a holomorphic function for $|s| \leq R$ with $F(0) = 0$. Let $A := \max_{|s|=R} \Re e\, F(s)$. Then we have*

$$\sup_{n\geq 1} \frac{R^n}{n!}|F^{(n)}(0)| \leq 2A. \tag{34}$$

Corollary 11.1. *Under the same assumptions, for any r with $0 < r < R$, we have that*

$$M(r) := \max_{|s|\leq r}|F(s)| \leq \frac{2Ar}{R-r}. \tag{35}$$

Proof of the theorem. Consider the Taylor series of F at $s = 0$,

$$F(s) = \sum_{n=1}^{\infty} a_n s^n \qquad (|s| \leq R).$$

Letting θ_n denote, for each n, the argument of a_n, we can write

$$\Re e\, F(Re^{i\theta}) = \sum_{n=1}^{\infty} |a_n| R^n \cos(n\theta + \theta_n),$$

the series being absolutely and uniformly convergent for $0 \le \theta \le 2\pi$. But the fact that F is holomorphic implies that the function $\Re e\, F(s)$ has the mean value property

$$\int_0^{2\pi} \Re e\, F(Re^{i\theta})\, d\theta = 0.$$

For each $n \ge 1$, it therefore follows that

$$\pi|a_n|R^n = \int_0^{2\pi} (1 + \cos(n\theta + \theta_n))\Re e\, F(Re^{i\theta})\, d\theta \le 2\pi A.$$

This completes the proof.

We omit the proof of Corollary 11.1, which is obvious. Another useful consequence of Theorem 11 is contained in the following statement.

Corollary 11.2. *Let F be a holomorphic function in the disc $|s| \le R$ with $F(0) = 1$ and such that $|F(s)| \le M$ for $|s| = R$. Let Z denote the finite sequence of zeros ρ of F, counted with multiplicity, in the disc $|s| \le \frac{1}{2} R$. For any real number r with $0 < r < \frac{1}{2} R$, we have*

$$\sup_{|s| \le r} \left| \frac{F'(s)}{F(s)} - \sum_{\rho \in Z} \frac{1}{s - \rho} \right| \le \frac{4R \log M}{(R - 2r)^2}. \tag{36}$$

Proof. Let $G(s) := F(s) \prod_{\rho \in Z} (1 - s/\rho)^{-1}$. Then $|G(s)| \le M$ for $|s| = R$, since each factor of the product over ρ has modulus at most 1. The function $H(s) := \log G(s)$ satisfies the conditions of Theorem 11, with $A = \log M$, in the disc $|s| \le \frac{1}{2} R$. For $|s| = r < \frac{1}{2} R$, we then have

$$\begin{aligned} |H'(s)| &\le \sum_{n=1}^{\infty} \frac{1}{(n-1)!} r^{n-1} |H^{(n)}(0)| \\ &\le 2 \log M \sum_{n=1}^{\infty} n r^{n-1} (R/2)^{-n} = \frac{4R \log M}{(R - 2r)^2}. \end{aligned}$$

§ 3.7 Global distribution of zeros

In the following section we shall develop, in the special case of the zeta function, the general factorisation method of Hadamard for entire functions of finite order. Here we establish the preliminary results. The function $(s-1)\zeta(s)$ is entire, but it is more convenient to consider instead

$$\xi(s) := s(s-1)\pi^{-s/2}\Gamma(\tfrac{1}{2} s)\zeta(s) = s(s-1)\Phi(s) \tag{37}$$

which, in addition, satisfies the simpler functional equation

$$\xi(s) = \xi(1 - s) \qquad (s \in \mathbb{C}). \tag{38}$$

The regularity of ξ for $\sigma \geq \frac{1}{2}$ follows from that of ζ, since the pole at $s = 1$ is compensated by the factor $(s - 1)$. By (38), $\xi(s)$ is thus an entire function. We note that $\xi(0) = \xi(1) = 1$.

From Theorem 9, $\xi(s) \neq 0$ for $\sigma \geq 1$. By (38), it follows that $\xi(s)$ does not vanish for $\sigma \leq 0$. Hence all the zeros of $\xi(s)$ lie in the critical strip $0 < \sigma < 1$. In this strip, $\xi(s)$ and $\zeta(s)$ have the same zeros. The situation is different in the half-plane $\sigma \leq 0$, since there the trivial zeros of $\zeta(s)$ are compensated by the poles of $\Gamma\left(\frac{1}{2}s\right)$.

Traditionally, the general notation for a zero of ξ (i.e. a non-trivial zero of ζ) is $\rho = \beta + i\gamma$, and we set

$$N(T) := \sum_{\rho\,:\,0 \leq \gamma \leq T} 1,$$

where, here and in the sequel, we adopt the convention that all zeros ρ are counted with multiplicity.

In this section, our aim is to establish an asymptotic formula for $N(T)$ as $T \to \infty$. With this in mind, it is clearly worthwhile to have information on the order of growth of $\xi(s)$. By Theorem 6, we have

$$\zeta(s) \ll \tau \qquad (\sigma \geq 0,\ |\tau| \geq 1). \tag{39}$$

Moreover, integrating by parts the last term in the complex Stirling formula (13), we obtain

$$\log \Gamma(s) = \left(s - \tfrac{1}{2}\right) \log s - s + \tfrac{1}{2} \log(2\pi) + \frac{1}{12s} - \tfrac{1}{2} \int_0^\infty \frac{B_2(t)}{(t+s)^2}\, \mathrm{d}t \tag{40}$$

where $B_2(t) = \{t\}^2 - \{t\} + \frac{1}{6}$ denotes the second Bernoulli function. This implies that

$$\log|\xi(s)| \ll |s| \log|s| \qquad (|s| \to \infty) \tag{41}$$

for $\sigma \geq \frac{1}{2}$, and hence in the whole plane, by (38).

We shall use an analogous upper bound to prove the following crucial result.

Theorem 12. *For $T \geq 2$, we have*

$$N(T+1) - N(T) \ll \log T. \tag{42}$$

Proof. We apply Jensen's formula (30) to the function

$$F(s) := \xi(s + 2 + iT)/\xi(2 + iT)$$

with $R = 3$.

In the first instance, for all r with $\sqrt{5} \le r \le 3$, the disc $|s| \le r$ contains the rectangle with vertices $-2, -2+i, -1, -1+i$, hence $n(r) \ge N(T+1) - N(T)$. Next, from (39) and (40), we readily deduce the estimate

$$\log|F(s)| \ll \log T \qquad (|s| \le 3).$$

Jensen's formula then implies that

$$\big(N(T+1) - N(T)\big)\log\big(3/\sqrt{5}\big) \le \int_{\sqrt{5}}^{3} \frac{n(r)}{r}\,\mathrm{d}r \ll \log T,$$

and (42) follows.

We are now in a position to establish the following fundamental theorem.

Theorem 13. *As T tends to infinity, we have*

$$N(T) = \frac{T}{2\pi}\log\frac{T}{2\pi} - \frac{T}{2\pi} + O(\log T). \tag{43}$$

Remark. This implies, in particular, that $\zeta(s)$ has infinitely many non-trivial zeros.

Proof. Without loss of generality, we may assume that T is not the ordinate of a zero of $\xi(s)$. Let $\mathcal{R}$ be the positively oriented rectangle with vertices $2 \pm iT, -1 \pm iT$. By a well-known formula,

$$2N(T) = \frac{1}{2\pi i}\int_{\mathcal{R}} \frac{\xi'(s)}{\xi(s)}\,\mathrm{d}s = \frac{1}{2\pi}\,\Im m \int_{\mathcal{R}} \frac{\xi'(s)}{\xi(s)}\,\mathrm{d}s.$$

Since $\mathcal{R}$ is symmetric with respect to the axes $\sigma = \frac{1}{2}$ and $\tau = 0$, and $\xi'(s)/\xi(s)$ respects this symmetry because of the functional equation, we deduce that

$$N(T) = \frac{1}{\pi}\,\Im m \int_{\mathcal{L}} \frac{\xi'(s)}{\xi(s)}\,\mathrm{d}s = \frac{1}{\pi}\big[\arg \xi(s)\big]_{\mathcal{L}} \tag{44}$$

where $\mathcal{L}$ is the polygonal line $2, 2+iT, \frac{1}{2}+iT$. Now, we have

$$\frac{\xi'(s)}{\xi(s)} = \frac{1}{s} + \frac{1}{s-1} - \tfrac{1}{2}\log\pi + \frac{\Gamma'(\frac{1}{2}s)}{2\Gamma(\frac{1}{2}s)} + \frac{\zeta'(s)}{\zeta(s)}. \tag{45}$$

The contribution to (44) of the first two terms is clearly $O(1)$. That of the third is exactly

$$-\frac{T}{2\pi}\log\pi.$$

By Stirling's formula (40), we have furthermore

$$\begin{aligned}\frac{1}{\pi}\Im m\int_{\mathcal{L}}\frac{\Gamma'(\frac{1}{2}s)}{\Gamma(\frac{1}{2}s)}\frac{\mathrm{d}s}{2} &= \frac{1}{\pi}\Im m\,\log\Gamma\left(\tfrac{1}{4}+\tfrac{1}{2}iT\right)\\ &= \frac{T}{2\pi}\log\frac{T}{2}-\frac{T}{2\pi}+O(1).\end{aligned}$$

Thus, it only remains to prove that

$$\Im m\int_{\mathcal{L}}\frac{\zeta'(s)}{\zeta(s)}\,\mathrm{d}s \ll \log T. \tag{46}$$

Since the vertical segment of $\mathcal{L}$ is located in the half-plane of absolute convergence, we can integrate termwise. Hence

$$\left|\int_2^{2+iT}\frac{\zeta'(s)}{\zeta(s)}\,\mathrm{d}s\right| \leq 2\sum_{n=2}^{\infty}\frac{\Lambda(n)}{\log n}n^{-2} < \infty.$$

In order to deal with the horizontal segment, we appeal to Corollary 11.2 with $F(s) = \zeta(2+iT+s)/\zeta(2+iT)$ and $R=4$, $r=3/2$. The numerator is of the order of a power of T, and the denominator satisfies

$$|\zeta(2+iT)| \geq 1-\sum_{n=2}^{\infty}n^{-2} = 2-\pi^2/6 > 0.$$

Letting Z denote the finite sequence of zeros of $\zeta(s)$ (counted with multiplicity) in the disc $|s-2-iT| \leq \frac{3}{2}$, we therefore obtain that, for each s in the horizontal segment $\left[\frac{1}{2}+iT, 2+iT\right]$,

$$\frac{\zeta'(s)}{\zeta(s)}-\sum_{\rho\in Z}\frac{1}{s-\rho} \ll \log T.$$

Theorem 12 implies that

$$\sum_{\rho\in Z}1 \ll \log T$$

and, clearly, for each ρ in Z, we have

$$\left|\Im m\int_{2+iT}^{\frac{1}{2}+iT}\frac{\mathrm{d}s}{s-\rho}\right| \leq \pi.$$

This implies (46), and completes the proof of Theorem 13.

§ 3.8 Expansion as a Hadamard product

We can now make explicit the Hadamard expansion in terms of the zeros for the entire function $\xi(s)$ (and consequently for $\zeta(s)$). As will be clear from the proof, only Theorem 12, which locally bounds the number of non-trivial zeros, is needed.

Theorem 14 (Hadamard product formula). *For suitable constants a, b, we have*

$$\xi(s) = e^{as} \prod_{\rho} (1 - s/\rho) e^{s/\rho} \qquad (s \in \mathbb{C}) \tag{47}$$

and

$$\zeta(s) = \frac{e^{bs}}{2(s-1)} \Gamma\big(\tfrac{1}{2}\, s + 1\big)^{-1} \prod_{\rho} (1 - s/\rho) e^{s/\rho} \qquad (s \neq 1). \tag{48}$$

Remark. In Exercise 2 we outline a proof of the formula

$$b = \log(2\pi) - 1 - \tfrac{1}{2}\, \gamma. \tag{49}$$

Proof. By (42), we have

$$\sum_{|\gamma| \leq T} |\rho|^{-2} \leq 2 \int_0^T t^{-2}\, \mathrm{d}N(t) \ll \sum_{k \leq T} k^{-2} \log(2k) \ll 1.$$

This implies the convergence of the infinite product

$$P(s) := \prod_{\rho} (1 - s/\rho) e^{s/\rho} \qquad (s \in \mathbb{C})$$

where the general term is, for each s, of the form $1 + O(|\rho|^{-2})$. $P(s)$ is therefore an entire function of s having the same zeros as $\xi(s)$ and, since $P(0) = \xi(0) = 1$, we can find a holomorphic determination $F(s)$ of $\log\big(\xi(s)/P(s)\big)$ taking the value 0 at $s = 0$. We shall see that

$$\max_{0 \leq \theta \leq 2\pi} \Re e\, F(Re^{i\theta}) \ll R(\log R)^2 \tag{50}$$

for a sequence of values of R tending to infinity. By the Borel–Carathéodory theorem, it follows from (50) that F is a linear function of s, which is precisely the required result.

We now prove that (50) holds for all R satisfying

$$\min_{\rho} \big|R - |\rho|\big| \geq (C \log R)^{-1} \tag{51}$$

where C is a suitable positive constant. Each interval $[k, k+2]$ with $k \geq 2$, contains at least one such R. Indeed, if C is sufficiently large, we have

$$\operatorname{card}\{\rho : k \leq |\rho| \leq k+2\} \leq 2\Big(N(k+2) - N(k-1)\Big) < J,$$

with $J := \big[C \log k\big]$. By the pigeonhole principle, there exists at least one interval $[k + 2j/J, k + (2j+2)/J]$ with $0 \leq j < J$, containing no $|\rho|$. We can thus choose $R = k + (2j+1)/J$.

Consider some R satisfying (51). For $|s| = R$, we have

$$\begin{aligned}\log|P(s)| &= \sum_\rho \log\big|(1 - s/\rho)e^{s/\rho}\big| \\ &\geq -\Bigg\{\sum_{|\rho|\leq R/2} \frac{R}{|\rho|} + \sum_{R/2<|\rho|\leq 2R} \Big(2 + \log(2CR\log R)\Big) + \sum_{|\rho|>2R} \frac{R^2}{|\rho|^2}\Bigg\},\end{aligned}$$

where we have used the inequalities

$$\log|(1-z)e^z| \geq \begin{cases} -|z| & (|z| \geq 2) \\ -2 + \log|1-z| & (\frac{1}{2} \leq |z| < 2) \\ -|z|^2 & (|z| < \frac{1}{2}). \end{cases}$$

Using (42), we deduce from the above lower bound that

$$\log(1/|P(s)|) \ll R(\log R)^2.$$

We omit the details, which are easy. Now, by (41), we also have

$$\log|\xi(s)| \ll R \log R.$$

This implies (50) and so completes the proof.

We conclude this section with a remark. The bound

$$N(T+1) - N(T) \ll \log T$$

sufficed to prove the product formula (47) for $\xi(s)$. Now this formula easily implies that $N(T) \to \infty$ as $T \to \infty$. Indeed, otherwise, $\xi(s)$ would have the form

$$\xi(s) = e^{cs}Q(s)$$

where c is a constant and $Q(s)$ is a polynomial in s. The functional equation $\xi(s) = \xi(1-s)$ would then imply that $c = 0$, but Stirling's formula applied to real, positive s prevents ξ from having polynomial growth. Actually, Titchmarsh (1951, §9.1) provides a simple argument which enables one to deduce the estimate

$$N(T+A) - N(T) \gg 1$$

(for some suitable absolute constant A), merely from the functional equation.

§ 3.9 Zero-free regions

It is clear that the proof of Theorem 9, leading to the assertion that $\zeta(s) \neq 0$ for $\sigma \geq 1$, could yield an explicit region with no zero of $\zeta(s)$ and yet intersecting the half-plane $\sigma < 1$. We provide the details in §4.2, for purpose of information only. Indeed, the results of the previous sections, and particularly the product formula, enable us to obtain rather easily a much larger region.

Theorem 15. *There exists some positive constant c such that $\zeta(s)$ has no zero in the region of the complex plane defined by the inequality*

$$\sigma \geq 1 - c/\log(2 + |\tau|). \tag{52}$$

Proof. The Dirichlet series

$$-\frac{\zeta'(s)}{\zeta(s)} = \sum_{n=1}^{\infty} \Lambda(n)n^{-s}$$

has non-negative coefficients. By Theorem 8, it hence follows that, for all $\sigma > 1$ and all real γ, we have

$$-3\frac{\zeta'(\sigma)}{\zeta(\sigma)} - 4\Re e\,\frac{\zeta'(\sigma + i\gamma)}{\zeta(\sigma + i\gamma)} - \Re e\,\frac{\zeta'(\sigma + 2i\gamma)}{\zeta(\sigma + 2i\gamma)} \geq 0. \tag{53}$$

We shall obtain the stated result by bounding each of the three terms on the left-hand side from above, for γ the ordinate of a zero $\rho = \beta + i\gamma$ of $\zeta(s)$.

In the first place, we have

$$-\frac{\zeta'(\sigma)}{\zeta(\sigma)} = \frac{1}{\sigma - 1} + O(1).$$

Then, logarithmically differentiating the product formula (48), for $s \neq 1$, $s \neq \rho$, we obtain

$$-\frac{\zeta'(s)}{\zeta(s)} = -b + \frac{1}{s-1} + \frac{\Gamma'(\tfrac{1}{2}s + 1)}{2\Gamma(\tfrac{1}{2}s + 1)} - \sum_{\rho}\Big(\frac{1}{\rho} + \frac{1}{s - \rho}\Big). \tag{54}$$

Estimating the term involving Γ by Stirling's formula (13), and noting that, for $\sigma > 1$, the numbers ρ and $(s - \rho)$ have positive real part, we have

$$-\Re e\,\frac{\zeta'(s)}{\zeta(s)} \leq O(\log|\tau|) \qquad (\sigma > 1,\ |\tau| \geq 2),$$

and also, taking into account the contribution of the zero $\beta + i\gamma$,

$$-\Re e\, \frac{\zeta'(\sigma + i\gamma)}{\zeta(\sigma + i\gamma)} \leq O(\log|\gamma|) - \frac{1}{\sigma - \beta}.$$

Substituting these estimates in (53), we then deduce the existence of a constant c_1 such that for $|\gamma| \geq 2$ we have

$$\frac{3}{\sigma - 1} - \frac{4}{\sigma - \beta} \geq -c_1 \log|\gamma|$$

from which we get

$$1 - \beta \geq \frac{1 - c_1(\sigma - 1)\log|\gamma|}{(3/(\sigma - 1)) + c_1 \log|\gamma|}.$$

Choosing $\sigma = 1 + 1/(2c_1 \log|\gamma|)$, it follows that

$$1 - \beta \geq c_2/\log|\gamma|$$

with $c_2 = 1/14c_1$. This implies the stated result, since the conclusion is trivial for $|\tau| < 2$ with sufficiently small c.

§ 3.10 Bounds for ζ'/ζ, $1/\zeta$ and $\log\zeta$

Theorem 15 determines a region in which the three functions in the title of this section are holomorphic. It is naturally desirable to have at hand corresponding explicit upper bounds in the same domain.

Theorem 16. *There exists a positive constant c such that, for $|\tau| \geq 3$ and $\sigma \geq 1 - c/\log|\tau|$, we have*

$$\zeta'(s)/\zeta(s) \ll \log|\tau|, \tag{55}$$

$$1/\zeta(s) \ll \log|\tau|, \tag{56}$$

$$|\log\zeta(s)| \leq \log_2|\tau| + O(1). \tag{57}$$

Proof. We can assume that $\tau > 0$. By Theorem 15, there exists a constant c with $0 < c < 1/16$, such that each non-trivial zero $\rho = \beta + i\gamma$ of $\zeta(s)$ satisfies

$$\beta < 1 - 8c/\log(|\gamma| + 2).$$

We shall see that this implies that

$$\min_{\rho} \Re e \left\{\frac{1}{\rho} + \frac{1}{s - \rho}\right\} \geq 0 \qquad (\tau \geq 4,\ \sigma \geq 1 - 4c/\log\tau). \tag{58}$$

When $|s-\rho| > \frac{1}{2}|\rho|$, the quantity to be minimised may be rewritten in the following form, setting $\theta := 2|s-\rho|/|\rho| \geq 1$,

$$\begin{aligned}\beta|\rho|^{-2} + (\sigma-\beta)|s-\rho|^{-2} &= |s-\rho|^{-2}\left(\tfrac{1}{4}\theta^2\beta + \sigma - \beta\right)\\ &\geq |s-\rho|^{-2}\left(\sigma - \tfrac{3}{4}\right) > 0.\end{aligned}$$

When $|s-\rho| \leq \frac{1}{2}|\rho|$, we have $|\tau-\gamma| \leq \frac{1}{2}(|\gamma|+1)$, so that $|\gamma| \leq 2\tau+2$ and hence

$$\beta < 1 - 8c/\log(2\tau+4) \leq 1 - 4c/\log\tau \leq \sigma.$$

This implies (58).

Substituting in (54), it follows that

$$-\Re e\,\frac{\zeta'(s)}{\zeta(s)} \leq K\log\tau \qquad (\tau \geq 4,\ \sigma \geq 1 - 4c/\log\tau) \tag{59}$$

where K is some suitable absolute constant.

We are now in a position to establish the bound (55). It is clear that we can assume τ to be sufficiently large. Let then $s = \sigma + i\tau$ be a fixed complex number with $\tau \geq 5$, $\sigma \geq 1 - c/\log\tau$. Write $\eta := c/\log\tau$, $s_0 := 1+\eta+i\tau$. Then for each w in the disc $|w| \leq 4\eta$, the point $s_0 + w = \sigma' + i\tau'$ satisfies $\tau' \geq 4$, $\sigma' \geq 1 - 4c/\log\tau'$, and as a result we can write, by (59),

$$-\Re e\,\frac{\zeta'(s_0+w)}{\zeta(s_0+w)} \leq 2K\log\tau.$$

This implies that the function

$$F(w) := \frac{\zeta'(s_0)}{\zeta(s_0)} - \frac{\zeta'(s_0+w)}{\zeta(s_0+w)}$$

satisfies the hypotheses of the Borel–Carathéodory theorem in the disc $|w| \leq 4\eta$, with $A := 2K\log\tau + |\zeta'(s_0)/\zeta(s_0)|$. Since $|s - s_0| \leq 2\eta$, Corollary 11.1 implies that

$$\left|\frac{\zeta'(s)}{\zeta(s)}\right| \leq 4K\log\tau + 3\left|\frac{\zeta'(s_0)}{\zeta(s_0)}\right|.$$

The upper bound (55) follows from this inequality, since

$$\left|\frac{\zeta'(s_0)}{\zeta(s_0)}\right| \leq \sum_{n=1}^{\infty}\Lambda(n)n^{-1-\eta} = \eta^{-1} + O(1) \ll \log\tau.$$

Since the estimate (56) is immediate from (57), it suffices to establish this last inequality. To this end, we use (55) in the form

$$\log\left(\frac{\zeta(s)}{\zeta(s_0)}\right) = \int_{s_0}^{s} \frac{\zeta'(w)}{\zeta(w)}\,dw \ll |s-s_0|\log\tau \ll 1. \tag{60}$$

This implies the required result upon noting that

$$\begin{aligned}|\log\zeta(s_0)| &= \left|\sum_{n=2}^{\infty}\frac{\Lambda(n)}{\log n}n^{-s_0}\right| \leq \sum_{n=2}^{\infty}\frac{\Lambda(n)}{\log n}n^{-1-\eta} = \log\zeta(1+\eta)\\ &= \log(1/\eta)+O(1) = \log_2\tau + O(1).\end{aligned}$$

Notes

§§ **3.2–3.3.** In the proofs of the fundamental Theorems 1 and 2, we have followed the original method of Riemann (1859). Titchmarsh (1951) gives several other proofs for the analytic continuation and the functional equation of $\zeta(s)$. See in particular Exercises 10 and 11.

§ **3.4.** Corollary 5.1 is due to Hardy & Littlewood (1921). In many applications, the fact that the sum over n contains at least $C|\tau|$ terms is prohibitive. Another result of Hardy & Littlewood (1923, 1929) allows one to reduce this number of terms to $O(\sqrt{|\tau|}\,)$. This is the famous *approximate functional equation:*

Theorem 17. *For each $\delta > 0$, and uniformly for $0 \leq \sigma \leq 1$, $x > \delta$, $y > \delta$, $\tau = 2\pi xy$, we have*

$$\zeta(s) = \sum_{n\leq x} n^{-s} + \chi(s)\sum_{n\leq y} n^{s-1} + O\big(x^{-\sigma} + \tau^{\frac{1}{2}-\sigma}y^{\sigma-1}\big)$$

where $\chi(s)$ is defined by (11).

For the proof of this result, its extension to powers of $\zeta(s)$, and numerous related remarks, see chapter 4 of Ivić (1985) and chapter IV of Titchmarsh–Heath-Brown (1986).

Improving on a result of Bombieri & Iwaniec (1986), Huxley (1993b) has shown that the exponent $\frac{1}{6}$ in the bound for $\zeta\left(\frac{1}{2}+i\tau\right)$ can be replaced by $\frac{89}{570}+\varepsilon$ for any $\varepsilon > 0$.

§ **3.5.** The formula (28) is a particular case of Ramanujan's formula

$$\frac{\zeta(s)\zeta(s-a)\zeta(s-b)\zeta(s-a-b)}{\zeta(2s-a-b)} = \sum_{n=1}^{\infty} \sigma_a(n)\sigma_b(n)n^{-s}$$

which is valid for $\sigma > 1+\max\{0, \Re e\, a, \Re e\, b, \Re e\, (a+b)\}$. Titchmarsh (1951) gives a proof of this result using Euler products. An alternative approach consists in noting that the left-hand side is the series associated with the arithmetic function

$$\begin{aligned}\sum_{h^2 k\ell m|n} h^{a+b}\mu(h)k^a\ell^b m^{a+b} &= \sum_{KLm|n} K^aL^bm^{a+b} \sum_{h|(K,L)} \mu(h)\\ &= \sum_{\substack{KLm|n\\(K,L)=1}} K^aL^bm^{a+b} = \sum_{d|n}\sum_{[\kappa,\lambda]=d} \kappa^a\lambda^b\\ &= \sum_{\kappa,\lambda|n} \kappa^a\lambda^b = \sigma_a(n)\sigma_b(n).\end{aligned}$$

§ **3.7.** Theorem 13, conjectured by Riemann in his memoir of 1859, was proved by von Mangoldt in 1895. Let $S(T) = (1/\pi)\big[\arg\zeta(s)\big]_{\mathcal{L}}$ where $\mathcal{L}$ is the polygonal line $\mathcal{L}$ joining 2, $2+iT$, $\frac{1}{2}+iT$. In his book, Titchmarsh (1951, §9.2) shows that

$$N(T) = \frac{T}{2\pi}\log\frac{T}{2\pi} - \frac{T}{2\pi} + \tfrac{7}{8} + S(T) + O\Big(\frac{1}{T}\Big).$$

§ **3.9.** Note that, in general, an upper bound for $\zeta(s)$ on the line $\sigma = 1$ provides, via Corollary 11.2, a corresponding upper bound for $-\Re e\big(\zeta'(s)/\zeta(s)\big)$ in terms of the abscissae of the zeros, by an argument analogous to that used in the proof of Theorem 15. Thus we can state that, conceptually, it is equivalent to bound $|\zeta(1+i\tau)|$ and to determine a zero-free region for $\zeta(s)$. Korobov (1958) and Vinogradov (1958) established the upper bound

$$\zeta(s) \ll \big(1+\tau^{A(1-\sigma)^{3/2}}\big)(\log\tau)^{2/3} \qquad (\sigma \geq 0,\ \tau \geq 2) \tag{61}$$

from which follows the best zero-free region known to date, namely

$$\sigma \geq 1 - C(\log\tau)^{-2/3}(\log_2\tau)^{-1/3} \qquad (\tau \geq 3). \tag{62}$$

For a detailed proof of (61) and the deduction of (62), see Ivić (1985), chapter 6.

§ **3.10.** The zero-free region (62) stated above easily implies improvements on the estimates in Theorem 16. One obtains

$$\zeta'(s)/\zeta(s) \ll (\log \tau)^{2/3}(\log_2 \tau)^{1/3},$$

$$1/\zeta(s) \ll (\log \tau)^{2/3}(\log_2 \tau)^{1/3},$$

$$|\log \zeta(s)| \le \tfrac{2}{3}\log_2 \tau + \tfrac{1}{3}\log_3 \tau + O(1),$$

for s satisfying (62).

Exercises

1. *Complex Stirling formula.*

We assume the Stirling formula for $N!$ (cf. Exercise I.0.3) and the infinite product expansion (cf., for example, Titchmarsh (1939), § 4.41)

$$\Gamma(z) = z^{-1}e^{-\gamma z}\prod_{n=1}^{\infty}\Big(1+\frac{z}{n}\Big)^{-1}e^{z/n}.$$

(a) Using the Euler–Maclaurin summation formula for $f(t) := \log(t+z)$, show that one has for $N \ge 1$, $z \in \mathbb{C}\smallsetminus\mathbb{R}^-$,

$$\sum_{n=1}^{N}\log\Big(1+\frac{z}{n}\Big) = z(1+\log N) - \big(z+\tfrac{1}{2}\big)\log z - \tfrac{1}{2}\log(2\pi) + \int_0^\infty \frac{B_1(t)}{t+z}\,\mathrm{d}t + R_N(z)$$

with $\lim_{N\to\infty} R_N(z) = 0$.

(b) Deduce the complex Stirling formula

$$\log\Gamma(z) = \big(z-\tfrac{1}{2}\big)\log z - z + \tfrac{1}{2}\log(2\pi) - \int_0^\infty \frac{B_1(t)}{t+z}\,\mathrm{d}t.$$

2. *Computation of the constant b in Theorem 14.*

(a) Show that $\Gamma'(1) = -\gamma$ by using any of the following: the infinite product expansion for $\Gamma(z)$, the complex Stirling formula, or the definition of $\Gamma(z)$ in integral form.

(b) Show that $\zeta'(0) = -\frac{1}{2}\log(2\pi)$ by considering a suitable Taylor approximation, in a neighbourhood of $s = 1$, of each side of the functional equation for $\zeta(s)$.

(c) Show that the constant b such that

$$\zeta(s) = e^{bs}\{2(s-1)\Gamma(\tfrac{1}{2}s+1)\}^{-1}\prod_{\rho}(1-s/\rho)e^{s/\rho}$$

has value $b = \log(2\pi) - 1 - \frac{1}{2}\gamma$. [*One may calculate the logarithmic derivative of* $\zeta(s)$ *at* $s = 0$.]

3. *Dirichlet L-functions.*

Let χ be a Dirichlet character to the modulus q, i.e. a multiplicative homomorphism of the group $(\mathbb{Z}/q\mathbb{Z})^*$ into that of the complex numbers of modulus 1. Extend χ to $\mathbb{N}^*$ by setting $\chi(n) := \chi(a)$ if $n \equiv a \pmod q$, $(n,q) = 1$, and $\chi(n) := 0$ if $(n,q) > 1$ [a runs through $(\mathbb{Z}/q\mathbb{Z})^*$]. We say that χ is *primitive* if there does not exist any proper divisor q^* of q such that χ coincides with a character modulo q^* on the set of integers prime to q.

(a) Show that χ defines a completely multiplicative arithmetic function.

(b) Establish the orthogonality relations

$$\varphi(q)^{-1}\sum_{\chi}\overline{\chi(m)}\chi(n) = \begin{cases} 1, & \text{if } n \equiv m \pmod q,\ (m,q) = 1, \\ 0, & \text{otherwise,} \end{cases}$$

where the χ-sum is taken over all Dirichlet characters to the modulus q.

(c) For each χ, one defines the Dirichlet series

$$L(s,\chi) = \sum_{n=1}^{\infty}\chi(n)n^{-s}.$$

Show that $L(s,\chi)$ can be expanded as an Euler product for $\sigma > 1$.

(d) Let χ_0 denote the *principal* character, i.e. $\chi_0(n) = 1$ for all n with $(n,q) = 1$. Show that

$$L(s,\chi_0) = \prod_{p|q}(1-p^{-s})\zeta(s).$$

(e) Show that for $\chi \neq \chi_0$, $L(s,\chi)$ is convergent for $\sigma > 0$.

(f) Show that $L(s,\chi)$ can be analytically continued as a meromorphic function in the whole complex plane.

(g) Let χ be a primitive character to the modulus q. Define $\alpha := \alpha(\chi) = 1$ (if $\chi(-1) = -1$), $\alpha := 0$ (if $\chi(-1) = 1$). Show that $L(s, \chi)$ satisfies the functional equation

$$\xi(s, \chi) = E(\chi)\xi(1 - s, \overline{\chi})$$

where

$$\xi(s, \chi) := \left(\frac{\pi}{q}\right)^{-\frac{1}{2}(s+\alpha)} \Gamma\left(\frac{s+\alpha}{2}\right) L(s, \chi), \ E(\chi) := i^{3\alpha} q^{-1/2} \sum_{h \,(\mathrm{mod}\, q)} \chi(h) e(h/q).$$

4. Let $\tau_k(n)$ denote the number of solutions in integers $m_1, m_2, \ldots, m_k \geq 1$ of the equation $n = m_1 m_2 \ldots m_k$.

(a) Show that $\sum_{n=1}^{\infty} \tau_k(n) n^{-s} = \zeta(s)^k$.

(b) Show that, for each $k \geq 1$, there exists some $\delta_k > 0$, which can be calculated explicitly, such that

$$\sum_{n \leq x} \tau_k(n) = x P_{k-1}(\log x) + O_\varepsilon(x^{1-\delta_k+\varepsilon}),$$

where P_{k-1} denotes a polynomial of degree $k-1$. Determine the coefficients for $k = 1, 2, 3$. What value for δ_k can be obtained assuming Lindelöf's hypothesis?

5. Show that, for $s = 1 + i\tau$, one has

$$\sum_{n \leq x} \mu(n) n^{-s} = \frac{1}{2\pi i} \int_{\kappa - iT}^{\kappa + iT} \zeta(s+w)^{-1} x^w w^{-1} \, dw + O(T^{-1} \log x + x^{-1}),$$

uniformly for $x \geq 2$, $T \geq 1$, and with $\kappa = 1/\log x$. Using the calculations in §3.10, deduce the estimate

$$\sum_{n \leq x} \mu(n) n^{-s} = \zeta(s)^{-1} + O_s\big(e^{-c\sqrt{(\log x)}}\big) \qquad (\sigma = 1, \ x > 1).$$

6. Use the effective Perron formula and some estimates from §3.10 to show that

$$\sum_{n \leq x} \Lambda(n) n^{-1-i\tau} = \begin{cases} -\dfrac{\zeta'(1+i\tau)}{\zeta(1+i\tau)} - \dfrac{x^{-i\tau}}{i\tau} + O_\tau\big(e^{-c\sqrt{(\log x)}}\big) & (\tau \neq 0) \\ \log x - \gamma + O\big(e^{-c\sqrt{(\log x)}}\big) & (\tau = 0). \end{cases}$$

Deduce that the following expressions hold for $\sigma = 1$, $\tau \neq 0$:

$$\log \zeta(s) = \sum_{n=2}^{\infty} \frac{\Lambda(n)}{\log n} n^{-s}, \qquad \zeta(s) = \prod_p (1 - p^{-s})^{-1}.$$

7. Let $\tau^*(n)$ be the number of odd divisors of an integer n. Determine the Dirichlet series associated with $\tau^*(n)$ and use complex integration to deduce the asymptotic expansion

$$T^*(x) := \sum_{n\leq x}\tau^*(n) = \tfrac{1}{2}\,x\{\log(2x) + 2\gamma - 1\} + O_\varepsilon(x^{\frac{1}{2}+\varepsilon}). \tag{63}$$

Set $T(x) := \sum_{n\leq x}\tau(n)$. Show that

$$T^*(x) = T(x) - T(\tfrac{1}{2}\,x)$$

and deduce an improvement in the error term of (63).

8. Let z be a complex number of modulus ≤ 1. Using Euler products, show that the Dirichlet series $F(s,z) := \sum_{n=1}^{\infty} z^{\Omega(n)}n^{-s}$ converges for $\sigma > 1$ and has an analytic continuation given by the convergent infinite product

$$F(s,z) = \prod_{k=1}^{\infty}\zeta(ks)^{a_k(z)},$$

with $a_k(z) := k^{-1}\sum_{d|k}\mu(d)z^{k/d}$, in any simply connected open subset of the half-plane $\sigma > 0$ which excludes all values of s such that ks ($k = 1, 2, \ldots$) is either a pole or a zero of the zeta function.

9. Show that the following asymptotic formulae are equivalent to Lindelöf's hypothesis:

(a) $\displaystyle\int_1^T |\zeta(\tfrac{1}{2} + i\tau)|^{2k}\mathrm{d}\tau \ll_{\varepsilon,k} T^{1+\varepsilon} \qquad (\varepsilon > 0,\ k = 1, 2, \ldots)$

(b) $\displaystyle\int_1^T |\zeta(\sigma + i\tau)|^{2k}\mathrm{d}\tau \ll_{\sigma,\varepsilon,k} T^{1+\varepsilon} \qquad (\sigma > \tfrac{1}{2},\ \varepsilon > 0,\ k = 1, 2, \ldots)$

(c) $\displaystyle\frac{1}{T}\int_1^T |\zeta(\sigma + i\tau)|^{2k}\mathrm{d}\tau \sim \sum_{n=1}^{\infty}\frac{\tau_k(n)^2}{n^{2\sigma}} \qquad (\sigma > \tfrac{1}{2},\ k = 1, 2, \ldots)$

[*cf. Titchmarsh (1951), chapter XIII*].

10. *The analytic continuation by the Euler–Maclaurin formula.*

Applying the Euler–Maclaurin formula to $f(t) := t^{-s}$, show that one has for $s > 1$, with the notation of Chapter I.0,

$$\zeta(s) = \frac{s}{s-1} + \sum_{r=0}^{k}\frac{B_{r+1}}{r+1}\binom{s+r-1}{r} - \binom{s+k}{k+1}\int_1^{\infty} B_{k+1}(t)t^{-s-k-1}\,\mathrm{d}t.$$

Deduce from this an alternative proof of Theorems 1 and 2.

11. *The functional equation by the theta function.*
Let $\theta(x) := \sum_{n\in\mathbb{Z}} e^{-\pi n^2 x}$ $(x > 0)$.

(a) Applying Poisson's formula to $f(t) := e^{-\pi t^2 x}$, show that

$$\theta(1/x) = \sqrt{x}\,\theta(x) \qquad (x > 0).$$

(b) Make the change of variables $x = \pi n^2 y$ in the integral $\Gamma(s/2) = \int_0^\infty e^{-x}x^{s/2-1}\,dx$ and deduce that one has for $\sigma > 1$

$$\zeta(s)\Gamma(s/2)\pi^{-s/2} = \int_0^\infty \psi(x)x^{s/2-1}\,dx, \tag{64}$$

where $\psi(x) := \frac{1}{2}\left(\theta(x) - 1\right)$.

(c) Show that the right-hand side of (64) also equals

$$\frac{1}{s(s-1)} + \int_1^\infty \left(x^{-(s+1)/2} + x^{s/2-1}\right)\psi(x)\,dx,$$

where the last integral defines an entire function of s which is invariant under the transformation $s \mapsto 1-s$. Deduce an alternative proof of Theorems 3 and 4.

12. Prove Corollary 5.1 without appealing to van der Corput's theory by employing the following strategy to obtain formula (20): apply the Euler–Maclaurin formula of order 0 to $f(t) := t^{-i\tau}$, expand $B_1(t)$ as a Fourier series, integrate termwise and estimate each of the resulting integrals by the second mean value theorem.

II.4

The prime number theorem and the Riemann hypothesis

§ 4.1 The prime number theorem

The information obtained in the previous chapter about the function $\zeta(s)$ allows us easily to obtain the prime number theorem in its classical form.

Theorem 1. *There exists a positive number c such that we have, as $x \to \infty$,*

$$\psi(x) = x + O\big(xe^{-c\sqrt{(\log x)}}\big), \tag{1}$$

$$\pi(x) = \mathrm{li}(x) + O\big(xe^{-c\sqrt{(\log x)}}\big). \tag{2}$$

Proof. The second formula follows easily from the first by partial summation. Let us prove (1). Since $\Lambda(n) \leq \log n$ and $|\zeta'(1+\sigma)/\zeta(1+\sigma)| \ll 1/\sigma$ $(\sigma > 0)$, the second effective Perron formula (cf. §2.1) allows us to write, for $x \geq 2$, $T \geq 2$,

$$\psi(x) = -\frac{1}{2\pi i}\int_{\kappa-iT}^{\kappa+iT} \frac{\zeta'(s)}{\zeta(s)}\frac{x^s}{s}\,\mathrm{d}s + O\left(\log x\left(1+\frac{x\log T}{T}\right)\right) \tag{3}$$

with $\kappa := 1 + 1/\log x$. By Theorem 3.15, there exists some positive constant c_0 such that the point $s = 1$ is the only singularity of the integrand in the rectangle $|\tau| \leq T$, $1 - c_0/\log T \leq \sigma \leq \kappa$. Since the residue at $s = 1$ equals x, we can write

$$-\frac{1}{2\pi i}\int_{\kappa-iT}^{\kappa+iT} \frac{\zeta'(s)}{\zeta(s)}\frac{x^s}{s}\,\mathrm{d}s = x - \frac{1}{2\pi i}\int_{\mathcal{G}} \frac{\zeta'(s)}{\zeta(s)}\frac{x^s}{s}\,\mathrm{d}s \tag{4}$$

where $\mathcal{G}$ is the polygonal line $\kappa - iT$, $1 - c_0/\log T - iT$, $1 - c_0/\log T + iT$, $\kappa + iT$. The upper bound from Theorem 3.16, namely

$$\frac{\zeta'(s)}{\zeta(s)} \ll \log T \qquad (s \in \mathcal{G}),$$

then implies that the contribution to the integral of the horizontal segments of $\mathcal{G}$ is $\ll x(\log T)/T$ whereas that of the vertical segment is

$$\ll x\exp\Big\{-c_0\frac{\log x}{\log T}\Big\}(\log T)^2.$$

Choosing $T := \exp\sqrt{(c_0 \log x)}$, we see that the integral over $\mathcal{G}$ is of order at most that of the remainder term of (1) provided $c < \sqrt{c_0}$. Substituting (4) in (3) then implies the desired result.

§ 4.2 Minimal hypotheses

It is conceptually interesting to investigate what are, in the theory of the zeta function, the minimal pieces of information which lead to a proof of the prime number theorem.

The fundamental (but easy) inequality

$$(5) \qquad \zeta(\sigma)^3|\zeta(\sigma+i\tau)|^4\,|\zeta(\sigma+2i\tau)| \geq 1 \qquad (\sigma > 1),$$

due to de La Vallée–Poussin (§3.5), led us to establish that $\zeta(s)$ is non-zero for $\sigma = 1$. Actually, it is almost sufficient on its own to yield the prime number theorem. Indeed, using the elementary upper bounds (cf. §3.4)

$$(6) \qquad \zeta^{(k)}(s) \ll (\log|\tau|)^{k+1} \qquad (|\tau| \geq 2,\ \sigma \geq 1 - c/\log|\tau|)$$

for $k = 0$ and 1, we can employ (5) to determine effectively a zero-free region of $\zeta(s)$ which intersects the half-plane $\sigma < 1$.

Let $s = \sigma + i\tau$, $0 < \eta < c/\log|\tau|$, $s_0 = 1 + \eta + i\tau$; then for $\sigma \geq 1 - \eta$ we have

$$|\zeta(s) - \zeta(s_0)| = \left|\int_{s_0}^{s} \zeta'(w)\,\mathrm{d}w\right| \leq C_0\eta(\log|\tau|)^2.$$

Now (5) and (6) imply

$$|\zeta(s_0)|^4 \geq C_1\eta^3(\log|\tau|)^{-1},$$

from which it follows that

$$|\zeta(s)| \geq C_1^{1/4}\eta^{3/4}(\log|\tau|)^{-1/4} - C_0\eta(\log|\tau|)^2.$$

Choosing η optimally, we obtain

$$(7) \qquad |\zeta(s)| \gg (\log|\tau|)^{-7} \qquad \left(|\tau| \geq 2,\ \sigma \geq 1 - C_2(\log|\tau|)^{-9}\right),$$

where C_2 is some suitable positive constant.

We leave the reader the easy task of checking that the proof of Theorem 1 given in the previous section can be adapted to these weakened assumptions. The sole modification resides in the choice of the parameter T, for which the optimal value becomes

$$T := \exp\left\{c_3(\log x)^{1/10}\right\}.$$

We thus obtain

$$\psi(x) = x + O\Big(x \exp\big\{ -c(\log x)^{1/10}\big\}\Big).$$

It is remarkable that *one can equally well arrive at the prime number theorem* (in its weakest form, i.e. $\psi(x) \sim x$) *using only the properties of* $\zeta(s)$ *in the half-plane* $\sigma \geq 1$.

Indeed, let us apply Perron's formula (2.10) to the Dirichlet series

$$F(s) = -\frac{\zeta'(s)}{\zeta(s)} - \zeta(s) = \sum_{n=1}^{\infty} (\Lambda(n) - 1)n^{-s}.$$

We obtain, for each $\kappa > 1$,

$$\int_0^x \big(\psi(t) - [t]\big)\,\mathrm{d}t = \frac{x^{\kappa+1}}{2\pi}\int_{-\infty}^{+\infty} \frac{F(\kappa + i\tau)x^{i\tau}}{(\kappa + i\tau)(\kappa + 1 + i\tau)}\,\mathrm{d}\tau. \tag{8}$$

Estimates (6) and (7) show that

$$F(\kappa + i\tau) \ll (\log(2 + |\tau|))^9 \qquad (\kappa \geq 1), \tag{9}$$

so that the theorem of dominated convergence applies to the integral occurring on the right-hand side of (8). It follows that

$$\int_0^x \big(\psi(t) - [t]\big)\,\mathrm{d}t = x^2 J(x)$$

with

$$J(x) := \frac{1}{2\pi}\int_{-\infty}^{+\infty} \frac{F(1 + i\tau)x^{i\tau}}{(1 + i\tau)(2 + i\tau)}\,\mathrm{d}\tau.$$

Thus $J(x)$ is the Fourier transform at the point $(-\log x)$ of an integrable function. By the Riemann–Lebesgue lemma, we then deduce that $J(x) = o(1)$ (as $x \to \infty$), from which it we derive that

$$\int_0^x \psi(t)\,\mathrm{d}t = \tfrac{1}{2}\,x^2 + x^2\varepsilon(x)$$

with $\varepsilon(x) = J(x) + O(1/x) = o(1)$.

Let us define $\eta(x) := \max_{\frac{1}{2}x \leq y \leq \frac{3}{2}x} |\varepsilon(y)|$. For all h, $0 < h \leq \frac{1}{2}x$, we can write

$$x - \tfrac{1}{2}\,h - \frac{2}{h}\eta(x)x^2 \leq \frac{1}{h}\int_{x-h}^{x} \psi(t)\,\mathrm{d}t \leq \psi(x)$$

$$\leq \frac{1}{h}\int_{x}^{x+h} \psi(t)\,\mathrm{d}t \leq x + \tfrac{1}{2}\,h + \frac{4}{h}x^2\eta(x).$$

Choosing $h := x\sqrt{\eta(x)}$, we obtain

$$\psi(x) = x\{1 + O(\sqrt{\eta(x)})\} = x(1 + o(1)). \tag{10}$$

Thus the prime number theorem follows easily from the fact that $\zeta(s)$ does not vanish for $\sigma = 1$ and from any "reasonable" upper bound of $F(s)$ for $\sigma \geq 1$. Indeed, it would be posssible to replace, without additional difficulty, the estimate (9) in the above proof by

$$F(s) \ll 1 + |\tau|^{1-\delta} \qquad (\sigma \geq 1).$$

We shall see in Chapter 7 that the *Wiener–Ikehara Tauberian theorem* yields (10) assuming only that $\zeta(1 + i\tau) \neq 0$, without needing *any* upper bound of $F(s)$.

The absence of zeros of $\zeta(s)$ on the line $\sigma = 1$ thus constitutes a *minimal* hypothesis for the prime number theorem, of which it is also an easy implication.

Indeed, let us assume that $\psi(x) \sim x$ $(x \to \infty)$ and that $s_0 = 1 + i\tau$ is a zero of order $m \geq 1$ of $\zeta(s)$. In these circumstances, we have

$$\lim_{\sigma \to 1+} (\sigma - 1)\frac{\zeta'(\sigma + i\tau)}{\zeta(\sigma + i\tau)} = m \tag{11}$$

whereas the following formula, proved by partial integration,

$$-\frac{\zeta'(s)}{\zeta(s)} = \frac{s}{s-1} + s\int_1^\infty (\psi(t) - t)t^{-s-1}\,dt$$

implies that

$$\begin{aligned}(\sigma - 1)\left|\frac{\zeta'(\sigma + i\tau)}{\zeta(\sigma + i\tau)}\right| &\leq \left|\frac{\sigma + i\tau}{\tau}\right|(\sigma - 1) + |\sigma + i\tau|(\sigma - 1)\int_0^\infty o(t)t^{-\sigma-1}\,dt \\ &= o(1) \qquad (\sigma \to 1+).\end{aligned}$$

This contradicts (11).

§ 4.3 The Riemann hypothesis

In 1859 Riemann conjectured that all non-trivial zeros of $\zeta(s)$ lie on the line $\sigma = \frac{1}{2}$. Having stated that the number of *complex* roots of the equation

$$\zeta(\tfrac{1}{2} + i\tau) = 0 \qquad (0 \leq \Re e\,\tau \leq T) \tag{12}$$

is "*approximately equal to* $\frac{T}{2\pi}\log\frac{T}{2\pi} - \frac{T}{2\pi}$" (cf. Theorem 3.13), he writes: "*indeed, one encounters between these limits a number of real roots approximately equal to this, and it is very probable that all the roots are real*". However, a note has been found in Riemann's personal papers, specifying that he has "*not yet completed the proof*" of the first point—cf. Riemann (1859) pp. 168–169 and 175.

Neither of Riemann's two claims has yet been proved or disproved. In 1914 Hardy established that $\zeta(s)$ *has infinitely many zeros on the critical line* $\sigma = \frac{1}{2}$, and in 1942 Selberg showed that a positive proportion of the zeros is definitely located on the critical line, in other words that, for some $c > 0$,

$$N_0(T) := \operatorname{card}\{\tau : 0 \leq \tau \leq T,\ \zeta(\tfrac{1}{2} + i\tau) = 0\} \geq cN(T) \quad (T \to \infty). \tag{13}$$

Today explicit values of c are available, following the work of Levinson (1974), who showed that for T sufficiently large one can take $c = 0.342$.

The statement that all the roots of (12) are real (that is to say that $N_0(T) = N(T)$) is known under the name of the *Riemann hypothesis*. It is one of the most famous conjectures of mathematics. It has profound implications throughout analytic number theory. In this section we shall develop two of them.

Theorem 2. *The Riemann hypothesis implies that of Lindelöf. More precisely, if all the non-trivial zeros of* $\zeta(s)$ *have real part equal to* $\frac{1}{2}$, *then we have*

$$\log \zeta(s) \ll_\varepsilon (\log|\tau|)^{2-2\sigma+\varepsilon} \qquad (\tfrac{1}{2} < \sigma \leq 1,\ |\tau| \geq 2). \tag{14}$$

The proof uses the following classical result.

Lemma 2.1 (Hadamard's three circles lemma). *Let* $F(s)$ *be a holomorphic function in the annulus* $R_1 \leq |s| \leq R_2$. *Then the function*

$$M(r) := \max_{|s|=r} |F(s)| \qquad (R_1 \leq r \leq R_2)$$

is a log-convex function of $\log r$ *in the interval* $R_1 \leq r \leq R_2$.

Proof of Lemma. For $R_1 < r < R_2$ the maximum modulus principle applied to $s^m F(s)^n$ immediately implies that

$$r^m M(r)^n \leq \frac{R_1^m M(R_1)^n}{r - R_1} + \frac{R_2^m M(R_2)^n}{R_2 - r}.$$

Let α be such that $R_1^\alpha M(R_1) = R_2^\alpha M(R_2)$. In the above inequality, let m and n tend to infinity in such a way that $m/n \to \alpha$. We then obtain

$$r^\alpha M(r) \leq R_1^\alpha M(R_1),$$

and hence, on replacing α by its actual value,

$$\log M(r) \leq \frac{\log(R_2/r)}{\log(R_2/R_1)} \log M(R_1) + \frac{\log(r/R_1)}{\log(R_2/R_1)} \log M(R_2). \tag{15}$$

This establishes the desired property.

Proof of Theorem 2. Observe first of all that the Riemann hypothesis implies that, for sufficiently large τ, the function

$$F(w) := \log \frac{\zeta(s_0 + w)}{\zeta(s_0)} \qquad (s_0 := 2 + i\tau)$$

is holomorphic in the disc $|w| < \frac{3}{2}$. Since it follows from Theorem 3.7 (for example) that, for all positive δ,

$$\Re e\, F(w) \leq O(\log \tau) \qquad (|w| \leq \tfrac{3}{2} - \delta),$$

we readily deduce from the Borel–Carathéodory theorem (cf. Corollary 3.11.1) that we have

$$\log \zeta(s) \ll_\sigma \log \tau \qquad (\sigma > \tfrac{1}{2},\ \tau \geq 2). \tag{16}$$

Then, let $s := \sigma + i\tau$ with $\sigma > \frac{1}{2}$, $\tau \geq \tau_0$, and consider parameters σ_1, ε, such that $\sigma_1 > 1$, $0 < \varepsilon < \min\{\sigma_1 - 1, \sigma - \frac{1}{2}\}$. Applying the three circles lemma to $\log \zeta(\sigma_1 + i\tau + w)$ for

$$R_1 := \sigma_1 - 1 - \varepsilon, \quad r := \sigma_1 - \sigma, \quad R_2 := \sigma_1 - \tfrac{1}{2} - \varepsilon$$

(so that, on these three circles, the point $\sigma_1 + i\tau + w$ passes respectively through $1 + \varepsilon + i\tau$, s, $\frac{1}{2} + \varepsilon + i\tau$) we obtain

$$|\log \zeta(s)| \leq M_1^{1-\alpha} M_2^{\alpha}$$

with

$$\alpha := \frac{\log(r/R_1)}{\log(R_2/R_1)} = \frac{\log\left(\dfrac{\sigma_1 - \sigma}{\sigma_1 - 1 - \varepsilon}\right)}{\log\left(\dfrac{\sigma_1 - \frac{1}{2} - \varepsilon}{\sigma_1 - 1 - \varepsilon}\right)} \sim 2(1 - \sigma + \varepsilon) \qquad (\sigma_1 \to \infty)$$

and

$$M_1 \leq \sup_{\sigma \geq 1+\varepsilon} |\log \zeta(\sigma + i\tau)| \ll_\varepsilon 1, \qquad M_2 \ll_\varepsilon \log \tau,$$

where the last estimate follows from (16). Letting σ_1 tend to infinity, we obtain the stated result.

Theorem 3. *Let θ be the lower bound of those real numbers ξ such that*

$$\psi(x) = x + O(x^{\xi}).$$

Then we have

$$\theta = \sup_{\rho} \beta \tag{17}$$

where the supremum is taken over all non-trivial zeros $\rho = \beta + i\gamma$ of $\zeta(s)$.

Corollary 3.1. *The Riemann hypothesis is equivalent to*

$$(\forall \varepsilon > 0) \qquad \psi(x) = x + O_{\varepsilon}(x^{1/2+\varepsilon}).$$

Proof of Theorem. Write $R(x) := \psi(x) - x$. For $\sigma > 1$ we have

$$\begin{aligned} -\frac{\zeta'(s)}{\zeta(s)} = \int_{1-}^{\infty} t^{-s}\,\mathrm{d}\psi(t) &= \frac{1}{s-1} + \int_{1-}^{\infty} t^{-s}\,\mathrm{d}R(t) \\ &= \frac{1}{s-1} + 1 + s\int_{1}^{\infty} R(t)t^{-s-1}\,\mathrm{d}t. \end{aligned}$$

By definition of θ, we may write $R(t) \ll_{\varepsilon} t^{\theta+\varepsilon}$, hence the formula above defines an analytic continuation of $\zeta'(s)/\zeta(s)$ as a meromorphic function for $\sigma > \theta$ having $s = 1$ as its sole singularity. In particular, this implies that $\zeta(s)$ does not vanish for $\sigma > \theta$.

Conversely, if $\zeta(s)$ does not vanish for $\sigma > \theta$, the upper bound

$$\log|\zeta(s)| \leq O(\log|\tau|) \qquad (\sigma > \tfrac{1}{2},\ |\tau| \geq 2)$$

and the Borel–Carathéodory theorem imply that

$$\log\zeta(s) \ll_{\sigma} \log|\tau| \qquad (\sigma > \theta,\ |\tau| \geq 2). \tag{18}$$

(We omit the details, the reasoning being identical with that leading to (16).) Cauchy's formula applied with a circle of radius $\delta < \sigma - \theta$ now yields

$$\frac{\zeta'(s)}{\zeta(s)} \ll_{\sigma} \log|\tau| \qquad (\sigma > \theta,\ |\tau| \geq 2). \tag{19}$$

By the effective Perron formula (3) and the residue theorem in the form (4)—choosing now $\mathcal{G}$ to be the polygonal line $\kappa - iT$, $\theta + \delta \pm iT$, $\kappa + iT$—, we deduce that

$$\psi(x) - x \ll \frac{x\log T}{T} + x^{\theta+\delta}(\log T)^2 + \log x\Big(1 + \frac{x\log T}{T}\Big).$$

Selecting $T := x$, this upper bound is $O\big(x^{\theta+2\delta}\big)$. Since δ is arbitrarily small, this yields the desired conclusion.

Notes

§ **4.1.** The best error term known for the prime number theorem is

$$O\big(x\exp\{-c(\log x)^{3/5}(\log_2 x)^{-1/5}\}\big).$$

It is obtained using the zero-free region of Korobov–Vinogradov—cf. Chapter 3, Notes. For a proof of this result see Ellison & Mendès France (1975) or Ivić (1985). Up to now, elementary methods have only been able to yield a weaker bound, namely

$$O_\varepsilon\big(x\exp\{-(\log x)^{(1/6)-\varepsilon}\}\big).$$

On this topic, one can consult the remarkable survey of Diamond (1982).

§ **4.2.** The proof given here of the prime number theorem, using only the properties of $\zeta(s)$ for $\sigma \geq 1$, essentially follows that of Widder (1971), chapter 4.

§ **4.3.** Titchmarsh (1951) gives several proofs of Hardy's theorem and also establishes that of Selberg. Here we limit ourselves to indicating how to prove briefly that

$$N_0(T) \gg \log T \qquad (T \to \infty).$$

The basic idea simply consists of making use of the fact that, if $Z(\tau)$ is a real continuous function and if

$$\int_T^{2T} Z(\tau)\,\mathrm{d}\tau = o\Big(\int_T^{2T} |Z(\tau)|\,\mathrm{d}\tau\Big) \qquad (T \to \infty) \tag{20}$$

then $Z(\tau)$ certainly vanishes on the interval $[T, 2T]$ for sufficiently large T. Titchmarsh's choice is

$$Z(\tau) := \chi\big(\tfrac{1}{2} + i\tau\big)^{-1/2}\zeta\big(\tfrac{1}{2} + i\tau\big)$$

where $\chi(s)$ denotes the "elementary" factor of the functional equation of $\zeta(s)$—cf. formula (3.11). The equation $\zeta(s) = \chi(s)\zeta(1-s)$ shows immediately that

$$\chi(s)\chi(1-s) = 1$$

(which can also be seen by noting that

$$\chi(s) = \pi^{s-\frac{1}{2}}\Gamma\big(\tfrac{1}{2}(1-s)\big)/\Gamma\big(\tfrac{1}{2}s\big)\big),$$

and hence that $\chi(\frac{1}{2}+i\tau)$ has modulus 1. Therefore

$$Z(\tau)=\chi(\tfrac{1}{2}+i\tau)^{-1/2}\zeta(\tfrac{1}{2}-i\tau)\chi(\tfrac{1}{2}+i\tau)=\overline{Z(\tau)},$$

that is, the function $Z(\tau)$ is real when τ is real.

In order to estimate the left-hand side of (20), we use the residue theorem in the form

$$\int_{\mathcal{R}}\chi(s)^{-1/2}\zeta(s)\,\mathrm{d}s=0$$

where $\mathcal{R}$ is the rectangle with vertices $\frac{1}{2}+iT$, $\frac{1}{2}+2iT$, $\frac{5}{4}+2iT$, $\frac{5}{4}+iT$. The integral on the left-hand side of $\mathcal{R}$ has the value

$$i\int_T^{2T}Z(\tau)\,\mathrm{d}\tau$$

and we bound this quantity by estimating the integrals on the other three sides.

The upper bounds of §3.4

$$\chi(s)^{-1}\ll\tau^{\sigma-1/2},\qquad\zeta(s)\ll_\varepsilon\tau^{((1-\sigma)/2)+\varepsilon}\qquad(0\le\sigma\le1)$$

immediately imply that

$$\chi(s)^{-1/2}\zeta(s)\ll_\varepsilon\begin{cases}\tau^{(1/4)+\varepsilon} & (\frac{1}{2}\le\sigma\le1)\\ \tau^{(3/8)+\varepsilon} & (1\le\sigma\le\frac{5}{4})\end{cases}.$$

The integrals on the horizontal segments of $\mathcal{R}$ are therefore $\ll T^{(3/8)+\varepsilon}$.

In order to estimate the contribution of the vertical segment $\left[\frac{5}{4}+iT,\frac{5}{4}+2iT\right]$, we evaluate $\chi(s)$ by the complex Stirling formula (3.13). This contribution turns out to be

$$i\int_T^{2T}\left(\frac{\tau}{2\pi}\right)^{(3/8)+i\tau}e^{-i\tau/2-i\pi/8}\{1+O(\tau^{-1})\}\zeta(\tfrac{5}{4}+i\tau)\,\mathrm{d}\tau.$$

We can treat the remainder term trivially. It gives a quantity $\ll T^{3/8}$. Replacing $\zeta(\frac{5}{4}+i\tau)$ by the series, and interchanging summations, we see that the main term has the value

$$ie^{-i\pi/8}\sum_{n=1}^{\infty}n^{-5/4}\int_T^{2T}\left(\frac{\tau}{2\pi}\right)^{3/8}e^{iF_n(\tau)}\,\mathrm{d}\tau\tag{21}$$

with $F_n(\tau):=\tau\big(\log(\tau/2\pi)-\frac{1}{2}+\log n\big)$. We have

$$F_n''(\tau)\asymp T^{-1}\qquad(T\le\tau\le2T)$$

from which, by Theorem I.6.3,

$$\int_T^\tau e^{iF_n(t)}\,\mathrm{d}t \ll T^{1/2} \qquad (T \le \tau \le 2T).$$

An integration by parts then allows us to deduce from the above that the expression (21) is $\ll T^{7/8}$, and so finally

$$\int_T^{2T} Z(\tau)\,\mathrm{d}\tau \ll_\varepsilon T^{7/8}. \tag{22}$$

Now let us estimate the right-hand side of (20). We have

$$\int_T^{2T} |Z(\tau)|\,\mathrm{d}\tau = \int_T^{2T} \left|\zeta(\tfrac12 + i\tau)\right|\mathrm{d}\tau \ge \left|\int_T^{2T} \zeta(\tfrac12 + i\tau)\,\mathrm{d}\tau\right|.$$

By the residue theorem, we can write

$$i\int_T^{2T} \zeta(\tfrac12 + i\tau)\,\mathrm{d}\tau = \int_{\mathcal{G}} \zeta(s)\,\mathrm{d}s$$

where $\mathcal{G}$ is the polygonal line joining $\frac12 + iT$, $2 + iT$, $2 + 2iT$, $\frac12 + 2iT$. Since $\zeta(s) \ll_\varepsilon T^{(1/4)+\varepsilon}$ on $\mathcal{G}$, the contribution of the horizontal segments is $\ll_\varepsilon T^{(1/4)+\varepsilon}$. The contribution of the vertical segment is

$$i\sum_{n=1}^{\infty} n^{-2}\int_T^{2T} n^{-i\tau}\,\mathrm{d}\tau = i\left\{T + O\Big(\sum_{n=2}^{\infty} n^{-2}(\log n)^{-1}\Big)\right\}.$$

We thus have

$$\int_T^{2T} |Z(\tau)|\,\mathrm{d}\tau \gg T,$$

which establishes (20).

A trivial modification of the argument presented above would give a lower bound of the type

$$N_0(T) \gg T^{(1/4)-\varepsilon}$$

(cf. Blanchard (1969), chapter IV.6).

The best numerical constant obtained so far via (a refinement of) the method introduced by Levinson—cf. (13)—is due to Conrey (1989) and exceeds 2/5.

A result more precise than Theorem 3 can be established, namely

$$\psi(x) = x + O(x^{\theta}(\log x)^2)$$

where $\theta = \sup_\rho \beta$—see for example Ellison & Mendès France (1975).

§ **5.5.** In the same order of ideas, Pintz (1984) established a precise link between the size of the remainder term $R(x) = \psi(x) - x$ and the localisation of the zeros of $\zeta(s)$: we have

$$\log \frac{x}{R(x)} \sim \min_\rho \{(1-\beta)\log x + \log|\gamma|\} \qquad (x \to \infty).$$

Another connection, already mentioned by Riemann and proved by von Mangoldt in 1895, appears in the "*explicit formula*"

$$\psi(x) = x - \sum_\rho \frac{x^\rho}{\rho} - \log 2\pi - \tfrac{1}{2}\log(1 - x^{-2})$$

valid for $x > 1$, $x \neq p^\nu$, where the sum over ρ converges in "principal value":

$$\sum_\rho \frac{x^\rho}{\rho} := \lim_{T\to\infty} \sum_{|\gamma| \leq T} \frac{x^\rho}{\rho}.$$

This result can be proved without difficulty by contour integration—see for example Ellison & Mendès France (1975), §5.5; see also chapter XIV of Titchmarsh (1951), and chapter 12 of Ivić (1985).

Exercises

1. Let k be an integer ≥ 1.

(a) Show that the complex numbers h_ν determined by the identity

$$1 + \sum_{\nu=1}^{\infty} h_\nu z^\nu = (1-z)^k \Big(1 + \frac{kz}{1-z}\Big)$$

are bounded by a function of k alone, viz

$$|h_\nu| \leq C(k) \qquad (\nu = 1, 2, \ldots).$$

(b) Show that $\sum_{n=1}^{\infty} k^{\omega(n)} n^{-s} = H_k(s)\zeta(s)^k$ where $H_k(s)$ is a Dirichlet series which converges absolutely for $\sigma > \frac{1}{2}$.

(c) From the above deduce the following relation

$$\sum_{n\leq x} k^{\omega(n)} = xP_{k-1}(\log x) + O_\varepsilon(x^{1-\delta_k+\varepsilon})$$

where P_{k-1} is a polynomial of degree $k-1$ (determine the leading coefficient!) and where δ_k is some constant > 0 (to be made explicit!).

(d) Calculate δ_k under the Riemann hypothesis.

2. *Bateman's theorem (1972).* Let φ denote Euler's totient function.

(a) Show that the number

$$a_n := \operatorname{card}\{m : m \geq 1, \varphi(m) = n\}$$

is finite for each integer $n \geq 1$.

(b) Establish the relation

$$\sum_{n=1}^{\infty} a_n n^{-s} = \zeta(s)G(s) \qquad (\sigma > 1)$$

with $G(s) := \prod_p (1 + (p-1)^{-s} - p^{-s})$.

(c) Show that the infinite product defining $G(s)$ is absolutely convergent for $\sigma > 0$.

(d) Show that

$$|(p-1)^{-s} - p^{-s}| \leq \min\left(2(p-1)^{-\sigma},\ |s|(p-1)^{-\sigma-1}\right)$$

and deduce the existence of an absolute constant A such that

$$G(s) \ll (\log|\tau|)^A \qquad (|\tau| \geq 2,\ \sigma \geq 1 - 1/\log|\tau|).$$

(e) Set $\Phi(x) := \operatorname{card}\{m : m \geq 1, \varphi(m) \leq x\}$. Show that $\Phi(x)$ is the summatory function for a_n and establish the asymptotic formula

$$\int_0^x \Phi(t)\,\mathrm{d}t = \tfrac{1}{2}\,\alpha x^2 + O\big(x^2 e^{-c_0\sqrt{(\log x)}}\big) \qquad (x \to \infty)$$

with $\alpha := G(1) = \zeta(2)\zeta(3)/\zeta(6)$.

(f) Using the monotonicity of Φ, show—by an argument analogous to that leading to (10)—that one has

$$\Phi(x) = \alpha x + O\big(xe^{-c\sqrt{(\log x)}}\big).$$

(g) Show that

$$a_n \ll ne^{-c\sqrt{(\log n)}} \qquad (n \geq 1).$$

3. Show that the Riemann hypothesis implies the convergence of the series $\sum \mu(n)n^{-s}$ for all s in the half-plane $\sigma > \frac{1}{2}$. [*Hint: use Theorem 2 and the Schnee–Landau theorem 2.4.*]

4. Put $\tau(n,\theta) := \sum_{d|n} d^{i\theta}$. Using Ramanujan's identity (3.28), show that, uniformly for $x \geq 2$, $|\theta| \leq 1$, $\theta \neq 0$, one has

$$\sum_{n \leq x} |\tau(n,\theta)|^2 = \frac{6}{\pi^2} x \Big\{ |\zeta(1+i\theta)|^2 \log x + O\big(|\theta|^{-3}\big) \Big\}.$$

5. Consider the arithmetic function

$$T(n) := \big|\{(d,d') : d|n,\ d'|n,\ |\log(d'/d)| \leq 1\}\big|.$$

Introducing the weight function $w(t) := \dfrac{1}{2\pi}\Big(\dfrac{\sin(t/2)}{t/2}\Big)^2$, with Fourier transform $\widehat{w}(\theta) = \big(1 - |\theta|\big)^+$, show that

$$w(1) \int_{-1}^{1} |\tau(n,\theta)|^2 \,\mathrm{d}\theta \leq T(n) \leq \frac{1}{2\pi w(1)} \int_{-1}^{1} |\tau(n,\theta)|^2 \,\mathrm{d}\theta.$$

Deduce, using the result of the previous exercise, that

$$\sum_{n \leq x} T(n) \asymp x(\log x)^2.$$

II.5

The Selberg–Delange method

§ 5.1 Complex powers of $\zeta(s)$

The analytic study of the Riemann zeta function enabled us to estimate the summatory function of certain arithmetic functions with Dirichlet series that can be simply expressed in terms of $\zeta(s)$. Thus the asymptotic formulae

$$\psi(x) := \sum_{n\leq x} \Lambda(n) = x + O\big(xe^{-c\sqrt{(\log x)}}\big)$$

$$M(x) := \sum_{n\leq x} \mu(n) = O\big(xe^{-c\sqrt{(\log x)}}\big)$$

$$T_k(x) := \sum_{n\leq x} \tau_k(n) = xP_{k-1}(\log x) + O_k(x^{1-\delta_k})$$

(where P_{k-1} is a polynomial of degree $k-1$ and δ_k a positive constant) are straightforward consequences of properties established in Chapter 3 for $\zeta(s)$: they are obtained by application of the Perron formulae to the meromorphic functions

$$-\zeta'(s)/\zeta(s), \qquad 1/\zeta(s), \qquad \zeta(s)^k.$$

In this chapter we consider a double extension of this method. On the one hand we want to be able to deal with Dirichlet series having singularities *which are not poles.* On the other hand we want to exhibit a kind of *stability* for the phenomenon, in the sense of an invariance in the nature of the asymptotic formulae obtained for two Dirichlet series whose ratio is a sufficiently regular analytic function. In this direction, we will consider the case of series admitting a representation of the type

$$F(s) = G(s;z)\zeta(s)^z \qquad (z \in \mathbb{C}) \tag{1}$$

for $\sigma > 1$, where $G(s;z)$ satisfies certain conditions to which we will return later.

Before developing this theory, due essentially to Selberg (1954) and Delange (1959, 1971), let us take a deeper look at the analytic nature of $\zeta(s)^z$ where z is a fixed complex number.

Defining the generalised binomial coefficient by

$$\binom{w}{\nu} := \frac{1}{\nu!}\prod_{j=0}^{\nu-1}(w-j) \qquad (w \in \mathbb{C},\ \nu \in \mathbb{N}),$$

we can write for $|\xi| < 1$, $z \in \mathbb{C}$,

$$(1-\xi)^{-z} = \sum_{\nu=0}^{\infty}\binom{z+\nu-1}{\nu}\xi^{\nu}.$$

When z is a negative integer, this formula reduces to the classical binomial formula. For $\sigma > 1$ we deduce that

$$\text{(2)} \qquad \zeta(s)^z = \prod_p (1-p^{-s})^{-z} = \prod_p\left(1+\sum_{\nu=1}^{\infty}\binom{z+\nu-1}{\nu}p^{-\nu s}\right),$$

the infinite product being absolutely convergent. By Theorem 1.2, it follows that $\zeta(s)^z$ is representable in the half-plane $\sigma > 1$ as the Dirichlet series of a multiplicative function, say $\tau_z(n)$, defined by

$$\text{(3)} \qquad \tau_z(p^{\nu}) := \binom{z+\nu-1}{\nu}.$$

This definition generalises that of the functions $\tau_k(n)$, corresponding to the case when $z = k$ is a positive integer. In these circumstances τ_k is the kth convolution power of the function $\mathbf{1}$, and $\tau_k(n)$ can be interpreted combinatorially as the number of decompositions of n as a product of k factors, viz.

$$\text{(4)} \qquad \tau_k(n) = \sum_{d_1 d_2 \ldots d_k = n} 1.$$

Of course, $\tau_2 = \tau$.

We shall see below that the function

$$\text{(5)} \qquad Z(s;z) := s^{-1}\{(s-1)\zeta(s)\}^z$$

plays a special role. It is defined on any simply connected domain which does not contain a zero of $\zeta(s)$. We shall always suppose that this domain includes the real half-line $[1, +\infty[$. We can then choose the principal value of the complex logarithm, so that

$$Z(1;z) = 1.$$

Theorem 1. *The function $Z(s;z)$ is holomorphic in the disc $|s-1|<1$, and can be represented there by the Taylor series*

$$(6) \qquad Z(s;z) = \sum_{j=0}^{\infty} \frac{1}{j!}\gamma_j(z)(s-1)^j$$

where the coefficients $\gamma_j(z)$ are entire functions of z satisfying, for all $A>0$ and $\varepsilon>0$, the upper bound

$$(7) \qquad \frac{1}{j!}\gamma_j(z) \ll_{A,\varepsilon} (1+\varepsilon)^j \qquad (|z|\le A).$$

Proof. All the assertions in the statement of the theorem follow immediately from the fact that $\zeta(s)$ does not vanish for $|s-1|<1$, via Cauchy's formula

$$\frac{1}{j!}\gamma_j(z) = \frac{1}{2\pi i}\oint_{|s-1|=r} Z(s;z)\frac{\mathrm{d}s}{(s-1)^{j+1}}.$$

It is indeed well-known—cf. for example Titchmarsh (1951), chapter XV—that the non-trivial zero of $\zeta(s)$ with smallest modulus in the half-plane $\tau \ge 0$ is

$$\rho = \tfrac{1}{2} + i\,14.13472\ldots.$$

For the sake of completeness, we give a quick proof of the fact that $\zeta(s)\ne 0$ for $|s-1|<1$. After integration by parts, formula (3.18) gives

$$\zeta(s) = \frac{1}{s-1} + \tfrac{1}{2} + \tfrac{1}{12}s - \tfrac{1}{2}s(s+1)\int_1^{\infty} B_2(t)t^{-s-2}\,\mathrm{d}t.$$

Suppose that $\zeta(\rho)=0$ where $\rho=\beta+i\gamma$ and $|\rho-1|<1$. Then, by symmetry, we may assume that $\beta\ge\frac{1}{2}$. Since $|B_2(t)|\le B_2=\frac{1}{6}$, we obtain, by setting $s=\rho$ in the above formula, that

$$1+\tfrac{1}{2}(\rho-1)\{1+\tfrac{1}{6}\rho-\tfrac{1}{6}\theta\rho(\rho+1)\int_1^{\infty} t^{-5/2}\,\mathrm{d}t\}=0$$

with $|\theta|\le 1$, which in turn implies

$$1\le\tfrac{1}{2}|\rho-1|\{1+\tfrac{1}{6}|\rho|+\tfrac{1}{9}|\rho(\rho+1)|\}<\tfrac{1}{2}\{1+\tfrac{1}{3}+\tfrac{2}{3}\}=1.$$

This is the required contradiction.

By Theorem 3.15, there exists an absolute positive constant c such that $\zeta(s)$ does not vanish in the region

$$\sigma \geq 1 - c/(1 + \log^+ |\tau|). \tag{8}$$

In the rest of this chapter *we let $\mathcal{D}$ denote the simply connected domain obtained by deleting the real segment* $[1 - c, 1]$ *from the region (8)*. We then have the analytic continuation

$$\zeta(s)^z = sZ(s; z)(s - 1)^{-z} \qquad (s \in \mathcal{D}). \tag{9}$$

Moreover, the upper bound $|\log \zeta(s)| \leq \log_2 |\tau| + O(1)$ from Theorem 3.16 shows that we have, for each constant $A > 0$,

$$\zeta(s)^z \ll_A (1 + \log^+ |\tau|)^A \qquad (|z| \leq A, \ s \in \mathcal{D}, \ |s - 1| \gg 1). \tag{10}$$

§ 5.2 Hankel's formula

This section is devoted to a classical result concerning the Γ function, which we shall need in the course of our study of Dirichlet series that are "close" to a complex power of the zeta function.

Given a positive parameter r, we designate by *Hankel contour* the path formed from the circle $|s| = r$ excluding the point $s = -r$, together with the half-line $]-\infty, -r]$ traced out twice, with respective arguments $+\pi$ and $-\pi$.

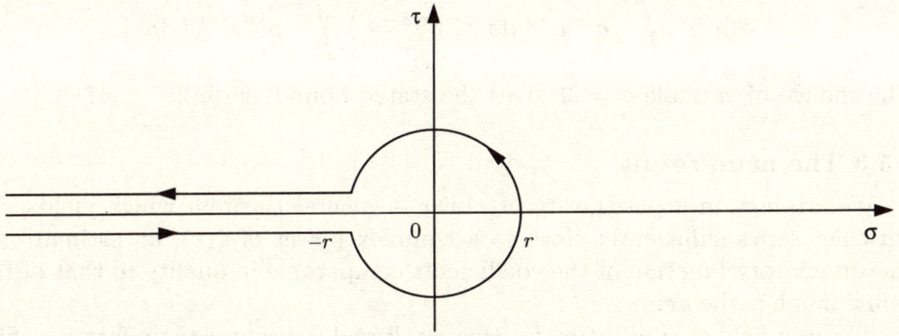

Theorem 2 (Hankel's formula). *Let $\mathcal{H}$ be a Hankel contour. For any complex number z, we have*

$$\frac{1}{\Gamma(z)} = \frac{1}{2\pi i} \int_{\mathcal{H}} s^{-z} e^s \, ds. \tag{11}$$

Proof. The integral is absolutely and uniformly convergent for each z. It thus defines an entire function of z which, by the residue theorem, is independent of r, since the sole singularity of the integrand occurs at $s = 0$. When $\Re e\, z < 1$, the integral round the circular part $|s| = r$ of the Hankel contour tends to 0 with r, whereas the integral along the doubled half-line tends to

$$\frac{1}{2\pi i}\int_0^\infty (e^{i\pi z} - e^{-i\pi z})\sigma^{-z}e^{-\sigma}\,\mathrm{d}\sigma$$
$$= \frac{\sin \pi z}{\pi}\int_0^\infty \sigma^{-z}e^{-\sigma}\,\mathrm{d}\sigma = \frac{\sin \pi z}{\pi}\Gamma(1-z) = \frac{1}{\Gamma(z)}.$$

This proves (11) when $\Re e\, z < 1$, and hence for all z by analytic continuation.

Corollary 2.1. *For each $X > 1$, let $\mathcal{H}(X)$ denote the part of the Hankel contour situated in the half-plane $\sigma > -X$. Then we have uniformly for $z \in \mathbb{C}$*

$$(12)\qquad \frac{1}{2\pi i}\int_{\mathcal{H}(X)} s^{-z}e^{s}\,\mathrm{d}s = \frac{1}{\Gamma(z)} + O\big(47^{|z|}\,\Gamma(1+|z|)\,e^{-\frac{1}{2}X}\big).$$

Proof. For $s = \sigma e^{\pm i\pi}$, $\sigma > 1$, we have

$$|s^{-z}e^{s}| \le (e^{\pi}\sigma)^{|z|}e^{-\sigma}.$$

The difference between the left-hand side of (12) and the integral (11) is therefore

$$\ll e^{\pi|z|}\int_X^\infty \sigma^{|z|}e^{-\sigma}\,\mathrm{d}\sigma \le e^{\pi|z|-\frac{1}{2}X}\int_0^\infty \sigma^{|z|}e^{-\frac{1}{2}\sigma}\,\mathrm{d}\sigma.$$

The change of variable $\sigma = 2t$ gives the stated bound, since $2e^{\pi} < 47$.

§ 5.3 The main result

We are now in a position to establish a general theorem which yields, for Dirichlet series sufficiently close to a complex power of $\zeta(s)$, an estimate for the summatory function of the coefficients comparable in quality to that of the prime number theorem.

The vast field of application for this result and the complete uniformity of its statement justify the explicit presence of rather numerous parameters. We have attempted to clarify the situation by introducing the following terminology.

Let $z \in \mathbb{C}$, $c_0 > 0$, $0 < \delta \le 1$, $M > 0$. We say that *a Dirichlet series $F(s)$ has the property $\mathcal{P}(z; c_0, \delta, M)$* if the Dirichlet series

$$G(s;z) := F(s)\zeta(s)^{-z}$$

may be continued as a holomorphic function for

$$\sigma \geq 1 - c_0/(1 + \log^+ |\tau|)$$

and, in this domain, satisfies the bound

$$|G(s;z)| \leq M(1 + |\tau|)^{1-\delta}. \tag{13}$$

If $F(s)$ has the property $\mathcal{P}(z; c_0, \delta, M)$ and if there exists a sequence of positive real numbers $\{b_n\}_{n=1}^{\infty}$ such that

$$|a_n| \leq b_n \qquad (n = 1, 2, \ldots)$$

and the series

$$\sum_{n=1}^{\infty} b_n n^{-s}$$

satisfies $\mathcal{P}(w; c_0, \delta, M)$ for a certain complex number w, we say that $F(s)$ has *type* $\mathcal{T}(z, w; c_0, \delta, M)$. It is worthwhile to bear in mind that a series with positive coefficients having property $\mathcal{P}(z; c_0, \delta, M)$ is trivially of type $\mathcal{T}(z, z; c_0, \delta, M)$.

In the domain where $G(s; z)$ is holomorphic we set

$$G^{(k)}(s; z) := \frac{\partial^k}{\partial s^k} G(s; z) \tag{14}$$

and

$$\lambda_k(z) := \frac{1}{\Gamma(z-k)} \sum_{h+j=k} \frac{1}{h!j!} G^{(h)}(1; z)\gamma_j(z) \tag{15}$$

where the $\gamma_j(z)$ are the entire functions appearing in Theorem 1.

Theorem 3. *Let*

$$F(s) := \sum_{n=1}^{\infty} a_n n^{-s}$$

be a Dirichlet series of type $\mathcal{T}(z, w; c_0, \delta, M)$. *For* $x \geq 3$, $N \geq 0$, $A > 0$, $|z| \leq A$, $|w| \leq A$, *we have*

$$\sum_{n \leq x} a_n = x(\log x)^{z-1} \Big\{ \sum_{k=0}^{N} \frac{\lambda_k(z)}{(\log x)^k} + O(MR_N(x)) \Big\} \tag{16}$$

with

$$R_N(x) := e^{-c_1\sqrt{(\log x)}} + \Big(\frac{c_2 N + 1}{\log x}\Big)^{N+1}.$$

The positive constants c_1, c_2 and the implicit constant in the Landau symbol depend at most on c_0, δ and A.

We shall see in the next chapter (§6.2) how one can take advantage of the uniformity in M in formula (16). Uniformity in N is equally significant. Consider for example the case when $z \in \mathbb{Z}$. Formula (15) then shows that $\lambda_k(z) = 0$ whenever $k \geq z$. We can hence choose N so as to minimise the error term in (16). For the value $N := [(\log x)/ec_2]$ we obtain

$$\sum_{n\leq x} a_n = x(\log x)^{z-1}\left\{P\Big(\frac{1}{\log x}\Big) + O\big(Me^{-c_1\sqrt{(\log x)}}\big)\right\}$$

where P is a polynomial of degree at most $z-1$.

Exercise 4.2 involved a direct study in a situation of this type. The corresponding result is actually an immediate consequence of Theorem 3. For the choice

$$a_n := \operatorname{card}\{m \geq 1 : \varphi(m) = n\}$$

we indeed have

$$F(s) = \sum_{m=1}^{\infty} \varphi(m)^{-s} = \zeta(s)G(s)$$

with

$$G(s) = \prod_p \big(1 + (p-1)^{-s} - p^{-s}\big) \qquad (\sigma > 0).$$

The easy upper bound

$$G(s) \ll (\log|\tau|)^{O(1)} \qquad (|\tau| \geq \tau_0,\ \sigma \geq 1 - 1/\log|\tau|)$$

shows that $F(s)$ has type $\mathcal{T}(1,1;1,\delta,M(\delta))$ for any fixed $\delta < 1$ and $M(\delta)$ depending only on δ. Since

$$\lambda_0(1) = G(1) = \zeta(2)\zeta(3)/\zeta(6) \approx 1.9436$$

is the only non-zero number among the $\lambda_k(1)$, we obtain the following statement.

Theorem 4 (Bateman, 1972). *There exists a positive constant c such that for $x \geq 1$ we have*

$$|\{n \geq 1 : \varphi(n) \leq x\}| = \frac{\zeta(2)\zeta(3)}{\zeta(6)}x + O\big(xe^{-c\sqrt{(\log x)}}\big). \tag{17}$$

Of course Theorem 3 provides an analogous result when $\varphi(n)$ is replaced by $\sigma(n)$ or, more generally, by any positive multiplicative function f such that

$$f(p^\nu) = p^\nu + O(p^{\nu-\delta}) \qquad (\nu \geq 1)$$

with $\delta > 0$.

§ 5.4 Proof of Theorem 3

Let c be a positive constant such that $c < c_0$ and such that $\zeta(s)$ has no zero in the region

$$\sigma \geq 1 - c/(1 + \log^+ |\tau|).$$

Then $F(s)$ is continuable to a holomorphic function in the domain $\mathcal{D}$ defined in §5.1, and we have by (10) and (13)

$$F(s) \ll_A M\big(1 + \log^+ |\tau|\big)^A \big(1 + |\tau|\big)^{1-\delta} \ll_{A,\delta} M(1 + |\tau|)^{1-\delta/2} \tag{18}$$

uniformly for $s \in \mathcal{D}$, $|s - 1| \gg 1$, $|z| \leq A$.

Set

$$A(x) := \sum_{n \leq x} a_n.$$

Then Perron's formula (2.10) allows us to write

$$\int_0^x A(t)\,\mathrm{d}t = \frac{1}{2\pi i} \int_{\kappa - i\infty}^{\kappa + i\infty} F(s) x^{s+1} \frac{\mathrm{d}s}{s(s+1)}$$

with $\kappa := 1 + 1/\log x$. Let $T > 1$ be a parameter whose value will be determined later. The residue theorem allows us to deform the segment of integration $[\kappa - iT, \kappa + iT]$ into some path joining the end-points and contained entirely in $\mathcal{D}$. We choose the path symmetrically with respect to the real axis (see the diagram below). Its upper part is made up of: the truncated Hankel contour Γ, surrounding the point $s = 1$, with radius $r = 1/(2\log x)$, and linear part joining $1 - r$ to $1 - \frac{1}{2}c$; the curve

$$\sigma = \sigma(\tau) := 1 - \tfrac{1}{2}c/(1 + \log^+ \tau)$$

for $0 \leq \tau \leq T$; and the horizontal segment $[\sigma(T) + iT, \kappa + iT]$.

We shall see that the main contribution arises from the integral over the truncated Hankel contour Γ.

Appealing to (18), we see immediately that the contribution from the vertical half-lines $[\kappa \pm iT, \kappa \pm i\infty[$ is $\ll_{A,\delta} Mx^2 T^{-\delta/2}$. This upper bound is equally valid for the contributions of the horizontal segments $[\sigma(T) \pm iT, \kappa \pm iT]$. Finally, that of the arcs $\sigma = \sigma(\tau)$ is

$$\ll_{A,\delta} Mx^{1+\sigma(T)} \int_0^T (1 + \tau)^{-1-\delta/2}\,\mathrm{d}\tau \ll_{A,\delta} Mx^{1+\sigma(T)}.$$

Selecting $T = \exp\sqrt{\big((c/\delta)\log x\big)}$, for $x \geq x_0$, it then follows that

$$\int_0^x A(t)\,\mathrm{d}t = \Phi(x) + O\Big(Mx^2 e^{-c_3\sqrt{(\log x)}}\Big), \tag{19}$$

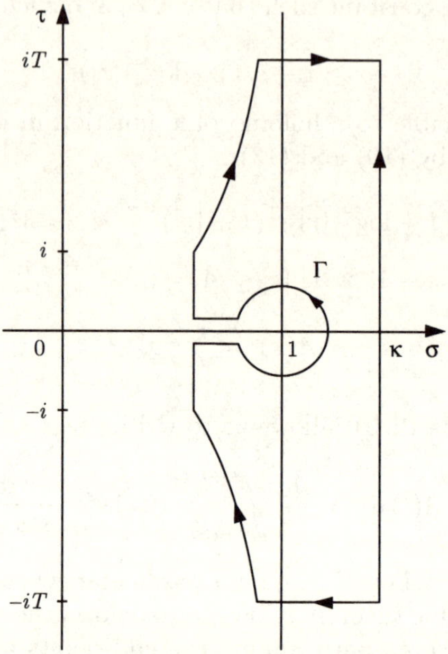

with

$$\Phi(x) := \frac{1}{2\pi i}\int_\Gamma F(s)x^{s+1}\frac{\mathrm{d}s}{s(s+1)}. \tag{20}$$

Here, and for the rest of the proof, we make the convention that all constants, explicit or implicit, depend at most on c_0, δ and A.

It remains to study the main term $\Phi(x)$ of (19). Clearly it is an infinitely differentiable function of x on $\mathbb{R}^+$, and in particular we have

$$\Phi'(x) = \frac{1}{2\pi i}\int_\Gamma F(s)x^s\frac{\mathrm{d}s}{s}, \qquad \Phi''(x) = \frac{1}{2\pi i}\int_\Gamma F(s)x^{s-1}\,\mathrm{d}s.$$

For $s \in \mathcal{D}$, we can write

$$F(s) = sG(s;z)Z(s;z)(s-1)^{-z} \tag{21}$$

with the result that by (6), (7) and (13)

$$F(s) \ll M|s-1|^{-A} \qquad (s \in \Gamma).$$

Since $r = 1/(2\log x)$, it follows that

$$\Phi''(x) \ll M(\log x)^A. \tag{22}$$

When $s \in \Gamma$, we have by (6)

$$G(s;z)Z(s;z) = \sum_{k=0}^{\infty} g_k(z)(s-1)^k$$

with

$$g_k(z) := \frac{1}{k!} \sum_{h+j=k} \binom{k}{j} G^{(h)}(1;z)\gamma_j(z) = \Gamma(z-k)\lambda_k(z).$$

In addition, since $G(s;z)Z(s;z)$ is holomorphic and $O(M)$ in the disc $|s-1| \le c$, the Cauchy formulae imply that

$$g_k(z) \ll Mc^{-k}.$$

Observing that Γ is contained in the disc $|s-1| \le \frac{1}{2}c$, we can write for $s \in \Gamma$, $N \ge 0$,

$$G(s;z)Z(s;z) = \sum_{k=0}^{N} g_k(z)(s-1)^k + O\Big(M\big(|s-1|/c\big)^{N+1}\Big).$$

Therefore

$$\Phi'(x) = \sum_{k=0}^{N} g_k(z) \frac{1}{2\pi i} \int_\Gamma x^s (s-1)^{k-z}\,\mathrm{d}s + O\big(Mc^{-N}R(x)\big), \tag{23}$$

with

$$\begin{aligned} R(x) &:= \int_\Gamma \big|x^s(s-1)^{N+1-z}\big|\,|\mathrm{d}s| \\ &\ll \int_{1-c/2}^{1-r} (1-\sigma)^{N+1-\Re e\, z} x^\sigma\,\mathrm{d}\sigma + x^{1+r} r^{N+2-\Re e\, z} \\ &\ll x(\log x)^{\Re e\, z-N-2} \left\{ \int_{\frac{1}{2}}^{+\infty} t^{N+1-\Re e\, z} e^{-t}\,\mathrm{d}t + 2^{-N} \right\} \\ &\ll x(\log x)^{\Re e\, z-N-2}\Gamma(N+A+2) \ll x(\log x)^{z-1}\Big(\frac{c_4 N+1}{\log x}\Big)^{N+1}. \end{aligned}$$

Using the change of variable $w = (s-1)\log x$, we can equally well write, with the notation of Corollary 2.1,

$$\begin{aligned}\frac{1}{2\pi i}\int_\Gamma x^s(s-1)^{k-z}\,\mathrm{d}s &= \frac{x}{2\pi i}(\log x)^{z-1-k}\int_{\mathcal{H}\left(\frac12 c\log x\right)} w^{k-z}e^w\,\mathrm{d}w\\ &= x(\log x)^{z-1-k}\Big\{\frac{1}{\Gamma(z-k)}+O\big((c_5k+1)^k x^{-c/4}\big)\Big\}.\end{aligned}$$

The main term of (23) thus has the value

$$x(\log x)^{z-1}\Big\{\sum_{k=0}^{N}\frac{\lambda_k(z)}{(\log x)^k}+O(E_N)\Big\}$$

with

$$\begin{aligned}E_N := x^{-c/4}\sum_{k=0}^{N}|g_k(z)|\Big(\frac{c_5k+1}{\log x}\Big)^k &\ll Mx^{-c/4}c_6^N\sum_{k=0}^{N}k!\Big(\frac{5}{c\log x}\Big)^k\\ &\le Mx^{-c/4}\Big(\frac{5c_6}{c\log x}\Big)^N\sum_{k=0}^{N}\frac{N!}{(N-k)!}\Big(\frac{c\log x}{5}\Big)^{N-k}\\ &\le Mx^{-c/20}N!\Big(\frac{5c_6}{c\log x}\Big)^N \le M\Big(\frac{c_7N+1}{\log x}\Big)^{N+1}.\end{aligned}$$

Substituting in (23), it follows that

$$\Phi'(x) = x(\log x)^{z-1}\left\{\sum_{k=0}^{N}\frac{\lambda_k(z)}{(\log x)^k}+O\left(M\Big(\frac{c_8N+1}{\log x}\Big)^{N+1}\right)\right\}. \tag{24}$$

We shall show, by using (19) and (22), that $\Phi'(x)$ is a suitable approximation for $A(x)$. To this end, let us take a parameter h, $0 < h < x/2$, and apply (19) for both x and $x+h$. Subtracting these estimates, we obtain

$$\int_x^{x+h} A(t)\,\mathrm{d}t = \Phi(x+h)-\Phi(x)+O\big(Mx^2e^{-c_3\sqrt{(\log x)}}\big) \tag{25}$$

while (22) implies that

$$\begin{aligned}\Phi(x+h)-\Phi(x) &= h\Phi'(x)+h^2\int_0^1(1-t)\Phi''(x+th)\,\mathrm{d}t\\ &= h\Phi'(x)+O\big(Mh^2(\log x)^A\big).\end{aligned} \tag{26}$$

We can therefore write

$$A(x) = \Phi'(x) + O\big(Mx^2h^{-1}e^{-c_3\sqrt{(\log x)}} + Mh(\log x)^A + h^{-1}L\big) \tag{27}$$

with

$$L := \int_x^{x+h} |A(t) - A(x)|\,\mathrm{d}t.$$

It is at this stage that we use the hypothesis that the series associated with $\{b_n\}_{n=1}^{\infty}$ has property $\mathcal{P}(w; c_0, \delta, M)$. Writing $B(t) := \sum_{n\leq t} b_n$ we have

$$L \leq \int_x^{x+h} (B(t) - B(x))\,\mathrm{d}t \leq \int_x^{x+h} B(t)\,\mathrm{d}t - \int_{x-h}^{x} B(t)\,\mathrm{d}t. \tag{28}$$

Now, our assumptions on the sequence $\{b_n\}_{n=1}^{\infty}$ imply the existence of an infinitely differentiable function Φ_1, satisfying (26), such that (25) is valid on replacing $A(t)$ by $B(t)$ and Φ by Φ_1. By (28) this implies that

$$L \ll Mx^2e^{-c_9\sqrt{(\log x)}} + Mh^2(\log x)^A.$$

In other words, up to modifying the value of c_3, we can suppress the term $h^{-1}L$ in the remainder of (27). Thus, choosing

$$h := xe^{-\frac{1}{2}c_3\sqrt{(\log x)}},$$

it follows that

$$A(x) = \Phi'(x) + O\big(Mxe^{-c_{10}\sqrt{(\log x)}}\big). \tag{29}$$

The conclusion of Theorem 3 follows by combining (24) and (29).

§ 5.5 A variant of the main theorem

We are now going to establish a result of the same type as Theorem 3, but in which the assumptions on the analytic continuation of the series $G(s; z)$ are replaced by convergence conditions on the successive derivatives at $s = 1$.

Theorem 5. *Let $F(s) := \sum_{n=1}^{\infty} a_n n^{-s}$ be a Dirichlet series converging for $\sigma > 1$. Suppose that there exists a complex number z and an integer $N \geq 0$ such that the derivatives of order $\leq N + 3 + \big[|z|\big]$ of the Dirichlet series*

$$G(s; z) := \zeta(s)^{-z}F(s) =: \sum_{n=1}^{\infty} b_z(n)n^{-s}$$

are absolutely convergent for $s = 1$. Set

$$H_N(z) := \sum_{n=1}^{\infty} |b_z(n)|(\log 3n)^{|z|+N+2}n^{-1}.$$

Then we have, for $|z| \le A$,

$$\sum_{n\le x} a_n = x(\log x)^{z-1}\Big\{\sum_{k=0}^{N}\frac{\lambda_k(z)}{(\log x)^k} + O_A\big(H_N(z)R_N(x)\big)\Big\} \tag{30}$$

with

$$R_N(x) := e^{-c_1\sqrt{(\log x)}} + \Big(\frac{c_2N+1}{\log x}\Big)^{N+1},$$

where c_1, c_2 *are positive constants depending at most on* A, *and the quantity* $\lambda_k(z)$ *is defined by* (15) *for* $0 \le k \le N$.

Proof. First of all, let us apply Theorem 3 to the Dirichlet series $\zeta(s)^z$. This is possible, since the upper bound

$$|\tau_z(n)| \le \tau_{|z|}(n) \qquad (n = 1, 2, \ldots)$$

guarantees that $\zeta(s)^z$ is of type $\mathcal{T}(z, |z|; c_0, 1, 1)$ for some positive absolute constant c_0. Setting

$$T_z(x) := \sum_{n\le x} \tau_z(n), \tag{31}$$

we obtain

$$\begin{aligned} T_z(x) = x(\log x)^{z-1}\Big\{&\sum_{k=0}^{N}\frac{\gamma_k(z)}{k!\,\Gamma(z-k)(\log x)^k} \\ &+ O_A\Big(e^{-c_3\sqrt{(\log x)}} + \Big(\frac{c_4N+1}{\log x}\Big)^{N+1}\Big)\Big\} \end{aligned} \tag{32}$$

where the $\gamma_k(z)$ are the entire functions from Theorem 1, and where c_3, c_4 are positive constants only depending on A.

Next, the convolution formula

$$a_n = \tau_z * b_z(n) \qquad (n = 1, 2, \ldots)$$

enables us to write

$$A(x) = \sum_{mn\le x} \tau_z(m)b_z(n) = \sum_{n\le x} b_z(n)T_z(x/n).$$

Since $T_z(x) = 1$ for $1 \le x < 2$, we further obtain

$$A(x) = \sum_{n\le x/2} b_z(n)T_z(x/n) + \sum_{x/2<n\le x} b_z(n). \tag{33}$$

The stated result follows from this formula, evaluating $T_z(x/n)$ by (32). At several points we will have occasion to use the following estimate, valid for $y \geq 1$ and $0 \leq h \leq N+2$,

$$\sum_{n \geq y} |b_z(n)|(\log n)^h n^{-1} \leq H_N(z)(\log 3y)^{h-2-N-|z|}. \tag{34}$$

This follows immediately from the definition of $H_N(z)$ and the trivial bound

$$(\log n)^h \leq \Big(\frac{\log 3n}{\log 3y}\Big)^{|z|+N+2} (\log 3y)^h \qquad (n \geq y).$$

Applied with $h = 0$, $y = x/2$, inequality (34) implies that

$$\Big| \sum_{x/2 < n \leq x} b_z(n) \Big| \leq x \sum_{n > x/2} |b_z(n)| n^{-1} \leq H_N(z) x (\log x)^{-N-|z|-2}.$$

On the other hand, formula (32) with $N = 0$ gives

$$T_z(x) \ll_A x(\log x)^{z-1} \qquad (x \geq 2),$$

so that, for $x \geq 4$, we can write

$$\begin{aligned}
\sum_{\sqrt{x} < n \leq x/2} b_z(n) T_z(x/n) &\ll_A x \sum_{\sqrt{x} < n \leq x/2} |b_z(n)| \Big(\log \frac{x}{n}\Big)^{\Re e\, z-1} n^{-1} \\
&\ll_A x(\log x)^{\Re e\, z-1} \sum_{\sqrt{x} < n \leq x/2} |b_z(n)| \Big(1 - \frac{\log n}{\log x}\Big)^{-|z|-1} n^{-1} \\
&\ll_A x(\log x)^{\Re e\, z+|z|} \sum_{n > \sqrt{x}} |b_z(n)| n^{-1} \ll_A 2^N H_N(z) x (\log x)^{z-N-2}
\end{aligned}$$

where the final bound follows from (34) with $h = 0$, $y = \sqrt{x}$. Substituting these estimates in (33) it follows that

$$A(x) = \sum_{n \leq \sqrt{x}} b_z(n) T_z(x/n) + O_A\big(2^N H_N(z) x (\log x)^{z-N-2}\big). \tag{35}$$

Formula (32) clearly implies the existence of positive constants $c_5 = c_5(A)$ and $c_6 = c_6(A)$ such that we have for $n \leq \sqrt{x}$

$$\begin{aligned}
T_z\Big(\frac{x}{n}\Big) = \frac{x}{n} \Big\{ &\sum_{k=0}^{N} \frac{\gamma_k(z)}{k!\,\Gamma(z-k)} \Big(\log \frac{x}{n}\Big)^{z-k-1} \\
&+ O_A\Big(e^{-c_5\sqrt{(\log x)}} + \Big(\frac{c_6 N + 1}{\log x}\Big)^{N+1}\Big)\Big\}.
\end{aligned} \tag{36}$$

The desired estimate for $A(x)$ will be obtained by expanding the terms $(\log x - \log n)^{z-k-1}$ in the above formula by means of the generalised binomial formula, and then substituting in (35).

For $|\xi| \leq \frac{1}{2}$, $0 \leq k \leq N$, we have

$$(1-\xi)^{z-k-1} = \sum_{h=0}^{N-k} \binom{z-k-1}{h} (-\xi)^h + O_A\left(3^N |\xi|^{N-k+1}\right).$$

This follows immediately by bounding the binomial coefficients using Cauchy's formula with a disc of radius 2/3, and we omit the details. Substituting $\xi = (\log n)/\log x$ and using the upper bound

$$\frac{\gamma_k(z)}{k!\,\Gamma(z-k)} \ll_A (c_7 N + 1)^N \qquad (|z| \leq A,\ 0 \leq k \leq N)$$

(which follows from (7) and the functional equation for the Γ function), we can rewrite the main term of (36) in the form

$$\frac{x}{n}\left\{\sum_{m=0}^{N} (\log x)^{z-m-1} \sum_{h+k=m} \frac{\gamma_k(z)}{k!\,\Gamma(z-k)} \binom{z-k-1}{h} (-\log n)^h \right.$$
$$\left. + O\Big((c_8 N + 1)^N (\log x)^{z-N-2} (\log n)^{N+1}\Big)\right\}.$$

Noting that, for $h + k = m$,

$$\frac{1}{\Gamma(z-k)} \binom{z-k-1}{h} = \frac{1}{h!\,\Gamma(z-m)},$$

we can then deduce from (36) that

$$\sum_{n \leq \sqrt{x}} b_z(n) T_z(x/n)$$
$$= x(\log x)^{z-1}\left\{\sum_{m=0}^{N} \frac{(\log x)^{-m}}{\Gamma(z-m)} \sum_{k+h=m} \frac{\gamma_k(z)}{k!h!} \sum_{n \leq \sqrt{x}} b_z(n)(-\log n)^h n^{-1} \right.$$
$$\left. + O\left(H_N(z)\left(e^{-c_5\sqrt{(\log x)}} + \left(\frac{c_9 N + 1}{\log x}\right)^{N+1}\right)\right)\right\}.$$

For each h the inner sum over n is equal to

$$G^{(h)}(1; z) + O_A(2^N H_N(z)(\log x)^{h-N-2}),$$

using (34). The contribution of the error term to the previous expression is thus

$$\ll_A 2^N H_N(z) x (\log x)^{z-N-2} \sum_{k+h \leq N} \frac{1}{h!\,|\Gamma(z-k-h)|} \left(\frac{2}{\log x}\right)^k$$
$$\ll_A H_N(z) x (\log x)^{z-1} \left(\frac{c_{10} N + 1}{\log x}\right)^{N+1}.$$

We have therefore shown that

$$\sum_{n\le\sqrt{x}} b_z(n)T_z(x/n) = x(\log x)^{z-1}\Bigg\{\sum_{m=0}^{N}\frac{\lambda_m(z)}{(\log x)^m} + O_A\bigg(H_N(z)\Big(e^{-c_5\sqrt{(\log x)}} + \Big(\frac{c_{11}N+1}{\log x}\Big)^{N+1}\Big)\bigg)\Bigg\}.$$

Inserting this in (35) completes the proof of the theorem.

Notes

§ **5.1.** The method of contour integration allows us to handle the case of Dirichlet series having singularities different in nature from those of $(s-1)^{-z}$ as considered here. Delange (1954) provided a formula for a singularity of type

$$(s-1)^{-\omega}\Big(\log\frac{1}{s-1}\Big)^k$$

where $\omega \in \mathbb{R}$ and k is zero or a positive integer—cf. Theorem 7.15. In Exercise 8 we sketch an argument adapted to a singularity of type

$$e^{\lambda/(s-1)}.$$

In all these cases it appears of crucial importance to be able to assume analytic continuation in some suitable neighbourhood (from which a half-line may be taken out) of the singularities.

Except in the case of $\gamma_0(z) \equiv 1$, it seems difficult to give a simple expression for $\gamma_j(z)$. Writing the analytic continuation of $\zeta(s)$ for $\sigma > 0$ in the form

$$\zeta(s) = \frac{s}{s-1} - s\int_1^\infty \{t\}t^{-s-1}\,\mathrm{d}t$$

readily gives for $j \ge 1$

$$\gamma_j(1) = (-1)^j\int_1^\infty \{t\}(\log t)^{j-1}\frac{\mathrm{d}t}{t^2}.$$

In particular we have $\gamma_1(1) = \gamma - 1$, where γ is Euler's constant.

§ **5.3.** The precise form obtained by Bateman for the error term in (17) is $O\big(x\exp\{-c_{12}\surd(\log x\log_2 x)\}\big)$ for any constant $c_{12} < \frac{1}{2}\surd 2$. The method of the present chapter gives the same result by slightly altering the contour of integration. Balazard & Smati (1990) have recently shown that this estimate can be precisely recovered by an elementary method.

It follows in particular from formula (17) that

$$a_n := \operatorname{card}\{m : \varphi(m) = n\} \ll ne^{-c\surd(\log n)} \qquad (n \geq 1).$$

In 1935, Erdős showed that there exists a constant $\delta > 0$ such that

$$a_n > n^{\delta}$$

for infinitely many integers n, and he conjectures that this result remains true for any $\delta < 1$. In the same direction, he ventures the hypothesis that the error term in (17) is

$$\Omega\left(x\exp\left\{-c\frac{\log x}{\log_2 x}\right\}\right)$$

for a suitable positive constant c.

§ **5.4.** The proof which we give here is largely inspired by that of Delange (1971) in which we have made explicit the dependence of the error term on M and N.

Formula (29) can yield a simpler approximation for $A(x)$ when the contribution to $\Phi'(x)$ from the circular part of the Hankel contour tends to zero with the radius. For example, under the hypotheses of Theorem 3 and assuming in addition that $\Re e\, z \leq 1 - \varepsilon$, we obtain in this way that

$$A(x) = \int_0^{c_{13}} x^{1-t}\alpha(t;z)\frac{\mathrm{d}t}{t^z} + O\big(xe^{-c_{14}\surd(\log x)}\big)$$

where c_{13} and c_{14} are positive constants, depending only on A and ε, and where $\alpha(t;z)$ is the continuous function on $[0, c_{13}]$ defined by

$$\alpha(t;z) = \frac{\sin\pi z}{\pi}Z(1-t;z)G(1-t;z).$$

Of course one would be able to recover this result from (16) by choosing $N = N(x)$ to tend to infinity suitably, writing

$$\begin{aligned}\frac{1}{\Gamma(z-k)} &= \frac{\sin\pi z}{\pi}(-1)^k\Gamma(k+1-z)\\ &= \frac{\sin\pi z}{\pi}(-1)^k\int_0^X e^{-t}t^{k-z}\,\mathrm{d}t + \text{remainder}\end{aligned}$$

with $X := c_{13}\log x$, and interchanging summation and integration in the main term of (16).

Exercises

1. Let $f(n)$ be a multiplicative arithmetic function such that $f(p) = \alpha$, $f(p^\nu) \ll p^{\delta\nu}$ $(\nu \geq 2)$ for some $\delta < \frac{1}{2}$. Evaluate the summatory function of $f(n)$.

2. Let $g(n)$ be the number of decompositions of n as a product of distinct numbers of the form $p - 1$, where p is prime. Estimate $\sum_{n \leq x} g(n)$.

3. Evaluate, uniformly for $x \geq 2$, $1 \leq q \leq x$, the sum $\sum_{n \leq x,\, (n,q)=1} 1/\tau(n)$.

4. Let $\lambda > 0$. Show that the asymptotic relation

$$\sum_{n \leq x} (-1)^n \lambda^{\omega(n)} = o\Big(\sum_{n \leq x} \lambda^{\omega(n)}\Big) \qquad (x \to \infty)$$

holds if, and only if, $\lambda = 1$.

5. Establish an asymptotic formula for $\sum_{1 < n \leq x} t^{\omega(n) - 1}$ valid uniformly for $0 \leq t \leq 1$. By integration, deduce a formula for

$$\sum_{1 < n \leq x} 1/\omega(n).$$

6. Use the method of the previous exercise to evaluate $\sum_{1 < n \leq x} \mu(n)^2/\omega(n)$.

7. Same question for $\sum_{1 < n \leq x} \varphi(n)/\omega(n)$.

8. Let $F(s) := \sum a_n n^{-s}$ be a Dirichlet series with non-negative coefficients. Suppose that $F(s)$ may be continued by continuity for $\sigma = 1$, $s \neq 1$, and in a punctured disc $0 < |s - 1| \leq c$. Assume also that the continuation satisfies the conditions

$$\text{(i)} \qquad F(s) \ll (1 + |\tau|)^{1-\delta}$$

$$\text{(ii)} \qquad F(s) = \exp\left\{\frac{\lambda}{s-1}\right\} \sum_{n=0}^{\infty} \alpha_n (s-1)^n \qquad (0 < |s - 1| \leq c)$$

where δ is fixed in $]0, 1[$, $\lambda > 0$, and $\sum \alpha_n \xi^n$ is a power series with radius of convergence at least equal to c. Set $A(x) = \sum_{n \leq x} a_n$.

(a) Show that

$$\int_0^x A(t)\,\mathrm{d}t = \frac{1}{2\pi i} \int_{\mathcal{C}} F(s) x^{s+1} \frac{\mathrm{d}s}{s(s+1)} + O(x^2)$$

where $\mathcal{C}$ is the half-circle $s = 1 + \frac{1}{2} c e^{i\theta}$, $-\frac{\pi}{2} \leq \theta \leq \frac{\pi}{2}$.

(b) In the above s-integral, use the change of variable $s = 1 + 1/w$ to show that there exists a Laurent series $L(w) := \alpha_0 + \sum_{n=1}^{\infty} \beta_n w^{-n}$, converging for sufficiently large $|w|$, and such that we have

$$\int_0^x A(t)\,\mathrm{d}t = \tfrac{1}{2} I(x) + O(x^2)$$

with

$$I(x) := \frac{x^2}{2\pi i} \int_{R-i\infty}^{R+i\infty} \exp\{w^{-1}\log x + \lambda w\} L(w) \frac{\mathrm{d}w}{w^2} \qquad (R \geq R_0).$$

(c) From now on, choose $R := \sqrt{(\lambda^{-1}\log x)}$. Show that $I(x)$ is infinitely differentiable, and that the contribution to $I'(x)$ from the domain $|\tau| \geq (\log x)^{1/3}$ is

$$\ll x \exp\{2\sqrt{(\lambda \log x)} - \lambda^{3/2}(\log x)^{1/6}\}.$$

(d) Expanding $w^{-1}\log x + \lambda w$ as a power series to order 4 for $|\tau| \leq (\log x)^{1/3}$, show that the contribution to $I'(x)$ from the domain $|\tau| \leq (\log x)^{1/3}$ is

$$\frac{\lambda^{1/4}}{\sqrt{\pi}}\, x \frac{e^{2\sqrt{(\lambda \log x)}}}{(\log x)^{3/4}} \Big\{\alpha_0 + O\Big(\frac{1}{\sqrt{(\log x)}}\Big)\Big\}.$$

(e) Show that $I''(x) \ll I'(x)/x$.

(f) Deduce from the preceding results, using the monotonicity of the function $A(t)$, that one has

$$A(x) = \frac{\lambda^{1/4}}{2\sqrt{\pi}}\, x \frac{e^{2\sqrt{(\lambda \log x)}}}{(\log x)^{3/4}} \Big\{\alpha_0 + O\Big(\frac{1}{\sqrt{(\log x)}}\Big)\Big\}.$$

9. *Oppenheim's problem on "factorisatio numerorum".* Let $Q(n)$ be the number of ways of decomposing n as a product of factors > 1, regardless of order. Thus

$$Q(n) := \big|\{(\nu_2, \nu_3, \ldots) : \textstyle\prod_{j=2}^{\infty} j^{\nu_j} = n\}\big|.$$

(a) Show that the Dirichlet series $F(s) := \sum_{n=1}^{\infty} Q(n) n^{-s}$ admits the product representation

$$F(s) = \prod_{j=2}^{\infty} (1 - j^{-s})^{-1}.$$

(b) Show that $F(s)$ satisfies the conditions of the previous exercise with $\lambda = 1$.

(c) Establish Oppenheim's formula (1927)

$$\sum_{n \leq x} Q(n) = x \frac{e^{2\sqrt{(\log x)}}}{2\sqrt{\pi}\,(\log x)^{3/4}} \Big\{1 + O\Big(\frac{1}{\sqrt{(\log x)}}\Big)\Big\}.$$

10. *A theorem of Diamond (1984).* Let $F_0(n) := \begin{cases} 1 & (n=1) \\ 0 & (n>1) \end{cases}$, and for $k \geq 1$ let

$$F_k(n) := \sum_{\substack{\nu_1\nu_2\ldots\nu_k=n \\ \nu_j>1\ (j=1,\ldots,k)}} \Big\{\prod_{j=1}^{k} \log \nu_j\Big\}^{-1}.$$

Let $F(n)$ denote the arithmetic function $F(n) = \sum_{k=0}^{\infty} F_k(n)/k! \quad (n \geq 1)$.

(a) Show that $\displaystyle\sum_{n=1}^{\infty} F(n)n^{-s} = \zeta(s) \exp\Big\{ - \sum_{n=2}^{\infty} \frac{\Lambda(n)-1}{n^s \log n}\Big\} \quad (\sigma > 1)$.

(b) Establish the asymptotic formula

$$\sum_{n \leq x} F(n) = Kx + O\big(xe^{-c\sqrt{(\log x)}}\big) \qquad (x \to \infty)$$

with $K := \exp\big\{\sum_{n=2}^{\infty} \big(1 - \Lambda(n)\big)/n \log n\big\} \approx 1.24292$.

11. Show that, uniformly for $x \geq 2$, $T > 0$, one has

$$\sum_{n \leq x,\, \omega(n) \leq T} 1 \ll x2^T/\sqrt{(\log x)}.$$

Using the canonical decomposition of an integer n in the form $n = qs$ with $\mu(q)^2 = 1$, $(q, s) = 1$, $p|s \Rightarrow p^2|s$, show that, for any fixed integer $k \geq 1$,

$$\sum_{n \leq x,\, k|\tau(n)} 1 = c_k x + O\big(x/\sqrt{(\log x)}\big).$$

For any prime number p, write $1 - c_p$ as an Eulerian product.

II.6

Two arithmetic applications

Here we intend to employ the results of the previous chapter, in particular Theorem 5.5, in order to investigate two concrete arithmetic problems.

§ 6.1 Integers having k prime factors

From the prime number theorem, it is easy to show by induction on the integer $k \geq 1$ that, for each fixed value of k, we have

$$N_k(x) := |\{n \leq x : \Omega(n) = k\}| \sim \frac{x}{\log x} \frac{(\log_2 x)^{k-1}}{(k-1)!} \qquad (x \to \infty)$$

— cf. Landau (1909). The same result holds for the function

$$\pi_k(x) := |\{n \leq x : \omega(n) = k\}|.$$

However, when k is allowed to tend to infinity with x, the study of the asymptotic behaviour of $\pi_k(x)$ or $N_k(x)$ by induction becomes very technical — cf. Sathe (1953, 1954). Another line of attack, devised by Selberg (1954), consists, for example in the case of $\omega(n)$, in identifying $\pi_k(x)$ with the coefficient of z^k in the expression

$$\sum_{n \leq x} z^{\omega(n)} \tag{1}$$

and then applying Cauchy's integral formula. This programme necessitates a good estimate for the sum (1), which will be provided by Theorem 5.3 or Theorem 5.5.

Indeed, let us consider the Dirichlet series

$$F_1(s;z) := \sum_{n=1}^{\infty} z^{\omega(n)} n^{-s} = \prod_p \Big(1 + \frac{z}{p^s - 1}\Big) \qquad (\sigma > 1).$$

Then the function

$$G_1(s;z) := F_1(s;z)\zeta(s)^{-z} = \prod_p \Big(1 + \frac{z}{p^s - 1}\Big)\Big(1 - \frac{1}{p^s}\Big)^z$$

is expandable as a Dirichlet series

$$G_1(s;z) = \sum_{n=1}^{\infty} b_{1z}(n)n^{-s}$$

where b_{1z} is the multiplicative function for which the values on prime powers are determined by the identity

$$1 + \sum_{\nu=1}^{\infty} b_{1z}(p^\nu)\xi^\nu = \left(1 + \frac{\xi z}{1-\xi}\right)(1-\xi)^z \qquad (|\xi| < 1).$$

In particular we have

$$(2) \qquad b_{1z}(p) = 0$$

and Cauchy's inequality implies that, for $|z| \le A$,

$$(3) \qquad \left|b_{1z}(p^\nu)\right| \le M \cdot 2^{\nu/2} \qquad (\nu \ge 2)$$

with

$$M = M(A) := \sup_{|z|\le A,\, |\xi| \le 1/\sqrt{2}} \left|\left(1 + \frac{\xi z}{1-\xi}\right)(1-\xi)^z\right|.$$

Relations (2) and (3) show that for $\sigma > \frac{1}{2}$

$$\sum_p \sum_{\nu=1}^{\infty} \left|b_{1z}(p^\nu)\right| p^{-\nu\sigma} \le 2M \sum_p \frac{1}{p^\sigma(p^\sigma - \sqrt{2})} \le cM/(\sigma - \tfrac{1}{2})$$

where c is an absolute constant. By Theorem 1.2, we deduce that $G_1(s;z)$ is absolutely convergent for $\sigma > \frac{1}{2}$ and that, for $\sigma \ge \frac{3}{4}$, we have

$$G_1(s;z) \ll_A 1.$$

The assumptions of Theorem 5.3 are therefore satisfied, and we can formulate the following result.

Theorem 1. *For any positive constant A, there exist positive constants $c_1 = c_1(A)$ and $c_2 = c_2(A)$ such that, uniformly for $x \ge 3$, $N \ge 0$, $|z| \le A$, we have*

$$(4) \qquad \sum_{n \le x} z^{\omega(n)} = x(\log x)^{z-1}\left\{\sum_{k=0}^{N} \frac{\lambda_k(z)}{(\log x)^k} + O_A(R_N(x))\right\}$$

with

$$(5) \qquad R_N(x) := e^{-c_1\sqrt{(\log x)}} + \left(\frac{c_2 N + 1}{\log x}\right)^{N+1}$$

and

$$(6) \qquad \lambda_k(z) := \frac{1}{\Gamma(z-k)} \sum_{h+j=k} \frac{1}{h!\,j!} G_1^{(h)}(1;z)\gamma_j(z),$$

where the $\gamma_j(z)$ are the entire functions defined in Theorem 5.1.

Note that $\lambda_k(0) = 0$ for all k. In particular, we write

$$\lambda_0(z) = z\lambda(z), \quad \text{with} \quad \lambda(z) := \frac{G_1(1;z)}{\Gamma(z+1)}. \tag{7}$$

Of course we can carry out a parallel study for

$$F_2(s;z) := \sum_{n=1}^{\infty} z^{\Omega(n)} n^{-s} = \prod_p \left(1 - z/p^s\right)^{-1}.$$

The coefficient $b_{2z}(n)$ of n^{-s} in

$$G_2(s;z) := F_2(s;z)\zeta(s)^{-z} = \prod_p \left(1 - z/p^s\right)^{-1} (1 - p^{-s})^z$$

is also a multiplicative function of n, determined by the identity

$$1 + \sum_{\nu=1}^{\infty} b_{2z}(p^\nu)\xi^\nu = (1 - \xi z)^{-1}(1 - \xi)^z.$$

As before we have $b_{2z}(p) = 0$, but in order to bound $|b_{2z}(p^\nu)|$ we must bear in mind that the right-hand side is only holomorphic in ξ inside the disc $|\xi| < \min\left(1, |z|^{-1}\right)$. For all δ, $0 < \delta < 1$, Cauchy's integral formula gives

$$|b_{2z}(p^\nu)| \leq M(\delta)(2-\delta)^\nu \qquad (|z| \leq 2 - 2\delta)$$

with

$$M(\delta) := \sup_{\substack{|\xi| \leq 1/(2-\delta) \\ |z| \leq 2-2\delta}} |(1 - \xi z)^{-1}(1-\xi)^z|.$$

This implies the absolute convergence of $G_2(s;z)$ for $\sigma \geq 1 - c_0(\delta)$, and Theorem 5.3 yields the result stated below.

Theorem 2. *For all δ, $0 < \delta < 1$, there exist positive constants $c_1 = c_1(\delta)$ and $c_2 = c_2(\delta)$ such that, uniformly for $x \geq 3$, $N \geq 0$, $|z| \leq 2 - \delta$,*

$$\sum_{n \leq x} z^{\Omega(n)} = x(\log x)^{z-1} \left\{ \sum_{k=0}^{N} \frac{\nu_k(z)}{(\log x)^k} + O_\delta(R_N(x)) \right\} \tag{8}$$

where $R_N(x)$ is defined by (5) and we have written

$$\nu_k(z) = \frac{1}{\Gamma(z-k)} \sum_{h+j=k} \frac{1}{h!\, j!} G_2^{(h)}(1;z)\gamma_j(z).$$

As before we have $\nu_k(0) = 0$ $(k \geq 0)$. We set

$$\nu_0(z) = z\nu(z), \quad \text{with} \quad \nu(z) := \frac{G_2(1;z)}{\Gamma(z+1)}. \tag{9}$$

The desired evaluations for $\pi_k(x)$ and $N_k(x)$, starting from formulae (4) and (8), result from the following general theorem.

Theorem 3. *Let $a_z(n)$ be an arithmetic function depending on a complex parameter z and with a power series expansion in the disc $|z| \le A$*

$$a_z(n) = \sum_{k=0}^{\infty} c_k(n) z^k \qquad (n = 1, 2, \ldots).$$

Let N be a non-negative integer. Suppose that there exist $N+1$ functions $h_j(z)$ $(0 \le j \le N)$, holomorphic for $|z| \le A$, and a quantity $R_N(x)$, independent of z, such that, for $x \ge 3$ and $|z| \le A$,

$$\sum_{n \le x} a_z(n) = x(\log x)^{z-1}\left\{\sum_{j=0}^{N} \frac{z h_j(z)}{(\log x)^j} + O_A\big(R_N(x)\big)\right\}. \tag{10}$$

Then uniformly for $x \ge 3$, $1 \le k \le A \log_2 x$, we have

$$\begin{aligned} C_k(x) &:= \sum_{n \le x} c_k(n) \\ &= \frac{x}{\log x}\left\{\sum_{j=0}^{N} \frac{Q_{j,k}(\log_2 x)}{(\log x)^j} + O_A\Big(\frac{(\log_2 x)^k}{k!} R_N(x)\Big)\right\}, \end{aligned} \tag{11}$$

with

$$Q_{j,k}(X) := \sum_{m+\ell=k-1} \frac{1}{m!\,\ell!} h_j^{(m)}(0) X^{\ell} \qquad (0 \le j \le N,\ k \ge 1). \tag{12}$$

If, in addition, we suppose that $|h_0''(z)| \le B$ for $|z| \le A$, then, uniformly for $x \ge 3$, $1 \le k \le A \log_2 x$, we have

$$C_k(x) = \frac{x}{\log x} \frac{(\log_2 x)^{k-1}}{(k-1)!}\left\{h_0\Big(\frac{k-1}{\log_2 x}\Big) + O_A\Big(\frac{B\cdot(k-1)}{(\log_2 x)^2} + \frac{\log_2 x}{k} R_0(x)\Big)\right\}. \tag{13}$$

Proof. First of all we prove formula (11). Since the main term is the coefficient of z^k in the main term of (10), it suffices to estimate the error term. By Cauchy's formula this is

$$\ll_A \oint_{|z|=r} \frac{x R_N(x)}{\log x} (\log x)^{\Re e z} |z|^{-k-1} |dz| \tag{14}$$

for all $r \le A$. Select $r := k/\log_2 x$. We have

$$\oint_{|z|=r} (\log x)^{\Re e z} |z|^{-k-1} |dz| = \Big(\frac{\log_2 x}{k}\Big)^k \int_0^{2\pi} e^{k\cos\theta}\, d\theta$$

and

$$\int_0^{2\pi} e^{k\cos\theta}\,\mathrm{d}\theta \le 2\int_0^{\pi/2} e^{k\cos\theta}\,\mathrm{d}\theta + \pi = 2\int_0^1 e^{kt}\frac{\mathrm{d}t}{\sqrt{(1-t^2)}} + \pi$$

$$\le 2e^k\int_0^1 e^{-k(1-t)}\frac{\mathrm{d}t}{\sqrt{(1-t)}} + \pi \le 2\Gamma(\tfrac{1}{2})e^k k^{-\frac{1}{2}} + \pi.$$

We thus obtain the stated result by substituting in (14).

When $k = 1$, formula (13) reduces to (11) for $N = 0$. Suppose therefore that $k \ge 2$, apply (11) with $N = 0$, and evaluate the main term by Cauchy's formula. For all $r \le A$, we have

$$Q_{0,k}(X) = \frac{1}{2\pi i}\oint_{|z|=r} h_0(z)e^{zX}z^{-k}\,\mathrm{d}z.$$

When $k \le AX$, we can choose $r = (k-1)/X$. Noting that

$$\frac{1}{2\pi i}\oint_{|z|=r}(z-r)e^{zX}z^{-k}\,\mathrm{d}z = \frac{X^{k-2}}{(k-2)!} - r\frac{X^{k-1}}{(k-1)!} = 0,$$

we see that

$$Q_{0,k}(X) = \frac{h_0(r)}{2\pi i}\oint_{|z|=r} e^{zX}z^{-k}\,\mathrm{d}z$$

$$+ \frac{1}{2\pi i}\oint_{|z|=r}\{h_0(z) - h_0(r) - (z-r)h_0'(r)\}e^{zX}z^{-k}\,\mathrm{d}z.$$

The first term on the right-hand side equals the stated main term $h_0(r)X^{k-1}/(k-1)!$. In order to evaluate the second, we note that the expression between curly brackets has the value

$$(z-r)^2\int_0^1 (1-t)h_0''(r+t(z-r))\,\mathrm{d}t$$

and that for each t, $0 \le t \le 1$,

$$|r + t(z-r)| = |r(1-t) + tz| \le r(1-t) + tr = r.$$

It follows that the modulus of the associated integral is

$$\le \frac{B}{2\pi}\int_0^{2\pi}|e^{i\theta}-1|^2 e^{rX\cos\theta}r^{3-k}\,\mathrm{d}\theta \le \frac{B}{\pi}r^{3-k}\int_0^{2\pi} e^{(k-1)\cos\theta}(1-\cos\theta)\,\mathrm{d}\theta.$$

Bounding the integral over θ by

$$2\int_0^1 e^{(k-1)t}\sqrt{(1-t)}\,\mathrm{d}t + 2\pi \le 2\Gamma(\tfrac{3}{2})e^{k-1}(k-1)^{-3/2} + 2\pi$$

and applying Stirling's formula, we deduce that

$$Q_{0,k}(X) = h_0(r)\frac{X^{k-1}}{(k-1)!} + O\Big(B\frac{X^{k-1}}{(k-1)!}\frac{k-1}{X^2}\Big) \qquad (k \le AX).$$

This proves (13).

Theorems 1, 2 and 3 now enable us to state explicit asymptotic formulae for $\pi_k(x)$ and $N_k(x)$. The role of the function $h_0(z)$ is played respectively by the functions

$$\lambda(z) := \frac{1}{\Gamma(z+1)} \prod_p \Big(1 + \frac{z}{p-1}\Big)\Big(1 - \frac{1}{p}\Big)^z \tag{15}$$

and

$$\nu(z) := \frac{1}{\Gamma(z+1)} \prod_p \Big(1 - \frac{z}{p}\Big)^{-1}\Big(1 - \frac{1}{p}\Big)^z. \tag{16}$$

Theorem 4. *Let $A > 0$. There exist positive constants $c_1 = c_1(A)$ and $c_2 = c_2(A)$ such that, uniformly for $x \geq 3$, $1 \leq k \leq A \log_2 x$, $N \geq 0$, we have*

$$\pi_k(x) = \frac{x}{\log x}\Big\{\sum_{j=0}^{N} \frac{P_{j,k}(\log_2 x)}{(\log x)^j} + O_A\Big(\frac{(\log_2 x)^k}{k!} R_N(x)\Big)\Big\} \tag{17}$$

where $P_{j,k}(X)$ is a polynomial of degree at most $k-1$ and $R_N(x)$ is defined by (5). In particular, we have

$$P_{0,k}(X) = \sum_{m+\ell=k-1} \frac{1}{m!\,\ell!}\lambda^{(m)}(0)X^\ell.$$

Moreover, under the same conditions, we have

$$\pi_k(x) = \frac{x}{\log x}\frac{(\log_2 x)^{k-1}}{(k-1)!}\Big\{\lambda\Big(\frac{k-1}{\log_2 x}\Big) + O\Big(\frac{k}{(\log_2 x)^2}\Big)\Big\}. \tag{18}$$

Theorem 5. *Let δ satisfy $0 < \delta < 1$. There exist positive constants $c_1 = c_1(\delta)$ and $c_2 = c_2(\delta)$ such that, uniformly for $x \geq 3$, $1 \leq k \leq (2-\delta)\log_2 x$, $N \geq 0$, we have*

$$N_k(x) = \frac{x}{\log x}\Big\{\sum_{j=0}^{N} \frac{Q_{j,k}(\log_2 x)}{(\log x)^j} + O_\delta\Big(\frac{(\log_2 x)^k}{k!} R_N(x)\Big)\Big\} \tag{19}$$

where $Q_{j,k}$ is a polynomial of degree at most $k-1$ and $R_N(x)$ is defined by (5). In particular, we have

$$Q_{0,k}(X) = \sum_{m+\ell=k-1} \frac{1}{m!\,\ell!}\nu^{(m)}(0)X^\ell.$$

Moreover, under the same conditions, we have

$$N_k(x) = \frac{x}{\log x}\frac{(\log_2 x)^{k-1}}{(k-1)!}\Big\{\nu\Big(\frac{k-1}{\log_2 x}\Big) + O\Big(\frac{k}{(\log_2 x)^2}\Big)\Big\}. \tag{20}$$

The condition $k \leq (2-\delta)\log_2 x$ is a natural restriction for the validity of (20), since $\nu(z)$ has a pole at $z = 2$. We can take advantage of this situation in order to evaluate $N_k(x)$ when $(2+\delta)\log_2 x \leq k \leq A\log_2 x$.

Theorem 6. *Let $0 < \delta < 1$, $A > 0$. Then, uniformly for $x \geq 3$, and for*

$$(2+\delta)\log_2 x \leq k \leq A\log_2 x,$$

we have

$$N_k(x) = C\frac{x\log x}{2^k}\Big\{1 + O_A\Big((\log x)^{-\delta^2/5}\Big)\Big\}, \tag{21}$$

with

$$C := \tfrac{1}{4}\prod_{p>2}\Big(1+\frac{1}{p(p-2)}\Big) \approx 0.378694.$$

Proof. Let $\varepsilon = \varepsilon(A)$ be a positive parameter to be specified later. First of all, we apply Cauchy's formula to (8) with $N = 0$, for the circle $|z| = 2-\varepsilon$. This gives

$$N_k(x) = \frac{x}{2\pi i}\oint_{|z|=2-\varepsilon}\nu(z)(\log x)^{z-1}z^{-k}\,\mathrm{d}z + O_\varepsilon\Big(x(2-\varepsilon)^{-k}(\log x)^{-\varepsilon}\Big), \tag{22}$$

where we have estimated trivially the contribution of the error term. For $k \leq A\log_2 x$, we have

$$(2-\varepsilon)^{-k}(\log x)^{-\varepsilon} \leq 2^{-k}(\log x)^{-A\log(1-\varepsilon/2)-\varepsilon} \leq 2^{-k}(\log x)^{1/2}$$

if $\varepsilon = \varepsilon(A)$ is sufficiently small.

It remains to evaluate the main term of (22). Since $\nu(z)$ has a pole at $z = 2$ with residue $-C$, we see that this quantity equals

$$C\frac{x\log x}{2^k} - \frac{x}{2\pi i}\oint_{|z|=2+\delta}\nu(z)(\log x)^{z-1}z^{-k}\,\mathrm{d}z.$$

The last integral is trivially

$$\ll_\delta (\log x)^{1+\delta}(2+\delta)^{-k} \leq (\log x)2^{-k}(\log x)^{-\eta(\delta)}$$

with

$$\eta(\delta) := 2\Big\{\big(1+\tfrac{1}{2}\delta\big)\log\big(1+\tfrac{1}{2}\delta\big) - \tfrac{1}{2}\delta\Big\} > \tfrac{1}{5}\delta^2.$$

This completes the proof.

In Exercise 3 we propose an extension of this result to large values of k, i.e. such that $A\log_2 x < k \leq (\log x)/\log 2$.

§ 6.2 The average distribution of divisors: the arcsine law

In Part I we obtained a certain amount of information about the average and extreme values of the total number of divisors $\tau(n)$ of an integer n. A natural question to ask is whether these divisors are distributed in the interval $[1, n]$ according to some definite law. Obviously, no result valid for all integers can be expected. Here we investigate a study on average.

For each integer n, let us define a random variable D_n taking the values $(\log d)/\log n$, as d runs through the set of the $\tau(n)$ of divisors of n, with uniform probability $1/\tau(n)$. The distribution function F_n of D_n is then defined by

$$F_n(u) := \text{Prob}\,(D_n \le u) = \tau(n)^{-1} \sum_{d|n,\, d\le n^u} 1 \qquad (0 \le u \le 1).$$

It is clear that the sequence $\{F_n\}_{n=1}^{\infty}$ does not converge pointwise on $[0, 1]$. However, we shall see the sequence of Cesàro means

$$G_N(u) := N^{-1} \sum_{n\le N} F_n(u)$$

is *uniformly* convergent on $[0, 1]$. Remarkably, the limit is the distribution function of a probability law well-known to specialists: the arcsine law, with density $1/\pi\sqrt{(u(1-u))}$. Large and small values have high probability: if D is a random variable with this distribution law, we have

$$\text{Prob}\,(D < 0.01 \text{ or } D > 0.99) \approx 0.128.$$

This indicates that, on average, an integer has many small (and correspondingly many large) divisors.

Theorem 7. *Uniformly for $x \ge 2$, $0 \le u \le 1$, we have*

$$x^{-1} \sum_{n\le x} F_n(u) = \frac{2}{\pi} \arcsin \sqrt{u} + O\Big(\frac{1}{\sqrt{(\log x)}}\Big). \tag{23}$$

This result is an easy consequence of the following theorem.

Theorem 8. *Let*

$$h := \prod_p \sqrt{(p(p-1))} \log\Big(\frac{1}{1-1/p}\Big) \approx 0.969.$$

Then, uniformly for $x \ge 2$, $d \ge 1$, we have

$$\sum_{n\le x} \frac{1}{\tau(nd)} = \frac{hx}{\sqrt{(\pi \log x)}} \Big\{g(d) + O\Big(\frac{(3/4)^{\omega(d)}}{\log x}\Big)\Big\}, \tag{24}$$

where g is an arithmetic function satisfying

$$\sum_{n\le x} g(d) = \frac{x}{h\sqrt{(\pi \log x)}} \Big\{1 + O\Big(\frac{1}{\log x}\Big)\Big\}. \tag{25}$$

Let us provisionally accept this theorem and see how it implies formula (23). The symmetry of the divisors of n about $\sqrt{n}$ allows us to write

$$\begin{aligned} F_n(u) &= \operatorname{Prob}(D_n \geq 1-u) = 1 - \operatorname{Prob}(D_n < 1-u) \\ &= 1 - F_n(1-u) + O\big(1/\tau(n)\big). \end{aligned}$$

Let $S(x,u)$ denote the left-hand side of (23). Summing the above equality for $n \leq x$ and evaluating the error term by (24) with $d = 1$, we obtain

$$S(x,u) + S(x,1-u) = 1 + O\Big(\frac{1}{\sqrt{(\log x)}}\Big) \qquad (0 \leq u \leq 1).$$

Since we also have $(2/\pi)\arcsin\sqrt{u} + (2/\pi)\arcsin\sqrt{(1-u)} = 1$ $(0 \leq u \leq 1)$, we infer that it suffices to establish (23) for $0 \leq u \leq \frac{1}{2}$.

With u in this range, we can write

$$\begin{aligned} S(x,u) &= \frac{1}{x}\sum_{n\leq x}\frac{1}{\tau(n)}\sum_{d|n,\, d\leq n^u} 1 \\ &= \frac{1}{x}\sum_{n\leq x}\frac{1}{\tau(n)}\Big\{\sum_{d|n,\, d\leq x^u} 1 - \sum_{d|n,\, n^u<d\leq x^u} 1\Big\} \\ &= \frac{1}{x}\sum_{d\leq x^u}\sum_{m\leq x/d}\frac{1}{\tau(md)} - R(x,u), \end{aligned} \tag{26}$$

with

$$\begin{aligned} R(x,u) &:= \frac{1}{x}\sum_{d\leq x^u}\sum_{m\leq x/d,\,(md)^u<d}\frac{1}{\tau(md)} \leq \frac{1}{x}\sum_{d\leq x^u}\sum_{m\leq x/d,\, m\leq d^{(1-u)/u}}\frac{1}{\tau(m)} \\ &\ll \frac{1}{x}\sum_{d\leq x^u} d^{(1-u)/u}\Big(1+\log d^{(1-u)/u}\Big)^{-1/2} \\ &\ll \Big(1+\log x^{1-u}\Big)^{-1/2} \ll (\log x)^{-1/2}. \end{aligned}$$

Using (24) to estimate the inner sum of (26), it follows that

$$S(x,u) = \frac{h}{\sqrt{\pi}}\sum_{d\leq x^u}\frac{1}{d\sqrt{(\log x/d)}}\Big\{g(d) + O\Big(\frac{(3/4)^{\omega(d)}}{\log x}\Big)\Big\} + O\Big(\frac{1}{\sqrt{(\log x)}}\Big)$$

where we have used the fact that $u \leq \frac{1}{2}$ to write $1/(\log x/d)$ as $O(1/\log x)$.

Bounding $(3/4)^{\omega(d)}$ by 1, we see that, in the summation over d, the contribution of the error terms is $\ll 1/\sqrt{(\log x)}$. The main term is handled by partial integration. Let us write

$$\mathcal{G}(t) := \sum_{n\le e^t} g(n) = \frac{e^t}{h\sqrt{(\pi t)}}\Big\{1+O\Big(\frac{1}{t}\Big)\Big\} \qquad (t>1).$$

We have

$$\begin{aligned}\frac{h}{\sqrt{\pi}}\sum_{d\le x^u}\frac{g(d)}{d\sqrt{(\log x/d)}} &= \frac{h}{\sqrt{\pi}}\int_{0-}^{u\log x}\frac{e^{-t}}{\sqrt{(\log x - t)}}\,\mathrm{d}\mathcal{G}(t)\\ &= \frac{h}{\sqrt{\pi}}\int_0^{u\log x}\mathcal{G}(t)\frac{e^{-t}}{\sqrt{(\log x-t)}}\Big\{1-\frac{1}{2(\log x - t)}\Big\}\,\mathrm{d}t + O\Big(\frac{1}{\sqrt{(\log x)}}\Big)\\ &= \frac{1}{\pi}\int_0^{u\log x}\frac{1+O(1/(t+1))}{\sqrt{(t(\log x - t))}}\,\mathrm{d}t + O\Big(\frac{1}{\sqrt{(\log x)}}\Big)\\ &= \frac{1}{\pi}\int_0^u\frac{dv}{\sqrt{(v(1-v))}} + O\Big(\frac{1}{\sqrt{(\log x)}}\Big) = \frac{2}{\pi}\arcsin\sqrt{u} + O\Big(\frac{1}{\sqrt{(\log x)}}\Big).\end{aligned}$$

This establishes (23).

Proof of Theorem 8. Introduce the generating series

$$f_d(s) := \sum_{n=1}^{\infty}\tau(nd)^{-1}n^{-s} \qquad (d=1,2,\ldots).$$

Although the function $n \mapsto \tau(nd)$ is not multiplicative, the formula

$$\tau(nd) = \prod_p (v_p(n)+v_p(d)+1)$$

enables us to express $f_d(s)$ in the form of a product of Eulerian type, viz.

$$f_d(s) = \prod_p\sum_{\nu=0}^{\infty}\frac{p^{-\nu s}}{\nu + v_p(d)+1} = g_d(s)f_1(s),$$

with

$$g_d(s) := \prod_{p^\alpha\| d}\Big(\sum_{\nu=0}^{\infty}\frac{p^{-\nu s}}{\nu+\alpha+1}\Big)\Big(\sum_{\nu=0}^{\infty}\frac{p^{-\nu s}}{\nu+1}\Big)^{-1}.$$

For each d, $g_d(s)$ is a finite product of ratios of series which are each absolutely convergent for $\sigma > 0$. For $\sigma \geq \frac{3}{4}$, the absolute values of the denominators are bounded from below by an absolute positive constant, indeed

$$\Big|\sum_{\nu=0}^{\infty} \frac{p^{-\nu s}}{\nu+1}\Big| \geq 1 - \sum_{\nu=1}^{\infty} \frac{p^{-\nu\sigma}}{\nu} = 1 + \log(1 - p^{-\sigma}) \geq 1 + \log(1 - 2^{-3/4}) > 0.$$

This implies that, for $\sigma \geq \frac{3}{4}$,

$$|g_d(s)| \leq \prod_{p^\alpha \| d} \Big\{\frac{1}{\alpha+1} + O(p^{-\sigma})\Big\} \leq C\big(\tfrac{3}{4}\big)^{\omega(d)}, \tag{27}$$

where C is an absolute constant.

Now, we can also write

$$h(s) := f_1(s)\zeta(s)^{-1/2} = \sum_{n=1}^{\infty} b(n)n^{-s}$$

where $b(n)$ is the multiplicative function determined by the identity

$$\sum_{\nu=0}^{\infty} b(p^\nu)\xi^\nu = \sum_{\nu=0}^{\infty} \frac{\xi^\nu}{\nu+1}(1-\xi)^{1/2}.$$

In particular $b(p) = 0$ for all p and

$$b(p^\nu) \ll \big(\tfrac{3}{2}\big)^\nu \qquad (\nu = 2, 3, \ldots).$$

This implies the absolute convergence of $h(s)$ for $\sigma > \frac{1}{2}$. The hypotheses of Theorem 5.3 are therefore fulfilled for $f_d(s)$, with $z = \frac{1}{2}$ and $M \ll (3/4)^{\omega(d)}$. From this, noting that $h(1) = h$, we deduce the validity of (24), with $g(d) = g_d(1)$.

It remains to show (25). To this end we appeal once more to the Selberg–Delange theorem 5.3 (or 5.5), noting that g is the multiplicative function defined by

$$g(p^\nu) := \Big(\sum_{j=0}^{\infty} \frac{p^{-j}}{j+\nu+1}\Big)\Big(\sum_{j=0}^{\infty} \frac{p^{-j}}{j+1}\Big)^{-1}.$$

For $\sigma > 0$, we can write

$$\sum_{n=1}^{\infty} g(n)n^{-s} = \zeta(s)^{1/2} \sum_{n=1}^{\infty} \beta(n)n^{-s}$$

where β is the multiplicative function determined by

$$\sum_{\nu=1}^{\infty}\beta(p^\nu)\xi^\nu=(1-\xi)^{1/2}\sum_{\nu=0}^{\infty}g(p^\nu)\xi^\nu\qquad(|\xi|<1). \tag{28}$$

Since $|g(p^\nu)|\le 1$, the right-hand side is holomorphic for $|\xi|<1$ and we have

$$\beta(p^\nu)\ll(\tfrac{3}{2})^\nu\qquad(\nu=1,2,\ldots).$$

In addition

$$\beta(p)=g(p)-\tfrac12=O\Big(\frac1p\Big).$$

This implies the absolute convergence of $\sum\beta(n)n^{-s}$ for $\sigma>\frac12$.

Theorem 5.3 then yields the formula

$$\sum_{n\le x}g(n)=\frac{Bx}{\surd(\pi\log x)}\Big\{1+O\Big(\frac{1}{\log x}\Big)\Big\}, \tag{29}$$

with

$$B:=\prod_p\sum_{\nu=0}^{\infty}\beta(p^\nu)p^{-\nu}.$$

Using (28), we can write the ν-sum in this product in the form

$$\begin{aligned}&\Big(\sum_{\nu=0}^{\infty}g(p^\nu)p^{-\nu}\Big)\Big(1-\frac1p\Big)\Big(1-\frac1p\Big)^{-1/2}\\ &=\Big(\sum_{j=0}^{\infty}\frac{p^{-j}}{j+1}\Big)^{-1}\Big(1-\frac1p\Big)^{-1/2}\sum_{\nu=0}^{\infty}\Big(1-\frac1p\Big)\sum_{j=0}^{\infty}\frac{p^{-\nu-j}}{\nu+j+1}\\ &=\Big(\sum_{j=0}^{\infty}\frac{p^{-j}}{j+1}\Big)^{-1}\Big(1-\frac1p\Big)^{-1/2}=\Big\{\surd(p(p-1))\log\Big(\frac{1}{1-1/p}\Big)\Big\}^{-1}.\end{aligned}$$

We therefore obtain

$$B=1/h,$$

which concludes the proof of (25).

Notes

§ **6.1.** For the study of $N_k(x)$ and $\pi_k(x)$ by induction on k for bounded k, see Hardy & Wright (1938), § 22.18. Theorems 1–5 are due to Selberg (1954) in the case $N = 0$. For $N \geq 0$, arbitrary but fixed, they are special cases of results of Delange (1971).

The asymptotic formula contained in the estimate of Theorem 6 was announced by Selberg (1954). Using an elementary method, Nicolas (1984) extended this result by showing that we have, uniformly for $x \geq 3$ and $(2+\delta)\log_2 x \leq k \leq \log x/\log 2$,

$$N_k(x) = Cy\log y\Big\{1 + O\big((\log y)^{-\eta}\big)\Big\} \tag{30}$$

with $y := x/2^k$, and $\eta = \eta(\delta) > 0$. Nicolas uses the fact that for these large values of k, a dominating contribution to $N_k(x)$ arises from integers of the form $2^\ell m$ with $\ell \geq \frac{1}{2}k$ and $\Omega(m) \sim 2\log_2 y$—see Exercise 3, where we outline an analytic proof based on the Selberg–Delange method.

Balazard, Delange & Nicolas (1988) make precise the behaviour of $N_k(x)$ when k is close to $2\log_2 x$: for $|k - 2\log_2 x| \leq A\surd(\log_2 x)$, we have

$$N_k(x) = C\Phi\Big(\frac{k - 2\log_2 x}{\surd(2\log_2 x)}\Big)\frac{x\log x}{2^k}\Big\{1 + O_A\Big(\frac{1}{\surd(\log_2 x)}\Big)\Big\}$$

where

$$\Phi(z) := \frac{1}{\surd(2\pi)}\int_{-\infty}^{z} e^{-t^2/2}\,\mathrm{d}t.$$

In his thesis, Balazard (1987) unified the study of $N_k(x)$ by providing an analytic approximation to $N_k(x)$, valid uniformly for $1 \leq k \leq (\log x)/\log 2$. See also Balazard, Delange & Nicolas (1988).

As might be expected, it is possible, under additional assumptions concerning the derivatives of higher order of $h_0(z)$, to make formula (13) more precise. For example, if $h^{(4)}(z) \ll B$ for $|z| \leq A$, we obtain (with $r := (k-1)/\log_2 x$) that

$$\begin{aligned} C_k(x) = &\frac{x}{\log x}\frac{(\log_2 x)^{k-1}}{(k-1)!} \\ &\times\left\{h_0(r) - \frac{rh_0''(r)}{2\log_2 x} + O\left(\frac{B(k-1)^2}{(\log_2 x)^4} + \frac{\log_2 x}{k}R_0(x)\right)\right\}. \end{aligned}$$

This immediately implies corresponding refinements for Theorems 4 and 5.

The problem of the asymptotic behaviour of $\pi_k(x)$ for $k > A\log_2 x$ is much more complicated than that of $N_k(x)$, since no single prime number plays a special role. It is only recently that the limit of validity of Selberg's formula

$$\pi_k(x) \sim \lambda\Big(\frac{k-1}{\log_2 x}\Big)\frac{x}{\log x}\frac{(\log_2 x)^{k-1}}{(k-1)!} \tag{31}$$

has been determined. Hensley showed in 1987 that (31) holds if, and only if,

$$k = o\big((\log_2 x/\log_3 x)^2\big).$$

By a different method, Hildebrand & Tenenbaum (1988) showed that, uniformly for $k \ll (\log_2 x)^2$,

$$\pi_k(x) = \lambda(r)\frac{x}{\log x}\frac{(\log_2 x)^{k-1}}{(k-1)!}e^{-kh/2}\Big\{1 + O\Big(\frac{1+r}{\log_2 x}\Big)\Big\} \tag{32}$$

with $r := (k-1)/\log_2 x$, and

$$h := \frac{\log(2 + Ar\log r)\log(2 + Br\log r)}{(\log_2 x)^2},$$

where A and B are absolute positive constants. The same article contains numerous other facts about the behaviour of $\pi_k(x)$ for large values of k, notably the relation

$$\frac{\pi_{k+1}(x)}{\pi_k(x)} \sim \frac{L}{k} \qquad \Big(1 \le k \ll \frac{\log x}{(\log_2 x)^2}\Big)$$

with

$$L := \log\Big(\frac{\log x}{k\log(k+1)}\Big).$$

For further information on this subject, in particular concerning "very large" values of k ($k \gg \log x/\log_2 x$) see Pomerance (1984). Balazard (1990) has recently shown that, for x sufficiently large, the sequence $k \mapsto \pi_k(x)$ is unimodal.

Theorems 7 and 8 are due to Deshouillers, Dress & Tenenbaum (1979). As these authors remark, the remainder term $O\big(1/\surd(\log x)\big)$ in formula (23) is optimal if uniformity in u is required: for $u \in [0, (\log 2)/\log x[$ we have $F_n(u) = 1/\tau(n)$ $(n \le x)$, so that

$$S(x,u) \sim hx/\surd(\pi\log x).$$

Exercises

1. *A formula of Delange* (1959). In this exercise, we put $\xi = \xi(x) = \log_2 x$, and set

$$T(x,u) := \sum_{n\le x} u^{\omega(n)} \qquad (x\ge 1,\ 0<u<\infty).$$

(a) Show that for all $\varepsilon>0$, $u\le 1\le v$, $x\ge 1$, one has

$$|\{n\le x : |\omega(n)-\xi|>\varepsilon\xi\}| \le u^{-(1-\varepsilon)\xi}T(x,u)+v^{-(1+\varepsilon)\xi}T(x,v).$$

Deduce that there exists an absolute constant c such that

$$\operatorname{card}\{n\le x : |\omega(n)-\xi|>\xi^{3/5}\} \ll x\exp\{-c\xi^{1/5}\} \qquad (x\ge 3).$$

(b) Show that, for $|k-\xi|\le\xi^{3/5}$, one has

$$e^{-\xi}\frac{\xi^k}{k!} \sim \frac{1}{\sqrt{(2\pi\xi)}}e^{-\theta^2/2} \qquad (\xi\to\infty)$$

with $\theta := (k-\xi)/\sqrt{\xi}$. Deduce the asymptotic estimate of Delange, valid for each integer $m\ge 0$,

$$x^{-1}\sum_{n\le x}\big(\omega(n)-\xi\big)^m = \{\mu_m+o(1)\}\xi^{m/2} \qquad (x\to\infty) \tag{33}$$

with

$$\mu_m := \frac{1}{\sqrt{(2\pi)}}\int_{-\infty}^{+\infty} t^m e^{-t^2/2}\,\mathrm{d}t = \begin{cases} 0 & (\text{if } m=2k+1)\\ \dfrac{(2k)!}{2^k k!} & (\text{if } m=2k).\end{cases}$$

2. Give an explicit expression for the remainder term in formula (33) from the previous exercise.

3. *Nicolas' theorem (1984).* Throughout this exercise, we write $y := x/2^k$. Let $N_k(x)$ (resp. $N_k'(x)$) denote the number of integers (resp. odd integers) not exceeding x and such that $\Omega(n)=k$.

(a) Show that for all $x\ge 1$, $k\ge 1$, one has $N_k(x)=\sum_{j=0}^k N_j'(2^j y)$.

(b) Show that for all δ, $0<\delta<1$, and uniformly for $|z|\le 3-\delta$, $x\ge 3$, one has

$$\sum_{\substack{n\le x\\ n\equiv 1 (\mathrm{mod}\, 2)}} z^{\Omega(n)} = x(\log x)^{z-1}\Big\{zh(z)+O_\delta\Big(\frac{1}{\log x}\Big)\Big\}$$

with

$$h(z) := \frac{2^{-z}}{\Gamma(z+1)}\prod_{p>2}\Big(1-\frac{z}{p}\Big)^{-1}\Big(1-\frac{1}{p}\Big)^z.$$

(c) Using the previous result with $z = \frac{5}{2}$, establish the upper bound

$$N'_j(x) \ll x(\log x)^{3/2}(\tfrac{2}{5})^j \qquad (x \geq 3,\ j \geq 1).$$

(d) Set $Y := \log_2 3y$, and assume that $y \geq 1$. Show that

$$\sum_{j > \frac{5}{2}Y} N'_j(2^j y) \ll y(\log 3y)^{1-c_1}$$

with $c_1 = \frac{5}{2}\log\frac{5}{4} - \frac{1}{2} > \frac{1}{20}$.

(e) Deduce from part (b) that, uniformly for $x \geq 3$, $j \leq \frac{5}{2}\log_2 x$, one has

$$N'_j(x) = \frac{x}{\log x}\Big\{Q_j(\log_2 x) + O\Big(\frac{(\log_2 x)^j}{j!\log x}\Big)\Big\}$$

with $Q_j(X) := \displaystyle\sum_{m+\ell=j-1} \frac{1}{m!\,\ell!}h^{(m)}(0)X^\ell.$

(f) Show that $\dfrac{1}{m!}h^{(m)}(0) \ll (\frac{3}{8})^m \quad (m \geq 0)$ and deduce that

$$Q_j(X) \ll (\tfrac{3}{8})^j e^{8X/3} \quad (j > \tfrac{5}{2}X), \qquad Q'_j(X) \ll \frac{X^{j-2}}{(j-2)!} \quad (2 \leq j \leq \tfrac{5}{2}X).$$

(g) Deduce from parts (e) and (f) that, for $y \geq 1$, $j \leq \frac{5}{2}Y$, one has

$$N'_j(2^j y) = \frac{2^j y}{\log 3y}\Big\{Q_j(Y) + O\Big(\frac{Y^{j+1}e^{-Y}}{j!}\Big)\Big\}.$$

(h) Show that $\sum_{j=1}^{\infty} 2^j Q_j(Y) = 2h(2)e^{2Y}$ and that

$$\sum_{j > \frac{5}{2}Y} 2^j Q_j(Y) \ll e^{(2-c_2)Y}$$

with $c_2 := \frac{5}{2}\log\frac{4}{3} - \frac{2}{3} > \frac{1}{20}$.

(i) Show that for all $k \in \big[\frac{5}{2}Y,\ (\log x)/\log 2\big]$, one has

$$N_k(x) = Cy\log 3y\Big\{1 + O\big((\log 3y)^{-1/20}\big)\Big\},$$

with $C := 2h(2) = \frac{1}{4}\displaystyle\prod_{p>2}\Big(1 + \frac{1}{p(p-2)}\Big).$

4. *The Erdős–Kac theorem (1939).*

(a) Deduce from the central limit theorem in probability theory that for each real number λ

$$\lim_{x\to\infty} \sum_{k\le x+\lambda\sqrt{x}} e^{-x}\frac{x^k}{k!} = \frac{1}{\sqrt{(2\pi)}}\int_{-\infty}^{\lambda} e^{-t^2/2}\,\mathrm{d}t.$$

(b) Put $\varphi_x(t) := e^{-x}x^t/\Gamma(t+1)$. Show that for $x \ge 1$, $|t-x| \ll \sqrt{x}$, one has

$$\varphi_x(t) = \frac{1}{\sqrt{(2\pi x)}}\exp\{-(t-x)^2/2x\}\bigl(1+O(1/\sqrt{x})\bigr).$$

Evaluate the Stieltjes integral $\int_{-\infty}^{\lambda} \varphi_x(t)\,\mathrm{d}[t]$ and thus recover the result of the previous question.

(c) Applying the estimates obtained by the Selberg–Delange method for $\pi_k(x) := \bigl|\{n \le x : \omega(n) = k\}\bigr|$, prove the Erdős–Kac theorem:

$$(\forall\lambda\in\mathbb{R})\ \lim_{x\to\infty}\frac{1}{x}\bigl|\{n\le x:\omega(n)\le \log_2 x+\lambda\sqrt{(\log_2 x)}\}\bigr| = \frac{1}{\sqrt{(2\pi)}}\int_{-\infty}^{\lambda} e^{-t^2/2}\,\mathrm{d}t.$$

What explicit error term can you obtain ?

II.7

Tauberian theorems

§ 7.1 Introduction: Abelian/Tauberian theorems duality

In the course of the previous chapters we have developed a variety of methods for evaluating the summatory function of an arithmetic sequence, based on the analytic continuation of its Dirichlet series. In each case the study of this continuation beyond the domain of convergence revealed itself to be critical. Here, on the other hand, we aim at results which yield the asymptotic behaviour of summatory functions using the generating series exclusively at points of convergence. This type of theorem has the remarkable feature of being easy to apply, due to the weakening of the hypotheses. This advantage, however, is balanced by the poor quality of the error terms in standard applications—although it can usually be shown that these error terms are essentially optimal at the level of generality of the statements.

Let us begin by stating a classical result of Abel.

Theorem 1 (Abel). *Let $f(z) := \sum_{n=0}^{\infty} a_n z^n$ be a power series with radius of convergence 1, converging for $z = 1$. For any real number θ, $0 \le \theta < \frac{\pi}{2}$, and any sector*

$$S(\theta) := \{z : |z| < 1, \, |\arg(1-z)| \le \theta\},$$

we have

$$\lim_{z\to 1,\, z\in S(\theta)} f(z) = f(1).$$

Proof. Let $z = 1 - re^{i\varphi} \in S(\theta)$, with $r > 0$, $|\varphi| \le \theta$. We have

$$|z|^2 = 1 - 2r\cos\varphi + r^2 < 1, \ \text{ so } r < 2\cos\varphi.$$

Since $r \to 0$ as $z \to 1$, we can suppose that $r \le \delta < 2\cos\theta$ and choose δ so small that

$$|\log z| = \left|\log|z| + i\arctan\left(\frac{r\sin\varphi}{1 - r\cos\varphi}\right)\right| \le 2r \qquad (|\varphi| \le \theta).$$

The assumption of convergence at $z = 1$ allows us to associate an integer $N = N(\varepsilon)$ to each $\varepsilon > 0$ such that

$$\sup_{t\ge N}\Big|\sum_{N<n\le t} a_n\Big| \le \varepsilon.$$

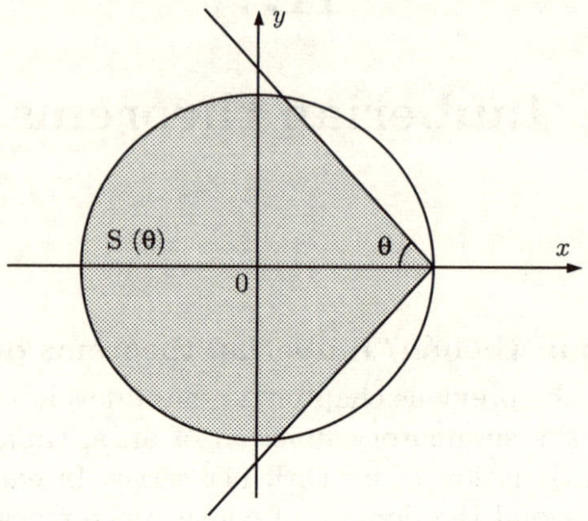

Let us write $A_N(t) := \sum_{N<n\leq t} a_n$. For all $M > N$, we have

$$\Big|\sum_{N<n\leq M} a_n z^n\Big| = \Big|\int_N^M z^t \,\mathrm{d}A_N(t)\Big| = \Big|A_N(M)z^M - \log z \int_N^M z^t A_N(t)\,\mathrm{d}t\Big|$$
$$\leq \varepsilon + 2r\varepsilon \int_N^M |z|^t \,\mathrm{d}t \leq \varepsilon(1 - 2r/\log|z|).$$

Noting that

$$\log|z| = \tfrac{1}{2}\log\left(1 - 2r\cos\varphi + r^2\right) \leq -r\cos\varphi + \tfrac{1}{2}r^2 \leq -\tfrac{1}{2}r(2\cos\theta - \delta),$$

we see that the preceding bound does not exceed

$$\varepsilon\Big(1 + \frac{4}{2\cos\theta - \delta}\Big).$$

This establishes the uniform convergence of the power series $f(z)$ in $S(\theta)$ and thereby completes the proof.

Theorem 1 is the prototype of a class of statements called *Abelian* because they share with it the following characteristic: they state that, if a sequence (or a function) is sufficiently regular, then a certain average of its values also has a regular behaviour. Thus the well-known implication

$$\lim_{n\to\infty} a_n = a \Rightarrow \lim_{n\to\infty} \frac{1}{n}\sum_{m=0}^{n} a_m = a$$

is an Abelian theorem. Setting $b_n := \sum_{m=0}^{n} a_m$ (so that $f(z) = \sum_{m=0}^{\infty} a_m z^m = (1-z)\sum_{n=0}^{\infty} b_n z^n$), Theorem 1 can equally well be written as

$$\lim_{n\to\infty} b_n = b \Rightarrow \lim_{z\to 1,\ z\in S(\theta)} (1-z)\sum_{n=0}^{\infty} b_n z^n = b.$$

The converse of an Abelian theorem is in general false. For example, the sum of the power series

$$f(z) = \sum_{n=0}^{\infty} (-1)^n z^n = (1+z)^{-1}$$

tends to $\frac{1}{2}$ as $z \to 1$, but the series is divergent at $z = 1$. A *Tauberian* theorem is one which provides a sufficient condition for such an inverse to be true. The first result of this type is due to the German mathematician A. Tauber (1897)—cf. §2.

We conclude this section with an Abelian theorem concerning Dirichlet series.

Theorem 2. *Let $F(s) := \sum_{n=1}^{\infty} a_n n^{-s}$ be a Dirichlet series convergent for $\sigma > a$. If there exist two constants c, ω, with $\omega > -1$, such that*

$$\sum_{n\le x} a_n = \left\{\frac{c}{\Gamma(\omega+1)} + o(1)\right\} x^a (\log x)^\omega \qquad (x\to\infty),$$

then we have

$$F(\sigma) = \frac{ca + o(1)}{(\sigma - a)^{\omega+1}} \qquad (\sigma \to a+).$$

Proof. Let us write $A(t) := \sum_{n\le e^t} a_n$ and

$$G(h) := \int_{0-}^{+\infty} e^{-(a+h)t}\,\mathrm{d}A(t) \qquad (h > 0).$$

Using the formula

$$\frac{c}{\Gamma(\omega+1)}\int_0^{+\infty} e^{-(a+h)t} t^\omega \,\mathrm{d}\{e^{at}\} = \frac{ca}{\Gamma(\omega+1)}\int_0^{+\infty} t^\omega e^{-ht}\,\mathrm{d}t = \frac{ca}{h^{\omega+1}},$$

we can write

$$\begin{aligned} G(h) - \frac{ca}{h^{\omega+1}} &= (a+h)\int_0^{+\infty} e^{-(a+h)t} A(t)\,\mathrm{d}t - \frac{ca}{h^{\omega+1}} \\ &= \int_0^{+\infty} e^{-(a+h)t}\left\{(a+h)A(t) - \frac{ac}{\Gamma(\omega+1)} e^{at} t^\omega\right\}\mathrm{d}t. \end{aligned}$$

By hypothesis there exists a function $\varepsilon(t)$, with $\lim_{t\to\infty}\varepsilon(t)=0$, such that the expression between curly brackets may be written as

$$\varepsilon(t)e^{at}t^{\omega}+O(he^{at}t^{\omega}).$$

This implies the required conclusion in the form

$$G(h)-\frac{ca}{h^{\omega+1}}=o\Big(\frac{1}{h^{\omega+1}}\Big)\qquad(h\to 0+).$$

§ 7.2 Tauber's theorem

In its original form, Tauber's theorem (1897) is the exact converse of Abel's theorem.

Theorem 3 (Tauber). *Let $f(z):=\sum_{n=0}^{\infty}a_nz^n$ be a power series with radius of convergence 1. Suppose that, for a suitable complex number ℓ, we have $\lim_{z\to 1,\,0\le z<1}f(z)=\ell$. Then the additional condition*

$$\text{(T)}\qquad \sum_{n\le x}na_n=o(x)\qquad(x\to\infty)$$

is sufficient to ensure that the series $\sum a_n$ is convergent and has sum ℓ.

Proof. Write $z:=e^{-\sigma}$ $(\sigma>0)$ and $A(t):=\sum_{n\le t}a_n$ $(t\in\mathbb{R})$. The hypotheses can be translated as

$$\text{(1)}\qquad f(e^{-\sigma})=\int_{0-}^{+\infty}e^{-\sigma t}\,\mathrm{d}A(t)\to\ell\qquad(\sigma\to 0+)$$

and

$$\text{(2)}\qquad \alpha(x):=x^{-1}\int_{0-}^{x}t\,\mathrm{d}A(t)\to 0\qquad(x\to+\infty).$$

For each pair of positive real numbers $\{\lambda,x\}$, we can write

$$\begin{aligned}A(\lambda x)-A(x)&=\int_x^{\lambda x}\mathrm{d}A(t)=\int_x^{\lambda x}t^{-1}\,\mathrm{d}\{t\alpha(t)\}\\&=\int_x^{\lambda x}t^{-1}\alpha(t)\,\mathrm{d}t+\alpha(\lambda x)-\alpha(x).\end{aligned}$$

This implies, for each *fixed* λ, that

$$\text{(3)}\qquad \lim_{x\to\infty}\{A(\lambda x)-A(x)\}=0.$$

In addition, since $\alpha(t)$ vanishes in the neighbourhood of 0, condition (2) implies the existence of a constant K such that $|\alpha(t)| \leq K$ $(t \geq 0)$. We deduce from this that

$$(4) \qquad |A(\lambda x) - A(x)| \leq K(2 + |\log \lambda|) \qquad (\lambda > 0,\, x > 0).$$

Now we have for all $\sigma > 0$

$$f(e^{-\sigma}) = \int_{0-}^{+\infty} e^{-\sigma t}\,\mathrm{d}A(t) = \sigma \int_0^{+\infty} e^{-\sigma t} A(t)\,\mathrm{d}t = \int_0^{+\infty} e^{-t} A(t/\sigma)\,\mathrm{d}t,$$

whence

$$f(e^{-\sigma}) - A(1/\sigma) = \int_0^{+\infty} e^{-t}\{A(t/\sigma) - A(1/\sigma)\}\,\mathrm{d}t.$$

The relations (3) and (4) show that, as $\sigma \to 0+$, the integrand tends to 0 pointwise on $\mathbb{R}^+$ and is bounded above independently of σ by an integrable function. Lebesgue's theorem then implies that

$$\lim_{\sigma \to 0+} \{f(e^{-\sigma}) - A(1/\sigma)\} = 0.$$

The desired result hence follows from (1).

Note that condition (T) is actually also *necessary* for the convergence of $\sum a_n$. Indeed, using the notation introduced in the course of the proof, we can write by partial summation

$$\alpha(x) = A(x) - x^{-1} \int_0^x A(t)\,\mathrm{d}t = x^{-1} \int_0^x \{A(x) - A(t)\}\,\mathrm{d}t.$$

Thus $A(x) \to \ell$ immediately forces $\alpha(x)$ to tend to 0.

Finally we can formulate Tauber's theorem in integral form as follows.

Theorem 4 (Tauber). *Let A be a function of bounded variation on any finite interval, with $A(0) = 0$. Suppose that the Laplace–Stieltjes integral*

$$F(\sigma) := \int_0^{+\infty} e^{-\sigma t}\,\mathrm{d}A(t)$$

converges for $\sigma > 0$ and satisfies

$$\lim_{\sigma \to 0+} F(\sigma) = \ell.$$

Then the two following assertions are equivalent:

(i) $\qquad A(x) = \ell + o(1) \qquad (x \to \infty)$

(ii) $\qquad x^{-1} \int_0^x t\,\mathrm{d}A(t) = o(1) \qquad (x \to \infty).$

Considering this statement, we can both enlarge and make more precise the notion of a Tauberian theorem. Given a real-valued function $\varphi(t,s)$ defined on $\mathbb{R}^+ \times S$ where $S \subseteq \mathbb{C}$, we define the φ-transform of a function A of bounded variation on any finite interval by

$$F(s) := \int_0^{+\infty} \varphi(t,s)\,\mathrm{d}A(t) \tag{5}$$

whenever the integral is convergent. Let us next assume that the following Abelian theorem holds for some number s_0 in the closure of S: *if* $\lim_{t\to\infty} A(t) = \ell$, *then the integral* (5) *converges for all* s *in* S, *and we have*

$$\lim_{s\to s_0,\, s\in S} F(s) = \ell. \tag{6}$$

In these circumstances, we shall say that a theorem is *Tauberian* if it provides a sufficient condition to deduce from (6) that $\lim_{t\to\infty} A(t) = \ell$.

The scope of the adjective "Tauberian" can be further extended to results which establish that $A(t)$ has a given behaviour at infinity (for example the existence of a limit for some ψ-transform) under an assumption of type (6)—see in particular chapter 4 of Bingham, Goldie & Teugels (1987).

§ 7.3 The theorems of Hardy–Littlewood and Karamata

Condition (T) of Theorem 3 readily follows from the assumption

$$a_n = o(1/n) \qquad (n \to \infty).$$

In 1913, Hardy & Littlewood showed that the one-sided condition

$$a_n \geq -K/n \qquad (n \geq 1),$$

where K is an arbitrary constant, suffices instead. The following theorem, due to Karamata (1931), allows a new, simpler proof of this result.

Theorem 5 (Karamata). *Let* $A(t)$ *be a non-decreasing function such that the integral*

$$F(\sigma) := \int_0^{\infty} e^{-\sigma t}\,\mathrm{d}A(t)$$

converges for all $\sigma > 0$. *Suppose that there exist two real numbers* $c \geq 0$, $\omega > 0$, *such that*

$$F(\sigma) = \{c + o(1)\}\sigma^{-\omega} \qquad (\sigma \to 0+).$$

Then we have

$$A(x) = \{c + o(1)\}\frac{x^{\omega}}{\Gamma(\omega+1)} \qquad (x \to \infty).$$

Proof. We can assume without loss of generality that $A(0) = 0$. Let n be a non-negative integer. We have

$$F\big((n+1)\sigma\big) = \int_0^\infty e^{-n\sigma t} e^{-\sigma t}\,\mathrm{d}A(t) = \frac{\{c+o(1)\}}{\Gamma(\omega)\sigma^\omega}\,\frac{\Gamma(\omega)}{(n+1)^\omega}$$
$$= \frac{\{c+o(1)\}}{\Gamma(\omega)\sigma^\omega}\int_0^\infty e^{-nt}e^{-t}t^{\omega-1}\,\mathrm{d}t \qquad (\sigma\to 0+).$$

From this it follows that, for any fixed polynomial P, we have

$$(7)\qquad \int_0^\infty P(e^{-\sigma t})e^{-\sigma t}\,\mathrm{d}A(t) = \frac{\{c+o(1)\}}{\Gamma(\omega)\sigma^\omega}\int_0^\infty P(e^{-t})e^{-t}t^{\omega-1}\,\mathrm{d}t \quad (\sigma\to 0+).$$

We shall show that this relation remains true when P is replaced by the function χ defined on $[0,1]$ by

$$\chi(e^{-t}) = \begin{cases} e^t & (0\le t\le 1) \\ 0 & (t>1) \end{cases}$$

and when σ tends to 0 in such a way that $1/\sigma$ avoids the points of discontinuity of A. Assuming this for the moment, it follows that

$$A\Big(\frac{1}{\sigma}\Big) = \int_0^\infty \chi(e^{-\sigma t})e^{-\sigma t}\,\mathrm{d}A(t) = \frac{\{c+o(1)\}}{\Gamma(\omega)\sigma^\omega}\int_0^\infty \chi(e^{-u})e^{-u}u^{\omega-1}\,\mathrm{d}u$$
$$= \frac{\{c+o(1)\}}{\Gamma(\omega)\sigma^\omega}\int_0^1 u^{\omega-1}\,\mathrm{d}u = \frac{\{c+o(1)\}}{\Gamma(\omega+1)\sigma^\omega}.$$

By our monotonicity hypothesis on A, this asymptotic relation is actually valid without restriction on the values of σ, and the stated conclusion follows.

In order to show that (7) is valid for χ, it is natural to approximate χ by polynomial functions. Let us consider the function $H(t)$ defined on $[0,1]$ by

$$(8)\qquad H(t) := \frac{\chi(t)-t}{t(1-t)} \quad (0<t<1), \quad H(0)=-1,\ H(1)=2.$$

It possesses a unique point of discontinuity at $t = 1/e$. For each $\varepsilon > 0$, there then exist continuous functions f and g satisfying

$$(9)\qquad \begin{cases} f(t)\le H(t)\le g(t) & (0\le t\le 1) \\ g(t)-f(t)\le\varepsilon & (|t-1/e|>\varepsilon) \\ g(t)-f(t)\le 12 & (|t-1/e|\le\varepsilon). \end{cases}$$

(The number 12 serves here only to exceed the value of the discontinuity of H at $t = 1/e$.) From the Weierstrass approximation theorem there exist polynomials p and q such that

$$\max_{0\le t\le 1} |p(t) - f(t)| \le \varepsilon, \qquad \max_{0\le t\le 1} |q(t) - g(t)| \le \varepsilon. \tag{10}$$

Setting

$$P(t) := t + t(1-t)\big(p(t) - \varepsilon\big), \qquad Q(t) := t + t(1-t)\big(q(t) + \varepsilon\big),$$

we deduce from the above that

$$P(t) \le \chi(t) \le Q(t) \qquad (0 \le t \le 1) \tag{11}$$

and, for $0 < \varepsilon < \frac{1}{2}$,

$$\int_0^1 \frac{Q(t) - P(t)}{t(1-t)}\,\mathrm{d}t \le \int_0^1 \big(q(t) - p(t)\big)\,\mathrm{d}t + 2\varepsilon.$$

It is easily deduced from (9) and (10) that the last integral does not exceed 27ε. By adjusting the value of ε in the preceding calculations, we therefore obtain the existence of two polynomials P and Q, which satisfy (11) and

$$\int_0^1 \frac{Q(t) - P(t)}{t(1-t)}\,\mathrm{d}t \le \varepsilon. \tag{12}$$

It is now easy to establish that (7) holds for the function χ and when σ tends towards $0+$ avoiding the discontinuities of $A(1/t)$. Under these conditions we indeed have

$$\begin{aligned} A(1/\sigma) &= \int_0^\infty \chi(e^{-\sigma t})e^{-\sigma t}\,\mathrm{d}A(t) \\ &\le \int_0^\infty Q(e^{-\sigma t})e^{-\sigma t}\,\mathrm{d}A(t) = \frac{\{c + o(1)\}}{\Gamma(\omega)\sigma^\omega} \int_0^\infty Q(e^{-t})e^{-t}t^{\omega-1}\,\mathrm{d}t. \end{aligned}$$

Similarly

$$A(1/\sigma) \ge \frac{\{c + o(1)\}}{\Gamma(\omega)\sigma^\omega} \int_0^\infty P(e^{-t})e^{-t}t^{\omega-t}\,\mathrm{d}t.$$

Now,

$$\int_0^\infty {}^{Q}_{P}(e^{-t})e^{-t}t^{\omega-1}\,\mathrm{d}t \lesseqgtr \int_0^\infty \chi(e^{-t})e^{-t}t^{\omega-1}\,\mathrm{d}t \pm R$$

with

$$R := \int_0^\infty \{Q(e^{-t}) - P(e^{-t})\} e^{-t} t^{\omega-1}\, dt$$

$$= \int_0^1 \Big\{\frac{Q(t) - P(t)}{t(1-t)}\Big\} t(1-t)\Big(\log\frac{1}{t}\Big)^{\omega-1} dt \le \varepsilon M$$

by (12), where $M := \sup_{0\le t\le 1} t(1-t)\{\log(1/t)\}^{\omega-1}$ is independent of σ and ε. By making successively σ and then ε tend to 0, we obtain the expected formula.

Remark. When $\omega = 0$, the conclusion of Theorem 5 remains true in the form

$$A(x) - A(0) \to c \qquad (x \to \infty).$$

This follows trivially from the partial integration formula

$$F(\sigma) = -A(0) + \sigma \int_0^\infty e^{-\sigma t} A(t)\, dt = \int_0^\infty e^{-t}\big\{A(t/\sigma) - A(0)\big\}\, dt.$$

Since A is non-decreasing, $A(t/\sigma)$ tends, for each t, to a limit $\alpha := \sup A(t)$. By Lebesgue's theorem, we then have

$$c = \lim F(\sigma) = \int_0^\infty e^{-t}\{\alpha - A(0)\}\, dt = \alpha - A(0).$$

In order to deduce the Hardy–Littlewood theorem from that of Karamata, we will need a lemma of Landau (1906), which is also of independent interest.

Theorem 6 (Landau). *Let f be a twice differentiable function on $\mathbb{R}^+$, α a real number, and M a positive constant. If we have*

(a) $\quad f(x) = o(x^\alpha) \qquad (x \to +\infty,$ *resp.* $0+)$

(b) $\quad f''(x) \le M x^{\alpha-2} \qquad (x > 0)$

then

$$f'(x) = o(x^{\alpha-1}) \qquad (x \to +\infty,\ \textit{resp. } 0+).$$

Proof. Let δ be a fixed real number, $0 < \delta \le \frac{1}{2}$. For all $x > 0$, Taylor's formula implies that

$$f(x \pm \delta x) - f(x) = \pm\delta x f'(x) + \delta^2 x^2 \int_0^1 (1-t) f''\big(x(1 \pm \delta t)\big)\, dt.$$

The integral does not exceed

$$M x^{\alpha-2}\{(1+\delta)^{\alpha-2} + (1-\delta)^{\alpha-2}\} \le K x^{\alpha-2}$$

where K is independent of δ. By (a) we can then write

$$o(x^\alpha) \le \pm\delta x f'(x) + K\delta^2 x^\alpha \qquad (x \to +\infty, \text{ resp. } 0+).$$

Dividing the two sides by δx^α and making x tend to $+\infty$ or $0+$, we obtain

$$\lim \begin{matrix}\sup\\ \inf\end{matrix}\, x^{1-\alpha} f'(x) \begin{matrix}\le\\ \ge\end{matrix} \pm K\delta.$$

The conclusion follows, since δ is arbitrarily small.

Theorem 7 (Hardy–Littlewood). *Let* $f(z) := \sum_{n=0}^{\infty} a_n z^n$ *be a power series with radius of convergence* 1 *and whose coefficients satisfy*

$$na_n \geq -K \qquad (n \geq 0) \tag{13}$$

for some suitable constant K*. If*

$$\lim_{z \to 1-} f(z) = \ell, \tag{14}$$

then we have $\sum_{n=0}^{\infty} a_n = \ell$*.*

Proof. By modifying a_0 if necessary, we can assume that $\ell = 0$. Consider the function

$$g(\sigma) = \sum_{n=0}^{\infty} a_n e^{-n\sigma}.$$

Hypothesis (14) implies that $g(\sigma) = o(1)$ as $\sigma \to 0+$ and, by (13), we have

$$g''(\sigma) = \sum_{n=0}^{\infty} n^2 a_n e^{-n\sigma} \geq -K \sum_{n=0}^{\infty} n e^{-n\sigma} = -Ke^{-\sigma}(1 - e^{-\sigma})^{-2}.$$

Applying Theorem 6 to $-g$, with $\alpha = 0$, it follows that

$$g'(\sigma) = \sum_{n=0}^{\infty} na_n e^{-n\sigma} = o(1/\sigma) \qquad (\sigma \to 0+),$$

from which

$$\sum_{n=0}^{\infty} (na_n + K)e^{-n\sigma} = \{K + o(1)\}/\sigma \qquad (\sigma \to 0+).$$

Karamata's theorem then gives the estimate

$$\sum_{n \leq x} (na_n + K) = \{K + o(1)\}x \qquad (x \to \infty),$$

that is to say

$$\sum_{n \leq x} na_n = o(x) \qquad (x \to \infty).$$

The desired conclusion then follows from Tauber's theorem.

In the context of Dirichlet series, we can formulate the following result, for which the reader will be able to supply the proof without difficulty.

Theorem 8 (Hardy–Littlewood–Karamata). *Let* $F(s) := \sum_{n=1}^{\infty} a_n n^{-s}$ *be a Dirichlet series convergent for* $\sigma > 1$. *Suppose that there exist real numbers* c, K, ω, *with* $\omega > 0$, *such that we have*

$$a_n \geq -K(\log n)^{\omega-1} \qquad (n \geq 2) \tag{15}$$

and

$$F(\sigma) = \frac{\{c + o(1)\}}{(\sigma - 1)^{\omega}} \qquad (\sigma \to 1+). \tag{16}$$

Then we have

$$\sum_{n \leq x} \frac{a_n}{n} = \frac{\{c + o(1)\}}{\Gamma(\omega + 1)} (\log x)^{\omega} \qquad (x \to +\infty). \tag{17}$$

Remark. Condition (15) can be replaced, more generally, by $a_n \geq -b_n$ $(n \geq n_0)$ where the b_n are the coefficients of a Dirichlet series satisfying both (16) and (17) for some suitable constant c, but with the same ω. For example, a condition of this type is

$$a_n \geq -K_1(\log n)^{\omega-1} - K_2\tau(n)^{\alpha} \qquad (n \geq 2) \tag{18}$$

with $\alpha := \log\omega/\log 2$.

§ 7.4 The remainder term in Karamata's theorem

During the nineteen fifties, Tauberian theory evolved in the direction of finding effective forms for the results, that is to say asymptotic formulae with explicit error terms. We give here a remainder version of Theorem 5 which is due to Freud (1952).

Theorem 9 (Karamata–Freud). *Let* $A(t)$ *be a non-decreasing function such that the integral*

$$F(\sigma) := \int_0^{\infty} e^{-\sigma t}\, \mathrm{d}A(t)$$

converges for all $\sigma > 0$. *Suppose that there exist two real numbers* $c \geq 0$, $\omega > 0$ *and a non-decreasing function* $\psi(t)$ *such that*

$$\psi(t) \to \infty, \qquad t^{-\omega}\psi(t) \text{ is non-increasing for large } t \tag{19}$$

and

$$F(\sigma) = \left\{c + O\left(\psi(1/\sigma)^{-1}\right)\right\}\sigma^{-\omega} \qquad (\sigma \to 0+). \tag{20}$$

Then we have

$$A(x) = \left\{c + O\left(\frac{1}{\log\psi(x)}\right)\right\}\frac{x^{\omega}}{\Gamma(\omega + 1)} \qquad (x \to +\infty). \tag{21}$$

It is important to note right from the outset that the error term in (21) cannot be improved without additional assumptions, as shown by the following counter-example, due to Karamata (1952). Consider the increasing function

$$A(x) := \int_0^x \left(1 + \cos\{(\log t)^2\}\right) \mathrm{d}t.$$

On the one hand, we have

$$A(x) = x + \Omega_\pm\Big(\frac{x}{\log x}\Big). \tag{22}$$

This follows by simple partial integration

$$\begin{aligned}\int_2^x \cos\left\{(\log t)^2\right\} \mathrm{d}t &= \int_2^x \frac{t}{2\log t}\,\mathrm{d}\sin\left\{(\log t)^2\right\}\\ &= \left[\frac{t}{2\log t}\sin\left\{(\log t)^2\right\}\right]_2^x - \int_2^x \Big(1-\frac{1}{\log t}\Big)\frac{\sin\{(\log t)^2\}}{2\log t}\,\mathrm{d}t\\ &= \frac{x}{\log x}\left(\tfrac12 \sin\left\{(\log x)^2\right\} + O\Big(\frac{1}{\log x}\Big)\right)\end{aligned}$$

where the last integral has been estimated by a second integration by parts.

On the other hand, we shall see that we have

$$F(\sigma) = \frac{1}{\sigma} + O(1) \qquad (\sigma \to 0+). \tag{23}$$

It is clear that $F(\sigma) = \dfrac{1}{\sigma} + \Re e\, J(\sigma) + O(1)$, with

$$J(\sigma) := \int_1^\infty \exp\left\{-\sigma t + i(\log t)^2\right\} \mathrm{d}t.$$

We estimate $J(\sigma)$ by replacing the integral over $[1, +\infty[$ by the complex integral over the contour formed by the circular arc $\Gamma := \left\{e^{i\theta} : 0 \le \theta \le \pi/4\right\}$ and the half-line $\Delta := \left\{re^{i\pi/4} : r \ge 1\right\}$. This transformation is justified for all $\sigma > 0$ by the bound

$$\left|\exp\left\{-\sigma z + i(\log z)^2\right\}\right| \le R^{-2\theta} e^{-\sigma R\cos\theta} \qquad \Big(z = Re^{i\theta},\ 0 \le \theta \le \frac{\pi}{4}\Big). \tag{24}$$

Now the integral over Γ is clearly $O(1)$, and the estimate (24) applied with $\theta = \pi/4 > \frac12$ shows that the integral over Δ is absolutely convergent and is bounded independently of $\sigma > 0$. This implies (23).

The idea behind the proof of Theorem 9 consists in making explicit the polynomial approximations for the function χ arising in the proof of Karamata's theorem. Precision is measured in the L^1-norm, and it is necessary to control the size of the coefficients. We define the *length* $\ell(P)$ of a polynomial $P(x) := \sum_{m=0}^n a_m x^m$ by

$$\ell(P) := \sum_{0\le m\le n} |a_m|.$$

We prove the following one-sided approximation result.

Theorem 10. *Let $f(t)$ be a function of bounded variation on $[0,1]$. There exist constants A_1, A_2 depending only on f, such that, for any integer $n \ge 1$, there exist two polynomials p and q, of degree at most n, satisfying*

$$\begin{cases} p(t) \le f(t) \le q(t) & (0 \le t \le 1) \\ \int_0^1 \big(q(t)-p(t)\big)\,\mathrm{d}t \le A_1/n \\ \ell(p)+\ell(q) \le A_2^n. \end{cases} \tag{25}$$

Initially let us assume this result and see how we can deduce Theorem 9 from it.

By the second of the assumptions (19), relation (20) implies that, for any integer $m \ge 1$,

$$\int_0^\infty e^{-m\sigma t}e^{-\sigma t}\,\mathrm{d}A(t) = \frac{1}{\sigma^\omega}\Big\{\frac{c}{\Gamma(\omega)}\int_0^\infty e^{-mt}e^{-t}t^{\omega-1}\,\mathrm{d}t + O\big(\psi(1/\sigma)^{-1}\big)\Big\}.$$

By linearity, we deduce from this that, for any polynomial P,

$$\begin{aligned} &\int_0^\infty P(e^{-\sigma t})e^{-\sigma t}\,\mathrm{d}A(t) \\ &\qquad = \frac{1}{\sigma^\omega}\Big\{\frac{c}{\Gamma(\omega)}\int_0^\infty P(e^{-t})e^{-t}t^{\omega-1}\,\mathrm{d}t + O\big(\ell(P)\psi(1/\sigma)^{-1}\big)\Big\}. \end{aligned} \tag{26}$$

We now apply Theorem 10 to the function H defined by (8), and we set $P(t) := t + t(1-t)p(t)$, $Q(t) := t + t(1-t)q(t)$. We have $P(t) \le \chi(t) \le Q(t)$ for all t in $[0,1]$ and from (26) it follows that, when $1/\sigma$ is not a discontinuity of A, we have

$$A(1/\sigma) \lesseqgtr \frac{1}{\sigma^\omega}\Big\{\frac{c}{\Gamma(\omega+1)} \pm R(\sigma)\Big\},$$

with

$$\begin{aligned} R(\sigma) &\ll \int_0^\infty \{Q(e^{-t})-P(e^{-t})\}e^{-t}t^{\omega-1}\,\mathrm{d}t + A_2^n\,\psi(1/\sigma)^{-1} \\ &= \int_0^1 \{q(t)-p(t)\}t(1-t)\Big(\log\frac{1}{t}\Big)^{\omega-1}\,\mathrm{d}t + A_2^n\,\psi(1/\sigma)^{-1} \\ &\ll (A_1/n) + A_2^n\,\psi(1/\sigma)^{-1}. \end{aligned}$$

For the choice $n := \left[\log\psi(1/\sigma)/(2\log A_2)\right]$, we obtain the required estimate (21).

Proof of Theorem 10. Let Y be the Heaviside function defined by

$$Y(t) = \begin{cases} 0 & (t < 0) \\ 1 & (t \geq 0). \end{cases}$$

We shall show the existence of absolute constants B_1, B_2 such that, for each $n \geq 1$, there exist polynomials R, S of degree at most n satisfying

$$(27) \qquad \begin{cases} R(t) \leq Y(t) \leq S(t) \qquad (-1 \leq t \leq 1) \\ \int_{-1}^{1} \left(S(t) - R(t)\right) \mathrm{d}t \leq B_1/n \\ \ell(R) + \ell(S) \leq B_2^n. \end{cases}$$

The general statement follows from this special case. Indeeed, if f_1 and f_2 are two non-decreasing functions such that $f = f_1 - f_2$, the polynomials defined on $[0,1]$ by

$$p(t) := f(0) + \int_0^1 R(t-\xi)\,\mathrm{d}f_1(\xi) - \int_0^1 S(t-\xi)\,\mathrm{d}f_2(\xi)$$

$$q(t) := f(0) + \int_0^1 S(t-\xi)\,\mathrm{d}f_1(\xi) - \int_0^1 R(t-\xi)\,\mathrm{d}f_2(\xi)$$

satisfy conditions (25) with

$$A_1 = B_1 \int_0^1 \mathrm{d}(f_1+f_2)(\xi), \qquad A_2 = 1 + 2|f(0)| + 2B_2 \int_0^1 \mathrm{d}(f_1+f_2)(\xi).$$

In order to construct R and S we appeal to the properties of the Chebyshev polynomials $T_r(x)$ defined on $[-1,1]$ by

$$T_r(x) := \cos ru = \sum_{2m \leq r} (-1)^m \binom{r}{2m} x^{r-2m}(1-x^2)^m \qquad (x = \cos u).$$

It is useful to note right away that

$$(28) \quad \ell(T_r) = \sum_{k \leq r} \Bigg| \sum_{2m \leq r,\, r-2m+2h=k} (-1)^m \binom{r}{2m}\binom{m}{h} \Bigg| \leq \sum_{\ell,h} \binom{r}{\ell}\binom{\ell}{h} = 3^r.$$

We denote the zeros of $T_r(x)$ by

$$x_\nu := \cos\left(\frac{\pi}{r}\left(\nu - \tfrac{1}{2}\right)\right) \qquad (1 \leq \nu \leq r),$$

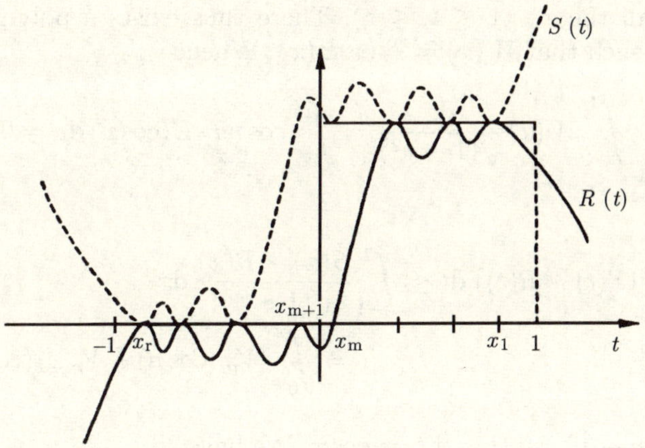

and set $m := \left[\frac{1}{2}(r+1)\right]$, so that $x_{m+1} < 0 \le x_m$. We then define $R(t)$ as the unique polynomial of degree $n = 2r - 2$ satisfying

$$R(x_\nu) = \begin{cases} 1 & (1 \le \nu \le m-1) \\ 0 & (m \le \nu \le r), \end{cases} \qquad R'(x_\nu) = 0 \quad (\nu \ne m).$$

Then $R'(t)$ vanishes at least once on each of the $(r-2)$ intervals $]x_{\nu+1}, x_\nu[$ $(1 \le \nu \le r,\, \nu \ne m-1)$. This demonstrates the existence of $r-1+r-2 = n-1$ zeros, so that $R'(t)$ does not vanish for $x_m \le t < x_{m-1}$ and $R'(x_m) > 0$ since $R(x_{m-1}) = 1 > R(x_m) = 0$. This implies that $R(t)$ has local maxima at all the x_ν $(\nu \ne m)$ and a minimum between each of these consecutive maxima. In particular, we do have $R(t) \le Y(t)$ $(-1 \le t \le 1)$, as requested in (27).

Analogously, we define $S(t)$ by the equations

$$S(x_\nu) = \begin{cases} 1 & (\nu \le m+1) \\ 0 & (m+2 \le \nu \le r), \end{cases} \qquad S'(x_\nu) = 0 \quad (\nu \ne m+1).$$

and check that this implies $S(t) \ge Y(t) \quad (-1 \le t \le 1)$.

In order to obtain the second of the properties (27), we introduce the polynomials

$$V_\nu(x) := T_r'(x_\nu)^{-2}\left(\frac{T_r(x)}{x - x_\nu}\right)^2 \qquad (1 \le \nu \le r). \tag{29}$$

We have

$$V_\nu(x_j) = \begin{cases} 1 & (j = \nu) \\ 0 & (j \ne \nu) \end{cases}, \qquad V_\nu'(x_j) = 0 \quad (j \ne \nu).$$

The polynomial $W(x) := S(x) - R(x) - V_m(x) - V_{m+1}(x)$ has degree n and vanishes at all the x_ν $(1 \le \nu \le r)$. There thus exists a polynomial $Z(x)$ of degree $r-2$ such that $W(x) = T_r(x)Z(x)$. Whence

$$\int_{-1}^{1} W(x)\frac{\mathrm{d}x}{\sqrt{(1-x^2)}} = \int_0^\pi \cos ru \cdot Z(\cos u)\,\mathrm{d}u = 0$$

and so

$$\int_0^1 \{S(t) - R(t)\}\,\mathrm{d}t \le \int_{-1}^{1} \frac{S(x)-R(x)}{\sqrt{(1-x^2)}}\,\mathrm{d}x = \int_0^\pi \{V_m(\cos u) + V_{m+1}(\cos u)\}\,\mathrm{d}u. \tag{30}$$

Set $\tau := \frac{\pi}{r}\left(m - \frac{1}{2}\right)$, so that $x_m = \cos\tau$. We have

$$T_r'(\cos\tau) = r\frac{\sin r\tau}{\sin\tau} = \pm\frac{r}{\sin\tau}.$$

Substituting in (29) with $\nu = m$ yields

$$\begin{aligned}\int_0^\pi V_m(\cos u)\,\mathrm{d}u &= \left(\frac{\sin\tau}{r}\right)^2 \int_0^\pi \left(\frac{\cos ru}{\cos u - \cos\tau}\right)^2 \mathrm{d}u \\ &= \left(\frac{\sin\tau}{r}\right)^2 \int_0^\pi \left\{\frac{\sin\left(\frac{1}{2}\,r(u-\tau)\right)}{\sin\left(\frac{1}{2}(u-\tau)\right)} \cdot \frac{\sin\left(\frac{1}{2}\,r(u+\tau)\right)}{\sin\left(\frac{1}{2}(u+\tau)\right)}\right\}^2 du.\end{aligned}$$

We can assume that $r \ge 3$. Then for u in $[0,\pi]$, we have

$$\tfrac{1}{4}\pi(1 - 2/r) \le \tfrac{1}{2}(u+\tau) \le \tfrac{3}{4}\pi.$$

The integral over u is thus

$$\ll \int_{-\pi}^{\pi} \left\{\frac{\sin\left(\frac{1}{2}rv\right)}{\sin\left(\frac{1}{2}v\right)}\right\}^2 \mathrm{d}v = 2\pi r,$$

from the classic form of the Fejér kernel

$$\left(\frac{\sin\frac{1}{2}rv}{\sin\frac{1}{2}v}\right)^2 = r + 2\sum_{j=1}^{r-1}(r-j)\cos(jv).$$

We have therefore shown that

$$\int_0^\pi V_m(\cos u)\,\mathrm{d}u \ll 1/r \ll 1/n.$$

A similar calculation provides the same estimate for the integral concerning V_{m+1}. Substituting in (30), we thus obtain the second property (27).

It remains to bound the lengths of R and S. Consider, for example, the case of R. We have

$$R(x) = \sum_{j=0}^{m-1} M_j(x)$$

where $M_j(x)$ is the unique polynomial of degree n satisfying

$$M_j(x_\nu) = \begin{cases} 1 & (\nu = j) \\ 0 & (\nu \neq j) \end{cases}, \qquad M_j'(x_\nu) = 0 \quad (\nu \neq m).$$

It thus suffices to estimate the length of the M_j. We have

$$(31) \qquad M_j(x) = r^{-2}\Big(\frac{u_j x + v_j}{x - x_m}\Big)\Big(\frac{T_r(x)}{x - x_j}\Big)^2$$

where u_j and v_j are determined by the equations

$$M_j(x_j) = 1, \qquad M_j'(x_j) = 0.$$

An easy calculation enables us to show that

$$u_j = 1 + x_j x_m - 2x_j^2, \qquad v_j = x_j^3 - x_m.$$

We therefore have in particular $\max(|u_j|, |v_j|) \leq 2$. From (28), we can write

$$T_r(x) = \theta_r \prod_{\nu=1}^{r} (x - x_\nu)$$

with $|\theta_r| \leq 3^r$. The desired upper bound for $\ell(M_j)$ then follows easily from the formula

$$M_j(x) = r^{-2}\theta_r(u_j x + v_j)(x - x_m) \prod_{\substack{1 \leq \nu \leq r \\ \nu \neq j,m}} (x - x_\nu)^2.$$

This completes the proof of Theorem 10.

§ 7.5 Ikehara's theorem

In terms of Dirichlet series, the Hardy–Littlewood–Karamata theorem might be described as a *limit Tauberian theorem*, in the sense that it links the behaviour of $\sum_{n=1}^{\infty} a_n n^{-\sigma}$ for $\sigma > 1$ to that of $\sum_{n \leq x} a_n n^{-1}$. It makes no assumptions about the values of the series at non-real points, and thus does not yield a proof of the prime number theorem—see the notes for this chapter for more detail concerning this last point.

This section is devoted to a Tauberian theorem different in nature. It concerns the obtaining of asymptotic information about the summatory function $\sum_{n \leq x} a_n$ using assumptions on the values of the Dirichlet series at non-real points $s = \sigma + i\tau$ with abscissa $\sigma > 1$. Schematically, one passes from the half-plane $\sigma > 1$ to the point $s = 0$, and this discontinuity between the hypothesis and the conclusion gives us ground for calling a result of this kind a *transcendental Tauberian theorem.*

Ikehara's theorem (1931) belongs to this category. We give a formulation below which reinforces it in two directions. On the one hand, we generalise the assumption by considering series having a singularity of type $s^{-\omega-1}$ $(\omega > -1)$ at $s = 0$—this form, which is today classical, is due to Ingham (1935). On the other hand, we give an effective version of the conclusion, with an explicit error term. Such a result does not seem to exist as such in the literature, but follows rather easily from the method of Ganelius (1971).

Theorem 11 ("Effective" Ikehara–Ingham). *Let $A(t)$ be a non-decreasing function such that the integral*

$$F(s) := \int_0^{\infty} e^{-st} \,\mathrm{d}A(t) \tag{32}$$

converges for $\sigma > a > 0$. Suppose that there exist constants $c \geq 0$, $\omega > -1$, such that the function

$$G(s) := \frac{F(s+a)}{s+a} - \frac{c}{s^{\omega+1}} \qquad (\sigma > 0) \tag{33}$$

satisfies

$$\eta(\sigma, T) := \sigma^{\omega} \int_{-T}^{T} |G(2\sigma + i\tau) - G(\sigma + i\tau)| \,\mathrm{d}\tau = o(1) \qquad (\sigma \to 0+) \tag{34}$$

for each fixed $T > 0$. Then we have

$$A(x) = \Big\{ \frac{c}{\Gamma(\omega+1)} + O\big(\rho(x)\big) \Big\} e^{ax} x^{\omega} \qquad (x \geq 1), \tag{35}$$

with

$$\rho(x) := \inf_{T \geq 32(a+1)} \{T^{-1} + \eta(1/x, T) + (Tx)^{-\omega-1}\}.$$

Furthermore, the implicit constant in (35) depends only on a, c and ω. An admissible choice for this constant is

$$52 + 1652c(a+1)(\omega+1) + 69c\big(1 + (\omega+1)e^{1-\omega}(\omega+1)^{\omega+2}\big)/\Gamma(\omega+1).$$

Ganelius' argument rests on a local version of Bohr's inequality (1935)

$$\|f\|_\infty \ll \|f'\|_\infty / T, \tag{36}$$

which is valid, when both sides are finite, for any $\mathbb{R}$-integrable function with Fourier transform $\widehat{f}(\tau)$ vanishing for $|\tau| \leq T$. Here we are content with a slightly weakened form of Ganelius' result, which is however sufficient for the applications envisaged. The proof depends on ideas similar to those in the classical inequality of Berry–Esseen—see for example Feller (1971).

Theorem 12 (Ganelius). *Let g be an integrable and bounded function on $\mathbb{R}$. Suppose there is a real positive number T such that*

$$\sup_{x \leq y \leq x+1/T} \{g(y) - g(x)\} \leq K < \infty, \tag{37}$$

and

$$\widehat{g}(\tau) := \int_{-\infty}^{+\infty} e^{-i\tau x} g(x)\,\mathrm{d}x = 0 \qquad (|\tau| \leq T). \tag{38}$$

Then we have

$$\|g\|_\infty := \sup_{x \in \mathbb{R}} |g(x)| \leq 16K. \tag{39}$$

Remark. We have made no attempt to optimise the constant 16 appearing in (39).

Proof. We can suppose that $T = 1$: the general case will follow by considering $g(x/T)$. Assumption (38) implies that

$$\widehat{g}(\tau)\widehat{\chi}(\tau) = 0 \qquad (\tau \in \mathbb{R}) \tag{40}$$

for any integrable function χ having Fourier transform with support contained in $[-1, 1]$. We choose

$$\chi(t) = \frac{1}{2\pi}\left(\frac{\sin \frac{1}{2}t}{\frac{1}{2}t}\right)^2$$

so that $\widehat{\chi}(\tau) = \max(1 - |\tau|, 0)$. Relation (40) implies that

$$\int_{-\infty}^{+\infty} g(x-t)\chi(t)\,\mathrm{d}t = 0 \qquad (x \in \mathbb{R}) \tag{41}$$

and the stated inequality will follow by using the fact that $\chi(t)$ "peaks" at $t = 0$. More precisely, we shall appeal to the inequality

$$I := \int_{|t|>5} \chi(t)\,\mathrm{d}t \le \frac{4}{\pi}\int_5^\infty \frac{\mathrm{d}t}{t^2} = \frac{4}{5\pi}. \tag{42}$$

Let ε, $0 < \varepsilon < 1$, be a real fixed number and let $\theta = \pm 1$ be so chosen that

$$\|g\|_\infty = \sup_{x \in \mathbb{R}}\{\theta g(x)\}.$$

There then exists an $x_0 = x_0(\varepsilon)$ such that $\theta g(x_0) \ge (1-\varepsilon)\|g\|_\infty$. Applying (41) with $x = x_0 - 5\theta$, and taking account of (42), we can write

$$\begin{aligned}
0 &= \theta\int_{-\infty}^{+\infty} g(x_0 - t - 5\theta)\chi(t)\,\mathrm{d}t \\
&= \theta\int_{-5}^{5} g(x_0)\chi(t)\,\mathrm{d}t - \theta\int_{-5}^{5}\{g(x_0) - g(x_0 - t - 5\theta)\}\chi(t)\,\mathrm{d}t \\
&\qquad + \theta\int_{|t|>5} g(x_0 - t - 5\theta)\chi(t)\,\mathrm{d}t \\
&\ge (1-\varepsilon)(1-I)\|g\|_\infty - 10K(1-I) - I\|g\|_\infty,
\end{aligned}$$

where we have used the fact that, for all t such that $|t| \le 5$, we have

$$\theta\{g(x_0) - g(x_0 - t - 5\theta)\} \le \sup_{x<y\le x+10}\{g(y) - g(x)\} \le 10K.$$

We therefore obtain

$$\{(1-\varepsilon)(1-I) - I\}\|g\|_\infty \le 10K(1-I)$$

and, letting ε tend to 0,

$$\|g\|_\infty \le \frac{10K(1-I)}{1-2I} \le 16K,$$

as stated.

An application of Theorem 12 on the lines of Ganelius' work would lead to an effective Tauberian theorem with rather unusual conditions, and which would not entirely generalise the Ikehara–Ingham theorem. We proceed in a slightly different manner by noting that Theorem 12 easily implies a result which simultaneously contains the Berry–Esseen inequality (cf. §7.6) and Theorem 11.

Theorem 13. *Let g be an integrable and bounded function on $\mathbb{R}$. Under assumption (37), we have*

$$\|g\|_\infty \leq 16K + 6\int_{-T}^{T} |\widehat{g}(\tau)|\,\mathrm{d}\tau.$$

Proof. Let $\varepsilon > 0$. We shall apply Theorem 12 to $g - f$ where f is the convolution of g with the integrable function

$$\alpha(t) := \frac{2}{\pi\varepsilon t^2}\sin\left(\varepsilon t/2\right)\sin\left((2T+\varepsilon)t/2\right)$$

whose Fourier transform is the trapezoidal function

$$\widehat{\alpha}(\tau) = \begin{cases} 1 & (|\tau| \leq T) \\ (T+\varepsilon-|\tau|)/\varepsilon & (T < |\tau| \leq T+\varepsilon) \\ 0 & (|\tau| > T+\varepsilon). \end{cases}$$

Since $\widehat{f} = \widehat{g}\,\widehat{\alpha}$ has compact support, we have

$$\|f\|_\infty \leq \frac{1}{2\pi}\|\widehat{f}\|_1 \leq \frac{1}{2\pi}\int_{-T-\varepsilon}^{T+\varepsilon} |\widehat{g}(\tau)|\,\mathrm{d}\tau. \tag{43}$$

It follows that

$$\sup_{x \leq y \leq x+1/T} \left\{g(y) - f(y) - \big(g(x) - f(x)\big)\right\} \leq K + 2\|f\|_\infty,$$

from which, by Theorem 12,

$$\|g\|_\infty \leq \|g - f\|_\infty + \|f\|_\infty \leq 16K + 33\|f\|_\infty\,.$$

Since $33/2\pi \leq 6$, the stated result follows by making ε tend to 0.

Proof of Theorem 11. Without loss of generality, we may assume that $A(0+) = 0$ and define $A(t) := 0$ for $t \leq 0$. Initially, we apply Theorem 13 to the function

$$g_\sigma(t) := A(t)e^{-(a+\sigma)t}(1 - e^{-\sigma t})$$

whose Fourier transform is given, for any positive value of the parameter σ, by the formula

$$\widehat{g}_\sigma(\tau) = G(\sigma + i\tau) - G(2\sigma + i\tau) + c\left\{(\sigma+i\tau)^{-\omega-1} - (2\sigma+i\tau)^{-\omega-1}\right\}.$$

The modulus of the last term does not exceed

$$c(\omega+1)\Big|\int_{\sigma+i\tau}^{2\sigma+i\tau} s^{-\omega-2}\,\mathrm{d}s\Big| \le c(\omega+1)\sigma|\sigma+i\tau|^{-\omega-2}.$$

In view of (34), we can thus write

$$\begin{aligned}\int_{-T}^{T}|\widehat{g}_\sigma(\tau)|\,\mathrm{d}\tau &\le \sigma^{-\omega}\Big\{\eta(\sigma,T)+c(\omega+1)\sigma^{\omega+1}\int_{-T}^{T}\max(\sigma,|\tau|)^{-\omega-2}\,\mathrm{d}\tau\Big\}\\ &\le \sigma^{-\omega}\big\{\eta(\sigma,T)+2c(\omega+2)\big\}.\end{aligned}$$

Then the monotonicity and the positivity of A imply that, for $x>0$, $y\ge 0$,

$$\begin{aligned}g_\sigma(x+y)-g_\sigma(x) &\ge A(x)e^{-(a+\sigma)x}(1-e^{-\sigma x})(e^{-(a+\sigma)y}-1)\\ &\ge -(a+\sigma)\|g_\sigma\|_\infty y.\end{aligned}\tag{44}$$

This inequality is also satisfied when $x\le 0$, since then $g_\sigma(x)=0$. Applying Theorem 13 to $-g_\sigma$, with $K=(a+\sigma)\|g_\sigma\|_\infty/T$, we deduce, for $0<\sigma\le 1$,

$$\|g_\sigma\|_\infty \le 6\sigma^{-\omega}\big\{\eta(\sigma,T)+2c(\omega+2)\big\}+16(a+1)\|g_\sigma\|_\infty/T.$$

Choosing $T=T_a:=32(a+1)$, we obtain

$$\|g_\sigma\|_\infty \le M_1(\sigma)\sigma^{-\omega}\qquad(0<\sigma\le 1),\tag{45}$$

with

$$M_1(\sigma):=12\big\{\eta(\sigma,T_a)+2c(\omega+2)\big\}.$$

Let us set

$$B(t):=\begin{cases}\dfrac{c}{\Gamma(\omega+1)}e^{-t}(1-e^{-t})t^\omega & (t>0)\\ 0 & (t\le 0).\end{cases}$$

The second step of the proof consists in a further application of Theorem 13, this time to the function

$$G_\sigma(t):=g_\sigma(t)-\sigma^{-\omega}B(\sigma t)=\Big(A(t)e^{-at}-\frac{c}{\Gamma(\omega+1)}t^\omega\Big)e^{-\sigma t}(1-e^{-\sigma t})$$

with Fourier transform

$$\widehat{G}_\sigma(\tau)=G(\sigma+i\tau)-G(2\sigma+i\tau).$$

For all $t > 0$, we have

$$\begin{aligned} B'(t) &= \frac{c}{\Gamma(\omega+1)} e^{-t} t^{\omega-1}\{2te^{-t} - t + \omega(1-e^{-t})\} \\ &\le \frac{c(\omega+1)}{\Gamma(\omega+1)} e^{-t}(1-e^{-t})t^{\omega-1} \le \frac{c(\omega+1)}{\Gamma(\omega+1)} e^{-t} t^{\omega}. \end{aligned}$$

Furthermore, B is continuous on $\mathbb{R}$ and so is equal to the integral of its derivative, defined, for example, by 0 at $t = 0$. Therefore, we have, on the one hand, for $x < 0$, $x + y \ge 0$,

$$B(x+y) - B(x) \le \int_0^y \frac{c(\omega+1)}{\Gamma(\omega+1)} e^{-t} t^{\omega} \, dt \le \frac{c}{\Gamma(\omega+1)} y^{\omega+1},$$

and on the other hand, for $x \ge 0$, $y \ge 0$,

$$\begin{aligned} B(x+y) - B(x) &\le \frac{c(\omega+1)}{\Gamma(\omega+1)} e^{-x} \int_x^{x+y} t^{\omega} \, dt \\ &\le \frac{c}{\Gamma(\omega+1)} e^{-x} \Big\{\theta(\omega)(\omega+1)(x+y)^{\omega} y + (1-\theta(\omega)) y^{\omega+1}\Big\} \end{aligned}$$

with $\theta(\omega) := 1$ for $\omega \ge 0$, and $\theta(\omega) := 0$ for $-1 < \omega < 0$. In the case $\omega < 0$, we have used the classical Minkowski inequality

$$(x+y)^{\omega+1} - x^{\omega+1} \le y^{\omega+1} \qquad (-1 < \omega < 0).$$

By an easy calculation, we may deduce that, for $x \in \mathbb{R}$, $0 \le y \le 1/T \le 1$, $0 < \sigma \le 1$, we have

$$B(\sigma x + \sigma y) - B(\sigma x) \le D\Big\{\frac{\sigma}{T} + \Big(\frac{\sigma}{T}\Big)^{\omega+1}\Big\},$$

with

$$D := \frac{c}{\Gamma(\omega+1)} \big(1 + e^{1-\omega}(\omega+1)^{\omega+2}\big).$$

Indeed, it can be checked that $D \ge \dfrac{c}{\Gamma(\omega+1)} \max\{1, \theta(\omega)(\omega+1)e \sup_{t \ge 0} e^{-t} t^{\omega}\}$.

By (44) and (45), we obtain under the same conditions

$$\begin{aligned} G_\sigma(x+y) - G_\sigma(x) &\ge -\{(a+\sigma)\|g_\sigma\|_\infty / T + \sigma^{-\omega}\big(B(\sigma x + \sigma y) - B(\sigma x)\big)\} \\ &\ge -\sigma^{-\omega}\Big\{M_2(\sigma)/T + D(\sigma/T)^{\omega+1}\Big\} \end{aligned}$$

with $M_2(\sigma) := (a+1)M_1(\sigma) + D = 12(a+1)\eta(\sigma, T_a) + 24c(a+1)(\omega+2) + D$. By then applying Theorem 13 to $-G_\sigma$, we infer that, for $T \geq T_a$,

$$\begin{aligned}|G_\sigma(x)| &\leq \sigma^{-\omega}\Big\{16M_2(\sigma)T^{-1} + 6\eta(\sigma,T) + 16D(\sigma/T)^{\omega+1})\Big\}\\ &\leq \sigma^{-\omega}\Big\{\frac{384c(a+1)(\omega+2)+16D}{T} + 12\eta(\sigma,T) + 16D\Big(\frac{\sigma}{T}\Big)^{\omega+1}\Big\}\\ &\leq M\sigma^{-\omega}\{T^{-1} + \eta(\sigma,T) + (\sigma/T)^{\omega+1}\},\end{aligned}$$

with

$$\begin{aligned}M &:= \max\big\{384c(a+1)(\omega+2) + 16D,\ 12\big\}\\ &\leq 12 + 384c(a+1)(\omega+2) + 16c\big(1 + (\omega+1)e^{1-\omega}(\omega+1)^{\omega+2}\big)/\Gamma(\omega+1).\end{aligned}$$

The conclusion follows by choosing $\sigma = 1/x$. The implicit constant in (35) does not exceed $eM/(1-e^{-1})$.

Ikehara's theorem implies the prime number theorem assuming only that $\zeta(s) \neq 0$ for $\sigma = 1$, without requiring an upper bound for $\zeta(s)^{-1}$—cf. Exercise 4. Of course any explicit upper bound for $\zeta'(s)/\zeta(s)$ or for $\zeta(s)^{-1}$ will provide, via Theorem 11, a corresponding effective version of the prime number theorem—see, in particular, Exercises 5 and 6.

§ 7.6 The Berry–Esseen inequality

As mentioned in the previous section, we now show how Theorem 13 implies the probabilistic inequality of Berry–Esseen. Conforming to usage, we say that the real function F of a real variable is a *distribution function* if it is non-decreasing and satisfies

$$F(-\infty) = 0, \qquad F(+\infty) = 1. \tag{46}$$

The *characteristic function* of F is the Fourier–Stieltjes transform

$$f(\tau) := \int_{-\infty}^{+\infty} e^{i\tau x}\,\mathrm{d}F(x).$$

Theorem 14 (Berry–Esseen). *Let F, G be two distribution functions with respective characteristic functions f, g. Suppose that G is differentiable and that G' is bounded on $\mathbb{R}$. Then we have for all $T > 0$*

$$\|F - G\|_\infty \leq 16\frac{\|G'\|_\infty}{T} + 6\int_{-T}^{T}\Big|\frac{f(\tau)-g(\tau)}{\tau}\Big|\,\mathrm{d}\tau. \tag{47}$$

Proof. Write $H := F - G$, and, for each value of the positive parameter ε, introduce the function

$$(48)\qquad H_\varepsilon(x) := -\int_0^\infty e^{-\varepsilon t}\,\mathrm{d}H(x-t) = e^{-\varepsilon x}\int_{-\infty}^x e^{\varepsilon t}\,\mathrm{d}H(t).$$

It is easily verified that H_ε is integrable and that its Fourier transform has the value

$$\widehat{H}_\varepsilon(\tau) = \int_{-\infty}^{+\infty} e^{-i\tau x} H_\varepsilon(x)\,\mathrm{d}x = \frac{f(-\tau) - g(-\tau)}{\varepsilon + i\tau}.$$

The second equality (48) gives, by partial integration,

$$(49)\qquad H_\varepsilon(x) = H(x) - \varepsilon e^{-\varepsilon x}\int_{-\infty}^x e^{\varepsilon t} H(t)\,\mathrm{d}t.$$

Since $\|H\|_\infty \leq 1$ and $H(-\infty) = 0$, we easily deduce that we have for any fixed x

$$\lim_{\varepsilon\to 0+} H_\varepsilon(x) = H(x).$$

Write $\alpha := \|G'\|_\infty$. Clearly we have

$$H(x+y) - H(x) \geq -\alpha y \qquad (x \in \mathbb{R},\ y > 0)$$

and

$$\frac{\mathrm{d}}{\mathrm{d}x}\Big\{e^{-\varepsilon x}\int_{-\infty}^x e^{\varepsilon t} H(t)\,\mathrm{d}t\Big\} = -\varepsilon e^{-\varepsilon x}\int_{-\infty}^x e^{\varepsilon t} H(t)\,\mathrm{d}t + H(x) \leq 2.$$

Substituting in (49), it follows that

$$H_\varepsilon(x+y) - H_\varepsilon(x) \geq -\frac{(\alpha + 2\varepsilon)}{T} \qquad (x \in \mathbb{R},\ 0 \leq y \leq \frac{1}{T}).$$

Applying Theorem 13 to $-H_\varepsilon$ with $K := \dfrac{\alpha + 2\varepsilon}{T}$, we obtain

$$|H_\varepsilon(x)| \leq 16\frac{(\alpha + 2\varepsilon)}{T} + 6\int_{-T}^T \left|\frac{f(\tau) - g(\tau)}{\tau}\right|\mathrm{d}\tau \qquad (x \in \mathbb{R}).$$

This yields (47) by letting ε tend to 0.

Remark. The numerical value of the constants in the Berry–Esseen inequality usually has little impact on the applications. Feller (1971) gives the values $24/\pi$ and $1/\pi$, both of which can be improved to respectively π and 0.16318 (cf. Vaaler (1985), theorem 13).

Notes

§ **7.1.** The restriction $z \in S(\theta)$ is necessary in Abel's theorem. B. de Mathan has communicated the following counter-example to the author. Let $\alpha_k := \exp(i/k)$ $(k = 1, 2, \ldots)$. The function

$$f(z) = \sum_{k=1}^{\infty} 2^{-k} \log(1 - z/\alpha_k)$$

is holomorphic for $|z| < 1$, since the series converges uniformly on every compact subset of the open disc. However we have

$$\lim_{z \to \alpha_k} |f(z)| = \infty \qquad (k = 1, 2, \ldots),$$

so $f(z)$ fails to have a limit when $z \to 1$. Now the Taylor series of $f(z)$ is

$$-\sum_{n=1}^{\infty} z^n n^{-1} \sum_{k=1}^{\infty} 2^{-k} \alpha_k^{-n},$$

which, by Dirichlet's test, converges at $z = 1$. Indeed, the partial sums of the double series $\sum\sum 2^{-k}\overline{\alpha}_k^n$ are uniformly bounded:

$$\begin{aligned}\Big| \sum_{1 \le n \le m} \sum_{k=1}^{\infty} 2^{-k} \overline{\alpha}_k^n \Big| &= \Big| \sum_{k=1}^{\infty} 2^{-k} \sum_{1 \le n \le m} \overline{\alpha}_k^n \Big| \\ &\le \sum_{k=1}^{\infty} 2^{-k} \Big| \frac{1 - \overline{\alpha}_k^m}{1 - \overline{\alpha}_k} \Big| \le 2 \sum_{k=1}^{\infty} 2^{-k} \big|1 - e^{-i/k}\big|^{-1} < \infty.\end{aligned}$$

§ **7.3.** It is possible to extend Karamata's theorem by considering the more general assumption

$$F(\sigma) = \{c + o(1)\}\sigma^{-\omega} L(1/\sigma) \qquad (\sigma \to 0+)$$

where $L(t)$ is a *slowly varying* function, i.e. satisfies

$$L(kt) \sim L(t) \qquad (t \to \infty)$$

for each fixed $k > 0$. The conclusion of Theorem 5 then holds on replacing x^ω by $x^\omega L(x)$—cf. Hardy (1949), theorem 108.

§ **7.4.** Theorem 10 is due to Freud (1952). See also Freud & Ganelius (1957). The self-contained proof which we give here essentially follows that of Korevaar (1954). Ingham (1965) gives a direct proof of the Karamata–Freud theorem which does not explicitly appeal to the lemma on the L^1-approximation by polynomials, but directly incorporates the use of "peak functions" in Karamata's proof.

§ **7.5.** For developments relating to Bohr's inequality (36), see Hörmander (1954).

As indicated in Exercise 5, Theorem 11 yields a remainder of order $\ll x(\log x)^{-2+\varepsilon}$ in the prime number theorem. Rieger (1983), also using the Wiener–Ikehara method, obtained a remainder of order

$$\ll x \exp\{-c(\log x)^{1/25}\}.$$

As Diamond remarked in his review (1988), it is not true *stricto sensu* that the Hardy–Littlewood–Karamata theorem cannot lead to a proof of the prime number theorem. Using a number of analytic properties of the Riemann zeta function, such as the functional equation, the product formula, the finiteness of the order, the absence of zeros on $\sigma = 1$ and a bound of type $N(T) \ll T^A$ for the number of zeros with ordinate not exceeding T, Littlewood indeed showed in 1971 that the Hardy–Littlewood–Karamata theorem can be used to obtain a "quick" proof of the prime number theorem. Since no known proof of the fact that $\zeta(s) \neq 0$ for $\sigma = 1$ furnishes an effective lower bound weaker than

$$|\zeta(s)| \gg \Big(\log(3+|\tau|)\Big)^{-A} \qquad (\sigma \geq 1), \tag{50}$$

such a procedure would seem, however, to have a rather limited theoretical interest. Under assumption (50), Perron's formula on its own enables one to recover the prime number theorem and no Tauberian theorem is needed whatsoever—cf. §4.2. The particular appeal of the Ikehara theorem is to yield the conclusion under the single minimal assumption $\zeta(s) \neq 0$ for $\sigma = 1$, without any additional analytic information.

Delange (1954) generalised Ikehara's theorem to the case of a singularity of mixed type, involving both monomial and logarithmic terms in its principal part.

Theorem 15 (Delange). *Let $F(s) := \sum_{n=1}^{\infty} a_n n^{-s}$ be a Dirichlet series with non-negative coefficients, converging for $\sigma > a > 0$. Suppose that $F(s)$ is holomorphic at all points of the line $\sigma = a$ other than $s = a$ and that, in the neighbourhood of this point and for $\sigma > a$, we have*

$$F(s) = (s-a)^{-\omega-1} \sum_{j=0}^{q} g_j(s)\Big(\log\Big(\frac{1}{s-a}\Big)\Big)^j + g(s),$$

where ω is some real number, and the $g_j(s)$ and $g(s)$ are functions holomorphic at $s = a$, the number $g_q(a)$ being non-zero. Then:

(i) *if ω is not a negative integer, we have as $x \to \infty$*

$$A(x) := \sum_{n \le x} a_n \sim \frac{g_q(a)}{a\Gamma(\omega+1)} x^a (\log x)^\omega (\log_2 x)^q,$$

(ii) *if $\omega = -m - 1$ for a non-negative integer m and if $q \ge 1$, we have as $x \to \infty$*

$$A(x) \sim (-1)^m m! \frac{q g_q(a)}{a} x^a (\log x)^{-m-1} (\log_2 x)^{q-1}.$$

Exercises

1. Deduce from Theorem 9 the estimate

$$\sum_{n \le x} \frac{\mu(n)}{n} \ll \frac{\log x}{\log_2 x} \qquad (x \to \infty).$$

[*This non-trivial estimate is of course much weaker than (I.3.15) but only necessitates a very fragmentary piece of information concerning the Dirichlet series $\zeta(s)^{-1}$ associated with the Möbius function.*]

2. Set $a_n := \sum_{d|n} (-1)^d 2^{\omega(n/d)}$, where $\omega(m)$ denotes the number of distinct prime factors of m. Determine the asymptotic behaviour of $\sum_{n=1}^{\infty} a_n n^{-\sigma}$ as $\sigma \to 1+$. Use the Karamata–Freud theorem to deduce that

$$\sum_{n \le x} \frac{a_n}{n} \ll \frac{(\log x)^3}{\log_2 x} \qquad (x \to \infty).$$

By using the method of Chapter II.5 show that one has

$$\sum_{n \le x} \frac{a_n}{n} \sim C (\log x)^2$$

with $C := -3\pi^{-2} \log 2$.

3. Let a_n be a bounded arithmetic function such that

$$\lim_{\sigma\to 1+}(\sigma-1)\sum_{n=1}^{\infty}a_n n^{-\sigma}=0.$$

Set $b_n := \sum_{d|n} a_d$. Show that $\sum_{n\le x} b_n = o(x\log x)$.

4. Set $Z(s) := -s^{-1}\zeta'(s)\zeta(s)^{-1}$. Show that the assumption $\zeta(1+i\tau)\neq 0$ for $\tau\neq 0$ implies that, for all fixed T,

$$\lim_{\sigma\to 0+}\int_{-T}^{T}\left|Z(1+2\sigma+i\tau)-Z(1+\sigma+i\tau)-\frac{1}{2\sigma+i\tau}+\frac{1}{\sigma+i\tau}\right|\mathrm{d}\tau=0.$$

Now apply Ikehara's theorem and deduce the prime number theorem.

5. Using the elementary bounds proved in §4.2

$$\zeta^{(k)}(s)\ll(\log|\tau|)^{k+1},\qquad \zeta(s)^{-1}\ll(\log|\tau|)^{7}\quad(|\tau|\ge 2,\ \sigma\ge 1),$$

show that the integral in Exercise 4 is $\ll \sigma(\log T)^{19}$ for $T\ge 2$, $\sigma>0$. Apply Theorem 11 and deduce the effective estimate

$$\psi(x):=\sum_{n\le x}\Lambda(n)=x+O\Big(x\frac{(\log_2 x)^{19}}{\log x}\Big).$$

6. Use the method of Exercise 5 to establish the bound

$$\sum_{n\le x}\mu(n)\ll x\frac{(\log_2 x)^{16}}{\log x}.$$

[*Hint: use the Dirichlet series* $\zeta(s)+\zeta(s)^{-1}$.]

7. *Berry–Esseen theorem.* Let X be a real random variable, with zero mean, variance equal to 1, and finite absolute moment of order 3 equal to ρ. Set $F(x) := \mathrm{Prob}\,(X\le x)$ $(x\in\mathbb{R})$ and

$$\varphi(\tau):=\int_{-\infty}^{+\infty}e^{i\tau x}\,\mathrm{d}F(x)\qquad(\tau\in\mathbb{R}).$$

Finally, let $F_n(x)$ denote the distribution function of $(1/\sqrt{n})\sum_{j=1}^{n}X_j$, where the X_j are independent variables with the same law as X.

(a) Show that $\rho\ge 1$.

(b) Using the inequalities

$$\left|e^{iy} - \sum_{j=0}^{m} \frac{1}{j!}(iy)^j\right| \le \frac{1}{(m+1)!}|y|^{m+1} \qquad (y \in \mathbb{R},\ m \ge 0)$$

and

$$|\log(1-y) + y| \le |y|^2 \qquad (|y| \le \tfrac{1}{2}),$$

show that

$$|\varphi(\tau) - 1| \le \tfrac{1}{2}\tau^2, \quad |\varphi(\tau) - 1 + \tfrac{1}{2}\tau^2| \le \tfrac{1}{6}\rho|\tau|^3 \quad (\tau \in \mathbb{R}),$$
$$\left|\log\varphi(\tau) + \tfrac{1}{2}\tau^2\right| \le \tfrac{1}{4}\tau^4 + \tfrac{1}{6}\rho|\tau|^3 \quad (\tau \in \mathbb{R}).$$

(c) Show that for all $n \ge 1$ one has

$$\left|\varphi\Big(\frac{\tau}{\sqrt{n}}\Big)^n - e^{-\tau^2/2}\right| \ll e^{-\tau^2/4}\Big\{\rho\frac{|\tau|^3}{\sqrt{n}} + \frac{\tau^4}{n}\Big\} \quad (\rho|\tau| \le \sqrt{n}).$$

(d) Deduce from the above that one has

$$\sup_{x\in\mathbb{R}}\left|F_n(x) - \frac{1}{\sqrt{(2\pi)}}\int_{-\infty}^{x} e^{-t^2/2}\,\mathrm{d}t\right| \le C\frac{\rho}{\sqrt{n}}$$

where C is some suitable absolute constant.
[*Feller (1971) shows that one can take* $C = 3$.]

8. Let $\{a_p\}$ be a bounded sequence indexed by the prime numbers. Show that if $\sum a_p p^{-\sigma}$ tends to a limit as $\sigma \to 1+$ then the series $\sum a_p/p$ converges. Show that the result fails for sequences indexed over the integers.

9. *On an idea of Wirsing (1956).* Let $\mathcal{P}$ be a set of prime numbers, and define $\theta(n)$ as 1 or 0 according to whether all prime factors of n belong to $\mathcal{P}$ or not. Set $T(x) := \sum_{n\le x}\theta(n)$. We assume there exist constants $\delta > 0$ and $K > 0$ such that, as $x \to \infty$,

$$\text{(i)} \quad \sum_{p\le x}\theta(p) \sim \frac{\delta x}{\log x}, \qquad \text{(ii)} \quad \prod_{p\le x}\Big(1 + \frac{\theta(p)}{p-1}\Big) \sim K(\log x)^{\delta}.$$

(a) Show that one has, as $x \to \infty$,

$$\text{(iii)} \quad \sum_{n\le x}\theta(n)\log n \sim \delta x\sum_{m\le x}\frac{\theta(m)}{m}, \qquad \text{(iv)} \quad T(x) \sim \frac{\delta x}{\log x}\sum_{m\le x}\frac{\theta(m)}{m}.$$

(b) Show that assumption (i) alone implies, as $\sigma \to 0+$,

$$\sum_{p \le \exp(1/\sigma)} \theta(p) p^{-1}(1 - p^{-\sigma}) - \sum_{p > \exp(1/\sigma)} \theta(p) p^{-1-\sigma} = \gamma\delta + o(1).$$

Deduce that $\sum_{m=1}^{\infty} \theta(m) m^{-1-\sigma} \sim e^{-\gamma\delta} \prod_{p \le \exp(1/\sigma)} \left(1 - \frac{\theta(p)}{p}\right)^{-1} \qquad (\sigma \to 0+).$

(c) Show that one has under hypotheses (i) and (ii)

$$T(x) \sim \frac{\delta K e^{-\gamma\delta}}{\Gamma(\delta + 1)} x (\log x)^{\delta - 1} \qquad (x \to \infty).$$

II.8

Prime numbers in arithmetic progressions

§ 8.1 Introduction: Dirichlet characters

In 1837 Dirichlet established that each arithmetic progression $n \equiv a(\bmod q)$, with $(a, q) = 1$, contains infinitely many prime numbers. The main novelty in his proof consisted in making use of characters modulo q, that is to say homomorphisms from the group $(\mathbb{Z}/q\mathbb{Z})^*$ of invertible residues $(\bmod q)$ into the multiplicative group of complex numbers of modulus 1.

Here we shall not develop the classical general theory of characters of finite Abelian groups, and we confine ourselves to recalling briefly the principal properties which will be useful in the present context. For a more comprehensive study the reader may consult for example Ayoub (1963) or Ellison & Mendès France (1975).

Let G be a finite Abelian group with decomposition as a direct product of cyclic groups

$$G \cong \prod_{j=1}^{k} G_j.$$

Let us write $n_j := |G_j|$ $(1 \leq j \leq k)$, so that

$$n := |G| = \prod_{j=1}^{k} n_j.$$

Denoting by γ_j $(1 \leq j \leq k)$ a generator of G_j, each element γ of G may be uniquely written as a product

$$\gamma = \prod_{j=1}^{k} \gamma_j^{r_j} \qquad \big(1 \leq r_j \leq n_j \ (1 \leq j \leq k)\big). \tag{1}$$

For any character χ of G, we can hence write

$$\chi(\gamma) = \prod_{j=1}^{k} \chi(\gamma_j)^{r_j}.$$

Thus, χ is entirely determined by the $\chi(\gamma_j)$ $(1 \le j \le k)$, and the fact that γ_j is of order n_j implies that $\chi(\gamma_j)$ is an n_jth root of unity. Conversely, for each choice

$$\zeta_j = e(h_j/n_j) \qquad (1 \le j \le k)$$

with $1 \le h_j \le n_j$, the map $\chi : G \to \mathbb{C}$ defined, for γ given by (1), by

$$\chi(\gamma) = \prod_{j=1}^{k} \zeta_j^{r_j}$$

is certainly a character of G, moreover all characters produced in this way are obviously distinct. There are thus n characters of G, which form, with respect to the ordinary product of maps, an Abelian group, labelled $\widehat{G}$.

A simple but fundamental property of characters is as follows. We let χ_0 denote the trivial character, defined by $\chi_0(\gamma) = 1$ for all $\gamma \in G$.

Theorem 1. *For any Abelian group G of order n, we have*

$$\sum_{\chi \in \widehat{G}} \chi(\gamma) = \begin{cases} n & \text{if } \gamma = 1 \\ 0 & \text{if } \gamma \ne 1 \end{cases} \qquad (\gamma \in G) \tag{2}$$

and

$$\sum_{\gamma \in G} \chi(\gamma) = \begin{cases} n & \text{if } \chi = \chi_0 \\ 0 & \text{if } \chi \ne \chi_0 \end{cases} \qquad (\chi \in \widehat{G}). \tag{3}$$

Proof. The proofs of these two dual relations being similar, we confine ourselves to establishing the first. Let S denote the left-hand side of (2). If $\gamma = 1$, we have $\chi(\gamma) = 1$ for all χ, hence

$$S = |\widehat{G}| = n.$$

If $\gamma \ne 1$, there exists some character, say χ_1, such that $\chi_1(\gamma) \ne 1$. Indeed, writing γ as in (1), we may suppose, by reordering the factors if need be, that $r_1 \ne n_1$. The map defined by $\chi_1(\gamma_1) = e(1/n_1)$, $\chi_1(\gamma_j) = 1$ $(2 \le j \le k)$ then has the required property. Since $\chi\chi_1$ runs over $\widehat{G}$ as χ runs over $\widehat{G}$, we may write

$$\chi_1(\gamma)S = \sum_{\chi \in \widehat{G}} \chi_1(\gamma)\chi(\gamma) = \sum_{\chi \in \widehat{G}} (\chi\chi_1)(\gamma) = S,$$

which implies that $S = 0$.

It follows from the above description of the character group that there are $\varphi(q)$ characters to the modulus q. They can be explicitly defined by decomposing $(\mathbb{Z}/q\mathbb{Z})^*$ as a direct product of cyclic groups. By the Chinese remainder theorem we have for any $q \geq 1$

$$(4) \qquad (\mathbb{Z}/q\mathbb{Z})^* \cong \prod_{p^\nu \| q} (\mathbb{Z}/p^\nu\mathbb{Z})^*.$$

Indeed, if we let q_j $(1 \leq j \leq \omega(q))$ denote the distinct prime powers dividing q, and if, for each j, a_j denotes the unique integer $(\operatorname{mod} q)$ such that

$$a_j \equiv 1 \,(\operatorname{mod} q_j), \qquad a_j \equiv 0 \,(\operatorname{mod} q_i) \quad (i \neq j),$$

then for each integer m we can write

$$m \equiv \sum_j a_j m_j \,(\operatorname{mod} q)$$

where m_j is the residue of m $(\operatorname{mod} q_j)$. Since $(m, q) = 1$ is equivalent to $(m_j, q_j) = 1$ for all j, we see that the canonical projection of the left-hand side of (4) onto the right is well-defined and surjective, and thus (4) follows.

When $p > 2$, the group $(\mathbb{Z}/p^\nu\mathbb{Z})^*$ is cyclic—cf. Exercises 1 and 2. When $p = 2$, there are three cases to consider for the structure of $(\mathbb{Z}/2^\nu\mathbb{Z})^*$. If $\nu = 1$, the group is trivial. If $\nu = 2$, it is cyclic of order 2. If $\nu \geq 3$, it is the direct product of a group of order 2 with a cyclic group of order $2^{\nu-2}$. This last point can be checked by first noting that a simple induction yields

$$(5) \qquad 5^{2^{m-2}} = 1 + 2^m h_m \qquad (m \geq 2)$$

where h_m is odd. Applying this relation with $m = \nu$, it is seen that the order of 5 modulo 2^ν divides $2^{\nu-2}$. But, choosing $m = \nu - 1$, we also see that this order does not divide $2^{\nu-3}$. It is thus exactly equal to $2^{\nu-2}$, and we conclude by noting that 5 is not a square modulo 8—thereby excluding the possibility that $(\mathbb{Z}/2^\nu\mathbb{Z})^*$ is cyclic.

We can now make the characters to the modulus q explicit. Write the prime power decomposition of q in the form

$$q = 2^\nu \prod_{j=1}^{k} p_j^{\nu_j}, \quad 2 < p_1 < \ldots < p_k, \quad \nu \geq 0, \quad \nu_j > 0 \quad (1 \leq j \leq k).$$

For each index j, let g_j be a primitive root modulo $p_j^{\nu_j}$. Then, for each integer m with $(m, q) = 1$, we can uniquely define $\varepsilon(m)$, $\eta(m)$, $\mu_j(m)$ $(1 \leq j \leq k)$ satisfying

$$m \equiv (-1)^{\varepsilon(m)} 5^{\eta(m)} \,(\operatorname{mod} 2^\nu), \quad \varepsilon(m) = 0 \text{ or } 1, \quad 0 \leq \eta(m) < 2^{\nu-2},$$

$$m \equiv g_j^{\mu_j(m)} \,(\operatorname{mod} p_j^{\nu_j}), \quad 0 \leq \mu_j(m) < \varphi(p_j^{\nu_j}) \quad (1 \leq j \leq k).$$

If $\nu = 0$ or 1, we choose $\varepsilon(m) = 0$. With this notation, the characters modulo q are the $\varphi(q)$ functions defined on $(\mathbb{Z}/q\mathbb{Z})^*$ by

$$\chi(m) = e\left(\frac{\lambda\varepsilon(m)}{2} + \frac{\lambda'\eta(m)}{2^{\nu-2}} + \sum_{j=1}^{k} \frac{\lambda_j\mu_j(m)}{\varphi(p_j^{\nu_j})}\right) \tag{6}$$

when the parameters λ, λ', λ_j are arbitrarily chosen so that

$$\lambda = 0 \text{ or } 1, \quad 0 \le \lambda' < 2^{\nu-2}, \quad 0 \le \lambda_j < \varphi(p_j^{\nu_j}) \quad (1 \le j \le k).$$

Definition. *A Dirichlet character to the modulus q is an arithmetic function extending a character χ of the group $(\mathbb{Z}/q\mathbb{Z})^*$ by means of the formula*

$$\chi(n) = \begin{cases} \chi(m), & \text{if } n \equiv m (\operatorname{mod} q),\ 1 \le m \le q,\ (m,q) = 1, \\ 0, & \text{if } (n,q) > 1. \end{cases} \tag{7}$$

The homomorphism property of the characters of $(\mathbb{Z}/q\mathbb{Z})^*$ translates into the fact that a Dirichlet character is a *completely multiplicative arithmetic function.* Theorem 1 then implies the following fundamental relations. As is customary, we call *principal character* and denote by χ_0 the extension of the trivial character, viz.

$$\chi_0(n) = \begin{cases} 1 & \text{if } (n,q) = 1, \\ 0 & \text{if } (n,q) > 1. \end{cases} \tag{8}$$

Theorem 2 (Orthogonality relations).

(a) *For all integers $n, m \ge 1$, we have*

$$\varphi(q)^{-1} \sum_{\chi\,(\operatorname{mod} q)} \chi(n)\overline{\chi(m)} = \begin{cases} 1 & \text{if } n \equiv m (\operatorname{mod} q) \text{ and } (m,q) = 1, \\ 0 & \text{otherwise.} \end{cases} \tag{9}$$

(b) *For all Dirichlet characters χ, χ', to the modulus q, we have*

$$\varphi(q)^{-1} \sum_{1 \le n \le q} \chi(n)\overline{\chi'(n)} = \begin{cases} 1 & \text{if } \chi = \chi', \\ 0 & \text{otherwise.} \end{cases} \tag{10}$$

§ 8.2 *L*-series. The prime number theorem for arithmetic progressions

For each character χ, the Dirichlet L-series is defined by the formula

$$L(s,\chi) := \sum_{n=1}^{\infty} \chi(n)n^{-s} \qquad (\sigma > 1), \tag{11}$$

with Euler expansion

$$L(s,\chi) = \prod_{p} \left(1 - \chi(p)p^{-s}\right)^{-1} \qquad (\sigma > 1). \tag{12}$$

In particular $L(s,\chi) \neq 0$ when $\sigma > 1$. For $\chi = \chi_0$, the above formula can be written in the form

$$L(s,\chi_0) = \prod_{p|q}(1 - p^{-s})\zeta(s). \tag{13}$$

This defines a meromorphic extension of $L(s,\chi_0)$ to a function having in the whole complex plane a single pole at $s = 1$, which is simple and has residue $\varphi(q)/q$. When $\chi \neq \chi_0$, the relation (10) applied with $\chi' = \chi_0$ implies by periodicity that

$$\max_{x \geq 1} \Big| \sum_{n \leq x} \chi(n) \Big| \leq \tfrac{1}{2}\, q. \tag{14}$$

By Dirichlet's test, we deduce that the series $L(s,\chi)$ converges for $\sigma > 0$, which actually determines the abscissa of convergence.

With the help of the orthogonality relations, the Euler expansion (12) allows us to investigate prime numbers in an arithmetic progression. As in the case of the Riemann zeta function, we proceed via the logarithmic derivative.

Theorem 3. *For all integers a, q with $(a,q) = 1$, we have*

$$\sum_{\substack{n=1 \\ n \equiv a \,(\mathrm{mod}\, q)}}^{\infty} \Lambda(n)n^{-s} = \varphi(q)^{-1} \sum_{\chi} \overline{\chi(a)} \cdot \frac{-L'}{L}(s,\chi) \qquad (\sigma > 1), \tag{15}$$

where the summation over χ is extended to the $\varphi(q)$ Dirichlet characters to the modulus q.

Proof. By (12), we have for any χ

$$\frac{-L'}{L}(s,\chi) = \sum_{p} \frac{\chi(p)\log p}{p^s - \chi(p)} = \sum_{p} \sum_{\nu=1}^{\infty} \frac{\chi(p^\nu)\log p}{p^{\nu s}} = \sum_{n=1}^{\infty} \chi(n)\Lambda(n)n^{-s}.$$

The result then follows from (9) with $m = a$, by interchanging the summations on the right-hand side of (15).

The Dirichlet series (15) is the Mellin–Stieltjes transform of the function

(16) $$\psi(x;a,q) := \sum_{\substack{n\leq x \\ n\equiv a(\bmod q)}} \Lambda(n).$$

As in the case of the classical Chebyshev function $\psi(x)$ $(= \psi(x;1,1))$, an integration by parts allows us to link the asymptotic behaviours of $\psi(x;a,q)$ and of the counting function of prime numbers in the corresponding arithmetic progression, i.e.

(17) $$\pi(x;a,q) := \sum_{\substack{p\leq x \\ p\equiv a(\bmod q)}} 1.$$

Dirichlet's theorem mentioned at the beginning of this chapter means that the expression (17) tends to infinity with x for all fixed a, q with $(a,q) = 1$. Actually, Dirichlet established a more precise result, which, in a certain sense, implies that the primes are evenly distributed among the $\varphi(q)$ possible residue classes modulo q. He showed that we have

(18) $$\sum_{\substack{p\leq x \\ p\equiv a(\bmod q)}} \frac{1}{p} \sim \frac{\log_2 x}{\varphi(q)} \qquad (x \to \infty).$$

This estimate readily follows from the assumption

(19) $$L(1,\chi) \neq 0 \qquad (\chi \neq \chi_0)$$

by means of the Hardy–Littlewood–Karamata theorem (Theorem 7.8). Indeed, we deduce from (15) and (19) that, for fixed a and q,

(20) $$\sum_{\substack{n=1 \\ n\equiv a(\bmod q)}}^{\infty} \frac{\Lambda(n)}{n^\sigma} = \frac{1}{\varphi(q)}\cdot\frac{-L'}{L}(\sigma,\chi_0) + O_q(1) \qquad (\sigma > 1),$$

and, by (13), we have under the same conditions,

(21) $$-\frac{L'}{L}(\sigma,\chi_0) = -\frac{\zeta'}{\zeta}(\sigma) + \sum_{p|q} \frac{\log p}{p^\sigma - 1} = \frac{1}{\sigma - 1} + O_q(1).$$

Applying Theorem 7.9 of Karamata–Freud, we even obtain the remainder formula

(22) $$\sum_{\substack{n\leq x \\ n\equiv a(\bmod q)}} \frac{\Lambda(n)}{n} = \frac{\log x}{\varphi(q)} + O_q\Big(\frac{\log x}{\log_2 x}\Big)$$

from which, by partial summation, an effective version of (18) may be extracted, viz.

$$\sum_{\substack{p\le x\\ p\equiv a(\mathrm{mod}\, q)}} \frac{1}{p} = \frac{\log_2 x}{\varphi(q)} + O_q(\log_3 x) \qquad (x\to\infty). \tag{23}$$

If the assumption (19) is strengthened to

$$L(1+i\tau,\chi)\neq 0 \qquad (\tau\in\mathbb{R},\ \chi\neq\chi_0) \tag{24}$$

then, by Ikehara's theorem, we obtain the stronger formula

$$\psi(x;a,q)\sim\frac{x}{\varphi(q)} \qquad (x\to\infty). \tag{25}$$

This result is commonly called the "prime number theorem for arithmetic progressions". One of the central questions in analytic number theory is to determine effective versions of this asymptotic formula with respect to the three variables involved.

In order to deduce (25) from (24), it suffices to apply Theorem 7.11 to the Dirichlet series (15), i.e.

$$F(s)=\int_{0-}^{\infty} e^{-st}\,\mathrm{d}\psi(e^t;a,q).$$

By (24), each function $\frac{-L'}{L}(s,\chi)$, $\chi\neq\chi_0$, may be continued to a function holomorphic for $\sigma\ge 1$ (that is to say in an open set containing this half-plane), and it then follows from (13) and (15) that

$$G(s):=\frac{F(s+1)}{s+1}-\frac{1}{\varphi(q)s} \tag{26}$$

can be continued to a function which is continuous for $\sigma\ge 0$. Ikehara's theorem then allows us to deduce the validity of (25).

We shall see in the following section that (19) and indeed (24) are satisfied. We shall even obtain explicit lower bounds for $|L(1+i\tau,\chi)|$. This will enable us to use the full force of our effective version 7.11 of Ikehara's theorem—and hence to derive some information about the dependence of the error term of (25) on x and q. We shall obtain the following result.

Theorem 4. *For any constant $A > 0$, and uniformly for*

$$x \geq 3, \quad 1 \leq q \leq (\log x)^A, \quad (a, q) = 1,$$

we have

$$\psi(x; a, q) = \frac{x}{\varphi(q)} + O\Big(2^{\omega(q)} x \frac{(\log_2 x)^{19}}{\log x}\Big). \tag{27}$$

In particular, the asymptotic formula (25) holds, for any $\varepsilon > 0$, uniformly in the domain

$$x \geq 3, \qquad 1 \leq q \leq (\log x)^{1-\varepsilon}. \tag{28}$$

A more careful study of the series $L(s, \chi)$ enables one to obtain a zero-free region analogous to that for the zeta function. Uniformity in q is crucial and particularly difficult—it incidentally raises an interesting open problem on the existence of the famous "Siegel zero", of which we give a very brief account in the Notes. The best results currently known yield the following statement, which considerably strengthens Theorem 4—cf. Siegel (1936).

Theorem 5 (Siegel–Walfisz). *Under the assumptions of Theorem 4, we have*

$$\psi(x; a, q) = \frac{x}{\varphi(q)} + O\big(xe^{-c\sqrt{(\log x)}}\big) \tag{29}$$

where c is an absolute positive constant.

The reader will find a detailed proof of this result in the book by Huxley (1972) or in that by Ellison & Mendès France (1975). A rather disappointing feature of Theorem 5 is its *ineffectiveness*: for $q \geq \log x$, one cannot, in the present state of knowledge, provide numerical values either to c or to the constant in the Landau symbol. Actually, the best effective results known, based on the Pólya–Vinogradov inequality (see the notes on this chapter), will only yield an asymptotic formula for $\psi(x; a, q)$ if $q \leq (\log x)^{2-\varepsilon}$—cf. Davenport (1980), p. 123. In this respect, the effective estimate (27) is hence of fairly good quality.

Before undertaking the proof of Theorem 4, a methodological remark is in order. For $q = 1$ the error term in (27) is notably less precise than the one which we were able to obtain for the prime number theorem, even without making use of delicate properties of the function $\zeta(s)$ in the critical strip—cf. §4.2. This suggests that we ought, *a priori*, to be able to improve Theorem 4 at least for small, or even bounded values of q. It indeed turns out that the lower bounds for $|L(1 + i\tau, \chi)|$ which we shall establish in the following section yield a zero-free region which in turn allows us to obtain, under assumption (28), a remainder term $O_A\bigl(x(\log x)^{-A}\bigr)$ for all $A > 0$. The details are made precise in Exercise 4. Here, we have however chosen to apply the effective Ikehara theorem: on the one hand because it provides, in a very simple way, a domain of validity for the asymptotic formula (25) which is practically identical to that obtained by contour integration and, on the other hand, because this will give us the opportunity for a detailed arithmetic use of a Tauberian theorem.

§ 8.3 Lower bounds for $|L(s,\chi)|$ when $\sigma \geq 1$. Proof of Theorem 4.

For ease of notation, we shall write

$$\mathcal{L} = \mathcal{L}(q,\tau) := \log(|\tau| + q + 1) \qquad (\tau \in \mathbb{R}). \tag{30}$$

Theorem 6. *For $k \geq 0$, $\chi \neq \chi_0$, we have*

$$L^{(k)}(s,\chi) \ll_k \mathcal{L}^{k+1} \qquad (\sigma \geq 1). \tag{31}$$

Proof. Let us write $K(t) := \sum_{n \leq t} \chi(n)$. It follows from (14) that

$$|K(t)| \leq \tfrac{1}{2} q \qquad (t \geq 0). \tag{32}$$

Let $x \geq 2$, $T = |\tau| + 2$. We have

$$\begin{aligned}
|L^{(k)}(s,\chi)| &= \Bigl|\sum_{n=1}^{\infty} \chi(n)(\log n)^k n^{-s}\Bigr| \\
&\leq \sum_{n \leq x} (\log n)^k n^{-1} + \Bigl|\int_x^{+\infty} (\log t)^k t^{-s}\,\mathrm{d}K(t)\Bigr| \\
&\ll_k (\log x)^{k+1} + (\log x)^k |K(x)| x^{-1} + T \int_x^{+\infty} |K(t)|(\log t)^k t^{-2}\,\mathrm{d}t \\
&\ll_k (\log x)^k \{\log x + qTx^{-1}\}.
\end{aligned}$$

The bound (31) follows from this estimate by choosing $x = qT$.

Theorem 7. *For all χ, we have*

$$L(\sigma,\chi_0)^3 |L(\sigma + i\tau,\chi)|^4 \; |L(\sigma + 2i\tau,\chi^2)| \geq 1 \qquad (\sigma > 1,\ \tau \in \mathbb{R}). \tag{33}$$

Proof. We proceed analogously to the proof of Theorem 3.8, using the inequality

$$V(\theta) = 3 + 4\cos\theta + \cos 2\theta \geq 0 \qquad (\theta \in \mathbb{R}) \tag{34}$$

which served to estimate $|\zeta(1+i\tau)|$ from below. For each χ, we define a function $\psi(n)$, with values in $[0, 2\pi[$, such that

$$\chi(n) = \chi_0(n)e^{i\psi(n)}.$$

By (12), we can write for $\sigma > 1$

$$\log L(s,\chi) = \sum_p \sum_{k=1}^{\infty} \frac{\chi(p)^k}{kp^{ks}} = \sum_{p\nmid q} \sum_{k=1}^{\infty} \frac{\exp\{i(\psi(p^k) - k\tau \log p)\}}{kp^{k\sigma}}$$

from which, taking real parts,

$$3\log L(\sigma,\chi_0) + 4\log|L(\sigma+i\tau,\chi)| + \log|L(\sigma+2i\tau,\chi^2)| = \sum_{p\nmid q} \sum_{k=1}^{\infty} \frac{V(\psi(p^k) - k\tau\log p)}{kp^{k\sigma}} \geq 0.$$

Formula (33) is then derived by exponentiation.

The inequality (34)—as actually all classical methods for estimating $|L(1+i\tau,\chi)|$ from below—gives a special role to *real characters*, that is to characters χ such that

$$\chi^2 = \chi_0. \tag{35}$$

In this case the factor $|L(\sigma+2i\tau,\chi^2)|$ becomes infinite as $\sigma \to 1+$, $\tau \to 0$, which has the consequence of reducing the effect of (33). The explicit description (6) shows that χ is real if and only if

$$\lambda' = \begin{cases} 0 & (\nu \leq 2) \\ 0,\ 2^{\nu-3} & (\nu \geq 3) \end{cases} \quad \text{and} \quad \lambda_j = 0 \text{ or } \tfrac{1}{2}\varphi(p_j^{\nu_j}) \quad (1 \leq j \leq k).$$

Since $\varepsilon(m) = 0$ when $\nu \leq 1$, we readily obtain that he number $r(q)$ of real characters modulo q is given by the formula

$$r(q) = \begin{cases} 2^{\omega(q)-1} & \text{if } q \equiv \pm 2 \pmod 8, \\ 2^{\omega(q)} & \text{if } q \equiv \pm 1, \pm 3, 4 \pmod 8, \\ 2^{\omega(q)+1} & \text{if } q \equiv 0 \pmod 8). \end{cases} \tag{36}$$

In all cases, we have $r(q) \leq \tau(q)$.

Theorem 8. *If* $\chi^2 \neq \chi_0$, *we have*

$$L(\sigma + i\tau, \chi)^{-1} \ll \mathcal{L}^7 \qquad (\sigma \geq 1). \tag{37}$$

Proof. We apply the method of §4.2. Write $s = \sigma + i\tau$, $s_0 = s + \eta$, where η is a parameter satisfying $0 \leq \eta \leq 1/\mathcal{L}$. It follows from (31) that we have

$$|L(s, \chi) - L(s_0, \chi)| \leq C_0\eta\mathcal{L}^2 \tag{38}$$

for a suitable absolute constant C_0. On the other hand, Theorem 6 implies that

$$L(\sigma_0 + 2i\tau, \chi^2) \ll \mathcal{L} \qquad (\sigma_0 := \sigma + \eta), \tag{39}$$

since $\chi^2 \neq \chi_0$. Substituting in (33) we obtain

$$|L(s_0, \chi)|^4 \gg \prod_{p|q}(1 - p^{-\sigma_0})^{-3}\zeta(\sigma_0)^{-3}\mathcal{L}^{-1} \gg \eta^3\mathcal{L}^{-1}. \tag{40}$$

Using (38), we infer that there exists some positive constant C_1 such that

$$|L(s, \chi)| \geq C_1\eta^{3/4}\mathcal{L}^{-1/4} - C_0\eta\mathcal{L}^2 = C_2\{C_1{\cdot}C_2^{-1/4} - C_0\}\mathcal{L}^{-7}$$

for the choice $\eta := C_2\mathcal{L}^{-9}$. When C_2 is sufficiently small, the expression in curly brackets is positive and we obtain (37).

Theorem 9. *If* $\chi^2 = \chi_0$, $\chi \neq \chi_0$, *we have*

$$L(\sigma, \chi) \geq \frac{1}{8q} \qquad (\sigma \geq 1). \tag{41}$$

Proof. We confine ourselves to the case $1 \leq \sigma \leq 1 + 1/7q$. Indeed, we otherwise have

$$L(\sigma, \chi) = \prod_p \left(1 - \chi(p)p^{-\sigma}\right)^{-1} \geq \prod_p (1 - p^{-\sigma}) = \frac{1}{\zeta(\sigma)} \geq \frac{\sigma - 1}{\sigma} > \frac{1}{8q}.$$

Consider the arithmetic function

$$f_\sigma(n) := \sum_{d|n} \chi(d)d^{1-\sigma}.$$

This is a multiplicative function. If $\sigma > 1$, we can write for each prime number p

$$f_\sigma(p^\nu) = \frac{1 - (\chi(p)p^{1-\sigma})^{\nu+1}}{1 - \chi(p)p^{1-\sigma}} \geq \frac{1 + (-1)^\nu p^{(1-\sigma)(\nu+1)}}{1 + p^{1-\sigma}}$$

where the inequality follows from the fact that $\chi(p) = \pm 1$ or 0. In particular

$$f_\sigma(p^\nu) \geq 0, \qquad f_\sigma(p^{2\nu}) \geq p^{2\nu(1-\sigma)} \qquad (\nu \geq 1).$$

Passing to the limit, these inequalities are still valid when $\sigma = 1$. We then obtain

$$f_\sigma(n) \geq 0, \qquad f_\sigma(n^2) \geq n^{2(1-\sigma)} \qquad (n \geq 1,\ \sigma \geq 1).$$

Let us now give ourselves a parameter α satisfying $\dfrac{1}{10q^2} \leq \alpha \leq 1$, and consider the expression

$$F(\sigma,\alpha) := \sum_{n=1}^{\infty} f_\sigma(n) e^{-\alpha n}.$$

On the one hand we have

$$\begin{aligned} F(\sigma,\alpha) &\geq \sum_{n=1}^{\infty} n^{2(1-\sigma)} e^{-\alpha n^2} \geq \int_1^\infty t^{2(1-\sigma)} e^{-\alpha t^2}\, \mathrm{d}t \\ &\geq \tfrac12 \alpha^{\sigma-3/2}\Gamma(\tfrac32 - \sigma) - (3-2\sigma)^{-1} \geq \tfrac12 \sqrt{\pi}\, \alpha^{\sigma-(3/2)} - \tfrac76. \end{aligned}$$

Noting that

$$\alpha^{\sigma-1} \geq \exp\Big\{-\frac{\log(10q^2)}{7q}\Big\} \geq 40^{-1/14},$$

we then obtain

$$F(\sigma,\alpha) \geq A\alpha^{-1/2} - \tfrac76 \tag{42}$$

with

$$A := \tfrac12 \sqrt{\pi} \cdot 40^{-1/14}.$$

On the other hand, using the definition of f_σ, we also have

$$\begin{aligned} F(\sigma,\alpha) &= \sum_{m,d=1}^{\infty} \chi(d) d^{1-\sigma} e^{-\alpha m d} = \sum_{d=1}^{\infty} \chi(d) d^{1-\sigma} (e^{\alpha d} - 1)^{-1} \\ &= \alpha^{-1} L(\sigma,\chi) + \sum_{d=1}^{\infty} \chi(d)\theta(d) \end{aligned}$$

with

$$\theta(d) := d^{1-\sigma}\Big\{\frac{1}{e^{\alpha d} - 1} - \frac{1}{\alpha d}\Big\}.$$

It is easily checked that θ is increasing on $[1, +\infty[$. The sum over d hence takes the value

$$\int_{1-}^{+\infty} \theta(t)\,\mathrm{d}K(t) = -\int_{1}^{\infty} K(t)\theta'(t)\,\mathrm{d}t \leq \tfrac{1}{2}\,q|\theta(1)|$$

where we have appealed to (32). Since

$$|\theta(1)| = \frac{1}{\alpha} - \frac{1}{e^{\alpha}-1} \leq \frac{1}{\alpha} - \frac{1}{\alpha(1+\frac{1}{2}\alpha)} \leq \tfrac{1}{2},$$

we obtain

$$F(\sigma,\alpha) \leq \frac{1}{\alpha}L(\sigma,\chi) + \tfrac{1}{4}q. \tag{43}$$

Substituting back in (42), it follows that

$$L(\sigma,\chi) \geq A\sqrt{\alpha} - \alpha(\tfrac{7}{6} + \tfrac{1}{4}q) \geq \frac{1}{8q}$$

for the optimal choice $\alpha := \dfrac{4A^2}{(q+14/3)^2}$.

Theorem 10. *There exists an absolute constant $c_0 > 0$ such that, for $\chi^2 = \chi_0$, $\chi \neq \chi_0$, $\sigma \geq 1$,*

$$L(s,\chi)^{-1} \ll \begin{cases} \mathcal{L}^6(\mathcal{L} + |\tau|^{-1}) & \text{if } |\tau| > c_0 q^{-1}(\log 2q)^{-2}, \\ q & \text{if } |\tau| \leq c_0 q^{-1}(\log 2q)^{-2}. \end{cases} \tag{44}$$

Proof. The second of these estimates follows immediately from Theorems 6 and 9. Indeed, a Taylor expansion at order 1 gives for all τ, $|\tau| \leq 1$,

$$|L(s,\chi)| \geq L(\sigma,\chi) + O\big(\tau(\log 2q)^2\big),$$

from which, taking account of (41), the required estimate follows.

We can therefore assume that $|\tau| \gg q^{-1}(\log 2q)^{-2}$. We then use the method of Theorem 8. Let η be a positive parameter. Set $\sigma_0 = \sigma + \eta$, $s_0 = \sigma_0 + i\tau$, $s_1 = \sigma_0 + 2i\tau$. We have

$$L(s_1,\chi_0) = \prod_{p|q}(1-p^{-s_1})\zeta(s_1) \ll \prod_{p|q}(1-p^{-\sigma_0})^{-1}(\mathcal{L} + |\tau|^{-1}).$$

It then follows from (33) that

$$L(s_0,\chi)^4 \gg \prod_{p|q}(1-p^{-\sigma_0})^{-2}\zeta(\sigma_0)^{-3}(\mathcal{L} + |\tau|^{-1})^{-1}.$$

By appealing to (31), we now deduce the existence of two absolute positive constants C_3 and C_4 such that

$$|L(s,\chi)| \geq C_3\eta^{3/4}(\mathcal{L}+|\tau|^{-1})^{-1/4} - C_4\eta\mathcal{L}^2.$$

Choosing $\eta = C_5(\mathcal{L}+|\tau|^{-1})^{-1}\mathcal{L}^{-8}$ for some suitable C_5, we obtain the desired bound.

Proof of Theorem 4.

We apply Theorem 7.11 to the series (15). If $G(s)$ is defined by (26), we obtain

$$\psi(x;a,q) = \frac{x}{\varphi(q)} + O\Big(x \min_{T\geq 64}\Big\{\frac{1}{T} + \eta\Big(\frac{1}{\log x}, T\Big)\Big\}\Big) \tag{45}$$

with

$$\eta(\sigma,T) := \int_{-T}^{T} \big|G(2\sigma+i\tau) - G(\sigma+i\tau)\big|\,\mathrm{d}\tau \quad (\sigma > 0). \tag{46}$$

We estimate this last quantity by bounding $|G'(s)|$. We have

$$G'(s) \ll \frac{1}{\varphi(q)}\sum_{\chi}|Z'(s,\chi)| \tag{47}$$

with

$$Z(s,\chi) := \begin{cases} \dfrac{-\zeta'(s+1)}{(s+1)\zeta(s+1)} - \dfrac{1}{s} + \dfrac{1}{(s+1)}\displaystyle\sum_{p|q}\frac{\log p}{p^s} & (\chi = \chi_0) \\ \dfrac{-L'(s+1,\chi)}{(s+1)L(s+1,\chi)} & (\chi \neq \chi_0). \end{cases}$$

From the estimates of §4.2 on $\zeta(s)$, and from Theorems 6, 8, 10 concerning $L(s,\chi)$, $\chi \neq \chi_0$, we obtain for $\sigma \geq 0$, after a routine calculation,

$$Z'(s,\chi) \ll \begin{cases} \mathcal{L}^{18}|s+1|^{-1}, & \text{if } \chi^2 \neq \chi_0 \text{ or } |\tau| > (\log 2q)^{-1}, \\ \mathcal{L}^{16}|\tau|^{-2}, & \text{if } \chi^2 = \chi_0 \text{ and } c_0(\log 2q)^{-2}q^{-1} < |\tau| \leq (\log 2q)^{-1}, \\ \mathcal{L}^4 q^2, & \text{if } \chi^2 = \chi_0 \text{ and } |\tau| < c_0(\log 2q)^{-2}q^{-1}. \end{cases}$$

Substituting first in (47) and then in (46), and taking account of (36), we deduce

$$\eta(\sigma,T) \ll \sigma\mathcal{L}(q,T)^{18}\Big\{\mathcal{L}(q,T) + \frac{2^{\omega(q)}q}{\varphi(q)}\Big\}.$$

Choosing $T = \log x$, we obtain from (45) the stated formula for $\psi(x;a,q)$.

Notes

§§ **8.2–3.** Siegel (1936) showed that for all $\varepsilon > 0$ we have

$$L(1,\chi) \gg_\varepsilon q^{-\varepsilon}. \tag{48}$$

Unfortunately the implied constant is *ineffective*, in the sense that the proof does not enable a numerical calculation for given $\varepsilon < \frac{1}{2}$. For a proof of (48), see Ellison & Mendès France (1975), § 8.4.

It can also be shown—cf. Davenport (1980)—that the product $\prod_\chi L(s,\chi)$ of the L-functions to a given modulus q possesses at most one zero (necessarily real and corresponding to a real character) in a region of the type

$$\sigma \geq 1 - c/\log\big(q(2+|\tau|)\big)$$

where c is absolute. Miech (1969) has shown that $c = 1/20$ is acceptable when $q \geq q_0$. Up to date, the best known value for c, valid for large enough q, is $c = 0{\cdot}10367$, due, according to Graham (1981b), to Schoenfeld in an unpublished work—see lemma 10 of Chen (1983) for a proof. Heath-Brown (1992) has recently shown that there is at most one zero in the region $\sigma \geq 1 - 0{\cdot}348/\log q$, $|\tau| \leq 1$. It is conjectured that the possible exceptional zero (commonly called the "Siegel zero") does not exist. The strongest conjecture in this direction is that all the zeros of L-functions are situated on the critical line $\sigma = \frac{1}{2}$: this is the *generalised Riemann hypothesis.*

Given the generalised Riemann hypothesis, the Siegel–Walfisz theorem can be improved to

$$\psi(x;a,q) = \frac{x}{\varphi(q)} + O\big(\sqrt{x}\cdot(\log x)^2\big). \tag{49}$$

The Bombieri–Vinogradov theorem—cf. Bombieri (1965), A.I. Vinogradov (1965, 1966)—shows unconditionally that (49) is satisfied on average.

Theorem 11 (Bombieri–Vinogradov). *Let A be a positive constant. Uniformly for $Q \geq 1$, $x \geq 1$, we have*

$$\sum_{q\leq Q} \max_{\substack{(a,q)=1\\ y\leq x}} \left|\psi(y;a,q) - \frac{y}{\varphi(q)}\right| \ll \frac{x}{(\log x)^A} + \sqrt{x}\cdot Q\big(\log(Qx)\big)^4. \tag{50}$$

For an elementary proof of this result, resting on a fertile general method, see Vaughan (1980). The exponent 4 can be improved to $\frac{5}{2} + \varepsilon$, cf. Dress, Iwaniec & Tenenbaum (1983). The Elliott–Halberstam conjecture states that the left-hand side of (50) is $o(x)$ for $Q \leq x^{1-\varepsilon}$. This would indeed imply even sharper results than those yielded by the generalised Riemann hypothesis in most arithmetic applications.

The trivial inequality (32) can be considerably improved. Pólya and I.M. Vinogradov discovered independently, in 1918, the inequality

(51) $$\max_{x\geq 1}\Big|\sum_{n\leq x}\chi(n)\Big| \leq 2\sqrt{q}\cdot\log q,$$

which is valid for all non-principal characters χ mod q—for a proof, see Ellison & Mendès France (1975), theorem 9.7. This inequality is clearly useless for small values of x, for example $x \leq \sqrt{q}$. Burgess (1962, 1963) studied this question. He shows, for example, that

(52) $$\sum_{n\leq x}\chi(n) \ll xq^{-\delta(\varepsilon)} \qquad (x \geq q^{3/8+\varepsilon})$$

where $\delta = \delta(\varepsilon) > 0$. In addition, (52) is valid for $x \geq q^{1/4+\varepsilon}$ if q is cubefree. See also Hildebrand (1986d).

Our proof of Theorem 9 essentially follows that of Ellison & Mendès France (1975), chapter 7, A1. Using the Pólya–Vinogradov inequality in place of (32), it can be shown that $L(\sigma,\chi)^{-1} \ll \sqrt{q}\cdot\log q$, thus obtaining a corresponding improvement in Theorem 10. The gain in Theorem 4 is not significant if we stick to the approach presented here, but the contour integration method does yield in this way an improvement on the uniformity in the variable q—see Exercise 4.

Paley (1932) showed that the inequality (51) is almost optimal. Indeed, we have for infinitely many q

$$\max_{\chi\neq\chi_0}\max_{1\leq x\leq q}\Big|\sum_{n\leq x}\chi(n)\Big| \gg \sqrt{q}\cdot\log_2 q.$$

Montgomery & Vaughan (1977) have shown, under the generalised Riemann hypothesis, that the right-hand side of (51) may be replaced by $O(\sqrt{q}\cdot\log_2 q)$.

It is easily seen that the left-hand side of (51) is $\geq c\sqrt{q}$ for some absolute positive constant c and any non-principal primitive character χ to the modulus q. This follows from the classical fact that the modulus of the Gaussian sum

$$\tau(\chi) := \sum_{1\leq n\leq q}\chi(n)e(n/q)$$

is equal to $\sqrt{q}$—see e.g. Davenport (1980), p. 66. Indeed we have

$$\tau(\chi) = -(2\pi i/q)\int_0^q K(x)e(x/q)\,\mathrm{d}x$$

with $K(x) := \sum_{1\leq n\leq x}\chi(n)$, so the required lower bound follows with $c = 1/2\pi$. If we use the relation $K(q-x) = -\chi(-1)K(x-)$, we obtain that $c = \frac{1}{4}$ is admissible. For further improvements on c, see Sárközy (1977a), Sokolovskii (1979). On the other hand, Montgomery & Vaughan (1979) have shown that, for any $\varepsilon > 0$, we have $\max_x |K(x)| \ll_\varepsilon \sqrt{q}$ for all but at most $\varepsilon\varphi(q)$ non-principal characters to the modulus q.

Exercises

1. (a) Let $p > 2$. For each $d \geq 1$, let $\psi(d)$ denote the number of elements in $(\mathbb{Z}/p\mathbb{Z})^*$ which are exactly of order d. Show that $\sum_{d|(p-1)} \psi(d) = p - 1$.

(b) Let y be an element of order d in $(\mathbb{Z}/p\mathbb{Z})^*$. Show that the equation $x^d - 1 \equiv 0 \pmod{p}$ has exactly d solutions, which can be made explicit. Deduce that each element of order d is one of the $\varphi(d)$ generators of the group generated by y.

(c) Show that $\psi(d) \leq \varphi(d)$ for all d, and hence that this inequality is an equality if $d|(p-1)$. Deduce that $(\mathbb{Z}/p\mathbb{Z})^*$ is cyclic.

2. Let $p > 2$, and g be a primitive root modulo p.

(a) Show that the order of $g \pmod{p^2}$ is either $p-1$ or $p(p-1)$. Show that, in the first case, $g+p$ does not have order $p-1$. Deduce that $(\mathbb{Z}/p^2\mathbb{Z})^*$ is cyclic.

(b) Let $\nu \geq 2$ and g be a primitive root $\bmod\, p^2$. Show that $g^{p-1} = 1 + hp$ with $(h, p) = 1$. Deduce that

$$g^{\varphi(p^\nu)} \not\equiv 1 \pmod{p^{\nu+1}}$$

and hence that g generates $(\mathbb{Z}/p^{\nu+1}\mathbb{Z})^*$.

3. Show, for $(a, q) = 1$, $\sigma > 1$, that

$$\sum_{p \equiv a \pmod q} p^{-s} = \varphi(q)^{-1} \sum_{\chi} \overline{\chi(a)} \log L(s, \chi) + h(s)$$

where $h(s)$ is a holomorphic function in the half-plane $\sigma > \frac{1}{2}$. Deduce, under the assumption that $L(s, \chi) \neq 0$ for $\sigma \geq 1$, that

$$\sum_{p \equiv a \pmod q} p^{-s} = \varphi(q)^{-1} \log\left(\frac{1}{s-1}\right) + h_1(s) \qquad (\sigma > 1)$$

where h_1 is holomorphic in the half-plane $\sigma \geq 1$. Establish the prime number theorem for arithmetic progressions by applying Delange's Tauberian theorem—cf. Chapter 7, Notes.

4. Write $Q = q(\log 2q)^2$. Prove the existence of absolute positive constants c_1, c_2 such that the estimates of Theorems 6, 8, 10 are in fact valid for

$$1 - \sigma \leq \begin{cases} c_1 \mathcal{L}^{-9} & (\chi^2 \neq \chi_0) \\ c_1 \mathcal{L}^{-8}(\mathcal{L} + |\tau|^{-1})^{-1} & (\chi^2 = \chi_0,\ |\tau| Q > c_2) \\ c_1 Q^{-1} & (\chi^2 = \chi_0,\ |\tau| Q \leq c_2). \end{cases}$$

Deduce the existence of an absolute constant $c > 0$ and of a quantity $\delta(q) \gg q^{-1}(\log q)^{-10}$ such that one has, uniformly for $q \leq \exp(\log x)^{1/10}$,

$$\psi(x; a, q) = \frac{x}{\varphi(q)} + O\Big(x \exp\{-c(\log x)^{1/10}\} + \frac{2^{\omega(q)}}{\varphi(q)} x^{1-\delta(q)}\Big).$$

Show by the same method, but now employing the Pólya–Vinogradov inequality, that, for each $\varepsilon > 0$, one has $\psi(x; a, q) \sim x/\varphi(q)$ uniformly for $q \leq (\log x)^{2-\varepsilon}$.

5. Set

$$Q(s, \chi) = \sum_{n=1}^{\infty} \mu(n)^2 \chi(n) n^{-s}.$$

Show that $Q(s, \chi) = L(s, \chi) H(s, \chi)$ where $H(s, \chi)$ is a bounded, holomorphic function for $\sigma \geq \sigma_0 > \frac{1}{2}$. Show that $L(s, \chi) \ll_\varepsilon (Tq)^{1-\sigma+\varepsilon}$ for $0 \leq \sigma \leq 1$, $|\tau| + 1 \leq T$ and deduce that, for all $\varepsilon > 0$ and uniformly for $x \geq 2$, $q \geq 1$, $(a, q) = 1$, one has

$$\sum_{\substack{n \leq x \\ n \equiv a \,(\mathrm{mod}\, q)}} \mu(n)^2 = \frac{x}{q} \prod_{p \nmid q} (1 - p^{-2}) + O_\varepsilon\Big(x^{2/3+\varepsilon} q^{1/3}\Big). \tag{53}$$

Show that the elementary convolution argument employed for the proof of Theorem I.3.9 yields the better bound $O(\sqrt{x})$ for the remainder term. Recover essentially this result (i.e. $\ll x^{1/2+\varepsilon}$) by the above analytic method using Montgomery's upper bound (1971), theorem 10.1,

$$\sum_{\chi \,(\mathrm{mod}\, q)} \int_{-T}^{T} |L(\tfrac{1}{2} + i\tau, \chi)|^4 \,\mathrm{d}\tau \ll (qT)^{1+\varepsilon}.$$

[*For more delicate, or average-type estimates, see Prachar (1958), Warlimont (1969, 1980). In particular Prachar shows by a simple modification of the elementary approach that the remainder term in (53) is* $\ll x^{1/2} q^{-1/4} + q^{1/2}$.]

6. Let χ denote the unique non-principal Dirichlet character to the modulus 4. Show that $L(1, \chi) = \pi/4$. Show that the set of prime numbers of the form $4m + 3$ satisfies hypotheses (i) and (ii) of Exercise 7.9, with $\delta = \frac{1}{2}$ and $K = \sqrt{(2/\pi)}\, e^{\gamma/2} A$, where we have set $A = \prod_{p \equiv 3 \,(\mathrm{mod}\, 4)} \left(1 - p^{-2}\right)^{-1/2}$. Deduce that the number $N_3(x)$ of those integers n not exceeding x and all of whose prime factors are of the form $4m + 3$ satisfies

$$N_3(x) \sim \frac{\sqrt{2} A x}{\pi \sqrt{(\log x)}}.$$

Give a remainder form of this result. Generalise. [*See Landau (1909), vol. 2, pp. 641-669, and Wirsing (1956).*]

Part III

Probabilistic methods

III.1

Densities

§ 1.1 Definitions. Natural density

Like other branches of mathematics (and perhaps even more so!) number theory is confronted with the problem of rigorously formalising intuitive notions. Foremost among these is the *probability that a number belongs to a sequence*. The point here is to give a mathematical meaning to statements of the type: one integer in two is even, almost no integer is the sum of two squares, etc.

The first approach which comes to mind is naturally to have recourse to the established theory of probability. The defining of a probability measure on $\mathbb{Z}^+$ does indeed allow us to associate a probability with each subsequence $\mathcal{A}$ of the sequence of integers. However the following result shows that such a framework fundamentally contradicts one of the strongest of our intuitions about numbers: that which suggests that the proportion of integers divisible by $a \geq 1$ is exactly $1/a$.

Theorem 1. *Let $a\mathbb{Z}^+$ denote the set of positive multiples of a. There exists no probability P on $\mathbb{Z}^+$ such that*

$$P(a\mathbb{Z}^+) = 1/a \qquad (a = 1, 2, \ldots).$$

Proof. Let us argue by contradiction. Since

$$a\mathbb{Z}^+ \cap b\mathbb{Z}^+ = ab\mathbb{Z}^+$$

whenever $(a, b) = 1$, we see that, under this assumption, the events $a\mathbb{Z}^+$ and $b\mathbb{Z}^+$ are independent. The same holds for their complements $\mathbb{Z}_a^+$ and $\mathbb{Z}_b^+$, with the notation $\mathbb{Z}_a^+ := \mathbb{Z}^+ \smallsetminus a\mathbb{Z}^+$. Therefore

$$P(\mathbb{Z}_a^+ \cap \mathbb{Z}_b^+) = \Big(1 - \frac{1}{a}\Big)\Big(1 - \frac{1}{b}\Big)$$

when $(a, b) = 1$. Inductively we immediately obtain for each pair of integers m, n, $m < n$, that

$$P(\{m\}) \leq P\Big(\bigcap_{m<p\leq n} \mathbb{Z}_p^+\Big) = \prod_{m<p\leq n} \Big(1 - \frac{1}{p}\Big).$$

Since n is arbitrarily large, we deduce from Mertens' theorem that $P(\{m\}) = 0$ for all $m \geq 1$, which yields the desired contradiction.

To define a probability law on $\mathbb{Z}^+$ is equivalent to considering a convergent series with sum 1, let us say

$$\sum_{n=1}^{\infty} \lambda_n = 1,$$

with $0 \leq \lambda_n \leq 1$ for all n. For any integer sequence $\mathcal{A} \subseteq \mathbb{Z}^+$, we then have

$$P(\mathcal{A}) = \sum_{a \in \mathcal{A}} \lambda_a$$

and we can appreciate the discrepancy between such a model and intuition by noting that the probability of a sequence virtually depends only on its initial terms. Indeed, for each $\varepsilon > 0$ there exists $N = N_\varepsilon$ such that

$$P(\{1, 2, \ldots, N\}) \geq 1 - \varepsilon.$$

We thus find ourselves in a typical situation, where the only theory at our disposal irrevocably invalidates two very natural "theorems", viz.

(i) $P(a\mathbb{Z}^+) = 1/a \qquad (a = 1, 2, \ldots)$,

(ii) $P(\mathcal{A}) = P(\mathcal{B}) \qquad (|\mathcal{A}\Delta\mathcal{B}| < \infty)$,

where $\mathcal{A}\Delta\mathcal{B}$ denotes the symmetric difference of $\mathcal{A}$ and $\mathcal{B}$. The choice (standard throughout the history of mathematics) of preferring intuitive theorems to established theories having been made, it is not hard to circumvent these difficulties. Introducing a *divergent* series of non-negative terms

$$\sum_{n=1}^{\infty} \lambda_n = \infty,$$

we define the *density* $d(\mathcal{A})$ of a sequence $\mathcal{A} \subseteq \mathbb{Z}^+$ as the limit, when it exists, of the ratio

$$d(\mathcal{A}; x) := \sum_{a \leq x,\, a \in \mathcal{A}} \lambda_a \Big/ \sum_{n \leq x} \lambda_n \tag{1}$$

as $x \to \infty$. It certainly follows that any finite sequence has zero density, but the concept thus introduced is not a measure on $\mathbb{Z}^+$:

(iii) sequences possessing a density do not form a σ-algebra,

(iv) density is not countably additive.

The simplest choice consists in taking $\lambda_n = 1$ for all $n \geq 1$. We thus obtain the notion of *natural density* or *asymptotic density*, both terms being commonly used. When it exists, the natural density of $\mathcal{A}$ is given by the formula

$$\mathbf{d}\mathcal{A} := \lim_{x \to \infty} x^{-1} \big|\{a \leq x : a \in \mathcal{A}\}\big|. \tag{2}$$

One denotes by upper (resp. lower) natural (or asymptotic) density the quantity $\overline{\mathbf{d}}\mathcal{A}$ (resp. $\underline{\mathbf{d}}\mathcal{A}$) obtained by replacing the symbol lim by lim sup (resp. lim inf) in (2).

Before proceeding further we make four simple observations.

(a) Any arithmetic progression $n \equiv a (\bmod q)$ has natural density, equal to $1/q$. This is immediate, since in this case we have

$$\big|\{a \leq x : a \in \mathcal{A}\}\big| = \big[x/q\big] + O(1).$$

The intuitive property (i) stated above is hence verified.

(b) A necessary and sufficient condition for the increasing sequence $a_1 < a_2 < \ldots$ to be of natural density α, $0 \leq \alpha \leq 1$, is that

$$\lim_{n\to\infty} n/a_n = \alpha. \tag{3}$$

The condition is obviously sufficient. We see that it is necessary by noting that, if $\{a_j\}_{j=1}^{\infty}$ has natural density α, then

$$n = \big|\{j : a_j \leq a_n\}\big| = \big(\alpha + o(1)\big)a_n \qquad (n \to \infty).$$

(c) There exist integer sequences failing to have natural density. Consider, for example, the sequence $\mathcal{A}$ of integers n with leading digit equal to 1 in the expansion to base 10. We have

$$\mathcal{A} := \bigcup_{k=0}^{\infty} \{n : 10^k \leq n < 2{\cdot}10^k\}. \tag{4}$$

Writing $A(x) := \big|\mathcal{A} \cap [1, x]\big|$, it follows that, for $m \geq 1$,

$$A(10^m - 1) = \sum_{k=0}^{m-1} 10^k = \tfrac{1}{9}(10^m - 1),$$
$$A(2{\cdot}10^m - 1) = \tfrac{1}{9}(10^m - 1) + 10^m = \tfrac{5}{9}(2{\cdot}10^m - 1) + \tfrac{4}{9}.$$

We easily deduce that

$$\underline{\mathbf{d}}\mathcal{A} = \tfrac{1}{9}, \qquad \overline{\mathbf{d}}\mathcal{A} = \tfrac{5}{9}.$$

(d) A formal link between the notion of density and probability theory can be defined. In the case of natural density, it suffices to observe that, if ν_N denotes the probability measure on $\mathbb{Z}^+$ obtained by assigning the uniform weight $1/N$ to each of the first N integers, then we have, assuming existence,

$$\mathbf{d}\mathcal{A} = \lim_{N\to\infty} \nu_N(\mathcal{A}).$$

This explains why natural density echoes intuitive criteria: the density of a sequence is the limit of its frequency in the first N integers.

§ 1.2 Logarithmic density

The density most used after natural density is that which is obtained by choosing $\lambda_n = 1/n$ for $n \geq 1$. The concept thus defined is called *logarithmic density*. The traditional notation is $\boldsymbol{\delta}\mathcal{A}$, so, assuming existence,

$$\boldsymbol{\delta}\mathcal{A} := \lim_{x\to\infty} \frac{1}{\log x} \sum_{a\leq x,\, a\in\mathcal{A}} \frac{1}{a}. \tag{5}$$

One defines in an obvious way the upper $\overline{\boldsymbol{\delta}}\mathcal{A}$ and lower $\underline{\boldsymbol{\delta}}\mathcal{A}$ logarithmic densities.

It is easy to construct examples of sequences failing to possess logarithmic density. The following theorem shows that they are to be found amongst sequences having no natural density.

Theorem 2. *For any integer sequence $\mathcal{A}$, we have*

$$\underline{\mathbf{d}}\mathcal{A} \leq \underline{\boldsymbol{\delta}}\mathcal{A} \leq \overline{\boldsymbol{\delta}}\mathcal{A} \leq \overline{\mathbf{d}}\mathcal{A}. \tag{6}$$

In particular, if $\mathcal{A}$ has natural density then it also has logarithmic density, and the two are equal.

Proof. Write $A(x) := \sum_{a\leq x} 1$, and $L(x) := \sum_{a\leq x} 1/a$, where the sums are over elements a of $\mathcal{A}$. By partial summation, we have

$$L(x) = \frac{A(x)}{x} + \int_1^x \frac{A(t)}{t^2}\,\mathrm{d}t \qquad (x \geq 1). \tag{7}$$

Let $\varepsilon > 0$. There exists some $t_0 = t_0(\varepsilon)$ such that

$$\underline{\mathbf{d}}\mathcal{A} - \varepsilon \leq A(t)/t \leq \overline{\mathbf{d}}\mathcal{A} + \varepsilon \qquad (t > t_0).$$

Substituting in (6) for $x > t_0$, we infer that

$$\big(\underline{\mathbf{d}}\mathcal{A} - \varepsilon\big) \log\big(x/t_0\big) \leq L(x) \leq 1 + \log t_0 + \big(\overline{\mathbf{d}}\mathcal{A} + \varepsilon\big) \log\big(x/t_0\big).$$

The stated result follows from these bounds, by allowing first x to tend to ∞ and then ε to 0.

The converse of Theorem 2 is false: the existence of logarithmic density says nothing about that of natural density. A counter-example is provided by the sequence (4). The following calculation shows that it has logarithmic density $\boldsymbol{\delta}\mathcal{A} = \log 2/\log 10$. Indeed

$$L(x) = \sum_{a\leq x} \frac{1}{a} = \sum_{0\leq k\leq \log x/\log 10} \sum_{\substack{10^k\leq n<2\cdot 10^k\\ n\leq x}} \frac{1}{n}$$

$$= \sum_{0\leq k\leq \log x/\log 10} \Big\{\log 2 + O\Big(\frac{1}{k+1}\Big)\Big\} + O(1) = \frac{\log 2}{\log 10}\log x + O(\log_2 x).$$

§ 1.3 Analytic density

The method for defining the density of an integer sequence described in Section 1 can be formally generalised. Instead of truncating a divergent series $\sum_{n=1}^{\infty} \lambda_n$, we consider a continuous family $(P_\sigma)_{\sigma \in S}$ of probability laws on $\mathbb{Z}^+$ and set

$$P_\sigma(\{n\}) =: \lambda_n(\sigma) \Big/ \sum_{m=1}^{\infty} \lambda_m(\sigma)$$

where each series $\sum_{m=1}^{\infty} \lambda_m(\sigma)$ $(\sigma \in S)$ is convergent. In addition, we consider a point σ_0, in the closure of S but not belonging to S, such that the series $\sum_{m=1}^{\infty} \lambda_m(\sigma_0)$, where each term is defined by continuity, is divergent. The density $d(\mathcal{A})$ is then defined as the possible limit of the ratio

$$P_\sigma(\mathcal{A}) = \sum_{n \in \mathcal{A}} \lambda_n(\sigma) \Big/ \sum_{m=1}^{\infty} \lambda_m(\sigma) \tag{8}$$

as $\sigma \to \sigma_0$ while remaining in S. The case of Section 1 corresponds to $S =]0,1]$, $\sigma_0 = 0$, and

$$\lambda_n(\sigma) = \begin{cases} \lambda_n & (n \leq 1/\sigma) \\ 0 & (n > 1/\sigma). \end{cases}$$

For each $\varepsilon > 0$, the convergence of the series $\sum_{m=1}^{\infty} \lambda_m(\sigma)$ when $\sigma \in S$ implies the existence of an integer $N = N(\varepsilon, \sigma)$, such that

$$P_\sigma(\mathcal{A}) = \Big(\sum_{n \leq N,\, n \in \mathcal{A}} \lambda_n(\sigma) \Big/ \sum_{m \leq N} \lambda_m(\sigma) \Big) + \theta\varepsilon,$$

with $|\theta| \leq 1$. By choosing $\varepsilon = \varepsilon(\sigma) \to 0$ as $\sigma \to \sigma_0$, we hence recover a definition analogous to that of Section 1, but where the function λ_n also depends on x. In certain instances, it may be established by a Tauberian argument that there exists an equivalent procedure which is strictly of type (1). Even then, such a framework can be useful: when the series $\sum \lambda_n(\sigma)$ are well-chosen, consideration of (8) rather than (1) can notably simplify the calculations, or favour the use of certain analytic techniques. This stems from replacing truncation by a smoother procedure.

A fundamental example is obtained from the choice $S =]1, +\infty[$, $\sigma_0 = 1$ and $\lambda_n(\sigma) = n^{-\sigma}$ $(n \geq 1)$. We then get

$$P_\sigma(\mathcal{A}) = \zeta(\sigma)^{-1} \sum_{n \in \mathcal{A}} n^{-\sigma}. \tag{9}$$

The presence of the Dirichlet series associated to the characteristic function of $\mathcal{A}$ opens the possibility of using all the analytic and algebraic properties of Dirichlet series for the computation of density—especially those connected to the convolution product.

The *analytic density* of a sequence $\mathcal{A}$ is defined as the limit (when it exists) of the quantity (9) as $\sigma \to 1+$. Instead of (9), we can equally well consider the expression

$$(\sigma - 1) \sum_{n \in \mathcal{A}} n^{-\sigma} \tag{10}$$

but it is often convenient to keep the factor $\zeta(\sigma)^{-1}$ when Euler products are involved.

The expected link between analytic density and a density of type (1) is made explicit in the following result.

Theorem 3. *Let $\mathcal{A}$ be an integer sequence. Then $\mathcal{A}$ has analytic density if and only if $\mathcal{A}$ has logarithmic density. In this case the two densities are equal.*

Proof. Let us retain the notation $L(x) = \sum_{a \leq x} 1/a$, where summation is restricted to elements $a \in \mathcal{A}$. By partial summation, we have

$$(\sigma - 1) \sum a^{-\sigma} = (\sigma - 1)^2 \int_0^\infty L(x) x^{-\sigma} \, \mathrm{d}x \qquad (\sigma > 1). \tag{11}$$

Suppose first that $\mathcal{A}$ has logarithmic density $\delta = \boldsymbol{\delta}\mathcal{A}$, that is to say

$$L(x) = \{\delta + o(1)\} \log x \qquad (x \to \infty).$$

Substituting in the right-hand side of (11), we deduce that

$$(\sigma - 1) \sum a^{-\sigma} = \delta + o(1) \qquad (\sigma \to 1+), \tag{12}$$

so $\mathcal{A}$ has analytic density, equal to δ.

Conversely, if (12) is satisfied, we can rewrite (11) in the following form, with $h := \sigma - 1$,

$$h \int_{0-}^{+\infty} e^{-ht} \, \mathrm{d}L(e^t) = h^2 \int_1^{+\infty} L(x) x^{-1-h} \, \mathrm{d}x = \delta + o(1) \qquad (h \to 0+).$$

Karamata's theorem II.7.5 then implies that

$$L(e^t) = \{\delta + o(1)\} t \qquad (t \to \infty),$$

which is equivalent to the required result.

§ 1.4 Probabilistic number theory

The concepts introduced in the previous sections form the basis of a new branch of number theory. The key idea is that of natural density, which enables us to view the behaviour of arithmetic functions in a fresh light.

Indeed, these functions usually have the specific property of varying in an intrinsically irregular and erratic manner, with the consequence that classical techniques of analysis are generally powerless to describe their behaviour appropriately. Probabilistic number theory corresponds to the challenge, which arises naturally in such circumstances, of seeking a statistical study of such behaviour. Together with the notions of extremal and average values, which allow a cursory classification, we shall also define (cf. Chapter 3) the notion of *normal order* of an arithmetic function, which reflects its "almost certain" behaviour. In practice, this involves a process of excluding a set of integers of zero density on which the function behaves aberrantly and studying the more "normal" behaviour elsewhere. Strikingly, this approach results in order and regularity suddenly emerging from apparent chaos. The illuminating focus of the "almost everywhere" concept has the effect of revealing a new field of investigation, which requires specific methods and yields distinctive results.

As has been remarked by Delange (1982), one ought rather to speak of "probabilistic theory of arithmetic functions". The typical viewpoint (cf. Remark (d) of §1) is that of considering an arithmetic function f as a random variable on the discrete space formed by the first N integers and equipped with the uniform law. The fundamental question then consists in determining in what sense(s) we can assert that the law of f tends towards a limit law as $N \to \infty$. We shall make this precise in the following chapters.

Notes

§§ **1.1–1.3.** Many other types of density of an integer sequence have been defined in the literature. Here we limit ourselves to mentioning three.

(a) The *multiplicative density* of Davenport & Erdős (1951) which is linked to the distribution of $\mathcal{A}$ among the different subsequences of $\mathbb{Z}^+$ obtained by removing those integers having a "large" prime factor—see Exercise 2.

(b) The *Schnirelmann density* (1930), defined by $\sigma(\mathcal{A}) = \inf_{n\geq 1} A(n)/n$.

It is related to the *addition of sequences*: if we define

$$\mathcal{A}+\mathcal{B} := \{a+b : a \in \mathcal{A},\ b \in \mathcal{B}\},$$

an important theorem of Mann (1942) states that

$$\sigma(\mathcal{A}+\mathcal{B}) \geq \min\big(1, \sigma(\mathcal{A})+\sigma(\mathcal{B})\big). \tag{13}$$

A fairly simple proof of (13) may be found in Halberstam & Roth (1983), Chapter I.4.

(c) The *divisor density* of Hall (1978). When this exists, it is the unique number $\mathbf{D}\mathcal{A}$ satisfying

$$\sum_{d|n,\, d\in\mathcal{A}} 1 = \big(\mathbf{D}\mathcal{A} + o(1)\big)\tau(n) \qquad \text{pp},$$

where the symbol "pp" (for the French *presque partout*, i.e. almost everywhere) indicates that the asymptotic formula thus described occurs as $n \to \infty$ inside some suitable sequence of natural density 1. This definition is quite different from those of §§1.1–1.3. For example, for any α, β in $[0,1]$ one can find a sequence $\mathcal{A}$ such that $\mathbf{d}\mathcal{A} = \alpha$, $\mathbf{D}\mathcal{A} = \beta$ (Hall, 1978). For other properties one can consult Hall (1981), Tenenbaum (1982), Dupain, Hall & Tenenbaum (1982), Hall & Tenenbaum (1986). [See also Exercises 3.9 and 3.10.]

§ **1.3.** A deeper study of the properties of arithmetic functions with respect to the laws $P_\sigma(\mathcal{A})$ has been undertaken by Nanopoulos (1975, 1977, 1982).

Exercises

We recall that $P^+(n)$ denotes the largest prime factor of an integer n, with the convention that $P^+(1) = 1$. For $y \geq 2$, $\mathcal{A} \subseteq \mathbb{Z}^+$, we set

$$\mathcal{A}_y := \mathcal{A} \cap \{n : P^+(n) \leq y\}.$$

For each integer $j \geq 1$, let $\mathcal{A}^{(j)}$ denote the finite sequence comprising the j smallest elements of $\mathcal{A}$.

1. For $n \geq 1$, write $n_y := \max\{d : d|n,\ P^+(d) \leq y\}$. Let $\mathcal{A} \subseteq \mathbb{Z}^+$ be a sequence of integers satisfying, for some $y \geq 2$, $\mathcal{A} = \{n : n_y \in \mathcal{A}_y\}$. Show that for $s \in \mathbb{C}$, $\sigma > 1$, we have

$$\sum_{n\in\mathcal{A}} n^{-s} = \zeta(s) \prod_{p\leq y} (1 - p^{-s}) \sum_{a\in\mathcal{A}_y} a^{-s}.$$

Deduce the existence of $\mathbf{d}\mathcal{A}$ and the formula

$$\mathbf{d}\mathcal{A} = \prod_{p\leq y} \Big(1 - \frac{1}{p}\Big) \sum_{a\in\mathcal{A}_y} \frac{1}{a}.$$

2. *Multiplicative density (Davenport & Erdős, 1951).*
For $y \geq 2$, $\mathcal{A} \subseteq \mathbb{Z}^+$, let us write

$$\mathbf{m}_y\mathcal{A} := \prod_{p\leq y} \Big(1 - \frac{1}{p}\Big) \sum_{a\in\mathcal{A}_y} \frac{1}{a}.$$

If $\mathbf{m}_y\mathcal{A}$ tends towards a limit $\mathbf{m}\mathcal{A}$ as $y \to \infty$, we say that $\mathbf{m}\mathcal{A}$ is the multiplicative density of $\mathcal{A}$. Now write

$$\overline{\mathbf{m}}\mathcal{A} := \limsup_{y\to\infty} \mathbf{m}_y\mathcal{A}, \qquad \underline{\mathbf{m}}\mathcal{A} := \liminf_{y\to\infty} \mathbf{m}_y\mathcal{A}.$$

(a) Show, for any $\mathcal{A}$, that $0 \leq \underline{\mathbf{m}}\mathcal{A} \leq \overline{\mathbf{m}}\mathcal{A} \leq 1$.

(b) Show that for any integer sequence $\mathcal{A}$ one has $\overline{\boldsymbol{\delta}}\mathcal{A} \leq e^\gamma \overline{\mathbf{m}}\mathcal{A}$.

(c) Let $\mathcal{A} := \{n \geq 1 : P^+(n) \leq \sqrt{n}\}$. Show that

$$\mathbf{d}\mathcal{A} = 1 - \log 2, \qquad \mathbf{m}\mathcal{A} = 1 - e^{-\gamma}.$$

3. *Sequential density of a set of multiples (Davenport & Erdős, 1951).*
Let $\mathcal{A}$ be an integer sequence. The *set of multiples* $\mathcal{M}(\mathcal{A})$ of $\mathcal{A}$ is defined as the sequence $\mathcal{M}(\mathcal{A}) = \{am : a \in \mathcal{A},\ m \geq 1\}$. [*See Halberstam & Roth (1966), Erdős, Hall & Tenenbaum (1994).*]

(a) Show that for each $j \geq 1$, $\mathcal{M}_j := \mathcal{M}(\mathcal{A}^{(j)}) \smallsetminus \mathcal{M}(\mathcal{A}^{(j-1)})$ has natural density $\mathbf{d}\mathcal{M}_j = \Delta_j(\mathcal{A})$ given by

$$\Delta_j(\mathcal{A}) := \frac{1}{a_j} + \sum_{1\leq k\leq j-1} (-1)^k \sum_{1\leq i_1<\ldots<i_k\leq j-1} \frac{1}{[a_{i_1}, \ldots, a_{i_k}, a_j]},$$

where a_j denotes the jth element of $\mathcal{A}$.

(b) Show that $\Delta(\mathcal{A}) := \sum_{j=1}^{\infty} \Delta_j(\mathcal{A})$ is finite and satisfies $0 \leq \Delta(\mathcal{A}) \leq 1$. We say that $\Delta(\mathcal{A})$ is the *sequential density* of the set of multiples $\mathcal{M}(\mathcal{A})$.

(c) Show that if $\sum_{j=1}^{\infty} a_j^{-1} < \infty$, then $\mathcal{M}(\mathcal{A})$ has natural density and $\mathbf{d}\mathcal{M}(\mathcal{A}) = \Delta(\mathcal{A})$.

4. Let $\mathcal{A} \subseteq \mathbb{Z}^+$, $y \geq 2$.

(a) Deduce from the results of Exercises 1 and 3 that

$$\mathbf{d}\mathcal{M}(\mathcal{A}_y) = \Delta(\mathcal{A}_y) = \mathbf{m}_y\mathcal{M}(\mathcal{A}).$$

(b) Show that $\mathbf{m}_y\mathcal{M}(\mathcal{A})$ is an increasing function of y.

(c) Let $j \geq 1$. Show that for sufficiently large y one has $\Delta_j(\mathcal{A}_y) = \Delta_j(\mathcal{A})$. Deduce that $\mathbf{m}\mathcal{M}(\mathcal{A}) \geq \Delta(\mathcal{A})$.

(d) Show that $\Delta(\mathcal{A}_y) \leq \Delta(\mathcal{A})$. Deduce that $\mathbf{m}\mathcal{M}(\mathcal{A}) = \Delta(\mathcal{A})$, in other words: *Any set of multiples has multiplicative density, which is equal to its sequential density* (Davenport–Erdős, 1951).

5. *The Davenport–Erdős theorem (1951).*

Let $\mathcal{A} \subseteq \mathbb{Z}^+$, $y \geq 2$.

(a) Deduce from Exercise 1 that $\mathbf{d}\mathcal{M}(\mathcal{A}_y) = \mathbf{m}_y\mathcal{M}(\mathcal{A})$.

(b) Let $x > y$. Show that

$$\sum_{\substack{n \in \mathcal{M}(\mathcal{A}) \smallsetminus \mathcal{M}(\mathcal{A}_y) \\ P^+(n) \leq x}} \frac{1}{n} = \prod_{p \leq x} \left(1 - \frac{1}{p}\right)^{-1} \{\mathbf{m}_x\mathcal{M}(\mathcal{A}) - \mathbf{m}_y\mathcal{M}(\mathcal{A})\}.$$

(c) By applying the result of Exercise 4 show that $\bar{\boldsymbol{\delta}}\mathcal{M}(\mathcal{A}) \leq \mathbf{m}\mathcal{M}(\mathcal{A})$.

(d) Show that $\mathbf{m}\mathcal{M}(\mathcal{A}) = \boldsymbol{\delta}\mathcal{M}(\mathcal{A}) = \Delta(\mathcal{A}) = \underline{\mathbf{d}}\mathcal{M}(\mathcal{A})$, in other words: *Any set of multiples has logarithmic density, equal to its lower natural density, and also to its multiplicative and sequential densities.* [A set of multiples does not necessarily have a natural density, cf. Exercise 3.8.]

6. *The Davenport–Erdős theorem (proof of 1937).*

Let $\mathcal{A} \subseteq \mathbb{Z}^+$. For $j \geq 1$, let $\theta_j(n)$ denote the set-theoretic characteristic function of $\mathcal{M}_j := \mathcal{M}(\mathcal{A}^{(j)}) \smallsetminus \mathcal{M}(\mathcal{A}^{(j-1)})$ and write $\Theta_j(n) := \sum_{1 \leq i \leq j} \theta_i(n)$. We consider the Dirichlet series

$$F_j(s) := \sum_{n=1}^{\infty} \theta_j(n) n^{-s}, \qquad F(s) := \sum_{j=1}^{\infty} F_j(s) = \sum_{n \in \mathcal{M}(\mathcal{A})} n^{-s},$$
$$G_j(s) := \zeta(s)^{-1} F_j(s), \qquad G(s) := \zeta(s)^{-1} F(s).$$

(a) Show, for $n \geq 1$, $j \geq 1$, that

$$\Theta_j(n) \log n \geq \sum_{d|n} \Theta_j(d) \Lambda(n/d).$$

Deduce that $\sum_{1 \leq i \leq j} G_i(\sigma)$ is a decreasing function of $\sigma > 1$.

(b) With the notation of Exercise 3 show that $G_j(1) = \Delta_j(\mathcal{A})$.

(c) Show that $\lim_{\sigma\to 1+} G(\sigma) = \Delta(\mathcal{A})$.

(d) Show that $\boldsymbol{\delta}\mathcal{M}(\mathcal{A}) = \Delta(\mathcal{A}) \ge \underline{\mathbf{d}}\mathcal{M}(\mathcal{A})$.

(e) Show that for each $j \ge 1$ one has $\underline{\mathbf{d}}\mathcal{M}(\mathcal{A}) \ge \sum_{1\le i\le j} \Delta_i(\mathcal{A})$. Deduce that $\underline{\mathbf{d}}\mathcal{M}(\mathcal{A}) = \Delta(\mathcal{A})$.

7. *A theorem of Rényi (1955).*

Let $z \in \mathbb{C}$. Write $\varphi_z(n) := z^{\Omega(n)-\omega(n)}$, $\lambda_z(n) := (\varphi_z * \mu)(n)$.

(a) Show, for $|z| < 2$, that $\sum_{n=1}^{\infty} |\lambda_z(n)|/n < \infty$.

(b) Deduce an asymptotic formula for $\sum_{n\le x} \varphi_z(n)$.

(c) Show that for each integer $k \ge 0$ the sequence $\{n : \Omega(n) - \omega(n) = k\}$ has natural density d_k, given by the formula

$$\sum_{k=0}^{\infty} d_k z^k = \frac{6}{\pi^2}\prod_p \Big(\frac{1 - z/(p+1)}{1 - z/p}\Big) \qquad (|z| < 2).$$

8. *Direct factors of* $\mathbb{Z}^+$.

Two subsets $\mathcal{A}$ and $\mathcal{B}$ of $\mathbb{Z}^+$ are said to form a pair of direct factors, if $1 \in \mathcal{A} \cap \mathcal{B}$ and if each integer $n > 1$ decomposes in a unique way as a product $n = ab$ with $a \in \mathcal{A}$, $b \in \mathcal{B}$. When this is the case, we write $a = \pi_A(n)$, $b = \pi_B(n)$, with the convention that $\pi_A(1) = \pi_B(1) = 1$. For $y \ge 2$, $n \ge 1$, let n_y denote the largest divisor of n free of prime factors $> y$, so that $n_y := \prod_{p^\nu \| n,\, p\le y} p^\nu$, and set

$$\mathcal{A}_y := \{n : n \in \mathbb{Z}^+,\ n_y \in \mathcal{A}\}, \quad A_y(x) := |\mathcal{A}_y \cap [1,x]|, \quad A(x) := |\mathcal{A} \cap [1,x]|.$$

Throughout the exercise a (resp. b) denotes generically an integer from $\mathcal{A}$ (resp. $\mathcal{B}$), where $(\mathcal{A}, \mathcal{B})$ is a given pair of direct factors of $\mathbb{Z}^+$.

(a) Show, for any $y \ge 2$, that

$$\sum_{P^+(a)\le y} \frac{1}{a} \sum_{P^+(b)\le y} \frac{1}{b} = \prod_{p\le y} \Big(1 - \frac{1}{p}\Big)^{-1}.$$

(b) Show the existence of $\mathbf{d}\mathcal{A}_y$ and the formula

$$\mathbf{d}\mathcal{A}_y = \Big(\sum_{P^+(b)\le y} \frac{1}{b}\Big)^{-1}.$$

(c) Let $\varphi_y : \mathcal{A} \to \mathcal{A}_y$ be the map defined by $\varphi_y(a) = \pi_A(a_y)a/a_y$. Show that if $\varphi_y(a) = \varphi_y(a')$, then $a\pi_B(a'_y) = a'\pi_B(a_y)$. Deduce that φ_y is injective and establish that $A(x) \le A_y(x)$ for $x \ge 1$, $y \ge 2$.

(d) Show that if $\sum 1/b = \infty$ then $\mathbf{d}\mathcal{A} = 0$.

(e) We now suppose that $\sum 1/b < \infty$. Let α, β denote the characteristic functions of $\mathcal{A}$, $\mathcal{B}$ respectively. Show that $\alpha = \mathbf{1} - \alpha * (\beta - \delta)$. Using the result of (c) above, deduce that one has, for $y \geq 2$,

$$\underline{\mathbf{d}}\mathcal{A} \geq \Big(\sum_{P^+(b) \leq y} \frac{1}{b} \Big)^{-1} \Big(1 - \sum_{P^+(b) > y} \frac{1}{b} \Big).$$

(f) Show that $\mathcal{A}$ has natural density, given by the formula

$$\mathbf{d}\mathcal{A} = \Big(\sum_{b \in \mathcal{B}} \frac{1}{b} \Big)^{-1}$$

where the right-hand side is interpreted as zero when the series diverges. [*This result is due to Saffari (1976) and Erdős, Saffari & Vaughan (1979). The proof indicated here is essentially that of Daboussi (1979).*]

III.2

Limiting distributions of arithmetic functions

§ 2.1 Definition—distribution functions

As we have seen in the previous chapter, probabilistic number theory can be regarded as the asymptotic study of the probability space

$$\{n : 1 \leq n \leq N\}$$

equipped with the uniform law ν_N. In this context, an arithmetic function may be viewed as a sequence of random variables

$$f_N = (f, \nu_N)$$

taking the values $f(n)$, $1 \leq n \leq N$, with probability $1/N$. We intend to investigate from this perspective the classical probabilistic notion of a distribution function.

A distribution function (d.f.) is a non-decreasing function $F : \mathbb{R} \to [0, 1]$ which is right-continuous and satisfies $F(-\infty) = 0$, $F(\infty) = 1$.

The set $\mathcal{D}(F)$ of *discontinuity points* of F is thus at most countable and only contains discontinuities of the first kind. We denote by $\mathcal{C}(F)$ the complement of $\mathcal{D}(F)$, that is to say *the set of continuity points* of F. A *point of increase* for F is a real number z such that $F(z + \varepsilon) - F(z - \varepsilon) > 0$ for any $\varepsilon > 0$. Each discontinuity point is a point of increase but the converse does not hold.

Write $\mathcal{D}(F) = \{z_\nu\}_{\nu=1}^{\infty}$ and $s_\nu := F(z_\nu) - F(z_\nu -)$. The function

$$\Phi(z) = \sum_{z_\nu \leq z} s_\nu$$

increases exclusively by jumps and is constant in any closed interval not containing some z_ν. It is called a step-function. The quantity s_ν is the saltus (or jump) of F at z_ν. If $\mathcal{D}(F)$ is not empty then Φ is, up to a constant multiple, a distribution function. Such d.f. is said to be *purely discrete* or *atomic*. It is immediately checked that $F - \Phi$ is continuous.

The d.f. F is called *improper* (or is said to be the d.f. of an improper law) if it equals a one-point step-function, say

$$F(z) = \begin{cases} 0 & (z < z_0) \\ 1 & (z \geq z_0). \end{cases}$$

A simple example of a continuous d.f. is a function of the type

$$F(z) = \int_{-\infty}^{z} h(t)\,\mathrm{d}t$$

where $h \geq 0$ is integrable in Lebesgue's sense and satisfies $\|h\|_1 = 1$. It is then said that F is *absolutely continuous.* The Radon–Nikodym Theorem (cf. for example, Rudin (1970), theorem 6.9) implies that each continuous d.f. may be written in the form

$$c_0 F_0 + c_1 F_1$$

where F_0 is absolutely continuous and F_1 is *purely singular*, that is to say continuous and such that

$$\int_{\mathcal{N}} \mathrm{d}F_1(z) = 1$$

where $\mathcal{N}$ is a subset of $\mathbb{R}$ with zero Lebesgue measure. We collect the preceding observations into the following statement.

Theorem 1 (Lebesgue decomposition theorem). *Each d.f. F can be uniquely written in the form*

$$F = \alpha_1 F_1 + \alpha_2 F_2 + \alpha_3 F_3$$

with α_1, α_2, $\alpha_3 \geq 0$, $\alpha_1 + \alpha_2 + \alpha_3 = 1$, and where the F_i are d.f.'s such that F_1 is absolutely continuous, F_2 is purely singular and F_3 is atomic.

A sequence $\{F_n\}_{n=1}^{\infty}$ of d.f.'s is said to *converge weakly* to a function F if we have

$$\lim_{n\to\infty} F_n(z) = F(z) \qquad (z \in \mathcal{C}(F)). \tag{1}$$

It is to be emphasised that the weak limit F is necessarily non-decreasing and bounded, but need not be a d.f. We may always suppose that it is right-continuous for (1) implies no constraint when $z \in \mathcal{D}(F)$.

Let us consider an arithmetic function f. For each $N \geq 1$ the function

$$F_N(z) := \nu_N\{n : f(n) \leq z\} = \frac{1}{N}|\{n \leq N : f(n) \leq z\}| \tag{2}$$

is an atomic d.f.

Definition. *An arithmetic function f is said to possess a (limiting) distribution function F (or: to have a limit law with d.f. F) if F is a d.f. and if the sequence F_N defined by (2) converges weakly to F.*

Thus the existence of a d.f. for f is equivalent to combining the two following assertions:

(i) The limit $F(z) := \lim_{N\to\infty} F_N(z)$ exists for all z belonging to a certain dense subset E of $\mathbb{R}$.

(ii) We have $\lim\limits_{\substack{z\to\pm\infty\\ z\in E}} F(z) = \begin{cases}1\\0\end{cases}$.

Indeed F can then be extended to a non-decreasing, right-continuous function. This implies the validity of (1).

The following result, which has often been implicitly used by Erdős, gives a practical sufficient condition for an arithmetic function to possess a limit law.

Theorem 2. *Let f be a real arithmetic function. Suppose that for any $\varepsilon > 0$ there exists a function with values in $\mathbb{Z}^+$, $n \mapsto a_\varepsilon(n)$, having the following properties:*

(i) $\lim\limits_{\varepsilon\to0} \limsup\limits_{T\to\infty} \overline{\mathbf{d}}\{n : a_\varepsilon(n) > T\} = 0;$

(ii) $\lim\limits_{\varepsilon\to0} \overline{\mathbf{d}}\{n : |f(n) - f(a_\varepsilon(n))| > \varepsilon\} = 0;$

(iii) *for each $a \geq 1$ the density $\mathbf{d}\{n : a_\varepsilon(n) = a\}$ exists.*

Then f has a limit law.

Proof. Let us choose, as $\eta \to 0+$, two functions $\varepsilon = \varepsilon(\eta) \to 0+$ and $T = T(\varepsilon(\eta)) \to \infty$ such that the upper density in (i) is $\leq \eta$. Denote by $\mathbf{d}(a,\varepsilon)$ the density mentioned in (iii) and write

$$F(z,\eta) := \sum_{a\leq T(\varepsilon),\, f(a)\leq z} \mathbf{d}(a,\varepsilon), \qquad F(z) := \limsup_{\eta\to0} F(z,\eta).$$

First let us show that $F_N(z)$, as defined by (2), converges weakly to F. For $z \in \mathcal{C}(F)$, we have

$$\begin{aligned} F_N(z) \leq N^{-1}\big|\{n \leq N : a_\varepsilon(n) \leq T(\varepsilon),\ f(a_\varepsilon(n)) \leq z+\varepsilon\}\big| \\ + N^{-1}\big|\{n \leq N : a_\varepsilon(n) > T(\varepsilon)\}\big| \\ + N^{-1}\big|\{n \leq N : |f(n) - f(a_\varepsilon(n))| > \varepsilon\}\big|. \end{aligned}$$

By (iii) the first term of this upper bound equals $F(z+\varepsilon,\eta) + o(1)$ as $N \to \infty$. The other two may be estimated using (i) and (ii). Letting N tend to infinity and then η tend to 0, we obtain

$$\limsup_{N\to\infty} F_N(z) \leq \limsup_{\eta\to0} F\big(z+\varepsilon(\eta),\eta\big) = F(z).$$

The last inequality results from the facts that $F(z',\eta)$ is a non-decreasing function of z' of and that $z \in \mathcal{C}(F)$. Similarly we obtain

$$\liminf_{N\to\infty} F_N(z) \geq \limsup_{\eta\to 0} F\big(z-\varepsilon(\eta),\eta\big) = F(z).$$

Thus F_N converges weakly to F, and we may normalise this function so that it becomes right-continuous.

It remains to show that $F(-\infty)=0$, $F(\infty)=1$. Since $F(z)=\lim F_N(z)$ for $z\in\mathcal{C}(F)$, we clearly have $0\leq F\leq 1$. Let $\varepsilon>0$. Choose z in $\mathcal{C}(F)$ so that $z>\max\{f(a): a\leq T(\varepsilon)\}+\varepsilon$. Then $f(n)>z$ implies either that $a_\varepsilon(n)>T(\varepsilon)$, or that $|f(n)-f(a_\varepsilon(n))|>\varepsilon$. By (i) and (ii) the corresponding density $1-F(z)$ tends to 0 as $\eta\to 0+$. This implies that $F(\infty)=1$. We obtain $F(-\infty)=0$ analogously.

In the probabilistic study of an arithmetic function, a natural normalisation is obtained by introducing the expectation and variance of f relative to ν_N, viz.

$$E_N(f) := \int_{-\infty}^{\infty} z\,\mathrm{d}F_N(z) = \frac{1}{N}\sum_{1\leq n\leq N} f(n), \tag{3}$$

and

$$\begin{aligned} V_N(f) = D_N(f)^2 &:= \int_{-\infty}^{\infty} \{z-E_N(f)\}^2\,\mathrm{d}F_N(z) \\ &= \frac{1}{N}\sum_{1\leq n\leq N}\{f(n)-E_N(f)\}^2. \end{aligned} \tag{4}$$

This suggests a different approach to the problem of the distribution of values of an arithmetic function. Instead of studying the asymptotic behaviour of $F_N(z)$, we consider that of

$$G_N(z) := \nu_N\big\{n : f(n)\leq E_N(f)+zD_N(f)\big\}. \tag{5}$$

This perspective may be regarded as being central in probabilistic number theory.

The sequence of d.f.'s $F_N(z)$ or $G_N(z)$ contains all the information concerning the arithmetic function f, and questions that are usually raised about f may be "translated" into equivalent ones regarding d.f.'s. Let us give two examples.

(a) It is equivalent to say that a sequence of integers $\mathcal{A}$ has a natural density α and to say that its set theoretic characteristic function $\chi(n)$ has an atomic d.f.

$$F(z) = \begin{cases} 0 & (z<0) \\ 1-\alpha & (0\leq z<1) \\ 1 & (z\geq 1)\end{cases} .$$

(b) If there exist real sequences A_N, B_N, such that

$$H_N(z) := F_N\big(A_N + zB_N\big)$$

converges weakly towards a d.f. $H(z)$ and if the convergence of the expectations $\int z\,\mathrm{d}H_N(z)$ is dominated in Lebesgue's sense, then we have

$$E_N(f) = A_N + \left\{\int_{-\infty}^{\infty} z\,\mathrm{d}H(z) + o(1)\right\}B_N \qquad (N \to \infty).$$

Thus a sufficiently precise knowledge of the d.f. of f provides an average estimate.

As a general rule we should only consider the study of an arithmetic function to be complete when its d.f. has been adequately described.

§ 2.2 Characteristic functions

By definition the characteristic function (c.f.) of a d.f. F is the Fourier transform of the Stieltjes measure $\mathrm{d}F(z)$, viz.

$$\varphi(\tau) := \int_{-\infty}^{\infty} e^{i\tau z}\,\mathrm{d}F(z) \qquad (\tau \in \mathbb{R}). \tag{6}$$

It is a uniformly continuous function on the real line, satisfying

$$|\varphi(\tau)| \le 1 = \varphi(0) \qquad (\tau \in \mathbb{R}).$$

The correspondence $F \leftrightarrow \varphi$ is one-to-one, as shown by the inversion formula

$$F(z+h) - F(z) = \lim_{T\to\infty} \frac{1}{2\pi}\int_{-T}^{T} \frac{1 - e^{-i\tau h}}{i\tau} e^{-i\tau z}\varphi(\tau)\,\mathrm{d}\tau, \tag{7}$$

valid for z and $z+h$ in $\mathcal{C}(F)$. The proof of (7) is analogous to that of the classical Fourier inversion formula, and presents no new difficulty. If F and G have the same c.f., it follows from (7) that $F(z+h) - F(z) = G(z+h) - G(z)$ for almost all z, h. Letting h tend to $-\infty$, we obtain that $F(z) = G(z)$ almost everywhere. Since F and G are non-decreasing and right-continuous, this implies that $F = G$.

The famous *continuity theorem* of Paul Lévy links the weak convergence of the d.f.'s to the pointwise convergence of c.f.'s.

Theorem 3 (Continuity theorem, Lévy, 1925). *Let $\{F_n\}_{n=1}^{\infty}$ be a sequence of d.f.'s and $\{\varphi_n\}_{n=1}^{\infty}$ the corresponding sequence of their c.f.'s. Then F_n converges weakly to a d.f. F if and only if φ_n converges pointwise on $\mathbb{R}$ to a function φ which is continuous at 0. In addition, in this case, φ is the c.f. of F, and the convergence of φ_n to φ is uniform on any compact subset.*

This classical result is established in detail in most books on probability—see for example Cramér (1970), Feller (1971), Loève (1963), Lukacs (1970). Here we limit ourselves to indicating the main steps. The first point is an identity which follows easily from (7).

Lemma 3.1. *Let F be a d.f. and φ its c.f. For $z \in \mathbb{R}$, $h > 0$, we have*

$$(8)\quad \frac{1}{h}\int_z^{z+h} F(t)\,\mathrm{d}t - \frac{1}{h}\int_{z-h}^{z} F(t)\,\mathrm{d}t = \frac{1}{2\pi}\int_{-\infty}^{\infty}\Big(\frac{\sin(\tau/2)}{\tau/2}\Big)^2 e^{-i\tau z/h}\varphi\Big(\frac{\tau}{h}\Big)\,\mathrm{d}\tau.$$

Proof of lemma. It suffices to apply (7) to the d.f. $F_0(z) := (1/h)\int_z^{z+h} F(t)\,\mathrm{d}t$. The c.f. of $F_0(z)$ is $\varphi(\tau)(1-e^{-i\tau h})/i\tau h$, as shown by partial integration. An obvious change of variable then yields the result.

Proof of the theorem. The necessity of the condition is easy. If F_n converges weakly to F we obtain, by a standard argument, that for every $\varepsilon > 0$, there exists a real number $X = X(\varepsilon)$ such that

$$\sup_{n\geq 1}\sup_{\tau\in\mathbb{R}}\Big|\int_{|z|>X} e^{i\tau z}\,\mathrm{d}F_n(z)\Big| \leq \sup_{n\geq 1}\int_{|z|>X}\mathrm{d}F_n(z) \leq \varepsilon.$$

Now X can be chosen in such a way that $\pm X \in \mathcal{C}(F)$, and recalling the definition of the Stieltjes integral we see that

$$\int_{-X}^{X} e^{i\tau z}\,\mathrm{d}F_n(z) \to \int_{-X}^{X} e^{i\tau z}\,\mathrm{d}F(z)$$

uniformly on any compact subset. Up to changing the value of X, the last integral equals $\varphi(\tau)+O(\varepsilon)$. This implies that $\varphi_n \to \varphi$ uniformly on any compact subset.

In order to establish the converse, it suffices to show that, if φ_n converges pointwise to φ and if φ is continuous at 0, then F_n converges weakly towards a d.f. F. Indeed it will then follow from the first part of the proof that φ is the c.f. of F and that the convergence $\varphi_n \to \varphi$ is uniform on any compact subset.

The first step consists in noting that we can extract from $\{F_n\}_{n=1}^{\infty}$ a subsequence $\{F_{n_j}\}_{j=1}^{\infty}$ converging weakly to a non-decreasing, right-continuous function F. This follows from a classical diagonal process, and we omit the details. We trivially have that $0 \leq F \leq 1$, and it remains to show that F is a d.f., that is to say that $F(\infty) - F(-\infty) = 1$. To this end, we apply (8) with $z = 0$, $F = F_{n_j}$, $\varphi = \varphi_{n_j}$, and let j tend to ∞. The conditions for the application of Lebesgue's theorem are trivially fulfilled. We thus obtain (8) where F is the weak limit of F_{n_j}, and φ is the pointwise limit of φ_{n_j}. As $h \to \infty$, the right-hand side tends to $F(\infty) - F(-\infty)$, for F is non-decreasing, and the left-hand side tends to $\varphi(0)$ because φ is continuous at 0 and bounded. Now $\varphi(0) = 1$ because $\varphi_n(0) = 1$ for all n. The weak limit of F_{n_j} is thus certainly a d.f.

This also holds for any other weak limit, say F^*. And since F^* necessarily still has φ for its c.f., we deduce that $F = F^*$. Therefore every weakly convergent subsequence of $\{F_n\}_{n=1}^{\infty}$ converges to F, and this means that $\{F_n\}_{n=1}^{\infty}$ itself converges weakly to F, thereby completing the proof.

When the limit law $F(z)$ is absolutely continuous with bounded density, say

$$F(z) = \int_{-\infty}^{z} h(t)\,\mathrm{d}t,$$

the Berry–Esseen inequality (Theorem II.7.14) gives a quantitative estimate for the approximation of F_n by F. For each $T > 0$ we have

$$\sup_{z\in\mathbb{R}} |F_n(z) - F(z)| \ll \frac{1}{T} \sup_{t\in\mathbb{R}} |h(t)| + \int_{-T}^{T} \left| \frac{\varphi_n(\tau) - \varphi(\tau)}{\tau} \right| \mathrm{d}\tau. \tag{9}$$

Since φ_n tends to φ uniformly on each compact subset, the right-hand side of (9) is bounded, for a suitable choice $T = T_n$, by a quantity ε_n depending only on n and such that $\lim_{n\to\infty} \varepsilon_n = 0$. This estimate is of major practical importance—cf. for example Theorem 4.8.

Bearing in mind the definition of the limiting d.f. of an arithmetic function, the continuity theorem immediately provides the following criterion.

Theorem 4. *Let f be a real arithmetic function. Then f possesses a d.f. F if, and only if, the sequence of functions*

$$\varphi_N(\tau) := \frac{1}{N} \sum_{n\le N} e^{i\tau f(n)} \tag{10}$$

converges pointwise on $\mathbb{R}$ to a function $\varphi(\tau)$ which is continuous at 0. In this case φ is the c.f. of F.

If $F_N(z)$ is defined by (2) we indeed have that

$$\varphi_N(\tau) = \int_{-\infty}^{\infty} e^{i\tau z}\,\mathrm{d}F_N(z).$$

When f is additive, the function $n \mapsto e^{i\tau f(n)}$ is, for each fixed τ, a multiplicative function with modulus 1. The problem of the existence of a limit law for f is thus equivalent to that of the existence of the mean value of a multiplicative function with values in the unit disc. We shall see in Chapter 4 how to exploit this duality.

Notes

§ **2.1.** For further details concerning the theory of d.f.'s see Feller (1970,1971), Loève (1963), or Lukacs (1970).

Theorem 2 is identical to Lemma A2 of Hall & Tenenbaum (1988).

The normalisation (5) for d.f.'s of arithmetic functions has an obvious theoretic interest. In practice, it is often preferable to keep a greater flexibility and to ask the more general question of determining those functions A_N, B_N for which the sequence of d.f.'s

$$H_N(z) := \nu_N\{n : f(n) \leq A_N + zB_N\}$$

converges weakly to a d.f. $H(z)$. This is the point of view adopted by Elliott (1979). Today his book is considered to be the indisputed reference for probabilistic number theory.

§ **2.2.** Our proof of Lévy's theorem essentially follows that of Cramér (1970).

The theory of c.f.'s is particularly suitable for the treatment of *convolutions* of d.f.'s. The convolution product of two d.f.'s F and G is the d.f. H defined by

$$H(z) := \int_{-\infty}^{\infty} F(z-y)\,\mathrm{d}G(y) = \int_{-\infty}^{\infty} G(z-y)\,\mathrm{d}F(y).$$

If $\varphi(\tau)$, $\gamma(\tau)$ are the c.f.'s for F, G respectively, then $\eta(\tau) = \varphi(\tau)\gamma(\tau)$ is the c.f. of H.

Parseval's formula reads

$$\int_{-\infty}^{\infty} g(z)\,\mathrm{d}F(z) = \frac{1}{2\pi}\int_{-\infty}^{\infty} \widehat{g}(\tau)\varphi(\tau)\,\mathrm{d}\tau \tag{11}$$

for all functions $g \in L^1(\mathbb{R})$ for which the Fourier transform

$$\widehat{g}(\tau) = \int_{-\infty}^{\infty} e^{-i\tau z} g(z)\,\mathrm{d}z$$

is also in $L^1(\mathbb{R})$.

Let $\mathcal{C}_b(\mathbb{R})$ denote the space of all continuous and bounded functions on $\mathbb{R}$. Equation (11) remains true in the form

$$(12) \qquad \int_{-\infty}^{+\infty} g(z)\,\mathrm{d}F(z) = \lim_{\lambda\to\infty} \frac{1}{2\pi}\int_{-\lambda}^{\lambda}\Big(1-\frac{|\tau|}{\lambda}\Big)\widehat{g}(\tau)\varphi(\tau)\,\mathrm{d}\tau$$

whenever $g \in L^2(\mathbb{R}) \cap \mathcal{C}_b(\mathbb{R})$. This can be quickly established using the properties of the Fejér kernel

$$w_\lambda(z) := \frac{\lambda}{2\pi}\Big(\frac{\sin \lambda z/2}{\lambda z/2}\Big)^2,$$

with Fourier transform $\widehat{w_\lambda}(\tau) = (1-|\tau|/\lambda)^+$. Indeed, under the above assumptions for g, it is classically known that $g_\lambda := g * w_\lambda$ converges towards g both pointwise and in quadratic mean. Furthermore $\|g_\lambda\|_\infty \leq \|g\|_\infty$. Now $\widehat{g_\lambda} = \widehat{g}\,\widehat{w_\lambda}$ lies in $L^1(\mathbb{R})$ as product of an element of $L^2(\mathbb{R})$ by a continuous function with compact support. Applying (11) to g_λ, and evaluating the limit of the left-hand side by Lebesgue's theorem of dominated convergence, we obtain (12).

Choosing in (12), for each fixed real y, $g(z) = \sin\{T(z-y)\}/\{T(z-y)\}$, so that $\widehat{g}(\tau) = \pi e^{-i\tau y}/T \quad (|\tau| \leq T)$, $\widehat{g}(\tau) = 0 \quad (|\tau| > T)$, we may write

$$\int_{-\infty}^{\infty} \frac{\sin\{T(z-y)\}}{T(z-y)}\,\mathrm{d}F(z) = \frac{1}{2T}\int_{-T}^{T} e^{-i\tau y}\varphi(\tau)\,\mathrm{d}\tau.$$

Integrating over $\mathbb{R}$ with respect to $\mathrm{d}F(y)$, it follows that

$$(13) \qquad \int_{-\infty}^{\infty}\int_{-\infty}^{\infty} \frac{\sin\{T(z-y)\}}{T(z-y)}\,\mathrm{d}F(z)\,\mathrm{d}F(y) = \frac{1}{2T}\int_{-T}^{T} |\varphi(\tau)|^2\,\mathrm{d}\tau.$$

From which, letting T tend to ∞, we infer that

$$(14) \qquad \sum_{n=0}^{\infty} s_n^2 = \lim_{T\to\infty} \frac{1}{2T}\int_{-T}^{T} |\varphi(\tau)|^2\,\mathrm{d}\tau$$

where $\{s_n\}_{n=1}^{\infty}$ is the sequence of jumps of F, ordered in such a way as to be decreasing.

In particular, *the relation*

$$(15) \qquad \lim_{T\to\infty} \frac{1}{2T}\int_{-T}^{T} |\varphi(\tau)|^2\,\mathrm{d}\tau = 0$$

is a necessary and sufficient condition for F to be continuous.

As we shall see in Chapter 4, the convergence of convolution products

$$F_1 * F_2 * \cdots * F_n \qquad (n \to \infty) \tag{16}$$

is of major importance for the study of limit laws of additive or multiplicative functions. In this context, we quote three fundamental results which we assemble in a single statement. We denote by $\varphi_j(\tau)$ the c.f. of F_j and let σ_j be its maximal saltus, i.e.

$$\sigma_j := \max_{z \in \mathbb{R}} \{F_j(z) - F_j(z-)\}.$$

Finally, we write $Y := \mathbf{1}_{[0,\infty[}$ for Heaviside's function.

Theorem 5. *The three following conditions are equivalent:*

(i) *The product (16) is weakly convergent, as $n \to \infty$, to a d.f. F,*

(ii) $\exists \delta > 0 : \lim_{m,n\to\infty} \prod_{m<j\leq n} \varphi_j(\tau) = 1 \quad (|\tau| \leq \delta)$,

(iii) $\lim_{m,n\to\infty} F_m * \cdots * F_n(z) = Y(z) \quad (z \neq 0)$.

Whenever any of these conditions is fulfilled, we have:

(a) (Lévy, 1931) *F is continuous if, and only if,*

$$\sum_{n=1}^{\infty} (1 - \sigma_n) = +\infty. \tag{17}$$

(b) (Jessen & Wintner, 1935) *If each F_n is atomic, then F is a pure law, i.e. F is either atomic, or continuous and purely singular, or absolutely continuous.*

The equivalence of conditions (i), (ii) and (iii) is an easy consequence of the continuity theorem. Put

$$F_{m,n} = F_m * \cdots * F_n, \qquad \varphi_{m,n}(\tau) = \prod_{m<j\leq n} \varphi_j(\tau).$$

If the product (16) converges weakly to F with c.f. φ, then φ is continuous and $\varphi(0) = 1$, hence φ does not vanish in a suitable neighbourhood of the origin, say $|\tau| \leq \delta$. Paul Lévy's theorem then immediately implies (ii). The implication (ii)$\Rightarrow$(iii) is a consequence of the general inequality

$$1 - \Re e\, \varphi(\tau) \geq \tfrac{1}{4}\big(1 - \Re e\, \varphi(2\tau)\big), \tag{18}$$

which holds for any c.f. φ and follows, by integration with respect to $\mathrm{d}F(z)$, from the elementary inequality $1 - \cos(z\tau) \geq \frac{1}{4}\big(1 - \cos(2z\tau)\big)$. Applying (18) to $\varphi_{m,n}$, we see that $\lim_{m,n\to\infty} \varphi_{m,n}(\tau) = 1$ for $|\tau| \leq 2\delta$ and finally, by iteration, for all τ. We deduce (iii) from this by the continuity theorem. It remains to show that (iii) implies (i). By a new application of the continuity theorem, we derive from (iii) that $\varphi_{m,n}(\tau) \to 1$ uniformly on every compact subset. This implies that the product $\prod_{j=1}^{\infty} \varphi_j(\tau)$ converges uniformly on every compact subset, its limit being *ipso facto* continuous at the origin. A last application of Theorem 3 yields the required conclusion.

A sufficient condition for the existence of an absolutely continuous limiting distribution is given by Babu (1992)—see Exercise 7 for a special case.

A complete proof of statements (a) and (b) of Theorem 5 may be found in Elliott (1979), lemma 1.22. The proof of criterion (17) rests upon the fruitful notion of *concentration function*, introduced by Paul Lévy in 1937. We give below a brief survey of some basic results and inform the reader that a deeper account on this topic may be found in the book of Hengartner & Theodorescu (1973). The concentration function Q_F of a d.f. F is defined on $\mathbb{R}^+$ by the formula

$$Q_F(\ell) := \sup_{z \in \mathbb{R}} \big(F(z+\ell) - F(z)\big).$$

We further define the *concentration* $Q(F) := Q_F(1)$ and observe that, for all $\ell > 0$, we have $Q_F(\ell) = Q(F_\ell)$, with $F_\ell(z) := F(\ell z)$. It is easy to see that F is continuous if, and only if, $Q_F(\ell) \to 0$ as $\ell \to 0+$, so the concentration function may be considered as a measure of the distance from F to the set of continuous d.f.'s. This notion fits particularly well with the study of convolutions of d.f.'s (or, what is the same, of sums of random variables), and indeed we have

$$Q(F * G) \le Q(F) \tag{19}$$

for any pair (F, G) of d.f.'s. Inequality (19) immediately follows from the definition of a convolution product:

$$\begin{aligned} Q(F * G) &= \sup_{z \in \mathbb{R}} \int\!\!\int_{z < x+y \le z+1} \mathrm{d}F(x)\,\mathrm{d}G(y) \\ &\le \int_{\mathbb{R}} \mathrm{d}G(y) \sup_{z \in \mathbb{R}} \int_{]z-y,z-y+1]} \mathrm{d}F(x) = Q(F). \end{aligned}$$

A deeper inequality between the concentration of a convolution product and that of its factors has beeen found by Kolmogorov (1956, 1958), and then refined by Rogozin (1961).

Theorem 6 (Kolmogorov–Rogozin inequality). *Let $\{F_j\}_{j=1}^n$ be a finite sequence of d.f.'s and set $F := F_1 * \cdots * F_n$. For $\ell > 0$, we have*

$$Q_F(\ell) \le C \Big/ \sqrt{1 + \sum_{j=1}^{n} (1 - Q_{F_j}(\ell))}, \tag{20}$$

where C is an absolute constant.

Prior to proving (20), let us see how this enables us to establish assertion (a) of Theorem 5.

Assume first that $\sum_{j=1}^{\infty}(1-\sigma_j)<\infty$. We may suppose without loss of generality that $\sigma_j \neq 0$ for all j and hence that $\prod_{j=1}^{\infty}\sigma_j > 0$. For each $j \geq 1$, we choose a real number u_j such that $\sigma_j = F_j(u_j)-F_j(u_j-)$. The series $\sum_{j=1}^{\infty} u_j$ is necessarily convergent: otherwise, there exists a number $\delta > 0$ and two integer sequences m_k, n_k tending to infinity such that $|\sum_{m_k<j\leq n_k} u_j| > \delta$ for all k, whence

$$\prod_{m_k<j\leq n_k} \sigma_j \leq \int_{|z|>\delta} \mathrm{d}F_{m_k,n_k}(z),$$

and this yields the desired contradiction, for the left-hand side approaches 1 as $k \to \infty$ whereas the right-hand side tends to 0. Setting $u := \sum_{j=1}^{\infty} u_j$, we hence obtain that

$$F(u) - F(u-) \geq \int_{\prod_{j=1}^{\infty}\{u_j\}} \prod_{j=1}^{\infty} \mathrm{d}F_j(z_j) = \prod_{j=1}^{\infty} \sigma_j > 0,$$

and F is not continuous at u. Conversely, if condition (17) is fulfilled, let $\varepsilon > 0$ be arbitrary and $m = m(\varepsilon)$ be such that $\sum_{1\leq j\leq m}(1-\sigma_j) \geq 1/\varepsilon^2$. For $\ell \leq \ell_0(\varepsilon)$ we have $Q_{F_j}(\ell) \leq \sigma_j + 1/m$ $(1 \leq j \leq m)$. Applying (19) and then (20) we get that $Q_F(\ell) \ll \varepsilon$. This shows that $Q_F(\ell)$ tends to 0 with ℓ and hence that F is indeed continuous.

The following lemma provides upper and lower bounds for the concentration $Q(F)$ in terms of mean values of $\varphi(\tau)$. We simply denote by $w(z)$ the Fejér kernel $w_1(z)$.

Lemma 6.1. *Set* $K_1 := w(1) > 0.146$ *and* $K_2 := 1/2\pi w(\frac{1}{2}) < 1.022$. *For any d.f.* F *with c.f.* φ, *we have*

$$K_1 \int_{-1}^{1} |\varphi(\tau)|^2 \,\mathrm{d}\tau \leq Q(F) \leq K_2 \int_{-1}^{1} |\varphi(\tau)| \,\mathrm{d}\tau. \tag{21}$$

Proof. For each $z \in \mathbb{R}$, we have

$$\begin{aligned} F(z+\tfrac{1}{2})-F(z-\tfrac{1}{2}) &= \int_{z-1/2}^{z+1/2} \mathrm{d}F(v) \\ &\leq \frac{1}{w(1/2)} \int_{-\infty}^{+\infty} w(z-v)\,\mathrm{d}F(v) = K_2 \int_{-1}^{1} (1-|\tau|)e^{-i\tau z}\varphi(\tau)\,\mathrm{d}\tau, \end{aligned}$$

where the last equality follows from (11). This implies the upper bound of (21). For the lower bound, we note that

$$Q(F) = Q(F) \int_{-\infty}^{+\infty} \int_{z-1/2}^{z+1/2} \mathrm{d}F(v)\,\mathrm{d}z \geq \int_{-\infty}^{+\infty} \big(F(z+\tfrac{1}{2}) - F(z-\tfrac{1}{2})\big)^2 \,\mathrm{d}z.$$

However, the inversion formula (7) in the form

$$F(z+\tfrac{1}{2})-F(z-\tfrac{1}{2})=\frac{1}{2\pi}\lim_{T\to\infty}\int_{-T}^{T}\frac{\sin(\tau/2)}{\tau/2}\varphi(\tau)e^{-i\tau z}\,\mathrm{d}\tau$$

being interpreted as a Fourier transform in $L^2(\mathbb{R})$, we may write Plancherel's formula

$$\int_{-\infty}^{+\infty}\big(F(z+\tfrac{1}{2})-F(z-\tfrac{1}{2})\big)^2\,\mathrm{d}z=\int_{-\infty}^{+\infty}w(\tau)|\varphi(\tau)|^2\,\mathrm{d}\tau\geq w(1)\int_{-1}^{1}|\varphi(\tau)|^2\,\mathrm{d}\tau.$$

This yields the required inequality.

We are now in a position to establish Theorem 6. The following proof is a variant of that of Esseen (1966) or Hengartner & Theodorescu (1973). Without loss of generality we may assume that $\ell=1$, and that $Q(F_j)<1$ for $1\leq j\leq n$. By (21), we may write

$$\text{(22)}\qquad \begin{aligned} Q(F)&\leq K_2\int_{-1}^{1}\prod_{j=1}^{n}|\varphi_j(\tau)|\,\mathrm{d}\tau\\ &\leq K_2\int_{-1}^{1}\exp\Big\{-\tfrac{1}{2}\sum_{1\leq j\leq n}(1-|\varphi_j(\tau)|^2)\Big\}\,\mathrm{d}\tau,\end{aligned}$$

as a consequence of the inequality $u\leq e^{-(1-u^2)/2}$, valid for $0\leq u\leq 1$.

Let $\check{F}_j(z):=1-F_j(-(z+))$ be the d.f. symmetric to $F_j(z)$, with c.f. $\overline{\varphi_j(\tau)}$. Define the symmetrised d.f. $G_j:=F_j*\check{F}_j$, with c.f. $|\varphi_j(\tau)|^2$. Thus

$$\text{(23)}\qquad 1-|\varphi_j(\tau)|^2=\int_{-\infty}^{+\infty}(1-\cos(\tau z))\,\mathrm{d}G_j(z).$$

Put $I:=[-\frac{1}{2},\frac{1}{2}[$ and

$$p_j:=\int_{\mathbb{R}\smallsetminus I}\mathrm{d}G_j(z)=\int_{-\infty}^{+\infty}\mathrm{d}F_j(y)\int_{x\notin y-I}\mathrm{d}F_j(x)\geq 1-Q(F_j)>0.$$

We now introduce the d.f. H_j defined by

$$H_j(z)=\begin{cases}G_j(z)/p_j & \text{if } z\notin I,\\ G_j(-\frac{1}{2})/p_j & \text{if } z\in I.\end{cases}$$

We have $\mathrm{d}G_j\geq p_j\,\mathrm{d}H_j$ and insert this bound in (23). Put $T:=\sum_{1\leq j\leq n}p_j$, $\alpha_j:=T/p_j$, so that $\sum_{1\leq j\leq n}1/\alpha_j=1$. By Hölder's inequality with exponents

α_j, we deduce from (22) and from the lower bound derived from (23) by introducing $\mathrm{d}H_j$ that

$$Q(F) \leq K_2 \prod_{j=1}^{n} \Big(\int_{-1}^{1} \exp\Big\{ -\tfrac{1}{2} \alpha_j p_j \int_{-\infty}^{+\infty} (1 - \cos(\tau z))\, \mathrm{d}H_j(z) \Big\}\, \mathrm{d}\tau \Big)^{1/\alpha_j}.$$

We have $\alpha_j p_j = T$ for all j. By Jensen's inequality, the exponential in the above formula does not exceed

$$\int_{-\infty}^{+\infty} e^{-T(1-\cos(\tau z))/2}\, \mathrm{d}H_j(z),$$

from which, permuting the order of integrations,

$$Q(F) \leq K_2 \prod_{j=1}^{n} \Big(\int_{-\infty}^{+\infty} \mathrm{d}H_j(z) \int_{-1}^{1} e^{-T(1-\cos(\tau z))/2}\, \mathrm{d}\tau \Big)^{1/\alpha_j}.$$

To bound the inner integral, we may assume that $|z| \geq \frac{1}{2}$ since I has zero measure for $\mathrm{d}H_j(z)$. It follows that

$$\begin{aligned}
\int_{-1}^{1} e^{-T(1-\cos(\tau z))/2}\, \mathrm{d}\tau &= \frac{2}{|z|} \int_{0}^{|z|} e^{-T(1-\cos v)/2}\, \mathrm{d}v \\
&\leq \frac{2}{|z|} \Big(1 + \frac{|z|}{\pi} \Big) \int_{0}^{\pi} e^{-T(1-\cos v)/2}\, \mathrm{d}v \\
&\leq \Big(4 + \frac{2}{\pi} \Big) \int_{0}^{+\infty} e^{-Tv^2/\pi^2}\, \mathrm{d}v \leq \frac{19}{\sqrt{T}}.
\end{aligned}$$

This yields the upper bound (20) when $T \geq 1$. Since the result is trivial otherwise, this completes the proof of Theorem 6.

For additional information on c.f.'s see the remarkable work of Lukacs (1970).

Exercises

1. *Distribution function of $\varphi(n)/n$.*
Let $\varepsilon > 0$, $y := \varepsilon^{-2}$, $a_\varepsilon(n) := \prod_{p^\nu \| n,\, p \leq y} p^\nu$.

(a) Show that $\mathbf{d}\{n : a_\varepsilon(n) = a\} = a^{-1} \prod_{p \leq y}(1 - p^{-1})$ for $a \geq 1$, $P^+(a) \leq y$. [*Use for example Exercise I.4.1.*]

(b) Establish the upper bound $\sum_{n \leq x} \log a_\varepsilon(n) \ll x \log(1/\varepsilon)$ $(x \geq 1)$.

(c) Write $f(n) = \varphi(n)/n$, where φ is Euler's function. Show that

$$|f(n) - f(a_\varepsilon(n))| \leq \sum_{p|n,\, p>y} \frac{1}{p}.$$

(d) Show that $f(n)$ has a limiting d.f. [*Deduce from (b) and (c) that the conditions (i) and (ii) of Theorem 2 are satisfied.*]

2. *A theorem of Schoenberg (1928).* For each $\tau \in \mathbb{R}$, determine an arithmetic function λ_τ satisfying $(\varphi(n)/n)^{i\tau} = \sum_{d|n} \lambda_\tau(d)$. Show that the series $\psi(\tau) := \sum_{d=1}^{\infty} \lambda_\tau(d)/d$ is absolutely convergent, and thus recover the result of Exercise 1. Show that the limit law F is given by

$$F(e^z) = \lim_{n\to\infty} F_{p_1} * F_{p_2} * \cdots * F_{p_n}(z)$$

where p_j denotes the jth prime number and $\mathrm{d}F_{p_j}$ is a linear combination of two Dirac measures. By applying Theorem 5(a), deduce that F is continuous. [*This property can also be established by directly appealing to criterion (15) of the Notes—cf. the end of the proof of Theorem 4.1.*] Show that F is pure by using Theorem 5(b).

3. Reconsider Exercises 1 and 2 but with $\sigma(n)/n$ (where $\sigma(n) := \sum_{d|n} d$) instead of $\varphi(n)/n$. Generalise.

4. *A sufficient condition for singularity (Erdős, 1939).* Denote by $P^+(n)$ the largest prime factor of an integer n, with the convention that $P^+(1) = 1$. Throughout the exercise assume as given a non-decreasing, integer-valued function $R(y)$ satisfying

(i) $\sum_p 1/pR(p) < \infty$;

(ii) for sufficiently large y, there exist $R(y)$ integers $m_1 < m_2 < \cdots < m_R$ such that

$$(\alpha) \quad P^+(m_j) \leq y, \ \mu(m_j)^2 = 1 \quad (1 \leq j \leq R(y))$$

$$(\beta) \quad \sum_{1\leq j\leq R(y)} 1/\varphi(m_j) \geq c \log y,$$

where c is an absolute positive constant.

Here we investigate the following result due to Erdős: *Let f be a strongly additive arithmetic function such that $f(p) \ll 1/R(p)^3$. Then f has a d.f., which is either atomic or purely singular.*

(a) Calculate $\lim_{x\to\infty} x^{-1}\sum_{n\leq x} e^{i\tau f(n)}$ by the method suggested in Exercise 2. Deduce the existence of the limit law and show that it is of pure type [*cf. Notes, Theorem 5*].

(b) Let $\lambda(I)$ denote the Lebesgue measure of a generic open subset I of $\mathbb{R}$. Show that, if the limit law of f is absolutely continuous, then

$$\lim_{\lambda(I)\to 0} \overline{\mathbf{d}}\{n : f(n)\in I\} = 0.$$

(c) Let $\varepsilon > 0$, and $y = y(\varepsilon)$ be sufficiently large. Set $a_\varepsilon(n) := \prod_{p\leq y, p|n} p$. Show that, for $a \geq 1$, $\mu(a)^2 = 1$, $P^+(a) \leq y$, we have

$$\mathbf{d}\{n : a_\varepsilon(n) = a\} = \varphi(a)^{-1}\prod_{p\leq y}(1-p^{-1})$$

and that Theorem 2 can be applied to provide a second proof of the existence of the d.f. of f.

(d) The number $y(\varepsilon)$ being chosen to be sufficiently large, write

$$I := \bigcup_{1\leq j\leq R(y)} \Big(f(m_j) + \big[-R(y)^{-2},\, R(y)^{-2}\big]\Big).$$

Show that $\overline{\mathbf{d}}\{n : f(n)\in I\} \gg 1$. Conclude.

5. Show that, for all $\delta > 0$, the function $R(y) = y^\delta$ satisfies the hypotheses of Exercise 4. Deduce that the d.f.'s of $\varphi(n)/n$ and $\sigma(n)/n$ are purely singular. [*Assume the results of Exercises* 2 *and* 3.]

6. Let $F(z)$ be a d.f. with c.f. $\varphi(\tau)$. Show that if $\varphi \in L^2(\mathbb{R})$, then F is absolutely continuous. [*One can appeal to Plancherel's theorem and to Parseval's formula (11) with an arbitrary function g of class C^2 with compact support.*]

7. *A result of Saffari (1979).*

Let f be the strictly additive function defined by $f(p) = (\log p)^{-\alpha}$, where $\alpha > 0$ is given.

(a) Show that f possesses a d.f. F, and express its c.f. $\varphi(\tau)$.

(b) By calculating $|\varphi(\tau)|^2$ show that

$$\varphi(\tau) \asymp \exp\Big\{-2\sum_p \frac{1}{p}\sin^2\big(\tfrac12\tau f(p)\big)\Big\}.$$

(c) Evaluate the sum over p by the prime number theorem and show that

$$|\varphi(\tau)| = |\tau|^{-1/\alpha+o(1)} \qquad (|\tau|\to\infty).$$

Appealing to the result of Exercise 6, deduce that F is absolutely continuous whenever $\alpha < 2$.

8. Let $h(n)$ be the number of prime factors p of n such that $n^{1/3} < p \le n^{1/2}$. Show that $h(n)$ has a distribution function.

9. *Concentration of an additive function on divisors.* Given a real additive function f, we define, for each integer n, a d.f. F_n by

$$F_n(z) := \frac{1}{\tau(n)} \sum_{d|n,\, f(d)\le z} 1.$$

(a) Compute the c.f. $\varphi_n(\tau)$ of F_n and deduce that $F_n = \mathop{\ast}_{p^\nu \| n} F_{p^\nu}$. Under the extra assumption that there is an absolute positive constant c such that $Q(F_{p^\nu}) \le 1 - c$, show that $Q(F_n) \ll (1+\omega(n))^{-1/2}$.

(b) For the case $f = \log$, we define the Δ-function of Hooley (1979) by

$$\Delta(n) = Q(F_n)\tau(n) = \max_{u\in\mathbb{R}} \sum_{d|n,\, e^u < d \le e^{u+1}} 1.$$

Show that $\Delta(n) \ll \tau(n)/\sqrt{(1+\omega(n))}$, and establish that this upper bound is optimal up to the value of the implicit constant.

10. *Chains of divisors.* For each integer n, we say that a set $\{d_1, \ldots, d_h\}$ of divisors of n is an n-chain if

(i) $\Omega(d_1) + \Omega(d_h) = \Omega(n)$,

(ii) d_{j+1}/d_j is a prime number for $1 \le j < h$.

(a) Let $n = mp^\nu$, with $p \nmid m$. Show that if $\{d_1, \ldots, d_h\}$ is an m-chain, then $\mathcal{C}_r := \{d_r, d_r p, \ldots, d_r p^{\nu+1-r}, d_{r+1}p^{\nu+1-r}, \ldots, d_h p^{\nu+1-r}\}$ is an n-chain for all r, $1 \le r \le R := \min(h, \nu+1)$. Show that the $\mathcal{C}_r$ induce a partition of $\{d_j p^\mu : 1 \le j \le h,\, 0 \le \mu \le \nu\}$. Deduce, by induction on $k = \omega(n)$, that for each n the set of all divisors of n is a disjoint union of n-chains. We denote by $\gamma(n)$ the maximal number of these n-chains.

(b) Show that each n-chain contains exactly one divisor d such that $\Omega(d) = [\Omega(n)/2]$.

(c) Show that the function Ω fulfills the condition of Exercise 9 with $c = \frac{1}{2}$. Deduce an inequality for $\gamma(n)$.

[*This result, as well as that of question (a) of the next exercise, is due to de Bruijn, van Tengbergen & Kruyswijk (1949–51).*]

11. *Primitive sequences.* We recall the notation $\gamma(n)$ from the preceding exercise. An integer sequence $\mathcal{A}$ is said to be *primitive* if none of its elements divides any other. For each integer n, we put $\tau(n, \mathcal{A}) := \sum_{d|n,\, d\in\mathcal{A}} 1$.

(a) Show that $\tau(n, \mathcal{A}) \le \gamma(n)$ $(n \ge 1)$ for any primitive sequence $\mathcal{A}$.

(b) By estimating $\sum_{n\le x}\tau(n,\mathcal{A})$, show Behrend's result (1935)

$$\sup_{\mathcal{A}\text{ primitive}} \sum_{a\le x,\, a\in\mathcal{A}} \frac{1}{a} \ll \frac{\log x}{\sqrt{(\log_2 x)}}. \tag{24}$$

(c) By considering $\mathcal{A}=\mathcal{A}_x=\{n\ge 1:\Omega(n)=[\log_2 x]\}$ and appealing to the results obtained by the Selberg–Delange method, show that the estimate (24) is optimal up to the value of the implicit constant.

[*Behrend's original proof, which rests on a theorem of Sperner (1928), is reproduced in Chapter 5 of Halberstam & Roth (1966). Erdős, Sárközy & Szemerédi (1967) have shown that the left-hand side of (24) is asymptotically* $(\log x)/\sqrt{(2\pi\log_2 x)}$.]

III.3

Normal order

§ 3.1 Definition

The concept of normal order of an arithmetic function corresponds, in probabilistic number theory, to that of almost sure equality of random variables in probability theory. More precisely, we say that *an arithmetic function f has normal order g if g is an arithmetic function such that, for any $\varepsilon > 0$, we have*

$$|f(n) - g(n)| \leq \varepsilon |g(n)|$$

on a set of integers n of density 1. A useful notation to express such a situation is

$$f(n) = \big(1 + o(1)\big)g(n) \qquad \text{pp}, \tag{1}$$

where the symbol pp (*presque partout*, i.e. almost everywhere) means that the relation thus designated holds on a suitable subset with natural density 1. Of course a given function can have several normal orders, which must all have the same asymptotic behaviour. However, the notion is pertinent only for those functions g whose behaviour is in a certain sense simpler than that of f. It was certainly with this idea in mind that Hardy & Ramanujan (1917), in an article which can be considered today as marking the birth of probabilistic number theory, added the restriction that a normal order should be *elementary and monotone.* It is always rather delicate to venture a definition of the word *elementary.* To fix ideas we may give it here the relatively wide meaning of: *that which can be expressed by means of the symbols of real analysis.* Thus we consider as elementary a function such as $\mathrm{li}(x)$, but reject the label in the case of a function defined exclusively by a Cauchy integral. Although they indeed hold in almost all known examples, we have not retained the limitations of Hardy & Ramanujan here, in order to keep the theoretical notion as flexible as possible.

In terms of distribution functions, the existence of a normal order can be interpreted, after suitable renormalisation, as a convergence to an improper law. For instance, it is immediate, in the case of postive functions, that (1) is equivalent to the weak convergence of the d.f.'s

$$H_N(z) := \nu_N\{n : f(n)/g(n) \leq z\}$$

to $H(z) := \mathbf{1}_{[1,\infty[}(z)$.

In the course of this and the following chapters we shall see that numerous arithmetic functions, with apparently chaotic variations, do indeed possess a normal order. Their behaviour is thus satisfactorily described "almost everywhere", and the collection of information of this type contributes to elaborating our current model of a "normal" number—i.e. a number which is "random", in a sense suited to natural density. That such a concept lends itself to quantitative assertions, and can ultimately serve as the basis of a complete mathematical theory, is by no means an insignificant attraction of the probabilistic point of view on the theory of numbers.

§ 3.2 The Turán–Kubilius inequality

In the study of an arithmetic function f, it is often reasonable to regard the expectation (or a suitable approximation of it)

$$g(N) := E_N(f) = \frac{1}{N} \sum_{1 \le n \le N} f(n) \tag{2}$$

as a good candidate for a normal order of f. When g is monotone increasing, a necessary condition for this is that its rate of growth be sufficiently slow so that

$$g(n) = \bigl(1 + o(1)\bigr) g(N) \tag{3}$$

holds for all but at most $o(N)$ integers $n \le N$. It is then natural to expect that a dispersion computation, i.e. an evaluation of the variance

$$V_N(f) := E_N\bigl(\{f(n) - g(N)\}^2\bigr), \tag{4}$$

will allow, via the Bienaymé–Chebyshev inequality, an effective proof that g is indeed a normal order for f.

This method is particularly efficient for additive functions. The Turán–Kubilius inequality (Theorem 1) provides in this case an upper bound for (4) which is often sufficient to determine the normal order. Before stating the result, it is interesting to explain the underlying probabilistic ideas.

Let f be an additive function. We have

$$f(n) = \sum_{p^\nu \le N} f(p^\nu) \xi_{p^\nu}(n) \qquad (n \le N) \tag{5}$$

with

$$\xi_{p^\nu}(n) = \begin{cases} 1 & \text{if } p^\nu \| n, \\ 0 & \text{otherwise.} \end{cases}$$

In other words, in the probability space $\{n : 1 \le n \le N\}$ equipped with the uniform law ν_N, we have an equality between random variables

$$f = \sum_{p^\nu \le N} f(p^\nu) \xi_{p^\nu}.$$

The ξ_{p^ν} are Bernoulli variables, determined by their expectations

$$\begin{aligned}E_N(\xi_{p^\nu}) = \nu_N\{n : \xi_{p^\nu}(n) = 1\} &= \frac{1}{N}\Big(\Big[\frac{N}{p^\nu}\Big] - \Big[\frac{N}{p^{\nu+1}}\Big]\Big)\\ &= (1-p^{-1})p^{-\nu} + O(N^{-1}).\end{aligned}$$

For $q \neq p$, we can further write

$$\begin{aligned}E_N(\xi_{p^\nu}\xi_{q^\mu}) &= \frac{1}{N}\Big(\Big[\frac{N}{p^\nu q^\mu}\Big] - \Big[\frac{N}{p^{\nu+1}q^\mu}\Big] - \Big[\frac{N}{p^\nu q^{\mu+1}}\Big] + \Big[\frac{N}{p^{\nu+1}q^{\mu+1}}\Big]\Big)\\ &= (1-p^{-1})p^{-\nu}(1-q^{-1})q^{-\mu} + O(N^{-1})\\ &= E_N(\xi_{p^\nu})E_N(\xi_{q^\mu}) + O(N^{-1}).\end{aligned}$$

Thus ξ_{p^ν} and ξ_{q^μ} are asymptotically independent when $p \neq q$, and for fixed p, the numbers $E_N(\xi_{p^\nu})$ are close to the probabilities of a geometric random variable with parameter $(1 - 1/p)$. This leads to the heuristic assumption that the distribution law of f, as a random variable on $\{n : 1 \leq n \leq N\}$, is close to that of

(6) $$\eta_N = \eta_N(f) := \sum_{p \leq N} \zeta_p$$

where the $\zeta_p = \zeta_p(f)$ are independent variables, on an abstract probability space, with law

(7) $$\operatorname{Prob}\big(\zeta_p = f(p^\nu)\big) = (1-p^{-1})p^{-\nu} \qquad (\nu = 0, 1, 2, \ldots).$$

[Note that $f(1) = 0$.] By familiar abuse of notation, we interpret (7) by agreeing that, if several values (possibly infinite in number) $f(p^\nu)$ are equal, the corresponding probability is the sum of the probabilities occurring on the right-hand side.

We have

(8) $$V(\eta_N) = \sum_{p \leq N} V(\zeta_p) \leq \sum_{p \leq N} E(\zeta_p^2) \leq \sum_{p \leq N} \sum_{\nu=1}^{\infty} f(p^\nu)^2 p^{-\nu},$$

where the first inequality is almost optimal, since it easily follows from the Cauchy–Schwarz inequality that $E(\zeta_p)^2 \leq E(\zeta_p^2)/p$ for each p, from which $V(\zeta_p) \geq (1 - 1/p)E(\zeta_p^2)$. The Turán–Kubilius inequality states in essence that the bound (8)—which follows, by a well-known probabilistic theorem, from the independence of the ζ_p—remains valid for $V_N(f)$ up to an *absolute* multiplicative constant.

Theorem 1 (Turán–Kubilius inequality). *There exists a function* $\varepsilon(x)$, *with* $\lim_{x\to\infty}\varepsilon(x) = 0$, *which has the following property. For each additive, complex-valued arithmetic function* f *we have*

$$x^{-1}\sum_{n\le x}\big|f(n) - A(x)\big|^2 \le \big(2+\varepsilon(x)\big)B(x)^2 \qquad (x\ge 2), \tag{9}$$

with

$$A(x) := \sum_{p^\nu\le x} f(p^\nu)p^{-\nu}(1-p^{-1}), \qquad B(x)^2 := \sum_{p^\nu\le x}\big|f(p^\nu)\big|^2p^{-\nu}. \tag{10}$$

Proof. We shall establish (9) with

$$\varepsilon(x) := 4x^{-1}\Bigg(\sum_{\substack{p^\nu q^\mu\le x\\ p\ne q}} p^\nu q^\mu\Bigg)^{1/2} + 4x^{-1}\Bigg(\sum_{\substack{p^\nu\le x\\ q^\mu\le x}} q^\mu p^{-\nu}\Bigg)^{1/2}.$$

By standard partial integration, the details of which we omit, the prime number theorem readily yields the estimate

$$\varepsilon(x) \ll \Big(\frac{\log_2 x}{\log x}\Big)^{1/2}.$$

We can assume without loss of generality that x is integral. Let us first consider the case when f is real and non-negative. The left-hand side of (9) is

$$x^{-1}\sum_{n\le x} f(n)^2 - 2A(x)x^{-1}\sum_{n\le x} f(n) + A(x)^2 = M_2 - 2A(x)M_1 + A(x)^2, \tag{11}$$

say. We have

$$\begin{aligned} M_2 &= x^{-1}\sum_{n\le x}\sum_{p^\nu\|n,\, q^\mu\|n} f(p^\nu)f(q^\mu)\\ &= \sum_{p^\nu\le x} f(p^\nu)^2x^{-1}\sum_{\substack{n\le x\\ p^\nu\|n}} 1 + \sum_{\substack{p^\nu q^\mu\le x\\ p\ne q}} f(p^\nu)f(q^\mu)x^{-1}\sum_{\substack{n\le x\\ p^\nu\|n,\, q^\mu\|n}} 1. \end{aligned}$$

The first inner sum does not exceed $xp^{-\nu}$, and the second is at most equal to $xp^{-\nu}q^{-\mu}(1-p^{-1})(1-q^{-1})+2$. Therefore

$$M_2 \le B(x)^2 + A(x)^2 + 2x^{-1}\sum_{p^\nu q^\mu\le x,\, p\ne q} f(p^\nu)f(q^\mu). \tag{12}$$

Similarly we can write

$$M_1 = x^{-1} \sum_{n \le x} \sum_{p^\nu \| n} f(p^\nu) = \sum_{p^\nu \le x} f(p^\nu) x^{-1} \sum_{n \le x,\, p^\nu \| n} 1,$$

from which, inserting the lower bound $xp^{-\nu}(1 - p^{-1}) - 1$ for the inner sum, we deduce that

$$M_1 \ge A(x) - x^{-1} \sum_{p^\nu \le x} f(p^\nu). \tag{13}$$

Substituting (12) and (13) in (11), we obtain that the left-hand side does not exceed

$$B(x)^2 + 2x^{-1} \sum_{p^\nu q^\mu \le x,\, p \ne q} f(p^\nu) f(q^\mu) + 2A(x) x^{-1} \sum_{p^\nu \le x} f(p^\nu).$$

We estimate the last two terms by the Cauchy–Schwarz inequality. We have

$$\sum_{p^\nu q^\mu \le x,\, p \ne q} \frac{f(p^\nu)}{p^{\nu/2}} \frac{f(q^\mu)}{q^{\mu/2}} \cdot p^{\nu/2} q^{\mu/2} \le B(x)^2 \Big(\sum_{p^\nu q^\mu \le x,\, p \ne q} p^\nu q^\mu \Big)^{1/2}$$

and

$$A(x) \sum_{p^\nu \le x} \frac{f(p^\nu)}{p^{\nu/2}} \cdot p^{\nu/2} \le \Big(B(x)^2 \sum_{p^\nu \le x} p^{-\nu} \cdot B(x)^2 \sum_{p^\nu \le x} p^\nu \Big)^{1/2}.$$

This gives exactly (9), with $1 + \frac{1}{2}\varepsilon(x)$ instead of $2 + \varepsilon(x)$.

When f is real and takes values of both signs, we introduce the additive functions f^+ and f^-, defined by $f^\pm(p^\nu) = \max\big(\pm f(p^\nu), 0\big)$. With obvious notation, we then have

$$\big(f(n) - A(x)\big)^2 \le 2\big(f^+(n) - A^+(x)\big)^2 + 2\big(f^-(n) - A^-(x)\big)^2 \quad (1 \le n \le x)$$

and $B(x)^2 = B^+(x)^2 + B^-(x)^2$. Thus (9) holds in this case.

When f is complex-valued, we just apply the above result to the real and imaginary parts of f. This completes the proof.

The constant 2 appearing in (9) is not optimal. In 1983 Kubilius showed that if

$$C(x) := \sup_f x^{-1} B(x)^{-2} \sum_{n \le x} |f(n) - A(x)|^2 \tag{14}$$

then we have

$$C(x) = \tfrac{3}{2} + O\big(1/\sqrt{(\log x)}\big). \tag{15}$$

This result gives a remarkable measure of the discrepancy between probabilistic number theory and the probability theory model: the fact that $C(x)$ has a finite limit exhibits the influence of the (partial) independence of the ξ_{p^ν}; the fact that this limit is not equal to 1 underlines the limitations of this influence.

In the definition of $C(x)$, we may consider indifferently the supremum over real or over complex additive functions f. Two other independent proofs of the validity and optimality of the constant $\frac{3}{2}$ are due to Hildebrand (1983) and Stein (1984). When f is restricted to strongly additive functions, an elegant result of J. Lee (1989) states that

$$C(x) = \tfrac{3}{2} - \frac{\alpha}{\log x} + O\Big(\Big(\frac{\log_2 x}{\log x}\Big)^2\Big) \tag{16}$$

where α is a positive absolute constant. In particular, this implies the existence of an absolute x_0 such that for any strongly additive function f we have

$$x^{-1} \sum_{n \leq x} |f(n) - A(x)|^2 \leq \tfrac{3}{2} B(x)^2 \qquad (x \geq x_0). \tag{17}$$

§ 3.3 Dual form of the Turán–Kubilius inequality

For $n \geq 1$, $r = p^\nu$, let us write

$$c_{nr} := \begin{cases} r^{1/2} - r^{-1/2}(1 - p^{-1}) & \text{if } r \| n, \\ -r^{-1/2}(1 - p^{-1}) & \text{otherwise,} \end{cases}$$

and

$$y_r := f(r) r^{-1/2}.$$

For $1 \leq n \leq N$ we then have

$$f(n) - A(N) = \sum_{r \leq N} c_{nr} y_r,$$

so that, with the notation (14), the Turán–Kubilius inequality can be written as

$$\sum_{n \leq N} \Big| \sum_{r \leq N} c_{nr} y_r \Big|^2 \leq N C(N) \sum_{r \leq N} |y_r|^2.$$

Since the y_r are arbitrary complex numbers, we can apply the duality principle of Lemma I.4.5.1. It follows that

$$\sum_{r \leq N} \Big| \sum_{n \leq N} c_{nr} x_n \Big|^2 \leq N C(N) \sum_{n \leq N} |x_n|^2$$

for all complex numbers $x_1, x_2, \ldots, x_N$. Noting that

$$\sum_{n \leq N} c_{nr} x_n = r^{1/2} \Big\{ \sum_{n \leq N,\, r \| n} x_n - r^{-1}(1 - p^{-1}) \sum_{n \leq N} x_n \Big\},$$

we can therefore state the following result.

Theorem 2. *For any complex sequence $\{a_n : 1 \le n \le N\}$, we have*

$$\sum_{p^\nu \le N} p^\nu \Big| \sum_{n \le N,\, p^\nu \| n} a_n - p^{-\nu}(1-p^{-1}) \sum_{n \le N} a_n \Big|^2 \le NC(N) \sum_{n \le N} |a_n|^2. \tag{18}$$

We recognise here a variant of Theorem I.4.7, obtained as a corollary to the large sieve inequality. In practice, this estimate is very useful owing to its complete generality and total uniformity. It expresses the fact that *on average* every sufficiently dense sequence a_n is well distributed among the congruence classes $n \equiv 0 \pmod{p^\nu}$.

§ 3.4 The Hardy–Ramanujan theorem and other applications

As mentioned in Section 3.2, the Turán–Kubilius inequality can be used to determine the normal order of an additive function. We have the following general result, where the notation is that of Theorem 1.

Theorem 3. *Let f be a complex-valued additive function. Under the assumption*

$$B(N) = o\big(A(N)\big) \qquad (N \to \infty), \tag{19}$$

$A(n)$ is a normal order for $f(n)$.

Proof. Let us show first of all that $A(n) = \big(1+o(1)\big)A(N)$ for all integers $n \le N$ except at most $o(N)$. This is a consequence of the following upper bound, valid for $\sqrt{N} < n \le N$,

$$\begin{aligned} |A(N) - A(n)| &= \Big| \sum_{n < p^\nu \le N} f(p^\nu) p^{-\nu}(1-p^{-1}) \Big| \\ &\le \Big(\sum_{\sqrt{N} < p^\nu \le N} p^{-\nu} \sum_{p^\nu \le N} |f(p^\nu)|^2 p^{-\nu} \Big)^{1/2} \\ &\ll B(N) = o\big(A(N)\big). \end{aligned}$$

The assertion of the theorem hence follows from the validity of

$$\nu_N\big\{n : |f(n) - A(N)| > \varepsilon |A(N)|\big\} = o(1) \tag{20}$$

for all $\varepsilon > 0$. However, the left-hand side does not exceed

$$E_N\Big(\Big| \frac{f(n) - A(N)}{\varepsilon A(N)} \Big|^2 \Big) \ll \Big| \frac{B(N)}{\varepsilon A(N)} \Big|^2 = o(1), \tag{21}$$

by Theorem 1 and assumption (19). This concludes the proof.

The classic example of application of Theorem 3 is to the functions $\omega(n)$ or $\Omega(n)$. In both cases it is easily shown that

$$A(N) = \log_2 N + O(1), \qquad B(N)^2 = \log_2 N + O(1), \tag{22}$$

so that (19) is certainly satisfied. We thus obtain the remarkable result that *the number of prime factors of an integer n, counted with or without multiplicity, is normally asymptotic to* $\log_2 n$. It is this statement which is usually referred to as the Hardy–Ramanujan theorem (1917). The following quantitative version is due to Turán (1934).

Theorem 4. *Let $\xi(N) \to \infty$. Then*

$$\left|\{n \le N : |\omega(n) - \log_2 N| > \xi(N)\sqrt{(\log_2 N)}\}\right| \ll N/\xi(N)^2. \tag{23}$$

Moreover the same bound holds for $\Omega(n)$ in place of $\omega(n)$.

Proof. Given (22) it suffices to choose $\varepsilon = \xi(N)/\sqrt{(\log_2 N)}$ in (21).

The original proof of Hardy & Ramanujan rested on an upper bound, uniform for N and $k \ge 1$, for the *local laws* $\nu_N\{n : \omega(n) = k\}$. This implied rather involved considerations—cf. Exercise 1. It was with the intention of simplifying the argument that Turán proved in 1934 the initial form of Theorem 1.

In view of the inequality $2^{\omega(n)} \le \tau(n) \le 2^{\Omega(n)}$, which is valid for any integer $n \ge 1$, we immediately deduce from Theorem 4 that

$$\tau(n) = (\log n)^{\log 2 + o(1)} \qquad \text{pp} \tag{24}$$

and even that

$$\tau(n) = (\log n)^{\log 2} \exp\left\{O(\xi(n)\sqrt{(\log_2 n)})\right\} \qquad \text{pp} \tag{25}$$

for any function $\xi(n) \to \infty$. *Stricto sensu,* these relations ought to be interpreted as saying that $\log \tau(n)$ has a normal order. However (24) and (25) are often referred to by the rather loose statement that $\tau(n)$ has normal order $(\log n)^{\log 2}$.

For the first time we meet here a common situation in probabilistic number theory: the mean value (here $\log n$) does not reflect the normal behaviour (here $(\log n)^{\log 2}$). This phenomenon can be explained by the fact that the sum

$$\sum_{n \le x} \tau(n)$$

is dominated by a small number of abnormal integers, for which $\tau(n)$ is large. We shall see later (Exercise 11(b)) that these are precisely those integers n such that $\omega(n) \sim 2\log_2 n$. In fact, for each $\varepsilon > 0$,

$$\sum_{\substack{n\le x \\ |\omega(n)-2\log_2 n|\le\varepsilon\log_2 n}} \tau(n) \sim x\log x \sim \sum_{n\le x}\tau(n) \qquad (x\to\infty).$$

Such a situation leads to a vast field of practical applications. Indeed it opens the possibility of considering, in certain summations, the arithmetic function $\tau(n)$ as a weight coefficient which emphasises those integers n for which $\omega(n) \sim 2\log_2 n$.

We now give without proof some results which are staightforward applications of the Turán–Kubilius inequality. We let $\xi(n)$ denote an arbitrary quantity such that $\xi(n)\to\infty$.

(a) For $a \ge 1$, $q \ge 1$, $(a,q) = 1$, let $\omega(n;a,q)$ be the number of distinct prime factors of n in the arithmetic progression $p \equiv a \pmod q$. Then

$$\text{(26)} \qquad \omega(n;a,q) = \frac{1}{\varphi(q)}\log_2 n + O\Big(\xi(n)\sqrt{(\log_2 n)}\Big) \qquad \text{pp.}$$

More generally, if E denotes a set of prime numbers such that

$$E(x) := \sum_{p\le x,\, p\in E}\frac{1}{p} \to \infty \qquad (x\to\infty),$$

we have

$$\text{(27)} \qquad \sum_{p|n,\, p\in E} 1 = E(n) + O(\xi(n)\sqrt{(E(n))}) \qquad \text{pp.}$$

(b) For fixed $k \ge 2$, consider the number $\tau_k(n)$ of representations of n as a product of k integers. Then

$$\text{(28)} \qquad \tau_k(n) = (\log n)^{\log k}\exp\left\{O(\xi(n)\sqrt{(\log_2 n)})\right\} \qquad \text{pp.}$$

Note the disparity between the "normal" value $(\log n)^{\log k}$ and the mean value $(\log n)^{k-1}/(k-1)!$ of $\tau_k(n)$—cf. Theorem II.5.3.

(c) For $k \ge 1$, let $\delta_k(n)$ be the number of representations of n in the form $n = [m_1,\dots,m_k]$ with $m_1 \ge 1,\dots,m_k \ge 1$. Then

$$\text{(29)} \qquad \delta_k(n) = (\log n)^{\log(2^k-1)}\exp\left\{O(\xi(n)\sqrt{(\log_2 n)})\right\} \qquad \text{pp.}$$

Here again the normal order is quite far from the mean value: using the convolution formula $\delta_k * \mathbf{1} = \tau^k$, it may be easily deduced from Theorem II.5.3 that we have

$$\sum_{n\le x} \delta_k(n) \sim C_k x (\log x)^{2^k-2} \tag{30}$$

with

$$C_k := \frac{1}{(2^k-2)!} \prod_p \left(1-p^{-1}\right)^{2^k} \sum_{\nu=0}^{\infty} (\nu+1)^k p^{-\nu}.$$

§ 3.5 Effective mean value estimates for multiplicative functions

Here we describe another useful tool which furnishes an alternative proof of the Hardy–Ramanujan theorem. This concerns uniform estimates for mean values of multiplicative functions which belong to a certain class. It is important to note right away, and to bear in mind, that the applications are secondary to the fact that the upper bounds obtained are independent of the particular function chosen in this class.

Theorem 5. *Let f be a multiplicative non-negative function which for suitable constants A and B satisfies*

$$\text{(i)} \qquad \sum_{p\le y} f(p)\log p \le Ay \qquad (y \ge 0),$$

$$\text{(ii)} \qquad \sum_p \sum_{\nu\ge 2} \frac{f(p^\nu)}{p^\nu} \log p^\nu \le B.$$

Then, for $x > 1$,

$$\sum_{n\le x} f(n) \le (A+B+1)\frac{x}{\log x} \sum_{n\le x} \frac{f(n)}{n}. \tag{31}$$

Proof. Let $S(x)$ be the left-hand side of (31) and $L(x) := \sum_{n\le x} f(n)/n$. We have trivially

$$S(x) \le xL(x) \qquad (x \ge 1). \tag{32}$$

Next we can write

$$\begin{aligned} S(x)\log x &= \sum_{n\le x} f(n)\log\frac{x}{n} + \sum_{n\le x} f(n)\sum_{p\| n}\log p + \sum_{n\le x} f(n) \sum_{\nu\ge 2,\, p^\nu \| n} \log p^\nu \\ &= S_1 + S_2 + S_3 \qquad \text{(say).} \end{aligned}$$

Clearly, $S_1 \leq xL(x)$. Making the change of variable $n = mp$ in S_2, we obtain

$$S_2 = \sum_{m \leq x} f(m) \sum_{p \leq x/m,\, p \nmid m} f(p) \log p \leq AxL(x).$$

Finally we get by interchanging summations that

$$\begin{aligned} S_3 &= \sum_p \sum_{\nu \geq 2} f(p^\nu) \log p^\nu \sum_{m \leq x/p^\nu,\, p \nmid m} f(m) \\ &\leq \sum_p \sum_{\nu \geq 2} f(p^\nu) \log p^\nu S\Big(\frac{x}{p^\nu}\Big) \\ &\leq \sum_p \sum_{\nu \geq 2} f(p^\nu) \log p^\nu \frac{x}{p^\nu} L\Big(\frac{x}{p^\nu}\Big) \leq BxL(x), \end{aligned}$$

where we have used (32) and the fact that L is non-decreasing. This completes the proof.

In practice Theorem 5 is often used in the following form.

Corollary 5.1. *Let λ_1, λ_2 be constants such that $\lambda_1 > 0$, $0 \leq \lambda_2 < 2$. For any multiplicative arithmetic function f satisfying*

$$0 \leq f(p^\nu) \leq \lambda_1 \lambda_2^{\nu-1} \qquad (p \geq 2,\, \nu = 1, 2, \ldots), \tag{33}$$

we have uniformly

$$\sum_{n \leq x} f(n) \ll x \prod_{p \leq x} (1 - p^{-1}) \sum_{\nu=0}^{\infty} f(p^\nu) p^{-\nu} \qquad (x \geq 2). \tag{34}$$

The constant implied in (34) does not exceed

$$4\big(1 + 9\lambda_1 + \lambda_1\lambda_2/(2 - \lambda_2)^2\big). \tag{35}$$

Remark. Here we have not sought to optimise the numerical constants appearing in (35).

Proof. It is clear that (33) implies conditions (i) and (ii) of Theorem 5. By Theorem I.1.4 we have $A \leq \lambda_1 \log 4$. In addition,

$$\begin{aligned} B \leq \lambda_1 \sum_p \frac{\log p}{p} \sum_{\nu=2}^{\infty} \nu \Big(\frac{\lambda_2}{p}\Big)^{\nu-1} &\leq 2\lambda_1\lambda_2 \sum_p \frac{\log p}{(p - \lambda_2)^2} \\ &\leq \frac{2\lambda_1\lambda_2 \log 2}{(2 - \lambda_2)^2} + 4\lambda_1 \sum_{n=3}^{\infty} \frac{\log n}{(n-2)^2}. \end{aligned}$$

It is easily shown that the series over n is $\leq \frac{5}{2}$. Noting that

$$\sum_{n\leq x}\frac{f(n)}{n}\leq\prod_{p\leq x}\sum_{\nu=0}^{\infty}f(p^\nu)p^{-\nu},$$

we obtain the stated estimate with a constant at most equal to

$$K\log 4\Big(\frac{1+10\lambda_1}{\log 4}+\lambda_1+\frac{\lambda_1\lambda_2}{(2-\lambda_2)^2}\Big)$$

where $K := \sup_{x\geq 2} K(x)$ with $K(x) := (1/\log x)\prod_{p\leq x}(1-1/p)^{-1}$.

We now prove that $K = K(2) = 2/\log 2$. To see this, we first check numerically that $K(x) \leq K(2)$ for $x \leq 300$. Next, we observe that for $x > y \geq 2$ we have

$$K(x)\leq\frac{\pi^2}{6\log x}\prod_{p\leq x}\Big(1+\frac{1}{p}\Big)\leq\frac{\pi^2}{6\log x}\prod_{p\leq y}\Big(1+\frac{1}{p}\Big)\exp\Big\{\sum_{y<p\leq x}\frac{1}{p}\Big\}.$$

Using Theorem I.1.7 in the form $F(t) := \sum_{p\leq t}(\log p)/p \leq \log(4t)$ $(t \geq 2)$ and partial summation, we readily get

$$\sum_{y<p\leq x}\frac{1}{p}\leq 1+\frac{\log 4-F(y)}{\log y}+\log_2 x-\log_2 y \qquad (x>y\geq 2),$$

from which it follows, taking $y = 300$, that $K(x) < 2.87 < K(2)$ for $x > 300$. This completes the proof.

The following result, which is an immediate consequence of Corollary 5.1, will help to establish a strong form of Theorem 4. For $t > 0$ we write

$$\text{(36)} \qquad \omega(n,t):=\sum_{p|n,\,p\leq t}1,\qquad \Omega(n,t):=\sum_{p^\nu\|n,\,p\leq t}\nu \qquad (n\geq 1).$$

Theorem 6. *Let $y_0 > 0$. Uniformly for $0 \leq y \leq y_0$, $x \geq t \geq 2$, we have*

$$\text{(37)} \qquad \sum_{n\leq x}y^{\omega(n,t)}\ll x(\log t)^{y-1}$$

Under the additional hypothesis $y_0 < 2$, the same estimate holds for $\Omega(n,t)$.

Proof. For each fixed t the function $n \mapsto y^{\omega(n,t)}$ is multiplicative and satisfies (33) with $\lambda_1 = 1 + y_0$, $\lambda_2 = 1$. This condition is still realised for $y^{\Omega(n,t)}$, with $\lambda_1 = 1 + y_0$, $\lambda_2 = \max(1, y_0)$. The bound (34) is then

$$x\prod_{p\leq t}(1-p^{-1})(1+yp^{-1}+O(p^{-2})).$$

From this, we deduce the stated result by Mertens' formula.

Theorem 7. *We have, uniformly for $x \geq t \geq 3$ and $0 \leq \xi < \sqrt{(\log_2 t)}$,*

$$|\{n \leq x : |\omega(n,t) - \log_2 t| > \xi\sqrt{(\log_2 t)}\}| \ll xe^{-\xi^2/3}. \tag{38}$$

Furthermore, if $\xi \ll \big(\log_2 t\big)^{1/6}$ we can replace $e^{-\xi^2/3}$ by $e^{-\xi^2/2}$. The same assertions hold for $\Omega(n,t)$.

When $t = x$, we have $\omega(n,t) = \omega(n)$ for all $n \leq x$. In its range of validity, the bound (38) hence constitutes a remarkable improvement of (23).

Proof. Let $\delta := \xi/\sqrt{(\log_2 t)}$, so that $0 \leq \delta < 1$, and let $\chi(n)$ denote the characteristic function of the set of those integers n such that

$$\big|\omega(n,t) - \log_2 t\big| > \delta \log_2 t. \tag{39}$$

Plainly,

$$\chi(n) \leq (1-\delta)^{\omega(n,t)-(1-\delta)\log_2 t} + (1+\delta)^{\omega(n,t)-(1+\delta)\log_2 t}.$$

Summing over $n \leq x$ and appealing to (37), we obtain

$$\sum_{n \leq x} \chi(n) \ll x(\log t)^{-Q(1-\delta)} + x(\log t)^{-Q(1+\delta)} \tag{40}$$

with

$$Q(y) := y\log y - y + 1 = \int_0^{y-1} \log(1+v)\,dv \qquad (y > 0). \tag{41}$$

For $0 \leq \delta \leq 1$, we have $Q(1 \pm \delta) \geq \frac{1}{3}\delta^2$, and $Q(1 \pm \delta) = \frac{1}{2}\delta^2 + O(\delta^3)$. This implies the stated bounds. The proof is unchanged in the case of $\Omega(n,t)$.

§ 3.6 Normal structure of the set of prime factors of an integer

The factorisation of an integer n is completely determined by the behaviour of the step-function $\Omega(n,t)$. Here we deal with the technically more manageable function $\omega(n,t)$. To all intents and purposes, this choice is irrelevant, since for all n, t, we have $0 \leq \Omega(n,t) - \omega(n,t) \leq \Omega(n) - \omega(n)$, and Theorems I.3.6 and I.3.7 imply that this difference has bounded mean value: we deduce from this that

$$\Omega(n,t) - \omega(n,t) \ll \xi(n) \qquad \text{pp} \tag{42}$$

holds uniformly in t for any given function $\xi(n) \to \infty$.

Theorem 7 describes the behaviour of $\omega(n,t)$ for fixed t. However only the variations as a function of t can yield information on the multiplicative structure of n. Thus, we need a pp approximation to $\omega(n,t)$ with respect to *the norm of uniform convergence*. The following result, which is a simple consequence of Theorem 7, provides the required estimate.

Theorem 8. *Let* $\varepsilon > 0$, *and* $\xi(n) \to \infty$. *We have*

$$\sup_{\xi(n)\le t\le n}\left|\frac{\omega(n,t)-\log_2 t}{\sqrt{(2\log_2 t\log_3 t)}}\right| \le 1+\varepsilon \qquad \text{pp.} \tag{43}$$

Proof. Let ξ be an arbitrarily large real number. Plainly, $\xi(n) > \xi$ pp. It is hence sufficient to show that the upper density of those integers n which do not satisfy (43) with $\xi(n)$ replaced by ξ tends to 0 as $\xi \to \infty$. To this end, we introduce the check-points

$$t_j := \exp\exp j \qquad (j \ge \log_2 \xi),$$

and note that, since $\log_2 t_{j+1} - \log_2 t_j \le 1$ and since $\omega(n,t)$ is a non-decreasing function of t, we have for sufficiently large ξ

$$\sup_{\xi\le t\le n}\left|\frac{\omega(n,t)-\log_2 t}{\sqrt{(2\log_2 t\log_3 t)}}\right| > 1+\varepsilon \;\Rightarrow\; \sup_{\xi\le t_j\le n}\left|\frac{\omega(n,t_j)-j}{\sqrt{(2j\log j)}}\right| > \sqrt{(1+\varepsilon)}.$$

Choosing, in Theorem 7, $t = t_j$ and $\xi = \sqrt{((2+2\varepsilon)\log j)}$ for all integers j such that $\log_2 \xi \le j \le \log_2 x$, we see that the upper density of those integers n satisfying the last inequality is

$$\ll \sum_{j\ge\log_2\xi} j^{-1-\varepsilon} \ll (\log_2 \xi)^{-\varepsilon}.$$

This completes the proof.

Theorem 8 may be directly formulated in terms of the standard factorisation of n. Let $p_j(n)$ denote the jth distinct prime factor of n, so that the canonical factorisation may be written as

$$n = \prod_{1\le j\le\omega(n)} p_j(n)^{\nu_j}$$

where the ν_j are positive integers. Then $\omega(n, p_j(n)) = j$. Making the change of variables $t \mapsto p_j(n)$ in (43), we obtain the following result.

Theorem 9. *Let* $\varepsilon > 0$, $\xi(n) \to \infty$. *Then*

$$\sup_{\xi(n)\le j\le\omega(n)}\left|\frac{\log_2 p_j(n)-j}{\sqrt{(2j\log j)}}\right| \le 1+\varepsilon \qquad \text{pp.} \tag{44}$$

This result was announced in a more precise form by Erdős (1946), but without proof. The details have been provided by Hall & Tenenbaum (1988), Chapter 1—see Theorem 14 in the Notes. The main device to keep in mind is that, for a normal integer, the jth prime factor is roughly of size $\exp\exp j$. This generates a valuable heuristic model for the normal multiplicative structure of the integers—the scope of which, however, shouldn't be overemphasized: taken too literally, this model leads to false conjectures; cf. Hall & Tenenbaum (1988) § 1.2; see also the Notes.

Notes

§ **3.2.** The distribution laws of the independent random variables ζ_p defined in (7) do not involve the parameter N. The dependence upon N of the probabilistic model η_N for the additive function f is thus relatively trivial, in the sense that N only appears only as a truncation parameter. It is possible, following for instance Hildebrand (1983), to mimic reality more closely by writing

$$f(n) = \sum_{p \le N} f_p(n) \qquad (1 \le n \le N)$$

with

$$f_p(n) := \sum_{\nu:\, p^\nu \le N} f(p^\nu)\xi_{p^\nu}(n) = \begin{cases} f(p^\mu) & \text{if } p^\mu \| n, \\ 0 & \text{otherwise.} \end{cases}$$

The random variables f_p are not independent. If they were, the variance $V_N(f)$ would be equal to

$$B_N^*(f)^2 := \sum_{p \le N} V_N(f_p).$$

As Hildebrand remarks, we have

$$\tfrac{1}{8}B(N)^2 \le B_N^*(f)^2 \le B(N)^2 \tag{45}$$

with, as in Theorem 1, $B(N)^2 := \sum_{p^\nu \le N} f(p^\nu)^2 p^{-\nu}$. (Here and in the remainder of this discussion we suppose that f is real.) The Turán–Kubilius inequality then implies that

$$V_N(f) \le \left(\tfrac{3}{2} + o(1)\right)B(N)^2 \le \big(12 + o(1)\big)B_N^*(f)^2 \tag{46}$$

where the $o(1)$ quantities are independent of f.

In order to prove (45) we consider the expression

$$w_N(p^\nu) := E_N(\xi_{p^\nu}) = N^{-1}\Big\{[N/p^\nu] - [N/p^{\nu+1}]\Big\},$$

which satisfies

$$\tfrac{1}{4}p^{-\nu} \le w_N(p^\nu) \le p^{-\nu} \qquad (p^\nu \le N). \tag{47}$$

The inequality on the right is obvious; we obtain the one on the left by considering separately the cases $p^{\nu+1} \le \frac{1}{2}N$ and $\frac{1}{2}Np^{-1} < p^\nu \le N$, applying the respective lower bounds $p^{-\nu}(1-3/2p)$ and $\frac{1}{2}p^{-\nu}(1-1/p)$. Relation (45) follows easily from (47) and the Cauchy–Schwarz inequality

$$E_N(f_p)^2 \le E_N(f_p^2) \sum_{\nu:\, p^\nu \le N} w_N(p^\nu) \le E_N(f_p^2)p^{-1}. \tag{48}$$

We actually have

$$V_N(f_p) = E_N(f_p^2) - E_N(f_p)^2 \ge (1-p^{-1})E_N(f_p^2)$$

and

$$\tfrac{1}{4} \sum_{\nu:\, p^\nu \le N} f(p^\nu)^2 p^{-\nu} \le E_N(f_p^2) = \sum_{p^\nu \le N} f(p^\nu)^2 w_N(p^\nu) \le \sum_{\nu:\, p^\nu \le N} f(p^\nu)^2 p^{-\nu}.$$

It is expected, in general, that the distribution law of f (with respect to ν_N) is close to that of η_N. In this context, Ruzsa (1982) introduced an interesting methodological distinction by defining for problems of this type:

(a) *the direct approach*, in which probability theory appears only as a heuristic model and for which the proof is purely arithmetic;

(b) *the indirect approach*, characterised by a comparison of (f, ν_N) and η_N as random variables, and in which the arithmetic result is derived in an additional step from a specifically probabilistic theorem.

In the case of the Turán–Kubilius inequality, the direct statement (neglecting the values of the constants) is

$$V_N(f) \ll B(N)^2 \tag{49}$$

while the indirect statement is

$$V_N(f) \ll V(\eta_N). \tag{50}$$

Since

$$V(\zeta_p) \le E(\zeta_p^2) \le \sum_{\nu=1}^{\infty} f(p^\nu)^2 p^{-\nu},$$

the bound (49) follows from (50) by appealing to the probabilistic theorem on the variance of a sum of independent random variables. Our proof of Theorem 1, which essentially follows that of Elliott (1979), is of direct type. The following remarkable result of Ruzsa provides an indirect approach, of which (50) is an immediate consequence.

Theorem 10 (Ruzsa, 1984). *Let f be a complex-valued additive function and let $F : \mathbb{R}^+ \to \mathbb{R}^+$ be non-decreasing. Uniformly for $A \in \mathbb{R}$ and $N \in \mathbb{Z}^+$, we have*

$$E_N\big(F(|f - A|)\big) \ll E\big(F(3|\eta_N - A|)\big). \tag{51}$$

To determine the best constant in (51) is an interesting open problem. Disproving a conjecture made in his 1984 article, Ruzsa showed that it is not in general possible to replace the constant 3 by a function of N tending to 1. Actually, he proved (private communication) that the optimal constant is $\geq 1 + 2/e \approx 1.73575$.

The Turán–Kubilius inequality does not always provide the correct order of magnitude of $V_N(f)$. In 1983, Ruzsa showed that for real f we have

$$V_N(f) \asymp \min_{\lambda \in \mathbb{R}} \big\{\lambda^2 + B_N^2(f - \lambda \log)\big\}. \tag{52}$$

Subsequently, Hildebrand obtained an asymptotic formula for $V_N(f)$ as a function depending solely on the numbers $f(p)$, $p \leq N$. He derived several interesting corollaries from this result. One of these is a new proof that the factor $\frac{3}{2} + o(1)$ is admissible in the Turán–Kubilius inequality. Another is the characterization of additive functions such that

$$V_N(f) \asymp B(N)^2. \tag{53}$$

Theorem 11 (Hildebrand, 1983). *Let f be an additive function. Then (53) holds if, and only if,*

$$\limsup_{N \to \infty} \frac{1}{B(N) \log N} \Big| \sum_{p \leq N} \frac{f(p) \log p}{p} \Big| < \frac{1}{\sqrt{2}}. \tag{54}$$

§ **3.3.** Elliott (1979), chapter 4, gives several variants of the inequality (18) of Theorem 2. For instance we have the following result.

Theorem 12 (Elliott, 1979). *For any sequence $\{a_n : 1 \leq n \leq x\}$ and all $x \geq 2$, we have*

$$\sum_{p \leq \sqrt{x}} p \Big| \sum_{n \leq x,\, p|n} a_n - \frac{1}{p} \sum_{n \leq x} a_n \Big|^2 \leq x\Big(1 + \frac{36}{\log x}\Big) \sum_{n \leq x} |a_n|^2.$$

Once again the constant 36 is not optimal here.

For a deeper study of the relations between the Turán–Kubilius inequality and the large sieve, see chapter 4 of Elliott (1979).

§ **3.4.** It was with the aim of proving Theorem 4 that, in his 1934 article, Turán stated the inequality of Theorem 1 for the function $\omega(n)$—without giving a precise value to the constant. He soon generalised the result to the case when $0 \le f(p) \le C$ (1936). The method was later systematically exploited by Kubilius (1956, 1964).

§ **3.5.** Theorem 5 appears in this form in Hall & Tenenbaum (1988), theorem 01. It may be seen as a simplified, and slightly weaker, version of a result of Halberstam & Richert, itself generalising a theorem of Hall (1974)—cf. Exercise 13.

Theorem 13 (Halberstam–Richert, 1979). *Let f be a multiplicative non-negative function, which for some suitable constant $\kappa > 0$ satisfies*

$$\text{(i)} \qquad \sum_{p\le y} f(p)\log p \le \kappa y + O\Big(\frac{y}{(\log y)^2}\Big) \qquad (y \ge 2)$$

$$\text{(ii)} \qquad \sum_{p\ge y}\sum_{\nu\ge 2} \frac{f(p^\nu)}{p^\nu}\log p^\nu \ll \frac{1}{\log y} \qquad (y \ge 2).$$

Then for $x \ge 2$ we have

$$\sum_{n\le x} f(n) \le \Big\{1 + O\Big(\frac{1}{\log x}\Big)\Big\}\frac{\kappa x}{\log x}\sum_{n\le x}\frac{f(n)}{n}. \tag{55}$$

The implicit constant depends at most on the constants implied in (i) and (ii).

Analogous results for upper and lower bounds are given by Hildebrand (1984a, 1987b).

The constant κ of (55) is optimal, as shown by the choice

$$f(p^\nu) := \begin{cases} 0 & (p \le x/\log x), \\ \kappa & (x/\log x < p \le x). \end{cases}$$

The technique employed to prove Theorem 7 is an example of an extremely fruitful general elementary method—see Hall & Tenenbaum (1988), chapter 0, and also Section III.5.1 of the present work. In the case of Theorem 7, and for $t = x$, the idea goes back to Turán—cf. Elliott (1980), pp. 18–20.

For other applications of Theorem 5, see Exercises 5–11.

§ **3.6.** The more precise version of Theorem 9, announced by Erdős (1946) and proved in detail in the book of Hall & Tenenbaum (1988), chapter 1, is the following.

Theorem 14 (Erdős, 1946). *Let $\varepsilon > 0$, and let $\xi(n)$ a function tending to ∞ sufficiently slowly with n. We have*

$$\sup_{\xi(n)\le j\le\omega(n)}\left|\frac{\log_2 p_j(n)-j}{\sqrt{(2j\log_2 j)}}\right|\le 1+\varepsilon \qquad \text{pp.} \tag{56}$$

Moreover, if $1+\varepsilon$ is replaced by $1-\varepsilon$ on the right-hand side, the density of integers n satisfying (56) is zero.

The general idea in the proof of this result consists in showing that the law of the iterated logarithm is asymptotically applicable to the arithmetic function

$$\omega(n,t)=\sum_{p\le t}\chi_p(n)$$

where $\chi_p(n)$ is the characteristic function of the set of integers n divisible by p. As we previously saw, the χ_p are not completely independent, and this is what generates the main difficulty.

For individual values $j=j(n)$, it is possible to obtain very precise information. Galambos has thus shown the following result.

Theorem 15 (Galambos, 1976).

(i) *Let $\varepsilon > 0$ and let $j=j(N)\to\infty$ in such a way that*

$$j(N)\le\log_2 N-(\log_2 N)^{(1/2)+\varepsilon}.$$

Then we have

$$\lim_{N\to\infty}\nu_N\{n:\log_2 p_j(n)\le j+z\sqrt{j}\}=\frac{1}{\sqrt{(2\pi)}}\int_{-\infty}^{z}e^{-t^2/2}\,dt. \tag{57}$$

(ii) *Let $\varepsilon > 0$ and let $j=j(N)\to\infty$ in such a way that*

$$j(N)\le(1-\varepsilon)\log_2 N.$$

Then we have

$$\lim_{N\to\infty}\nu_N\{n:\log_2 p_{j+1}(n)-\log_2 p_j(n)\le z\}=1-e^{-z}\quad(z>0). \tag{58}$$

Theorems 8, 9 and 14 are consistent with the prediction that

$$\omega(n;s,t):=\omega(n,t)-\omega(n,s)$$

is normally of order $\log\big(\log t/\log s\big)$. In 1969 Erdős showed that

$$\omega(n;s,t)\sim\log\big(\log t/\log s\big)\qquad\text{pp}$$

uniformly for $3\le s\le t\le n$, provided that the left-hand side tends to ∞ faster than $\log_3 n$. This condition is indeed necessary. Other results on the normal distribution of prime numbers are available in the article of Bovey (1977).

Theorem 14 may be applied to determine the normal order of the jth divisor of an integer. Let $\{d_j(n):1\le j\le\tau(n)\}$ denote the increasing sequence of divisors of n. We have the following result (a weaker form of which is described in Exercise 6).

Theorem 16 (Hall & Tenenbaum, 1988). *Let $\varepsilon > 0$, and let $\xi(n)$ a function tending sufficiently slowly to ∞ with n. We have*

$$\sup_{\xi(n)\le j\le\tau(n)}\left|\frac{\log_2 d_j(n)-\log j/\log 2}{\sqrt{((2/\log 2)\log j\log_3 j)}}\right|\le 1+\varepsilon \qquad \text{pp.} \tag{59}$$

Moreover, if $1+\varepsilon$ is replaced by $1-\varepsilon$ in the right-hand side, the density of the set of integers n satisfying (59) is zero.

Thus the heuristic model for the structure of the set of divisors of a normal integer is

$$d_j(n)\approx\exp\left\{j^{1/\log 2}\right\}. \tag{60}$$

This agrees well with the normal order $(\log n)^{\log 2}$ of $\tau(n)$, but can lead to false conclusions if considered as an asymptotic formula with an excessively local interpretation. Thus an old conjecture of Erdős, established by Maier & Tenenbaum (1984), states that

$$\min_{1\le j\le\tau(n)}\frac{d_{j+1}(n)}{d_j(n)}\to 1 \qquad \text{pp.} \tag{61}$$

This implies that the above minimum is achieved as $j\to\infty$, which contradicts (60) if we define the symbol $\approx$ with too precise a meaning.

Another model which also clashes with an excessively strict interpretation of (60) as a description of the structure of the set of divisors of a normal number is that of a fractal object with dimension $\log 2$. Such a model may be shown to be consistent with some rather strong arithmetical results concerning arithmetic functions closely linked to the local structure of the set of divisors; see Mendès France & Tenenbaum (1993).

Theorems 14–16 do not provide any information on the sequences of $p_j(n)$ or $d_j(n)$ for small values of j. In 1979, Erdős introduced the densities $\Lambda_j(d)$, $\lambda_j(p)$, of the sequences of integers n satisfying $d_j(n)=d$ or $p_j(n)=p$, respectively. An asymptotic study of these densities was undertaken by Erdős & Tenenbaum (1989a). In particular, $\Lambda_j(d)>0$ if and only if $\tau(d)\le j\le d$.

Exercises

1. *The Hardy–Ramanujan inequality (1917).*

(a) Show that for $x > 0$, $0 < \alpha < 1 < \beta$ we have

$$\sum_{k \leq \alpha x} \frac{e^{-x} x^k}{k!} < \frac{e^{-Q(\alpha)x}}{(1-\alpha)\sqrt{(\alpha x)}}$$

and

$$\sum_{k \geq \beta x} \frac{e^{-x} x^k}{k!} < \frac{\sqrt{(\beta)} e^{-Q(\beta)x}}{(\beta-1)\sqrt{(2\pi x)}}$$

with $Q(y) := y \log y - y + 1$. [*See Norton, 1976, and, for more precise bounds, 1978.*]

(b) Consider the function $\pi_k(x) := \left|\{n \leq x : \omega(n) = k\}\right|$. Show that for $k \geq 1$, $x > 0$,

$$(k+1)\pi_{k+1}(x) \leq \sum_{p^\nu \leq x} \pi_k(x/p^\nu).$$

Deduce the existence of two absolute constants c_0, c_1, such that

$$\pi_k(x) \leq \frac{c_0 x}{\log x} \frac{(\log_2 x + c_1)^{k-1}}{(k-1)!} \qquad (x \geq 3,\ k \geq 1).$$

(c) Recover in this way Theorem 7 for the function $\omega(n) = \omega(n, x)$.

(d) Adapt the method to handle the case of the function $\Omega(n)$.

[*For generalizations and lower bounds, see Norton (1976, 1979, 1982), Balazard (1987), and the references given in Exercise 4.5.*]

2. Let $A(n)$ be the additive arithmetic function defined by $A(p^\nu) = \nu p$—cf. Alladi & Erdős (1977, 1979).

(a) Show that

$$\sum_{n \leq x} A(n) = \frac{\pi^2 x^2}{12 \log x}\left\{1 + O\left(\frac{1}{\log x}\right)\right\}.$$

(b) Show that for all $n \geq 1$ we have $P^+(n) - 1 \leq A(n) \leq \Omega(n) P^+(n)$.

(c) Show that the sequence of integers n such that $P^+(n) \geq \sqrt{n}$ has natural density equal to $\log 2$.

(d) Let $\chi(n)$ be the characteristic function of the set $\mathcal{A}$ of integers n such that $P^+(n) \leq n^{2/5}$. Show that $\chi(n) \geq 1 - \sum_{p|n,\, p>n^{2/5}} 1$, and deduce that $\mathcal{A}$ has positive lower density.

(e) Show that $A(n)$ does not have a monotonic normal order.

3. Let $g(n)$ be the strongly additive function such that $g(p^\nu) = \log p$.

(a) Show, with the notation of §2, that $B(x) \sim A(x)/\sqrt{2}$.

(b) Show that $\sum_{n\le x}\big(\log n - g(n)\big) \ll x$.

(c) Deduce that g has a normal order, *not accessible by the Turán–Kubilius inequality.*

4. Let $\delta > 1$, $g_\delta(n) := \sum_{p|n}(\log p)^\delta$. Show that

$$\alpha_n^\delta(\log n)^\delta \le g_\delta(n) \le \alpha_n^{\delta-1}(\log n)^\delta \qquad (n \ge 1)$$

with $\alpha_n := \log P^+(n)/\log n$. Show that for each α, $\frac{1}{2} \le \alpha \le 1$, the sequence of integers n such that $\alpha_n \ge \alpha$ has a natural density, and compute this density. Deduce that g_δ does not have a monotonic normal order. [*This also holds when* $0 < \delta < 1$, *see Exercise 5.4.*]

5. For $y \ge 2$, let $\chi(n;y)$ denote the characteristic function of the set of integers n such that $P^+(n) \le y$ and let $\Psi(x,y)$ be its summatory function.

(a) Show that for each $\alpha > 0$, we have

$$\Psi(x,y) \le \sqrt{x} + \sum_{n\le x}(n/\sqrt{x})^\alpha \chi(n;y).$$

(b) By applying Theorem 5 to the function $n \mapsto n^\alpha\chi(n;y)$ for some suitable choice of α, deduce that we have

$$\Psi(x,y) \le c_1 x \exp\Big\{-c_2\frac{\log x}{\log y}\Big\} \qquad (x \ge y \ge 2)$$

where c_1 and c_2 are absolute positive constants.

6. Let $\xi(n)$ denote a function tending sufficiently slowly to ∞. Write

$$L(t) := \sqrt{(2\log_2 t\log_3 t)}, \qquad \omega(n,t) := \sum_{p|n,\, p\le t} 1,$$

$$\Omega(n,t) := \sum_{p^\nu\|n,\, p\le t} \nu, \qquad \tau(n,t) := \sum_{d|n,\, d\le t} 1.$$

(a) Show that, for each $\varepsilon > 0$, one has $\displaystyle\sup_{\xi(n)\le t\le n}\left|\frac{\Omega(n,t)-\log_2 t}{L(t)}\right| \le 1+\varepsilon$, pp.

(b) Deduce that

$$\tau(n,t) \le (\log t)^{\log 2}2^{(1+\varepsilon)L(t)} \quad (\xi(n) \le t \le n) \qquad \text{pp.}$$

(c) Let $\tau(n;t,u) := \big|\{d : d|n,\, d > t,\, P^+(d) \le u\}\big|$. Show by using the result of the previous exercise that one has

$$\sum_{n\le x} \tau(n;t,u) \ll x \log u \exp\Big\{-c_2 \frac{\log t}{\log u}\Big\} \qquad (2 \le t,\, u \le x).$$

(d) Let $\alpha := 1/\log 2$, $t_j := \exp j^\alpha$ $(j = 1, 2, \ldots)$, $u_j := \exp\{c_2 j^\alpha/(10 \log j)\}$ $(j = 1, 2, \ldots)$. Show that $\tau(n; t_j, u_j) = 0 \quad (\xi(n) \le j \le (\log n)^{\log 2})$ pp.

(e) Show that $2^{\omega(n,u)} \le \tau(n,t) + \tau(n;t,u)$ and deduce that for each $\varepsilon > 0$ one has

$$\tau(n,t_j) \ge j 2^{-(1+\varepsilon)L(t_j)} \quad (\xi(n) \le j \le (\log n)^{\log 2}) \quad \text{pp.}$$

(f) Establish that for each $\varepsilon > 0$ one has

$$\sup_{\xi(n)\le j\le\tau(n)} \left| \frac{\log_2 d_j(n) - \alpha \log j}{\surd(2\alpha \log j \log_2 j)} \right| \le 1 + \varepsilon \qquad \text{pp.}$$

7. *Set of multiples of* $]y, 2y]$.
Let $y \ge 2$, $\mathcal{A} := \mathbb{Z}^+ \cap]y, 2y]$, and $\mathcal{M}_y := \mathcal{M}(\mathcal{A})$ be the set of multiples of $\mathcal{A}$, i.e. the set of integers n having at least one divisor d such that $y < d \le 2y$.

(a) Show the existence of $\mathbf{d}\mathcal{M}_y =: \varepsilon_y$ and derive an expression for this quantity from the inclusion–exclusion principle.

(b) Let $\mathcal{B}_y := \{n : \Omega(n,y) \ge (\log_2 y)/\log 2\}$. Show the existence of $\mathbf{d}\mathcal{B}_y$ and establish the formula

$$\mathbf{d}\mathcal{B}_y = \prod_{p\le y}\Big(1 - \frac{1}{p}\Big) \sum_{b\in\mathcal{B}_y,\, P^+(b)\le y} \frac{1}{b}.$$

[*Cf. Exercise 1.1.*]

(c) Let $\chi_y(n)$ be the characteristic function of $\mathcal{B}_y$. Show that for all $z \ge 1$

$$\chi_y(n) \le (\log y)^{-(\log z)/\log 2} z^{\Omega(n,y)} \qquad (n \ge 1).$$

By choosing the parameter z suitably, deduce that $\mathbf{d}\mathcal{B}_y \le K(\log y)^{-\delta}$ where K is an absolute constant and $\delta := 1 - (1 + \log_2 2)/\log 2 \approx 0.08607$.

(d) Let $\mathcal{B}'_y := \mathcal{M}_y \smallsetminus \mathcal{B}_y$. Show that $\mathcal{B}'_y$ has a natural density. Denote the characteristic function of $\mathcal{B}'_y$ by $\chi'_y(n)$. Show that for all z, $0 < z \le 1$, one has

$$\chi'_y(n) \le (\log y)^{-(\log z)/\log 2} z^{\Omega(n,y)} \sum_{d|n,\, y<d\le 2y} 1.$$

By choosing z suitably, deduce that $\mathbf{d}\mathcal{B}'_y \le K'(\log y)^{-\delta}$, where K' is absolute, and thus $\varepsilon_y \le K_0(\log y)^{-\delta}$ with $K_0 = K + K'$. [*In fact one has* $(\log y)^{-\delta} e^{-c_0\surd(\log_2 y \log_3 y)} \ll \varepsilon_y \ll (\log y)^{-\delta}(\log_2 y)^{-1/2}$. *See Hall & Tenenbaum (1988), chapter 2.*]

8. *A theorem of Besicovitch (1934).* With the notation $\mathcal{M}_y$ and ε_y from the previous exercise, let $H(x,y) := |\{n \le x : n \in \mathcal{M}_y\}|$. Let $\varepsilon > 0$. Show the existence of a sequence $y_k \to \infty$ such that

(i) $\varepsilon_{y_k} \le \varepsilon/2^{k+1} \qquad (k \ge 0)$

(ii) $H(x, y_k) \le 2\varepsilon_{y_k} x \qquad (k \ge 0,\ x \ge y_{k+1})$.

Set $\mathcal{A} = \mathbb{Z}^+ \cap \bigcup_{k\ge 0}]y_k, 2y_k]$, and let $\mathcal{M} = \mathcal{M}(\mathcal{A})$ be the set of multiples of $\mathcal{A}$. Show that $\underline{\mathbf{d}}\mathcal{M} \le \varepsilon$, $\overline{\mathbf{d}}\mathcal{M} \ge \frac{1}{2}$, and hence that *a set of multiples does not necessarily have a natural density.*

Using the method of the previous exercise, find a uniform upper bound for $H(x,y)$ in the range $2 \le y \le \sqrt{x}$. Deduce that an admissible choice is $y_k = \exp\{c(2^k\varepsilon^{-1})^{1/\delta}\}$ with some sufficiently large absolute constant c.

9. *Divisor density (Hall, 1978).* Let $\mathcal{A}$ be a sequence of integers with characteristic function χ. We say that $\mathcal{A}$ has a divisor density z, denoted $\mathbf{D}\mathcal{A} = z$, if we have

$$\tau(n, \mathcal{A}) := \sum_{d|n} \chi(d) = \big(z + o(1)\big)\tau(n) \qquad \text{pp.}$$

Throughout this exercise, let the integers a, q, with $q \ge 2$, be given and put

$$\mathcal{A} := \{d : \omega(d) \equiv a \,(\mathrm{mod}\, q)\}.$$

(a) Show that $\chi(d) = (1/q)\sum_{j=0}^{q-1} e^{2\pi i j(\omega(d)-a)/q}$ $(d \ge 1)$. Deduce the existence of $\mathbf{d}\mathcal{A}$ and indeed that $\mathbf{d}\mathcal{A} = 1/q$.

(b) Show that $\tau(n, \mathcal{A}) = (1/q)\sum_{j=0}^{q-1} e^{-2\pi i ja/q} \prod_{p^\nu \| n}(1 + \nu e^{2\pi i j/q})$.

(c) With the notation $m := \prod_{p\|n} p$, establish that

$$\big|\tau(n, \mathcal{A}) - \tau(n)/q\big| \le \tau(n/m)\{2\cos\pi/q\}^{\omega(m)} = \tau(n)\big(\cos\pi/q\big)^{\omega(m)}.$$

(d) Show that the arithmetic function $n \mapsto \omega(m)$ is additive and has normal order $\log_2 n$.

(e) Deduce that $\mathbf{D}\mathcal{A} = 1/q$.

10. *Another result on divisor density.* Let δ be a fixed real number, $0 < \delta < \frac{1}{2}$. Set

$$\mathcal{A}(\delta) := \{m : m \ge 3,\ \omega(m) \le (\tfrac{1}{2} + \delta)\log_2 m\}.$$

(a) Show that $\mathbf{d}\mathcal{A}(\delta) = 0$.

(b) Show that for each ε, $0 < \varepsilon < 1$, one has

$$|\{p : p|n,\ \log_2 p > (1-\varepsilon)\log_2 n\}| \sim \varepsilon\omega(n) \quad \text{pp}$$

and deduce that $|\{d : d|n,\ \log_2 d > (1-\varepsilon)\log_2 n\}| \sim \tau(n)$ pp.

(c) Show that $\sum_{d|n} y^{\omega(d)} \leq (1+y)^{\Omega(n)}$ $(y \geq 0,\ n \geq 1)$ and use this inequality to show that, for all η, $0 < \eta < \frac{1}{2}$, one has

$$|\{d : d|n, \omega(d) \leq (\tfrac{1}{2} + \eta)\Omega(n)\}| \sim \tau(n) \quad \text{pp.}$$

(d) Show that $\mathbf{D}\mathcal{A}(\delta) = 1$.

11. *Method of vanishing moments.* Show, for $-\frac{1}{2} \leq z \leq \frac{1}{2}$, that

$$\sum_{n \leq x} \tau(n)^z \ll x(\log x)^{2^z - 1}.$$

By choosing z sufficiently close to 0, deduce that if $\alpha < \log 2 < \beta$ then

$$\left|\{n \leq x : \tau(n) \notin [(\log x)^\alpha,\ (\log x)^\beta]\}\right| = o(x).$$

Sharpen the result by choosing $\alpha = \alpha(x) \to \log 2-$, $\beta = \beta(x) \to \log 2+$.

(b) Establish that the upper bound $\sum_{n \leq x} \tau(n) y^{\omega(n)} \ll x(\log x)^{2y-1}$ holds uniformly for $x \geq 2$, $0 \leq y \leq y_0$. Deduce that, for $0 < \varepsilon < 1$,

$$\sum_{n \leq x,\ |\omega(n) - 2\log_2 x| > \varepsilon \log_2 x} \tau(n) \ll x(\log x)^{1-2\eta}$$

with $\eta := (1 + \frac{1}{2}\varepsilon)\log(1 + \frac{1}{2}\varepsilon) - \frac{1}{2}\varepsilon > 0$.

12. *A theorem of Erdős & Hall (1974).*

Let $\alpha > \log 2$, $\varepsilon_d \ll \big(\log(d+1)\big)^{-\alpha}$ $(d \geq 1)$, and $f(n) := \sum_{d|n} \varepsilon_d$.

(a) Show that for each $\varepsilon > 0$,

$$\lim_{z \to \infty} \overline{\mathbf{d}}\{n : \sup_{d > z} \Omega(n,d)/\log_2 d \geq 1 + \varepsilon\} = 0.$$

(b) Show that

$$\lim_{z \to \infty} \limsup_{x \to \infty} x^{-1} \sum_{n \leq x} \sum_{d|n,\ d > z} |\varepsilon_d| 2^{-\Omega(n,d)} (\log d)^{\log 2} = 0.$$

(c) Establish, by making suitable use of Theorem 2.2, that $f(n)$ has a limiting d.f. [*Erdős & Hall actually prove that this d.f. is continuous.*]

13. *A theorem of Hall (1974).* Let f be a multiplicative function with $0 \leq f \leq 1$. Set $S(x) := \sum_{n \leq x} f(n)$.

(a) Let $k(n) := \prod_{p|n} p$. Show that

$$\sum_{n \leq x} f(n) \log k(n) \leq \sum_{m \leq x} f(m) \psi(x/m)$$

and deduce that the left-hand side does not exceed $x \sum_{m \leq x} f(m)/m + O(x)$.

(b) Let $N(x, y)$ be the number of integers $n \le x$ such that $k(n) \le y$. Show that for all y, $1 \le y \le x$, one has

$$\sum_{n \le x} f(n) \log\bigl(x/k(n)\bigr) \le N(x,y)\log x + S(x)\log(x/y).$$

(c) Appealing to Theorem II.1.13, deduce that

$$S(x) \le \Bigl\{1 + O\Bigl(\frac{\log_2 x}{\log x}\Bigr)\Bigr\}\frac{x}{\log x}\sum_{m \le x}\frac{f(m)}{m},$$

where the implied constant is absolute. [*One may choose* $y = x/\log^3 x$ *in (b).*]

14. Let f be an arithmetic function with a non-decreasing normal order. Show that, for all $\varepsilon > 0$, we have

$$\bigl|\{(m,n) : 1 \le m \le n \le x,\ f(n) \le (1-\varepsilon)f(m)\}\bigr| = o(x^2).$$

15. Let f be a multiplicative arithmetic function such that the set

$$\mathcal{A} := \{f(p) : p \text{ prime}\}$$

has at least two elements. We assume furthermore that, for some fixed $\delta > 0$ and all primes p, we have $f(p) = 1 + O(p^{-\delta})$. Put $f_y(n) := \prod_{p^\nu \| n,\, p>y} f(p^\nu)$. For large y, find an upper bound for the mean value of $\mu(n)^2|\log f_y(n)|$ and deduce that, for any $\varepsilon > 0$ and any $a \in \mathcal{A}$, the sequence of those integers n such that $|f(n) - a| \le \varepsilon$ has positive lower asymptotic density. Using the result of Exercise 14, show that f does not have a non-decreasing normal order. Applications: $f(n) = n/\varphi(n)$, $f(n) = \sigma(n)/n$.

16. *On the mean value of Hooley's Δ-function.*

Put $\Delta(n) := \max_{u \in \mathbb{R}} \sum_{d|n,\, e^u < d \le e^{u+1}} 1$ and $\tau(n,\theta) := \sum_{d|n} d^{i\theta}$. By appealing to Lemma 2.6.1, show that

$$\tau(n)^{-1}\int_0^1 |\tau(n,\theta)|^2\,\mathrm{d}\theta \ll \Delta(n) \ll \int_0^1 |\tau(n,\theta)|\,\mathrm{d}\theta.$$

Deduce the estimates

$$x\log_2 x \ll \sum_{n \le x}\Delta(n) \ll x(\log x)^{4/\pi - 1} \tag{62}$$

by proceeding as follows. For the upper bound, use Theorem 5. For the lower bound, first show that $\Delta(n) \gg \sum^*_{dd'|n} 2^{-\Omega(dd')}$, where the asterisk indicates that summation is restricted to those pairs of divisors $\{d, d'\}$ with $(d, d') = 1$ and $d < d' \le 2d$. The inner sum may then be estimated by the Selberg–Delange method. [*The lower estimate in (62) is due to Hall & Tenenbaum (1982) and the upper estimate to Hooley (1979). Tenenbaum (1985) has shown that the exponent* $4/\pi - 1$ *may be replaced by* $o(1)$*; see also Hall & Tenenbaum (1988), chapter 7.*]

III.4

Distribution of additive functions and mean values of multiplicative functions

§ 4.1 The Erdős–Wintner theorem

The following result gives a complete solution to the problem of the existence of a limiting distribution for an additive function. Both its statement and its proof place it in the realm of comparison theorems between additive functions and sums of independent random variables. The underlying probabilistic result is Kolmogorov's three series theorem—cf. for example Feller (1971), IX.9.

Theorem 1 (Erdős–Wintner, 1939). *A necessary and sufficient condition for a real additive function $f(n)$ to have a limiting distribution is that the following three series converge simultaneously for at least one value of the positive real number R:*

$$(a)\quad \sum_{|f(p)|>R} \frac{1}{p}; \qquad (b)\quad \sum_{|f(p)|\leq R} \frac{f(p)^2}{p}; \qquad (c)\quad \sum_{|f(p)|\leq R} \frac{f(p)}{p}.$$

When these conditions are satisfied, the characteristic function of the limit law is given by the convergent product

$$\varphi(\tau) = \prod_p \left(1 - p^{-1}\right) \sum_{\nu=0}^{\infty} e^{i\tau f(p^\nu)} p^{-\nu}. \tag{1}$$

The limit law is necessarily pure. It is continuous if, and only if,

$$\sum_{f(p)\neq 0} \frac{1}{p} = \infty. \tag{2}$$

Remarks. (i) Should the three series above converge for some value of $R > 0$, they do so for all values of $R > 0$. Therefore, in practice, there is no loss of generality in taking $R = 1$.

(ii) We emphasise that the existence of a limiting distribution for $f(n)$ does not depend on the values $f(p^\nu)$ with $\nu \geq 2$.

The original proof of this result, due solely to Erdős for sufficiency, is direct, and rests upon delicate estimates for the frequencies $\nu_x\{n : f(n) \le z\}$. We shall give here a simpler proof, due to Delange (1961), which depends in an essential way on the continuity theorem of Paul Lévy (Theorem 2.4). Delange establishes the following fundamental result, which has an intrinsic interest, and of which Theorem 1 is an easy consequence.

Theorem 2 (Delange). *Let g be a multiplicative function with values in the unit disc.*

(i) *If g possesses a non-zero mean value*

$$M(g) := \lim_{x\to\infty} \frac{1}{x} \sum_{n\le x} g(n)$$

then we have:

(a) *the series $\sum_p (1 - g(p))/p$ converges,*

(b) *there exists some integer $\nu \ge 1$ such that $g(2^\nu) \ne -1$.*

(ii) *If condition (a) above is satisfied, then g has a mean value, given by the formula*

$$M(g) = \prod_p (1 - p^{-1}) \sum_{\nu=0}^{\infty} g(p^\nu) p^{-\nu}. \tag{3}$$

We shall give the proof of this result in the next section. For the moment, let us show how Theorem 1 may be deduced from it—apart from the assertions concerning the purity and continuity of the limit law.

The following simple lemma will be useful.

Lemma 2.1. *Let $\{u_n\}_{n=1}^{\infty}$ and $\{v_n\}_{n=1}^{\infty}$ be two complex sequences such that*

$$\sum_{n=1}^{\infty} |u_n|^2 + |v_n| \le H < \infty. \tag{4}$$

Then the infinite product $\prod_{n=1}^{\infty}(1+u_n+v_n)$ converges if, and only if, the series $\sum_{n=1}^{\infty} u_n$ converges. In this case we have

$$\Big|\prod_{n=1}^{\infty}(1 + u_n + v_n)\Big| \le \exp\Big\{6H + \sum_{n=1}^{\infty} \Re e\, u_n\Big\}. \tag{5}$$

Proof. If $|u_n| + |v_n| \ge \frac{1}{2}$, then $|u_n|^2 + |v_n| \ge |u_n|^2 - |u_n| + \frac{1}{2} \ge \frac{1}{4}$. By (4), this implies that the set $\mathcal{E}$ of integers n for which $|u_n| + |v_n| \ge \frac{1}{2}$ has at most $4H$ elements.

When $n \notin \mathcal{E}$, we use the estimate

$$|\log(1+z) - z| \le |z|^2 \qquad (|z| \le \tfrac{1}{2}) \tag{6}$$

with $z = u_n + v_n$. We obtain that, for all m, M with $0 \le m < M$,

$$\begin{aligned}\Bigg|\sum_{\substack{m<n\le M\\ n\notin\mathcal{E}}} (u_n+v_n) - \log \prod_{\substack{m<n\le M\\ n\notin\mathcal{E}}} (1+u_n+v_n)\Bigg| &\le \sum_{\substack{m<n\le M\\ n\notin\mathcal{E}}} \big(|u_n|+|v_n|\big)^2 \\ &\le 2 \sum_{\substack{m<n\le M\\ n\notin\mathcal{E}}} \big(|u_n|^2+|v_n|^2\big) \le \sum_{m<n\le M} \big(2|u_n|^2+|v_n|\big).\end{aligned}$$

This shows that the convergence of the infinite product and the convergence of the series $\sum u_n$ are equivalent. Moreover, it follows from this calculation that

$$\begin{aligned}&\Bigg|\prod_{n=1}^{\infty}(1+u_n+v_n)\Bigg| \\ &\qquad\le \prod_{n\in\mathcal{E}} \exp\{|u_n|+|v_n|\} \prod_{n\notin\mathcal{E}} \exp\{\Re e(u_n+v_n)+2|u_n|^2+|v_n|\} \\ &\qquad\le \exp\Big\{\sum_{n=1}^{\infty} \Re e\, u_n + \sum_{n\in\mathcal{E}}(2|u_n|+|v_n|) + \sum_{n\notin\mathcal{E}}(2|u_n|^2+2|v_n|)\Big\}.\end{aligned}$$

We obtain (5) by noting that

$$\sum_{n\in\mathcal{E}} |u_n| \le \Big(4H\sum_{n=1}^{\infty}|u_n|^2\Big)^{1/2} \le 2H.$$

We are now in a position to tackle the proof of Theorem 1.

For each real number τ, let us consider the multiplicative function of modulus 1 defined by

$$g_\tau(n) := e^{i\tau f(n)} \qquad (n = 1, 2, \ldots). \tag{7}$$

Suppose initially that f has a limit law. By the continuity theorem (Theorem 2.4), it follows that g_τ has a mean value for each τ. Write $\varphi(\tau) := M(g_\tau)$. Then φ is the characteristic function of the limit law, and in particular $\varphi(0) = 1$, and φ is continuous at $\tau = 0$. Thus there exists some $T > 0$ such that $|\varphi(\tau)| \ge \frac{1}{2}$ for $|\tau| \le T$. We shall prove the convergence of the three series of Theorem 1 when $R = 2/T$.

For $|\tau| \le T$, assertion (i) of Delange's theorem implies the convergence of the series

$$\sum_p \frac{1 - g_\tau(p)}{p}, \tag{8}$$

and assertion (ii) then yields the representation

$$\varphi(\tau) = \prod_p (1 - p^{-1}) \sum_{\nu=0}^{\infty} g_\tau(p^\nu) p^{-\nu}. \tag{9}$$

The general term in this product equals $1 - (1 - g_\tau(p))/p + h_\tau(p)/p(p-1)$, say, with $|h_\tau(p)| \le 2$. By Lemma 2.1 we then have

$$|\varphi(\tau)| \ll \exp\Big\{ - \sum_p \frac{1 - \cos\big(\tau f(p)\big)}{p} \Big\}, \tag{10}$$

and, since $|\varphi(\tau)| \ge \frac{1}{2}$ when $|\tau| \le T$,

$$\sum_p \frac{1 - \cos\big(\tau f(p)\big)}{p} \ll 1 \qquad (|\tau| \le T). \tag{11}$$

Given the inequality $1 - \cos\theta \ge 2\theta^2/\pi^2$ ($|\theta| \le \pi$), it follows that

$$\sum_{|f(p)| \le 2/T} \frac{f(p)^2}{p} < \infty. \tag{12}$$

Averaging (11) over $[0, T]$, we further obtain that

$$\sum_p \frac{1}{p} \Big\{ 1 - \frac{\sin\big(Tf(p)\big)}{Tf(p)} \Big\} \ll 1,$$

with an obvious convention when $f(p) = 0$. The expression between curly brackets is ≥ 0 for all p and is $> \frac{1}{2}$ when $|f(p)| > R = 2/T$. Therefore

$$\sum_{|f(p)| > R} \frac{1}{p} < \infty. \tag{13}$$

This allows us to deduce from the convergence of $\sum_p \sin\big(Tf(p)\big)/p$, which follows from that of (8), the convergence of

$$\sum_{|f(p)| \le R} \frac{\sin\big(Tf(p)\big)}{p}.$$

Using (12) and the inequality $|\sin\theta - \theta| \le \frac{1}{6}|\theta|^3 \le \frac{1}{3}\theta^2$ ($|\theta| \le 2$), we obtain the convergence of

$$\sum_{|f(p)| \le R} \frac{f(p)}{p}. \tag{14}$$

This completes the proof that the conditions of Theorem 1 are necessary.

Conversely, let us assume that the three series in the statement are convergent for some positive real number R.

The trivial bound $|1-e^{i\tau f(p)}| \leq 2$ then implies, for each $T > 0$, the uniform convergence of the series

$$\sum_{|f(p)|>R} \frac{1-e^{i\tau f(p)}}{p} \qquad (|\tau| \leq T).$$

In addition, when $|f(p)| \leq R$, we can write

$$\begin{aligned} 1-e^{i\tau f(p)} &= -i\tau f(p) + \{1-\cos\left(\tau f(p)\right)\} + i\{\tau f(p) - \sin\left(\tau f(p)\right)\} \\ &= -i\tau f(p) + O\Big(T^2 f(p)^2 + T^3 R f(p)^2\Big), \end{aligned}$$

uniformly for $|\tau| \leq T$. This yields that the series

$$\sum_{|f(p)|\leq R} \frac{1-e^{i\tau f(p)}}{p}$$

is uniformly convergent on $[-T,T]$. Thus $\sum_p (1-g_\tau(p))/p$ is uniformly convergent on each compact subset. By assertion (ii) of Theorem 2 we may deduce, for each $\tau \in \mathbb{R}$, the existence of the mean value $M(g_\tau) = \varphi(\tau)$, with $\varphi(\tau)$ given by (9). Since the general term of the product (9) equals $1-(1-g_\tau(p))/p + O(1/p^2)$, Lemma 2.1 shows that the convergence of the infinite product is uniform on each compact subset. Consequently, $\varphi(\tau)$ is continuous. By Theorem 2.4, f has a limiting distribution.

The assertion concerning the purity and the continuity of the limit law follow from Theorem 2.5, since this law is that of the infinite convolution product

$$\mathop{\ast}_{p} F_p$$

where F_p is the atomic distribution function given by

$$F_p(z) = \sum_{f(p^\nu)\leq z} p^{-\nu}(1-p^{-1}) \qquad (z \in \mathbb{R}).$$

In particular, the largest saltus of F_p equals $1-1/p+O(1/p^2)$ if $f(p) \neq 0$ and $1+O(1/p^2)$ if $f(p)=0$.

For the reader's convenience we give here a direct proof of the fact that the condition

$$\sum_{f(p)\neq 0} \frac{1}{p} = \infty \tag{15}$$

is necessary and sufficient for the continuity of the limit law.

Let us show first of all that the condition is necessary. If the series (15) converges, then the sequence $\mathcal{A}$ of squarefree integers n such that $p|n \Rightarrow f(p) = 0$ has positive density. This follows, for instance, from Theorem I.3.11, since the characteristic function of $\mathcal{A}$ is multiplicative. (We obtain incidentally that $\mathbf{d}\mathcal{A} = 6\pi^{-2}\prod_{f(p)\neq 0}(1 + 1/p)^{-1}$, but this explicit value is not needed here.) Since $f(n) = 0$ for all n in $\mathcal{A}$, the limit law is not continuous at the origin.

In order to prove the sufficiency of condition (15), we shall establish the continuity of the limit law in the equivalent form (cf. Chapter 2, Notes)

$$\text{(16)} \qquad \lim_{T\to\infty} \frac{1}{2T}\int_{-T}^{T} |\varphi(\tau)|^2 \,\mathrm{d}\tau = \lim_{T\to\infty} \tfrac{1}{2}\int_{-1}^{1} |\varphi(Ty)|^2 \,\mathrm{d}y = 0.$$

Let N be a sufficiently large integer, and let $a_1 < a_2 < \cdots < a_J$ denote the distinct non-zero values of $f(p)$, $p \le N$. We write

$$\varepsilon_j := \sum_{p\le N,\, f(p)=a_j} \frac{1}{p} \quad (1 \le j \le J), \qquad S_N := \sum_{1\le j\le J} \varepsilon_j = \sum_{p\le N,\, f(p)\neq 0} \frac{1}{p}.$$

By Hölder's inequality, we have for each integer $h \ge 1$ and any real number y, $|y| \le 1$,

$$\Big(\sum_{1\le j\le J} \varepsilon_j \cos(a_j T y)\Big)^{2h} \le S_N^{2h-1} \sum_{1\le j\le J} \varepsilon_j \cos^{2h}(a_j T y),$$

from which

$$\int_{-1}^{1}\Big(\sum_{1\le j\le J} \varepsilon_j \cos(a_j T y)\Big)^{2h} \mathrm{d}y \le S_N^{2h-1} \sum_{1\le j\le J} \varepsilon_j \int_{-1}^{1} \cos^{2h}(a_j T y)\,\mathrm{d}y.$$

As $T \to \infty$, the last integral tends to

$$\frac{4}{\pi}\int_0^{\pi/2} \cos^{2h} w \,\mathrm{d}w \ll \frac{1}{\sqrt{h}},$$

where the bound follows from a classical evaluation of Wallis integrals—cf. Exercise I.0.3(b). For each fixed N, and sufficiently large T, we can thus assert that the measure of the set of those y in $[-1, 1]$ such that

$$\Big|\sum_{1\le j\le J} \varepsilon_j \cos(a_j T y)\Big| \ge (1 - 1/h) S_N$$

is $\ll h^{-1/2}$. Using (10), it follows that

$$\begin{aligned}\int_{-1}^{1} |\varphi(Ty)|^2 \,\mathrm{d}y &\ll \int_{-1}^{1} \exp\Big\{-2\sum_{1\le j\le J} \varepsilon_j\big(1 - \cos(a_j T y)\big)\Big\}\,\mathrm{d}y \\ &\ll h^{-1/2} + \exp\{-2S_N/h\} \ll \sqrt{(\log S_N/S_N)}\end{aligned}$$

for the choice $h := [S_N/\log S_N]$. Since (15) implies that $S_N \to \infty$ as $N \to \infty$, we obtain (16).

§ 4.2 Delange's theorem

Here we give a proof of Theorem 2 also due to Delange, but different from his original proof, and, following an idea of Rényi (1965), depending on the Turán–Kubilius inequality.

It rests fundamentally on the following result.

Theorem 3 (Delange). *Let g be a multiplicative function with values in the unit disc. Under the assumption that*

$$\sum_p \frac{1 - \Re e\, g(p)}{p} < \infty, \tag{17}$$

we have

$$\frac{1}{x}\sum_{n\le x} g(n) = \prod_{p\le x}(1-p^{-1})\sum_{\nu=0}^{\infty} g(p^\nu)p^{-\nu} + o(1) \qquad (x\to\infty). \tag{18}$$

Proof. For each $y \geq 2$, let us define a multiplicative function g_y by

$$g_y(p^\nu) = \begin{cases} g(p^\nu) & (p\le y) \\ 1 & (p > y). \end{cases}$$

The behaviour of g_y is simpler than that of g, since g_y is only non-trivial on a finite set of prime numbers. However, one can hope that g_y is a "good" approximation for g when y is large: this is the main thread of the proof.

Let us consider $h_y := \mu * g_y$. We have $h_y(p^\nu) = g(p^\nu) - g(p^{\nu-1})$ if $\nu \geq 1$, $p \leq y$, and $h_y(p^\nu) = 0$ if $\nu \geq 1$, $p > y$. This implies that

$$\begin{aligned} H_y(x) &:= \sum_{m\le x}|h_y(m)| + x\sum_{m>x}\frac{1}{m}|h_y(m)| \\ &\le \sqrt{x}\sum_{m=1}^{\infty}\frac{|h_y(m)|}{\sqrt{m}} \le \sqrt{x}\prod_{p\le y}\Big(1+\frac{2}{\sqrt{(p)}-1}\Big). \end{aligned}$$

We easily deduce that g_y has a mean value, viz.

$$\begin{aligned} \sum_{n\le x} g_y(n) &= \sum_{m\le x} h_y(m)\Big[\frac{x}{m}\Big] = x\sum_{m=1}^{\infty}\frac{1}{m}h_y(m) + O\big(H_y(x)\big) \\ &= x\{M(g_y)+o(1)\}, \end{aligned}$$

with

$$M(g_y) := \prod_{p\le y}(1-p^{-1})\sum_{\nu=0}^{\infty} g(p^\nu)p^{-\nu}.$$

Now let us define a multiplicative function $r(n)$ by $r(n) := |g(n)|$, and an additive function $\theta(n)$ by

$$\theta(p^\nu) := \begin{cases} \arg g(p^\nu), & \text{if } g(p^\nu) \neq 0, \\ 0, & \text{if } g(p^\nu) = 0, \end{cases}$$

the argument being chosen to lie in $]-\pi, \pi]$. Setting $A(x) := \sum_{p \le x} \theta(p)/p$, we use Cauchy's criterion to show that the quantity

$$M(g_y)e^{-iA(y)} \tag{19}$$

tends to a finite limit M, as $y \to \infty$. For $2 \le y \le z \le x$, we have

$$\begin{aligned} S(x,y,z) &:= \frac{1}{x}\Big|\sum_{n \le x} g_z(n)e^{-iA(z)} - \sum_{n \le x} g_y(n)e^{-iA(y)}\Big| \\ &\le \frac{1}{x}\sum_{n \le x}\Big|g_z(n)e^{-iA(z)} - g_y(n)e^{-iA(y)}\Big| \\ &\le \frac{1}{x}\sum_{n \le x}\Big|e^{-i\{A(z)-A(y)\}} \prod_{p^\nu \| n,\, y<p\le z} g(p^\nu) - 1\Big|. \end{aligned}$$

From the bound $|u_1 \ldots u_m - 1| \le \sum_{j=1}^m |u_j - 1|$, valid for all complex numbers $u_1, \ldots, u_m$ in the unit disc, we obtain

$$S(x,y,z) \le \frac{1}{x}\sum_{n \le x} \sum_{\substack{p^\nu \| n \\ y<p\le z}} \left(1 - r(p^\nu)\right) + \frac{1}{x}\sum_{n \le x}\left|e^{i\{A(z)-A(y)-\theta_{y,z}(n)\}} - 1\right|$$

where $\theta_{y,z}$ is the additive function defined by $\theta_{y,z}(p^\nu) := \theta(p^\nu)$ if $y < p \le z$, and $\theta_{y,z}(p^\nu) := 0$ otherwise.

We obtain an upper bound for the second sum over n by using the inequality $|e^{iu} - 1| = \left|\int_0^u e^{it}\,\mathrm{d}t\right| \le |u|$ $(u \in \mathbb{R})$. After interchanging summations in the first sum and applying the Cauchy–Schwarz inequality in the second, we can write

$$\begin{aligned} S(x,y,z) \le &\sum_{y<p\le z} \frac{1-r(p)}{p} + 2\sum_{y<p\le z} \frac{1}{p(p-1)} \\ &+ \left\{\frac{1}{x}\sum_{n \le x}\Big(\theta_{y,z}(n) - \big(A(z) - A(y)\big)\Big)^2\right\}^{1/2}. \end{aligned}$$

Estimating the last sum by the Turán–Kubilius inequality, we obtain

$$\begin{aligned} S(x,y,z) &\le \sum_{y<p\le z} \frac{1-r(p)}{p} + O\left(\frac{1}{y} + \sqrt{\sum_{y<p\le z} \frac{\theta(p)^2}{p} + \frac{1}{y}}\right) \\ &\ll \sum_{y<p\le z} \frac{1-\Re e\, g(p)}{p} + \frac{1}{\sqrt{y}} + \sqrt{\sum_{y<p\le z} \frac{\theta(p)^2}{p}}. \end{aligned}$$

Next observe that for each p we have

$$\theta(p)^2 \le \pi^2\big(1 - \Re e\, g(p)\big). \tag{20}$$

Indeed, either $|\theta(p)| > \frac{1}{2}\pi$, and the inequality is satisfied since $\Re e\, g(p) \le 0$, or $|\theta(p)| \le \frac{1}{2}\pi$, and we have

$$1 - \Re e\, g(p) = 1 - r(p)\cos\theta(p) \ge 1 - \cos\theta(p) \ge 2\pi^{-2}\theta(p)^2.$$

Substituting (20) in the above upper bound for $S(x,y,z)$, we obtain

$$S(x,y,z) \ll \eta(y) \qquad (2 \le y \le z \le x) \tag{21}$$

with

$$\eta(y) := 1/\sqrt{y} + \Big(\sum_{p>y} \big(1 - \Re e\, g(p)\big)/p\Big)^{1/2} = o(1) \quad (y \to \infty).$$

This shows that Cauchy's criterion is satisfied for (19), and hence there exists an M such that

$$M(g_y)e^{-iA(y)} = M + o(1) \qquad (y \to \infty). \tag{22}$$

Choosing $z = x$ in (21) and noting that $g_x(n) = g(n)$ for $n \le x$, it follows that

$$\frac{1}{x}\sum_{n\le x} g(n)e^{-iA(x)} = \frac{1}{x}\sum_{n\le x} g_y(n)e^{-iA(y)} + O\big(\eta(y)\big).$$

Letting, successively, x and y tend to ∞, we obtain

$$e^{-iA(x)}\frac{1}{x}\sum_{n\le x} g(n) = M + o(1) = M(g_x)e^{-iA(x)} + o(1),$$

as required.

We are now in a position to prove Theorem 2 of Delange. Suppose first that g has a non-zero mean value $M(g)$. By partial summation, we infer that

$$(\sigma-1)\sum_{n=1}^{\infty} g(n)n^{-\sigma} \sim \zeta(\sigma)^{-1}\sum_{n=1}^{\infty} g(n)n^{-\sigma} \sim M(g) \qquad (\sigma\to 1+)$$

and further that

$$\lim_{\sigma\to 1+}\prod_p (1-p^{-\sigma})\sum_{\nu=0}^{\infty} g(p^\nu)p^{-\nu\sigma} = M(g).$$

Noting that the general term of this product has modulus ≤ 1 and that

$$\Big|\sum_{\nu=0}^{\infty} g(p^\nu)(p^{-\nu}-p^{-\nu\sigma})\Big| \leq \frac{1-p^{-(\sigma-1)}}{(p-1)(1-p^{-\sigma})} \ll \frac{(\sigma-1)\log p}{p},$$

we can write

$$\begin{aligned}
|M(g)| &\ll \Big|\prod_{p\leq\exp\{1/(\sigma-1)\}} (1-p^{-\sigma})\sum_{\nu=0}^{\infty} g(p^\nu)p^{-\nu\sigma}\Big| \\
&\ll \Big|\prod_{p\leq\exp\{1/(\sigma-1)\}} \Big\{(1-p^{-\sigma})\sum_{\nu=0}^{\infty} g(p^\nu)p^{-\nu} + O\big((\sigma-1)\log p/p\big)\Big\}\Big| \\
&\ll \Big|\prod_{p\leq\exp\{1/(\sigma-1)\}} \Big\{1-(1-g(p))/p + O\big(1/p^2+(\sigma-1)\log p/p\big)\Big\}\Big| \\
&\ll \exp\Big\{-\sum_{p\leq\exp\{1/(\sigma-1)\}} (1-\Re e\, g(p))/p\Big\},
\end{aligned}$$

by Lemma 2.1.

This implies the convergence of the series $\sum_p \big(1-\Re e\, g(p)\big)/p$. By Theorem 3, it then follows that the product

$$\prod_p (1-p^{-1})\sum_{\nu=0}^{\infty} g(p^\nu)p^{-\nu} \tag{23}$$

converges with value $M(g)$. Its general term is $1-\big(1-g(p)\big)/p+O(1/p^2)$ and, by a further application of Lemma 2.1, we then obtain the convergence of the series $\sum_p(1-g(p))/p$. Finally, since $M(g)\neq 0$, no term of the infinite product (23) vanishes, and in particular $g(2^\nu)\neq -1$ must hold for some $\nu\geq 1$.

Let us now prove assertion (ii) of Theorem 2. If the series (a) converges, then, by Lemma 2.1, so does the product (23). The conclusion hence follows immediately from Theorem 3.

§ 4.3 Halász' theorem

Theorem 2 of Delange gives a necessary and sufficient condition for a multiplicative function of modulus ≤ 1 to have a non-zero mean value. Wirsing (1967) and Halász (1968) subsequently completed the study of the behaviour in the mean of a multiplicative function with modulus ≤ 1, the former for real and the latter for complex functions.

Theorem 4 (Halász). *Let g be a multiplicative function with values in the unit disc. If there exists some real number τ such that the series*

$$\sum_p \frac{1 - \Re e\big(g(p)p^{-i\tau}\big)}{p} \tag{24}$$

converges, then we have

$$\frac{1}{x}\sum_{n\leq x} g(n) = \frac{x^{i\tau}}{1+i\tau}\prod_{p\leq x}\big(1-p^{-1}\big)\sum_{\nu=0}^{\infty} g(p^\nu)p^{-\nu(1+i\tau)} + o(1) \qquad (x\to\infty). \tag{25}$$

If there exists no real number τ such that the series (24) converges, then we have

$$\frac{1}{x}\sum_{n\leq x} g(n) = o(1). \tag{26}$$

The first assertion of this theorem follows easily from Theorem 3. We indeed deduce from the convergence of (24) that

$$\frac{1}{x}\sum_{n\leq x}\frac{g(n)}{n^{i\tau}} = \prod_{p\leq x}\big(1-p^{-1}\big)\sum_{\nu=0}^{\infty} g(p^\nu)p^{-\nu(1+i\tau)} + o(1) \qquad (x\to\infty). \tag{27}$$

Moreover, formula (22) applied to $g(n)n^{-i\tau}$ allows us to replace the product on the right-hand side by

$$M_\tau e^{iA_\tau(x)} + o(1)$$

where M_τ is a suitable constant, and

$$A_\tau(x) := \sum_{p\leq x}\theta_\tau(p)/p$$

with $\theta_\tau(p) := \arg\big\{g(p)p^{-i\tau}\big\} \in]-\pi,\pi]$. By (20), the series $\sum_p \theta_\tau(p)^2/p$ must converge, and so, by applying the Cauchy–Schwarz inequality, we obtain

$$A_\tau(x) - A_\tau(y) = o\Big(1 + \log\Big(\frac{\log x}{\log y}\Big)\Big)$$

as $y \to \infty$, $x \geq y$. In particular, the function L_τ defined by $L_\tau(\log x) := e^{iA_\tau(x)}$ is *slowly varying*, i.e. satisfies

$$L_\tau(u) \sim L_\tau(v)$$

as u, $v \to \infty$ and $u \asymp v$. This property allows us to deduce (25) from (27), using a simple integration by parts. For convenience of later reference, we note that (25) may be rewritten in the form

$$\frac{1}{x}\sum_{n\leq x} g(n) = K_\tau x^{i\tau} L_\tau(\log x) + o(1) \tag{28}$$

where K_τ is a suitable constant and L_τ is a slowly varying function of modulus 1.

We shall establish later the second assertion of Halász' theorem. Before this, we derive the following important corollary, initially proved by Wirsing (1967).

Theorem 5 (Wirsing). *Let g be a real multiplicative function with values in $[-1, 1]$. Then we have*

$$\lim_{x\to\infty} \frac{1}{x}\sum_{n\leq x} g(n) = \prod_p \left(1 - p^{-1}\right) \sum_{\nu=0}^{\infty} g(p^\nu)p^{-\nu} \tag{29}$$

where the infinite product is to be taken as zero when it is divergent.

Proof. Under the assumptions of this statement the series

$$\sum_p \frac{1 - \Re e\big(g(p)p^{-i\tau}\big)}{p}$$

diverges for all $\tau \neq 0$. This follows immediately from the relation

$$\sum_{|\cos(\tau \log p)|\leq \frac{1}{2}} \frac{1}{p} = \infty$$

which is easily obtained by means of the prime number theorem. (Actually an estimate of the type $\sum_{x<n\leq cx} \Lambda(n) \gg x$ $(x \to \infty)$ valid for any arbitrary constant $c > 1$ would suffice.) Thus we need only consider the value $\tau = 0$ in Halász' theorem.

Remark. Wirsing's theorem lies at least as deep as the prime number theorem, since the function $g(n) = \mu(n)$ is within its range of application.

For the "divergent" case of Halász' theorem, we shall give an effective estimate due to Montgomery (1978b). The argument is essentially the same as that employed by Halász in his 1971 article, partially devoted to making his 1968 results effective, but Montgomery also introduces certain refinements and numerous simplifications.

Let us write

$$G(x) := \sum_{n\le x} g(n), \qquad F(s) := \sum_{n=1}^{\infty} g(n)n^{-s} \qquad (\sigma > 1)$$

and

$$H(\alpha)^2 := \sum_{k\in\mathbb{Z}} \frac{1}{k^2+1} \max_{\substack{\sigma=1+\alpha \\ |\tau-k|\le\frac12}} |F(s)|^2 \qquad (\alpha > 0). \tag{30}$$

Theorem 6 (Montgomery). *With the above notation, we have for each multiplicative function g with values in the unit disc*

$$G(x) \ll \frac{x}{\log x}\int_{1/\log x}^{1} \frac{H(\alpha)}{\alpha}\,\mathrm{d}\alpha. \tag{31}$$

Let us first show how the second part of Halász' theorem may be deduced from this statement. We need the following lemma.

Lemma 4.1. *Let $\{\varphi_n(\tau)\}_{n=1}^{\infty}$ be a sequence of continuous functions such that, for each τ with $|\tau|\le 1$, $\varphi_n(\tau)$ increases to ∞. Then the convergence is uniform.*

Proof. We argue by contradiction. If the conclusion fails, then $\inf_{|\tau|\le 1}\varphi_n(\tau)$ does not tend to infinity with n, and there exist some constant A and some subsequence with indices $\{n_j\}_{j=1}^{\infty}$, $n_j\to\infty$, such that $\inf_{|\tau|\le1}\varphi_{n_j}(\tau)\le A$ for $j\ge 1$. Since the φ_{n_j} are continuous, for each j there exists a $\tau_j\in[-1,1]$ such that $\varphi_{n_j}(\tau_j)=\inf_{|\tau|\le1}\varphi_{n_j}(\tau)$. By extracting some new subsequence if necessary, we may assume that $\tau_j\to\tau_0$ as $j\to\infty$. Now let n be some arbitrary fixed integer. For sufficiently large j, we have

$$\varphi_{n_j}(\tau_j) \ge \varphi_n(\tau_j) \ge \tfrac12\,\varphi_n(\tau_0)$$

since $\varphi_n(\tau_j)\to\varphi_n(\tau_0)$. It follows that $\varphi_n(\tau_0)\le 2A$, contradicting the hypothesis that $\varphi_n(\tau_0)\to\infty$ as $n\to\infty$.

We can now complete the proof of Halász' theorem. By Lemma 2.1, we have

$$\begin{aligned} F(s)\zeta(\sigma)^{-1} &= \prod_p \Big(1 - \frac{1-g(p)p^{-i\tau}}{p^\sigma} + \sum_{\nu=2}^{\infty}\frac{g(p^\nu)-g(p^{\nu-1})p^{i\tau}}{p^{\nu s}}\Big) \\ &\ll \exp\Big\{-\sum_p \frac{1-\Re e\big(g(p)p^{-i\tau}\big)}{p^\sigma}\Big\} \end{aligned} \tag{32}$$

and, under the assumption that the series (24) diverges for all τ, Lemma 4.1 implies that the right-hand side converges to 0 uniformly in τ on each compact subset, as $\sigma \to 1+$. We deduce that $H(\alpha) = o(1/\alpha)$ as $\alpha \to 0+$, from which, by substituting in (31), we get that $G(x) = o(x)$ as $x \to \infty$.

For the proof of Theorem 6, we need two auxiliary results. The first is a general inequality, also due to Montgomery (1971), p. 158, concerning quadratic means of almost-periodic functions. We state it in the context of Dirichlet series.

Lemma 6.1 (Montgomery).
Let $A(s) := \sum_{n=1}^{\infty} a_n n^{-s}$ and $B(s) := \sum_{n=1}^{\infty} b_n n^{-s}$ be two Dirichlet series which are convergent for $\sigma > 1$ and satisfy $|a_n| \le b_n$ for $n \ge 1$. Then for all $T \ge 0$ and all $\sigma > 1$ we have

$$\int_{-T}^{T} |A(s)|^2 \,\mathrm{d}\tau \le 3 \int_{-T}^{T} |B(s)|^2 \,\mathrm{d}\tau. \tag{33}$$

Proof. Consider the function $\chi(\tau) := \max(0, 1 - |\tau|/T)$ with Fourier transform

$$\widehat{\chi}(t) = \int_{-\infty}^{\infty} e^{i\tau t} \chi(\tau) \,\mathrm{d}\tau = T\Big(\frac{\sin(tT/2)}{tT/2}\Big)^2.$$

For all $\tau_0 \in \mathbb{R}$ we have

$$\begin{aligned}\int_{-\infty}^{\infty} \chi(\tau - \tau_0)|A(s)|^2 \,\mathrm{d}\tau &= \sum_{m,n=1}^{\infty} \frac{a_m \overline{a_n}}{(mn)^{\sigma}} \Big(\frac{n}{m}\Big)^{i\tau_0} \widehat{\chi}\Big(\log\frac{m}{n}\Big) \\ &\le \sum_{m,n=1}^{\infty} \frac{b_m b_n}{(mn)^{\sigma}} \widehat{\chi}\Big(\log\frac{m}{n}\Big) = \int_{-\infty}^{\infty} \chi(\tau)|B(s)|^2 \,\mathrm{d}\tau.\end{aligned}$$

Letting $h(\tau)$ denote the characteristic function of $[-T, T]$, we have

$$h(\tau) \le \chi(\tau - T) + \chi(\tau) + \chi(\tau + T),$$

from which

$$\int_{-T}^{T} |A(s)|^2 \,\mathrm{d}\tau \le 3 \int_{-\infty}^{\infty} \chi(\tau)|B(s)|^2 \,\mathrm{d}\tau \le 3 \int_{-T}^{T} |B(s)|^2 \,\mathrm{d}\tau.$$

The second auxiliary result is a convenient decomposition of the Euler product $F(s)$.

Lemma 6.2. *Let g be a multiplicative function of modulus ≤ 1. We have*

$$(34)\qquad \sum_{n=1}^{\infty} g(n)n^{-s} = \big(1 + D(s)\big)F_1(s)J(s) \qquad (\sigma > 1)$$

with

$$D(s) := \sum_{\nu=1}^{\infty} g(2^\nu)2^{-\nu s}, \qquad F_1(s) := \exp \sum_{p>2} g(p)p^{-s},$$

and where $J(s)$ is holomorphic for $\sigma > \frac{1}{2}$ and satisfies

$$(35)\qquad 1 \ll J(s) \ll 1, \qquad J'(s) \ll 1 \qquad (\sigma \geq 1).$$

Proof. We have

$$J(s) = \prod_{p>2} \exp\{-g(p)p^{-s}\} \sum_{\nu=0}^{\infty} g(p^\nu)p^{-\nu s}.$$

A typical factor of this product has value $1 + O(p^{-2\sigma})$, and so $J(s)$ is certainly holomorphic for $\sigma > \frac{1}{2}$ and uniformly bounded for $\sigma \geq \frac{3}{4}$. In particular, this implies the required upper bounds for $J(s)$ and $J'(s)$ in (35). Furthermore, since

$$\Big|\sum_{\nu=1}^{\infty} g(p^\nu)p^{-\nu s}\Big| \leq \frac{1}{p-1} \leq \tfrac{1}{2} \qquad (p > 2,\ \sigma \geq 1)$$

we deduce from (6) that

$$|\log J(s)| \leq \sum_{p>2} \frac{2}{(p-1)^2}$$

from which follows the lower bound for $J(s)$ in (35).

We can now tackle the proof of Theorem 6. It will be convenient to have at our disposal the following lower bound

$$(36)\qquad H(\alpha) \gg 1 \qquad (\alpha > 0).$$

For this, we first observe that, putting $\theta_\nu := \arg g(2^\nu)$ with $|\theta_\nu| \leq \pi$, we have

$$|1 + D(1+\alpha-i\tau)| \geq 1 - \sum_{\nu=1}^{\infty} 2^{-\nu}|\cos(\theta_\nu - \tau\nu\log 2)| \geq \tfrac{1}{2}\,(1 - |\cos(\theta_1 - \tau\log 2)|).$$

Let k_0 be one of the possibly two closest integers to $(\theta_1 - \frac{1}{2}\pi)/\log 2$, so that

$$|1 + D(s)| \geq \tfrac{1}{2}(1 - \sin\log 2) > 0 \qquad (\sigma \geq 1,\ |\tau - k_0| \leq \tfrac{1}{2}).$$

Plainly, $|k_0| \leq 8$. Furthermore

$$\int_{k_0-\frac{1}{2}}^{k_0+\frac{1}{2}} \log F_1(s)\,\mathrm{d}\tau = \sum_{p>2} \frac{g(p)}{p^\sigma} \int_{k_0-\frac{1}{2}}^{k_0+\frac{1}{2}} p^{-i\tau}\,\mathrm{d}\tau \ll \sum_p \frac{1}{p\log p} \ll 1.$$

This implies that $\max_{|\tau-k_0|\leq\frac{1}{2}} |F_1(s)| \gg 1$, and hence

$$\max_{\substack{\sigma=1+\alpha \\ |\tau-k_0|\leq\frac{1}{2}}} |F(s)| \geq \min\big|\big(1+D(s)\big)J(s)\big| \max|F_1(s)| \gg 1,$$

from which (36) follows.

The first step of the proof consists in establishing an upper bound for $G(x)$ in terms of an average of itself, namely

$$G(x) \ll \frac{x}{\log x}\int_1^x \frac{|G(t)|}{t^2}\,\mathrm{d}t + \frac{x\log_2 x}{\log x}. \tag{37}$$

Write $K(x) := \sum_{n\leq x} g(n)\log n$. Then

$$G(x)\log x - K(x) = \sum_{n\leq x} g(n)\log(x/n) \ll \sum_{n\leq x}\log(x/n) \ll x.$$

Hence, in order to establish (37), it suffices to show that

$$K(x) \ll x\int_1^x \frac{|G(t)|}{t^2}\,\mathrm{d}t + x\log_2 x. \tag{38}$$

Now we can write

$$\begin{aligned} K(x) &= \sum_{p^\nu\leq x} g(p^\nu)\log p^\nu \sum_{m\leq x/p^\nu,\, p\nmid m} g(m) \\ &= \sum_{p^\nu\leq x} g(p^\nu)\log p^\nu\big\{G(x/p^\nu) + O\big(x/p^{\nu+1}\big)\big\} \\ &\ll \sum_{p^\nu\leq x}\log p^\nu\big|G(x/p^\nu)\big| + x. \end{aligned} \tag{39}$$

In addition, note that for $1 \leq y \leq x$, $|t-x| \leq y$, we have $|K(x)-K(t)| \ll y\log x$. Therefore

$$|K(x)| \leq \frac{1}{y}\int_{x-y}^{x+y} |K(t)|\,\mathrm{d}t + y\log x.$$

Choose $y = x/\log x$, and in the last integral use (39) to evaluate $K(t)$. We obtain

$$\begin{aligned}
\int_{x-y}^{x+y} |K(t)|\,\mathrm{d}t &\ll \int_{x-y}^{x+y} \sum_{p^\nu \le t} \log p^\nu \big|G\big(t/p^\nu\big)\big|\,\mathrm{d}t + xy \\
&\ll \sum_{p^\nu \le 2x} \log p^\nu \int_{x-y}^{x+y} \big|G\big(t/p^\nu\big)\big|\,\mathrm{d}t + xy \\
&= \sum_{p^\nu \le 2x} p^\nu \log p^\nu \int_{(x-y)/p^\nu}^{(x+y)/p^\nu} |G(t)|\,\mathrm{d}t + xy \\
&= \int_1^x |G(t)| \sum_{(x-y)/t < p^\nu \le (x+y)/t} p^\nu \log p^\nu\,\mathrm{d}t + xy \\
&\ll x \int_1^x \frac{|G(t)|}{t} \sum_{(x-y)/t < p^\nu \le (x+y)/t} \log p^\nu\,\mathrm{d}t + xy.
\end{aligned}$$

By the Brun–Titchmarsh theorem (I.4.9), the inner sum is $\ll \dfrac{y}{t}\,\dfrac{\log(2x/t)}{\log(y/t)}$ whenever $y \ge 2t$. For $t \le y/\log x = x/\log^2 x$, this upper bound is $\ll y/t$, and the corresponding contribution to the last integral is

$$\ll y \int_1^x |G(t)| \frac{\mathrm{d}t}{t^2}.$$

For $y/\log x < t \le x$, we bound $|G(t)|$ trivially by t. The corresponding contribution is

$$\ll \sum_{p^\nu \le (x+y)/(y/\log x)} \log p^\nu \int_{(x-y)/p^\nu}^{(x+y)/p^\nu} \mathrm{d}t \ll y \sum_{p^\nu \le 2\log^2 x} \frac{\log p^\nu}{p^\nu} \ll y \log_2 x.$$

Rearranging the estimates we obtain (38).

The second step involves deriving the upper bound

$$\int_1^x \frac{|G(t)|\log t}{t^2}\,\mathrm{d}t \ll H\Big(\frac{2}{\log x}\Big)\log x \qquad (x \ge 2). \tag{40}$$

In (40), we may replace $|G(t)|\log t$ by $|K(t)|$. The implied error is $O(\log x)$, which is acceptable given (36). Next, the Cauchy–Schwarz inequality

$$\int_1^x \frac{|K(t)|}{t^2}\,\mathrm{d}t \le \Big\{\int_1^x \frac{|K(t)|^2}{t^3}\,\mathrm{d}t \int_1^x \frac{\mathrm{d}t}{t}\Big\}^{1/2}$$

enables us to reduce the proof of (40) to that of

$$\int_1^{\infty} \frac{|K(t)|^2}{t^{3+2\alpha}}\, dt \ll \frac{H(\alpha)^2}{\alpha} \qquad (\alpha > 0). \tag{41}$$

Indeed the desired upper bound follows from the choice $\alpha = 2/\log x$. The relation

$$\int_0^{\infty} K(e^u)e^{-u\sigma}e^{-iu\tau}\, du = -\frac{F'(s)}{s} \qquad (\sigma > 1)$$

allows us to write Plancherel's formula

$$\int_1^{\infty} \frac{|K(t)|^2}{t^{3+2\alpha}}\, dt = \int_0^{\infty} |K(e^u)|^2 e^{-2u(1+\alpha)}\, du = \frac{1}{2\pi}\int_{-\infty}^{\infty} \left|\frac{F'(1+\alpha+i\tau)}{1+\alpha+i\tau}\right|^2 d\tau$$

from which we infer that

$$\int_1^{\infty} \frac{|K(t)|^2}{t^{3+2\alpha}}\, dt \ll \sum_{k\in\mathbb{Z}} \frac{1}{k^2+1}\int_{k-\frac12}^{k+\frac12} |F'(1+\alpha+i\tau)|^2\, d\tau. \tag{42}$$

The last integral does not exceed

$$\max_{\substack{\sigma=1+\alpha\\ |\tau-k|\le\frac12}} |F(s)|^2 \int_{k-\frac12}^{k+\frac12} \left|\frac{F'}{F}(1+\alpha+i\tau)\right|^2 d\tau. \tag{43}$$

At this point we appeal to Lemma 6.2 in the form

$$\frac{F'}{F}(s) = \frac{D'(s)}{1+D(s)} + \frac{F_1'}{F_1}(s) + \frac{J'}{J}(s).$$

By (35), the last term is uniformly bounded. It thus contributes $O(1)$ to the integral in (43). In addition, we have

$$\frac{F_1'}{F_1}(s) = -\sum_{p>2} \frac{g(p)\log p}{p^s}.$$

By Lemma 6.1, it hence follows that

$$\begin{aligned}\int_{k-\frac12}^{k+\frac12} \left|\frac{F_1'}{F_1}(1+\alpha+i\tau)\right|^2 d\tau &= \int_{-\frac12}^{\frac12} \left|\frac{F_1'}{F_1}(1+\alpha+ik+i\tau)\right|^2 d\tau\\ &\le 3\int_{-\frac12}^{\frac12} \left|\frac{\zeta'}{\zeta}(1+\alpha+i\tau)\right|^2 d\tau \ll \int_{-\frac12}^{\frac12} \frac{d\tau}{\alpha^2+\tau^2} \ll \frac{1}{\alpha}.\end{aligned}$$

Furthermore

$$\frac{1}{1+D(s)} = 1 + \sum_{\nu=1}^{\infty} (-D(s))^{\nu} \qquad (\sigma > 1).$$

As a Dirichlet series, this expression has coefficients which do not exceed in absolute value those of

$$1 + \sum_{\nu=1}^{\infty} \Big(\sum_{m=1}^{\infty} 2^{-ms} \Big)^{\nu} = \frac{2^s - 1}{2^s - 2} \ll \frac{1}{|s-1|}.$$

Given that $D'(s) \ll 1$ uniformly for $\sigma \geq 1$, we hence deduce from Lemma 6.1 that

$$\begin{aligned}\int_{k-\frac{1}{2}}^{k+\frac{1}{2}} \left| \frac{D'(1+\alpha+i\tau)}{1+D(1+\alpha+i\tau)} \right|^2 \mathrm{d}\tau &\ll \int_{-\frac{1}{2}}^{\frac{1}{2}} \frac{\mathrm{d}\tau}{|1+D(1+\alpha+ik+i\tau)|^2} \\ &\ll \int_{-\frac{1}{2}}^{\frac{1}{2}} \left| \frac{2^{1+\alpha+i\tau}-1}{2^{1+\alpha+i\tau}-2} \right|^2 \mathrm{d}\tau \ll \frac{1}{\alpha}.\end{aligned}$$

It follows from the previous calculations that the integral in (43) is $O(1/\alpha)$ uniformly in k. Substituting in (43), this gives (41) and hence (40).

We can now complete the argument. Since $H(\alpha) \gg 1$, the second term of the bound (37) is of the required order of magnitude. In order to evaluate the first, we use (40) in the form

$$\int_{\sqrt{y}}^{y} \frac{|G(t)|}{t^2} \mathrm{d}t \ll H\Big(\frac{2}{\log y}\Big) \qquad (y \geq e).$$

It follows that

$$\begin{aligned}\int_{e^2}^{x} \frac{|G(t)|}{t^2} \mathrm{d}t &\ll \int_{e^2}^{x} \frac{|G(t)|}{t^2} \int_{t}^{t^2} \frac{\mathrm{d}y}{y \log y} \mathrm{d}t \ll \int_{e^2}^{x^2} \frac{\mathrm{d}y}{y \log y} \int_{\sqrt{y}}^{y} \frac{|G(t)|}{t^2} \mathrm{d}t \\ &\ll \int_{e^2}^{x^2} H\Big(\frac{2}{\log y}\Big) \frac{\mathrm{d}y}{y \log y} = \int_{1/\log x}^{1} \frac{H(\alpha)}{\alpha} \mathrm{d}\alpha.\end{aligned}$$

This concludes the proof of Theorem 6.

Corollary 6.3. *Let g be a multiplicative function of modulus ≤ 1. For $x \geq 2$, $T \geq 2$, set*

$$m(x,T) := \min_{|\tau| \leq T} \sum_{p \leq x} \frac{1 - \Re e\big(g(p)p^{-i\tau}\big)}{p}, \qquad R(x,T) := e^{-m(x,T)/2} + T^{-1/2}.$$

Then we have

(44) $$\sum_{n \leq x} g(n) \ll x\, R(x,T).$$

Remark. It can be shown that $R(x,T) = m(x,T)e^{-m(x,T)} + T^{-1/2}$ is admissible: see Exercise 6.

Proof. By (32), we have

$$\begin{aligned}\alpha F(1+\alpha+i\tau) &\ll \exp\Big\{-\sum_{p\le\exp(1/\alpha)}\frac{1-\Re e\big(g(p)p^{-i\tau}\big)}{p^{1+\alpha}}\Big\}\\ &\ll \exp\Big\{-\sum_{p\le\exp(1/\alpha)}\frac{1-\Re e\big(g(p)p^{-i\tau}\big)}{p}\Big\},\end{aligned}\tag{45}$$

since $\sum_{p\le\exp(1/\alpha)}(p^{-1}-p^{-1-\alpha})\ll 1$.

Hence, we have uniformly for $|\tau|\le T$, $1/\log x\le\alpha\le 1$

$$F(1+\alpha+i\tau)\ll e^{-m(\exp(1/\alpha),T)}\alpha^{-1}\ll e^{-m(x,T)}\alpha(\log x)^2,\tag{46}$$

since

$$\sum_{p\le z}\frac{2}{p}-m(z,T)=\max_{|\tau|\le T}\sum_{p\le z}\frac{1+\Re e\big(g(p)p^{-i\tau}\big)}{p}$$

is a non-decreasing function of z. Using the trivial bound $F(1+\alpha+i\tau)\ll 1/\alpha$ for $|\tau|>T$, it follows that

$$H(\alpha)\ll \alpha e^{-m(x,T)}(\log x)^2+\alpha^{-1}T^{-1/2}.\tag{47}$$

Employing (47) for $1/\log x\le\alpha\le e^{m(x,T)/2}/\log x$ and the trivial bound $H(\alpha)\ll 1/\alpha$ elsewhere, we obtain

$$\int_{1/\log x}^{1}\frac{H(\alpha)}{\alpha}\,\mathrm{d}\alpha\ll R(x,T)\log x.$$

Inserting this in (31), we deduce (44).

We now derive another consequence of Theorem 6, which concerns the mean value of real multiplicative functions. We shall make use of the following lemma.

Lemma 7.1. *Let h be a 2π-periodic function of bounded variation on $[0,2\pi]$, and with mean value*

$$\overline{h}:=\frac{1}{2\pi}\int_0^{2\pi}h(t)\,\mathrm{d}t.$$

For all real numbers τ, w, z such that $\tau\ne 0$, $1<w<z$, we have

$$\begin{aligned}\sum_{w<p\le z}\frac{1}{p}h(\tau\log p) &= \overline{h}\log\Big(\frac{\log z}{\log w}\Big)\\ &\quad+O\Big(\frac{V(h)}{|\tau|\log w}+\big\{M(h)+(1+|\tau|)V(h)\big\}e^{-\sqrt{(\log w)}}\Big),\end{aligned}\tag{48}$$

where $M(h):=\sup_t|h(t)|$, $V(h):=\int_0^{2\pi}|\mathrm{d}h(t)|$.

Proof. We may suppose that $\tau > 0$. We use the prime number theorem in the form $R(t) := \pi(t) - \mathrm{li}(t) \ll t \exp\{-2\sqrt{(\log t)}\}$. By partial summation, we can write the left-hand side of (48) as

$$\int_w^z \frac{h(\tau \log t)}{t \log t}\,\mathrm{d}t + \left[\frac{R(t)h(\tau \log t)}{t}\right]_w^z - \int_w^z R(t)\,\mathrm{d}\Big(\frac{h(\tau \log t)}{t}\Big)$$
$$= \overline{h} \log\Big(\frac{\log z}{\log w}\Big) + \int_{\tau \log w}^{\tau \log z} \big(h(t) - \overline{h}\,\big)\frac{\mathrm{d}t}{t}$$
$$- \int_w^z \frac{R(t)}{t}\,\mathrm{d}h(\tau \log t) + O\Big(M(h)e^{-\sqrt{(\log w)}}\Big).$$

Noting that we have for all a and b

$$\Big|\int_a^b \big(h(t) - \overline{h}\,\big)\,\mathrm{d}t\Big| \ll V(h)$$

and using the second mean value theorem, it follows that the second term in the above sum is $\ll V(h)/(\tau \log w)$. The third is

$$\ll e^{-\sqrt{(\log w)}} \int_{\tau \log w}^{\tau \log w + 2\pi} \sum_{k=0}^{\infty} e^{-\sqrt{((v+2\pi k)/\tau)}}\,|\,\mathrm{d}h(v)| \ll (1 + |\tau|)V(h)e^{-\sqrt{(\log w)}},$$

since the sum over k is $\ll (1 + |\tau|)$. This implies (48).

Theorem 7 (Hall & Tenenbaum, 1991). *Let φ_0 be the unique solution on $]0, 2\pi[$ of the equation $\sin\varphi_0 + (\pi - \varphi_0)\cos\varphi_0 = \frac{1}{2}\pi$. Put $K = \cos\varphi_0 \approx 0.32867$. Then, uniformly for all real multiplicative functions g with $-1 \leq g \leq 1$ and all $x \geq 2$, we have*

$$\sum_{n \leq x} g(n) \ll x \exp\Big\{-K \sum_{p \leq x} \frac{1 - g(p)}{p}\Big\}. \tag{49}$$

Remark. Hall & Tenenbaum in fact show that the quoted constant K is optimal.

Proof. Let $h(\theta) := |\cos(\theta) - K|$. We shall first show that, uniformly for $0 < \alpha \leq 1$, $\tau \in \mathbb{R}$, we have

$$\sum_{p \leq \exp(1/\alpha)} h(\tau \log p)/p \leq (1 - K)\log(1/\alpha) + O\big(\log_2\big(|\tau| + 3\big)\big). \tag{50}$$

We may plainly assume $\tau > 0$. For $0 < \tau \leq \alpha$, relation (50) follows immediately by summation over p from the estimate

$$h(\tau \log p) = 1 - K + O(\tau \log p). \tag{51}$$

When $\alpha < \tau \leq 1$, write $w := \exp(1/\tau)$. We deduce from (51) that

$$\sum_{p \leq w} \frac{h(\tau \log p)}{p} \leq (1 - K) \log_2 w + O(1). \tag{52}$$

Then, putting $z := \exp(1/\alpha)$, and applying Lemma 7.1 to the function h which has mean value

$$\begin{aligned}\overline{h} &= \frac{1}{2\pi} \int_0^{2\pi} |\cos\theta - \cos\varphi_0| \, \mathrm{d}\theta = \frac{2}{\pi}\{\sin\varphi_0 - \varphi_0 \cos\varphi_0\} + \cos\varphi_0 \\ &= 1 - \cos\varphi_0 = 1 - K,\end{aligned}$$

we infer that

$$\sum_{w < p \leq z} \frac{h(\tau \log p)}{p} \leq (1 - K)\big(\log(1/\alpha) - \log_2 w\big) + O(1), \tag{53}$$

so that, given (52), relation (50) is also valid in this case.

Suppose now that $1 < |\tau| \leq \exp(1/\sqrt{\alpha}) - 3$, and choose, in Lemma 7.1, $w := \exp\{\log^2(3 + |\tau|)\} \leq z = \exp(1/\alpha)$. Inequality (53) remains valid, and we deduce (50) by estimating the left-hand side of (52) trivially.

When $|\tau| > \exp(1/\sqrt{\alpha}) - 3$, we have that $\log(1/\alpha) \ll \log_2(3 + |\tau|)$ and the complete sum in (50) can be estimated trivially. We thus obtain that (50) is valid for $0 < \alpha \leq 1$ and any $\tau \in \mathbb{R}$.

Consider now the quantity $\lambda = \lambda(\alpha) \in [0, 2]$, defined by

$$\sum_{p \leq \exp(1/\alpha)} \frac{1 - g(p)}{p} = \lambda \sum_{p \leq \exp(1/\alpha)} \frac{1}{p}. \tag{54}$$

The inequality

$$\begin{aligned}\Re e\,\big(g(p)p^{i\tau}\big) &= g(p)\big(\cos(\tau \log p) - K\big) + K g(p) \\ &\leq |\cos(\tau \log p) - K| + K g(p) = h(\tau \log p) + K g(p)\end{aligned}$$

yields, after dividing by p, summing over $p \leq \exp(1/\alpha)$ and using (50) and (54), that

$$\Re e \sum_{p \leq \exp(1/\alpha)} g(p)/p^{1+i\tau} \leq (1 - K\lambda) \log(1/\alpha) + O\big(\log_2(|\tau| + 3)\big). \tag{55}$$

We now embark on the last phase of the proof. If $F(s)$ denotes the Dirichlet series appearing in (30), then for $s = 1 + \alpha + i\tau$ we have that

$$F(s) \ll \exp\Big\{\Re e \sum_p g(p)p^{-s}\Big\} \ll \exp\Big\{\Re e \sum_{p \leq \exp(1/\alpha)} g(p)p^{-1-i\tau}\Big\}, \tag{56}$$

by virtue of the estimate

$$\sum_{p \leq \exp(1/\alpha)} (1 - p^{-\alpha})/p + \sum_{p > \exp(1/\alpha)} p^{-1-\alpha} \ll 1,$$

which follows from the prime number theorem by partial summation.

We deduce from (55) and (56) that

$$F(s) \ll \alpha^{K\lambda-1} \log^B(|\tau|+3)$$

where B is some absolute constant. Montgomery's function $H(\alpha)$ defined by (30) thus satisfies

$$H(\alpha)^2 \ll \alpha^{2K\lambda-2} \sum_{k\in\mathbb{Z}} \frac{\log^B(|k|+4)}{k^2+1}$$

and hence

$$H(\alpha) \ll \alpha^{K\lambda-1}. \tag{57}$$

Let $S := \sum_{p\le x} (1-g(p))/p$. For $1/\log x \le \alpha \le 1$, we have

$$\begin{aligned}\lambda \log(1/\alpha) + O(1) = \sum_{p\le \exp(1/\alpha)} \frac{1-g(p)}{p} &\ge \sum_{p\le x} \frac{1-g(p)}{p} - \sum_{\exp(1/\alpha)<p\le x} \frac{2}{p} \\ &\ge S - 2\log_2 x + 2\log(1/\alpha) + O(1)\end{aligned}$$

from which

$$\alpha^\lambda \ll \alpha^2 e^{-S} \log^2 x.$$

Substituting in (57), we obtain

$$H(\alpha) \ll \alpha^{2K-1} e^{-KS} (\log x)^{2K},$$

and thus, by Theorem 6,

$$G(x) \ll \frac{x}{\log x} e^{-KS} (\log x)^{2K} \int_{1/\log x}^{1} \alpha^{2K-2}\, \mathrm{d}\alpha \ll x e^{-KS},$$

as required.

§ 4.4 The Erdős–Kac theorem

The problem of the weak convergence of the frequencies

$$\nu_N\{n : f(n) \le A(N) + zB(N)\} \qquad (N \to \infty) \tag{58}$$

is more delicate than that of the existence of a limit law. In the case of an additive function f, it is easy to appreciate the reason for this since the corresponding characteristic functions take the values

$$\frac{1}{N} \sum_{n\le N} \exp\{i\tau(f(n) - A(N))/B(N)\} \tag{59}$$

and we need to estimate the average over the first N integers of a multiplicative function *which depends on* N. The uniformity required for such a purpose does not follow from Theorems 2, 3 or 4, and the application of Theorem 6 depends on sometimes delicate estimates for the function $H(\alpha)$—cf. the Notes to §4.3 for references to the literature.

Here we shall treat the historically first example of a normalised distribution function. It concerns the arithmetic function $\omega(n)$, but an identical result also holds for $\Omega(n)$. The required effective mean value estimate is provided by Theorem II.5.3, of Selberg–Delange type. The link between characteristic functions and distribution functions is made explicit by the Berry–Esseen inequality (Theorem II.7.14). We let

$$\Phi(y) := \frac{1}{\sqrt{(2\pi)}} \int_{-\infty}^{y} e^{-t^2/2}\,\mathrm{d}t$$

denote the normal distribution function, with characteristic function

$$\varphi(\tau) := \int_{-\infty}^{\infty} e^{i\tau y}\,\mathrm{d}\Phi(y) = e^{-\tau^2/2}.$$

Theorem 8 (Erdős & Kac, 1939; Rényi & Turán, 1958). *We have uniformly for $N \geq 2$, $y \in \mathbb{R}$,*

$$\nu_N\Big\{n : \omega(n) \leq \log_2 N + y\sqrt{(\log_2 N)}\Big\} = \Phi(y) + O\Big(\frac{1}{\sqrt{(\log_2 N)}}\Big). \tag{60}$$

Proof. Let $F_N(y)$ denote the left-hand side of (60), and let $\varphi_N(\tau)$ be the corresponding characteristic function, viz.

$$\varphi_N(\tau) := \frac{1}{N}\sum_{n \leq N} \exp\Big\{\frac{i\tau}{\sqrt{(\log_2 N)}}\big(\omega(n) - \log_2 N\big)\Big\}.$$

By Theorem II.6.1, we have uniformly for $N \geq 2$, $t \in \mathbb{R}$,

$$\frac{1}{N}\sum_{n \leq N} e^{it\omega(n)} = A(e^{it})\big(\log N\big)^{e^{it}-1} + O\big((\log N)^{\cos t - 2}\big) \tag{61}$$

where $A(z)$ is an entire function of z with $A(1) = 1$.

Write $T := \sqrt{(\log_2 N)}$, and choose $t = \tau/T$. Noting that $\cos t - 1 \leq -2(t/\pi)^2$ for $|t| \leq 1$, we obtain

$$\varphi_N(\tau) \ll e^{-2\tau^2/\pi^2} \qquad (|\tau| \leq T). \tag{62}$$

We shall use this estimate when $T^{1/3} < |\tau| \leq T$.

Since

$$A(e^{it}) = 1 + O(t), \qquad e^{it} - 1 = it - \tfrac{1}{2}t^2 + O(t^3) \qquad (|t| \leq 1),$$

we can write, on the other hand, using (61),

$$\frac{1}{N}\sum_{n\le N} e^{i\tau\omega(n)/T} = \big(1+O(\tau/T)\big)\exp\big\{i\tau T - \tfrac12\tau^2 + O(\tau^3/T)\big\} + O\Big(\frac{1}{\log N}\Big)$$

from which

$$\varphi_N(\tau) = e^{-\tau^2/2}\Big\{1+O\Big(\frac{|\tau|+|\tau|^3}{T}\Big)\Big\} + O\Big(\frac{1}{\log N}\Big) \qquad (|\tau|\le T^{1/3}). \tag{63}$$

We shall use this estimate for $1/\log N < |\tau| \le T^{1/3}$.

When $|\tau|\le 1/\log N$, we simply insert the trivial estimate $e^{iy} = 1 + O(y)$ $(y\in\mathbb{R})$ in the definition of $\varphi_N(\tau)$. This yields

$$\varphi_N(\tau) = 1 + O\Big(\frac{|\tau|}{TN}\sum_{n\le N}|\omega(n)-\log_2 N|\Big) = 1 + O(|\tau|), \tag{64}$$

bounding the sum over n by successive applications of the Cauchy–Schwarz and Turán–Kubilius inequalities.

Now consider the Berry–Esseen inequality

$$\sup_{y\in\mathbb{R}}|F_N(y)-\Phi(y)| \ll \frac{1}{T} + \int_{-T}^{T}\Big|\varphi_N(\tau) - e^{-\tau^2/2}\Big|\frac{\mathrm{d}\tau}{|\tau|}.$$

We split the integral into three parts I_1, I_2, I_3 corresponding to the respective ranges of integration

$$T^{1/3} < |\tau| \le T, \qquad 1/\log N < |\tau| \le T^{1/3} \quad\text{and}\quad 0\le|\tau|\le 1/\log N.$$

By (62), (63), (64), we obtain

$$I_1 \ll \int_{T^{1/3}}^{\infty} e^{-2\tau^2/\pi^2}\frac{\mathrm{d}\tau}{\tau} \ll \frac{1}{T},$$

$$I_2 \ll \int_{-\infty}^{\infty}\Big(\frac{1+\tau^2}{T}\Big)e^{-\tau^2/2}\,\mathrm{d}\tau + \frac{1}{\log N}\int_{1/\log N}^{T^{1/3}}\frac{\mathrm{d}\tau}{\tau} \ll \frac{1}{T},$$

and

$$I_3 \ll \int_{-1/\log N}^{1/\log N}\mathrm{d}\tau \ll \frac{1}{T}.$$

This completes the proof.

Note that the error term in formula (60) is optimal. Indeed, we deduce from Theorem II.6.4 for $k := \left[\log_2 N\right]$

$$\nu_N\{n : \omega(n) = k\} \sim \frac{1}{\log N}\frac{(\log_2 N)^{k-1}}{(k-1)!} \sim \frac{1}{\sqrt{(2\pi \log_2 N)}}. \tag{65}$$

Now, if $R(N)$ denotes the supremum over $y \in \mathbb{R}$ of the error term in (60), we obtain, expressing (65) as a difference of two frequencies of type (60), that

$$\frac{1+o(1)}{\sqrt{(2\pi \log_2 N)}} = \Phi\Big(\frac{\theta}{\sqrt{(\log_2 N)}}\Big) - \Phi\Big(\frac{\theta - 1/2}{\sqrt{(\log_2 N)}}\Big) + O\big(R(N)\big)$$

with $\theta := k - \log_2 N$. Hence

$$R(N) \gg 1/\sqrt{(\log_2 N)}.$$

Notes

§ **4.1.** For a purely probabilistic proof (using Kolmogorov's theorem) of the Erdős–Wintner theorem, see Novoselov (1964) and Babu (1973). Another probabilistic interpretation is given in the excellent review by Galambos (1970).

The original proof of Erdős for the sufficient condition of Theorem 1 was obtained in three stages (1935/37/38). The necessary condition was obtained by Erdős & Wintner in 1939.

The argument presented here to obtain the continuity criterion for the limit law directly (i.e. without recourse to Theorem 2.5(a)) is a variant of a proof of Szüsz (1974), completed by himself in Elliott's book (1985), pp. 437–439. Another proof was given by Elliott (1979), pp. 220–224.

§ **4.2.** Theorem 3 has not been published by Delange but has been the object of several oral expositions.

Another proof of part (i) of Theorem 2 is due to Daboussi (1982, 1989).

§ **4.3.** Theorem 6 was stated and proved by Montgomery (1978b) for a *completely* multiplicative function g. Of course the extension to the general case is purely technical. In the same article Montgomery remarks that Theorem 6 enables one to establish estimates such as that in Theorem 7. See also Montgomery & Vaughan (1994). Evaluations of this type are referred to as *effective mean value estimates*, or *quantitative theorems* for mean values. They are, characteristically, uniform with respect to the summed function (in a certain class), which in particular can depend upon x. Theorem 3.5 is a simple example of such a result. The estimate of Theorem 7 can be extended to complex-valued functions g such that $|g| \leq 1$ and, for all p, $g(p)$ lies in the ellipse

$$\Im m\,(e^{-i\varphi}z)^2 \leq \delta^2\big(1 - \Re e\,(e^{-i\varphi}z)^2\big)$$

where δ, φ are arbitrary parameters such that $0 \leq \delta < 1$, $0 \leq \varphi < \pi$. Hall & Tenenbaum (1991) give, for each pair (δ, φ), the best constant $K(\delta, \varphi) > 0$ such that

$$G(x) \ll x \exp\Big\{ - K(\delta, \varphi) \sum_{p \leq x} \frac{1 - \Re e\, g(p)}{p} \Big\}.$$

For more general statements and the techniques necessary to prove them, see Elliott (1980), chapter 19, where the estimates of Halász (1971) are described and refined. Elliott proves the two following results; the first corresponds to Theorem 7, the second to an effective version of Delange's theorem (Theorem 2).

Theorem 9. *Let g be a completely multiplicative function, which for some suitable constant $\lambda > 0$, satisfies $g(p) = 0$ or $\lambda \leq |g(p)| \leq 2 - \lambda$ for all p. Let θ_p be the argument of $g(p)$ when $g(p) \neq 0$. Suppose that there exist real numbers θ_0, δ, $|\theta_0| \leq \pi$, $\delta > 0$, such that $|e^{i\theta_p} - e^{i\theta_0}| \geq \delta$ $\big(g(p) \neq 0\big)$. Then we have*

$$\Big|\sum_{n \leq x} g(n)\Big| \ll x \exp\Big\{ \sum_{p \leq x} \frac{|g(p)| - 1}{p} - c \sum_{p \leq x} \frac{|g(p)| - \Re e g(p)}{p} + 2\lambda \sum_{\substack{p \leq x \\ g(p)=0}} \frac{1}{p} \Big\}$$

where c is a constant only depending on δ and λ.

Theorem 10. *Let g be a completely multiplicative function which for all p satisfies $|g(p) - 1| \leq \eta \leq \eta_0 < 1$. Then we have*

$$\Big|\sum_{n \leq x} g(n) - Ax\Big| \ll \eta x \exp\Big\{ - \sum_{p \leq x} \frac{1 - \Re e\big(g(p)\big)}{p} \Big\} + x\Big(e^{-c_1/\eta} + (\log x)^{-c_2}\Big) \exp\Big\{ \sum_{p \leq x} \frac{|g(p)| - 1}{p} \Big\}$$

where $A := \exp\big\{ \sum_{p \leq x} (g(p) - 1)/p \big\}$ and c_1, c_2 are constants depending only on η_0.

Theorem 4 yields a complete solution to the problem of the limiting distribution of translated additive functions, namely the weak convergence of the d.f.'s

$$\nu_N\{n : f(n) \le A(N) + z\} \qquad (N \to \infty). \tag{66}$$

The following result, independently obtained by Elliott & Ryavec (1971), Levin & Timofeev (1971), Delange, and Kubilius (unpublished), is proved in Elliott (1979), chapter 7.

Theorem 11. *Let f be a real additive function. A necessary and sufficient condition for there to exist a function $A(N)$ such that the frequencies (66) converge weakly to a d.f. is that, for a suitable constant c,*

$$\sum_p \frac{1}{p} \min(1, |h(p)|^2) < \infty,$$

with $h(n) := f(n) - c\log n$ $(n \ge 1)$. In this case, we can choose

$$A(N) := \sum_{p\le N,\, |h(p)|\le 1} \frac{h(p)}{p} + c\log N.$$

For this function $A(N)$ the characteristic function of F is

$$\frac{1}{1+ic\tau} \prod_{|h(p)|>1} w_p(\tau) \prod_{|h(p)|\le 1} w_p(\tau) e^{-i\tau h(p)/p}$$

where $w_p(\tau) := \left(1 - p^{-1}\right) \sum_{\nu=0}^{\infty} e^{-i\tau h(p^{-\nu})} p^{-\nu}$. Moreover, F is of pure type. It is continuous if and only if

$$\sum_{f(p)\ne 0} \frac{1}{p} = \infty.$$

§ **4.4.** It is clear that the Erdős–Kac theorem can also be deduced from estimates of the frequencies $\nu_N\{n : \omega(n) = k\}$ by summing over appropriate values of k, as suggested in Exercise II.6.4. These estimates have been obtained in Chapter II.6, using the same basic tool as the one used here: the Selberg–Delange method.

The original proof of Erdős & Kac (1939) did not give an explicit error term. LeVeque conjectured the bound $O\big((\log_2 N)^{-1/2}\big)$ in 1949, and showed that $O\big(\log_3 N(\log_2 N)^{-1/4}\big)$ was admissible. This result was improved by Kubilius in 1956, and LeVeque's conjecture was finally established by Rényi & Turán in 1958, employing the same method as in this chapter. By using the estimates for $\sum_{n\le x} z^{\omega(n)}$ for complex z (cf. Theorem II.6.1), Delange (1959) obtained a more precise result: we have

$$\nu_N\{n:\omega(n)\le \log_2 N + y\surd(\log_2 N)\}$$
$$= \Phi(y) + \frac{e^{-y^2/2}}{\surd(2\pi\log_2 N)}\{a - \tfrac{1}{6}y^2 + g\big(\log_2 N + y\surd(\log_2 N)\big)\} + O\Big(\frac{1}{\log_2 N}\Big)$$

where $a \approx 0.40516$ and where $g(t)$ denotes the fractional part of t.

Exercises

1. Show that the three types of distribution function for an additive function can actually occur. [*Hint: see Exercises 2.4–2.7.*]

2. *On the distribution of multiplicative functions.*

Throughout this exercise, $g(n)$ denotes a real multiplicative function.

(a) Show (using for example Theorem I.3.11) that we have

$$\mathbf{d}\{n:g(n)=0\} = 1 - \prod_p (1-p^{-1}) \sum_{\nu\ge 0,\, g(p^\nu)\ne 0} p^{-\nu},$$

where the infinite product is to be interpreted as 0 when it is divergent. In particular, deduce that $g(n)$ vanishes pp if, and only if, $\sum_{g(p)=0} 1/p = \infty$.

(b) Suppose that $g(n)$ does not vanish pp and set

$$g^*(n) := \prod_{p^\nu \| n,\, g(p^\nu)\ne 0} g(p^\nu).$$

Show that g and g^* simultaneously possess limiting distributions.

(c) Suppose that $g(n)\ne 0$ for $n\ge 1$. Let E be a set of prime numbers such that $\sum_{p\in E} 1/p < \infty$. Set $h(n) := \prod_{p^\nu\|n,\, p\notin E} g(p^\nu)$, $A := \{a : p|a \Rightarrow p\in E\}$, and let a denote generically an integer which belongs to A. Show that $\sum_{a\in A} 1/a < \infty$. Show also that for all $N\ge 1$, $z\in\mathbb{R}$,

$$\nu_N\{n:g(n)\le z\} = \frac{1}{N}\sum_{a\le N}\sum_{a'\le N/a}\mu(a')\sum_{m\le N/aa',\, g(a)h(m)\le z} 1.$$

Deduce that if h has a limit distribution function H, then g has a limit distribution function G, given almost everywhere by

$$G(z) = \prod_{p\in E} (1 - p^{-1}) \sum_{a\in A} a^{-1} H_a(z/g(a))$$

where $H_a(z) := H(z)$ if $g(a) > 0$, $H_a(z) := 1 - H(z)$ if $g(a) < 0$.

(d) As a special case of the result proved in the previous question, recover the theorem of Galambos & Szüsz (1986): *Let g be a multiplicative function such that $g(n) \neq 0$ for $n \geq 1$, and $\sum_{g(p)<0} 1/p < \infty$. Then g has a limit law which is continuous at the origin, if, and only if, the additive function* $\log|g(n)|$ *satisfies the conditions of the Erdős–Wintner theorem.*
[*A necessary and sufficient condition for the existence of a limiting d.f. for strongly multiplicative real functions has been given by Galambos (1971).*]

3. *Daboussi's theorem (1974).* Write $e(t) = e^{2\pi i t}$ $(t \in \mathbb{R})$. An arithmetic function f is said to have the property (D) if $n \mapsto f(n)e(\alpha n)$ has zero mean value for all $\alpha \in \mathbb{R}\smallsetminus\mathbb{Q}$.

(a) Show that if g has property (D) and satisfies $\sum_{n\leq x} |g(n)| \ll x$, and $\sum_{n=1}^{\infty} |h(n)|/n < \infty$, then $h * g$ has property (D).

(b) Let f be a multiplicative function with modulus ≤ 1. Show that if $g(n) = \mu(n)^2 f(n)$ has property (D), then the same is true for f.

(c) For the rest of this exercise, define, for $y \geq 2$, the completely multiplicative functions u_y, v_y by

$$u_y(p) := 1 \quad (p > y), \qquad u_y(p) := 0 \quad (p \leq y), \qquad v_y(p) := 1 - u_y(p).$$

Show that $u_y := \mathbf{1} * v_y\mu$, and deduce that u_y has property (D).

(d) Let f be a multiplicative function of modulus ≤ 1 satisfying $f(p^\nu) = 0$ whenever $\nu \geq 2$. Show that $f = fu_y * fv_y$. Deduce by a suitable application of the Cauchy–Schwarz inequality that, for $x \geq 1$, we have

$$\Big|\frac{1}{x} \sum_{n\leq x} f(n)e(\alpha n)\Big|^2 \leq A(x)\{B_1(x) + B_2(x)\}$$

with $A(x) := (1/x)\sum_{n\leq x} u_y(n)$ and

$$B_1(x) := \frac{1}{x} \sum_{d\leq x} |f(d)v_y(d)|^2 \sum_{n\leq x/d} u_y(n)$$

$$B_2(x) := \sum_{1\leq d,d'\leq x,\, d\neq d'} fv_y(d)\overline{fv_y(d')}\,\frac{1}{x} \sum_{n\leq x/d',\, n\leq x/d} u_y(n)e\big(\alpha n(d - d')\big).$$

(e) Show that $\lim_{x\to\infty} A(x) = \prod_{p\le y}(1-1/p)$. Show that $B_1(x) \le 1$. Using the fact that $fv_y(d)$ is non-zero for at most a finite number of integers d, show that $B_2(x) \to 0$ as $x \to \infty$.

(f) Letting $y \to \infty$, prove Daboussi's theorem: *Every multiplicative function of modulus ≤ 1 has property (D).*

[*The proof described here, though different from Daboussi's original proof (see Daboussi & Delange, 1982), is also due to Daboussi (1989).*]

4. *An estimate of Montgomery & Vaughan (1994).*

Let $M(n,x) := \sum_{d|n,\, d\le x} \mu(d)$.

(a) Using Lemma 7.1, show that uniformly for $0 < \alpha \le 1$, $\tau \ne 0$, one has

$$-\sum_{p>\exp(1/|\tau|)} \frac{\cos(\tau \log p)}{p^{1+\alpha}} \le 2\log_2(3+|\tau|) + O(1),$$

$$\sum_{p>\exp(1/|\tau|)} \frac{|\cos(\tau \log p)|}{p^{1+\alpha}} \le \frac{2}{\pi}\log\frac{1}{\alpha} + (2-\frac{4}{\pi})\log_2(3+|\tau|) + O(1).$$

(b) Show, for each integer $n \ge 1$, that

$$\Big|\prod_{p|n}(1-p^{-1-\alpha-i\tau})\Big| \ll \alpha^{-1/\pi}\Big(\log(3+|\tau|)\Big)^{2-2/\pi} \qquad (0<\alpha\le 1,\ \tau\ne 0).$$

(c) Using Theorem 6, deduce that

$$\max_{n\ge 1} |M(n,x)| \ll x\big(\log x\big)^{(1/\pi)-1} \qquad (x\ge 2). \tag{67}$$

(d) Prove that the left-hand side of (67) is $\gg x/\log x$. For $x \ge 2$, evaluate $\int_1^\infty M(n_x,t)\,\mathrm{d}t/t^{2+i}$ with $n_x := \prod_{p\le x,\, \cos\log p \le 0} p$, and deduce that the exponent $1/\pi$ in (67) is sharp.

[*By the method of Hall & Tenenbaum (1991), it can actually be shown that $\ll$ in (67) may be replaced by $\asymp$.*]

5. *A theorem of Halász (1971).*

Let E be a set of prime numbers. Write $E(x) := \sum_{p\le x,\, p\in E} 1/p$, and consider the arithmetic function $\Omega(n;E) := \sum_{p^\nu \| n,\, p\in E} \nu$.

(a) Show, using Theorem 9, that if $\delta > 0$, $\delta \le |z| \le 2-\delta$, then one has, for some constant $c_1 > 0$ depending on δ only,

$$\sum_{n\le x} z^{\Omega(n;E)} \ll x\exp\big\{\big(|z|-1-c_1(|z|-\Re e z)\big)E(x)\big\}.$$

(b) Show, using Theorem 10, that if $|z-1| \leq \frac{1}{2}$, then one has, for absolute constants $c_2 > 0$ and $c_3 > 0$, that

$$\Big|\frac{1}{x}\sum_{n\leq x} z^{\Omega(n;E)} - e^{(z-1)E(x)}\Big|$$
$$\ll |z-1|e^{(\Re e z-1)E(x)} + e^{(|z|-1)E(x)}\Big\{e^{-c_2/|z-1|} + (\log x)^{-c_3}\Big\}.$$

(c) Deduce from (a) and (b), by employing Cauchy's formula in a manner similar to that in the proof of Theorem II.6.3, that one has uniformly for $E(x) \geq 2$, $\delta E(x) \leq m \leq (2-\delta)E(x)$,

$$\sum_{n\leq x,\, \Omega(n;E)=m} 1 = x\frac{E(x)^m}{m!}e^{-E(x)}\Big\{1 + O\Big(\frac{1}{\surd(E(x))} + \frac{|m-E(x)|}{E(x)}\Big)\Big\}.$$

[*For more precise results, see Halász (1972), Sárközy (1977b).*]

6. *On the effective Halász' theorem.* Let g be a multiplicative function with modulus at most 1. Put

$$\mu(x,\tau) := \sum_{p\leq x} \Re e\,\big(g(p)/p^{1+i\tau}\big), \qquad \mu^*(x,T) := \sup_{|\tau|\leq T} \mu(x,\tau),$$

so that, with the notation of Corollary 6.3, $m(x,T) = \log_2 x - \mu^*(x,T) + O(1)$.

(a) Show that, for any $\tau \in [0,T]$, one has

$$\mu(x,\tau) = \tfrac{1}{2}\log x \int_{\tau-1/\log x}^{\tau+1/\log x} \mu(x,\theta)\,\mathrm{d}\theta + O(1).$$

Deduce that $\mu^*(y,T) \leq \mu^*(x,T) + O(1)$ for $0 \leq y \leq x$.

(b) Show that $(1/x)\sum_{n\leq x} g(n) \ll m(x,T)e^{-m(x,T)} + 1/\surd T$.

7. *Convergence to the Gaussian law.*

Throughout this exercise, we set $y = y(x) = x^{1/\log_2 x}$.

(a) Let $\mathcal{A}(x)$ denote the class of those multiplicative functions G which take values in the unit disc and which satisfy $G(p^\nu) = 1$ for all $p > y$ and any $\nu \geq 1$. Show that, uniformly for $x \geq 2$, $G \in \mathcal{A}(x)$, we have

$$\sum_{n\leq x} G(n) = x\prod_{p\leq y}(1-p^{-1})\sum_{\nu=0}^{\infty} G(p^\nu)p^{-\nu} + O(x/\log x).$$

(b) Let $\{g_x : x \geq 2\}$ be a family of multiplicative functions of modulus at most 1 which satisfies $\lim_{x\to\infty} \sum_{y<p\leq x} |g_x(p) - 1|/p = 0$. Introducing the functions $G_x(n) := \prod_{p\leq y,\, p^\nu \| n} g_x(p^\nu)$ and using question (a), show that

$$\sum_{n\leq x} g_x(n) = x \prod_{p\leq x} (1 - p^{-1}) \sum_{\nu=0}^{\infty} g_x(p^\nu)p^{-\nu} + o(x).$$

(c) Let f be a real additive function which satisfies the following properties as $x \to \infty$:

(i) $D(x)^2 := \sum_{p\leq x} f(p)^2/p \to \infty$, (ii) $\sum_{p\leq x} f(p)/p = o\big(D(x)\big)$,

(iii) $\sum_{y<p\leq x} |f(p)|/p = o\big(D(x)\big)$, (iv) $\max_{p\leq x} |f(p)| = o\big(D(x)\big)$.

For $\tau \in \mathbb{R}$, put $g_x(n, \tau) := \exp\{i\tau f(n)/D(x)\}$. Show that for fixed τ and all $p \leq x$ we have

$$1 + \frac{g_x(p,\tau) - 1}{p} = \exp\Big\{\frac{i\tau f(p)}{pD(x)} - \frac{\tau^2 f(p)^2}{2pD(x)^2} + O\Big(\frac{|f(p)|^3}{pD(x)^3} + \frac{|f(p)|^2}{p^2 D(x)^2}\Big)\Big\}.$$

Deduce the existence of $\lim_{x\to\infty} x^{-1} \sum_{n\leq x} g_x(n,\tau)$ and find its value. Show that for all $z \in \mathbb{R}$ we have

$$\lim_{x\to\infty} x^{-1}\big|\{n \leq x : f(n) \leq zD(x)\}\big| = \frac{1}{\sqrt{(2\pi)}} \int_{-\infty}^{z} e^{-t^2/2}\,\mathrm{dt}.$$

III.5

Integers free of large prime factors. The saddle-point method

§ 5.1 Introduction. Rankin's method

In this chapter we shall describe the various methods which have been developed in the literature for the evaluation of the quantity

$$\Psi(x,y) := \big|\{n \le x : P^+(n) \le y\}\big|. \tag{1}$$

This is indeed crucial in numerous arithmetical problems, and the techniques that have been implemented to tackle it are instructive from several points of view.

$\Psi(x,y)$ is the summatory function of the multiplicative function

$$\chi(n,y) := \begin{cases} 1 & (P^+(n) \le y) \\ 0 & (P^+(n) > y). \end{cases} \tag{2}$$

The associated Dirichlet series is the leading factor of the Euler product for the Riemann zeta function, i.e.

$$\zeta(s,y) := \prod_{p \le y} \left(1 - p^{-s}\right)^{-1} = \sum_{n=1}^{\infty} \chi(n,y) n^{-s}. \tag{3}$$

In 1938, in an article devoted to large differences between consecutive primes, Rankin devised a technique for finding an upper bound for $\Psi(x,y)$ which rests essentially on the fact that the abscissa of convergence for $\zeta(s,y)$ is $\sigma = 0$. Indeed, for any $\sigma > 0$, we have that

$$\Psi(x,y) \le \sum_{n \ge 1,\, P^+(n) \le y} \left(\frac{x}{n}\right)^{\sigma} = x^{\sigma} \zeta(\sigma, y). \tag{4}$$

We then obtain an explicit upper bound by choosing σ optimally. This procedure, simple but remarkably efficient, can be generalised in several ways. It is now known under the title of *Rankin's method*. We apply it, in a rudimentary form, in the proof of Theorem 1 below.

The quantity

$$u := \frac{\log x}{\log y} \qquad (x \geq y \geq 2) \tag{5}$$

plays a pivotal role in the study of the asymptotic behaviour of $\Psi(x, y)$. In the course of this chapter we make systematic use of the notation (5).

Theorem 1. *We have*

$$\Psi(x, y) \ll xe^{-u/2} \qquad (x \geq y \geq 2). \tag{6}$$

Proof. We may assume that $y \geq 11$: otherwise $\Psi(x, y) \leq \Psi(x, 7) \ll (\log x)^4$, whereas the upper bound in (6) has order a power of x. This being the case, we have for all $\alpha \geq 0$ that

$$\Psi(x, y) \leq x^{3/4} + \sum_{n \leq x} \left(\frac{n}{x^{3/4}}\right)^{\alpha} \chi(n, y). \tag{7}$$

For the choice $\alpha := 2/(3\log y)$, the multiplicative function $n \mapsto n^{\alpha}\chi(n, y)$ satisfies the conditions of Corollary 3.5.1. We obtain

$$\sum_{n \leq x} n^{\alpha}\chi(n, y) \ll x \prod_{p \leq y} \left(1 + \frac{p^{\alpha} - 1}{p}\right) \leq x \exp\left\{O\left(\sum_{p \leq y} \frac{\alpha \log p}{p}\right)\right\} \ll x.$$

Substituting in (7) and noting that $\frac{1}{4}\log x > \frac{1}{2}u$ for $y \geq 11$, the result follows.

The above proof is clearly not optimal. A more thorough study of Rankin's bound (4) enabled de Bruijn (1966) to obtain the following result, which provides an asymptotic formula for $\log \Psi(x, y)$ as $y \to \infty$. We set

$$\begin{aligned} Z &:= \frac{\log x}{\log y} \log\left(1 + \frac{y}{\log x}\right) + \frac{y}{\log y} \log\left(1 + \frac{\log x}{y}\right) \\ &= u \int_0^1 \log\left(1 + \frac{y}{v \log x}\right) dv. \end{aligned} \tag{8}$$

Theorem 2 (de Bruijn). *We have, uniformly for $x \geq y \geq 2$,*

$$\log \Psi(x, y) = Z\left\{1 + O\left(\frac{1}{\log y} + \frac{1}{\log_2 2x}\right)\right\}. \tag{9}$$

Proof. We may plainly assume x to be sufficiently large. Denote by

$$\varphi_y(\sigma) := \sum_{p \leq y} \frac{\log p}{p^{\sigma} - 1} \tag{10}$$

the quantity $-\zeta'(\sigma, y)/\zeta(\sigma, y)$ $(0 \leq \sigma \leq 1)$. The optimal value of the parameter σ in (4) is the unique solution $\alpha = \alpha(x, y)$ to the equation

$$\varphi_y(\alpha) = \log x.$$

The first step of the proof consists in finding an explicit approximation for $\alpha(x, y)$. To this end, we appeal to the prime number theorem so as to evaluate $\varphi_y(\sigma)$ uniformly for $y \geq 2$, $0 < \sigma \leq 1$. We obtain

$$\varphi_y(\sigma) = \frac{y}{y^\sigma - 1} \cdot \frac{1 - y^{-(1-\sigma)}}{1 - \sigma} \Big\{1 + O\Big(\frac{1}{\log y}\Big)\Big\}. \tag{11}$$

We omit the proof, somewhat technical, of this estimate, but indicate the main steps of the reasoning in Exercise 1. We can deduce from (11) the estimate

$$\alpha(x, y) = \beta(x, y)\Big\{1 + O\Big(\frac{\log_2 y}{\log y}\Big)\Big\} \qquad (x \geq y \geq 2) \tag{12}$$

where $\beta = \beta(x, y)$ is the solution of the equation $y/(y^\beta - 1) = \log x$, namely

$$\beta(x, y) := \frac{\log(1 + y/\log x)}{\log y}. \tag{13}$$

To show (12), we first observe that $\frac{1}{2} \leq (1 - y^{-(1-\sigma)})/(1 - \sigma) \leq \log y$ $(0 \leq \sigma \leq 1)$, from which we get, inserting in (11), that $\alpha = \beta + O(\log_2 y/\log y)$. This yields (12) when $\beta > \frac{1}{2}$. If now $\beta \leq \frac{1}{2}$, we substitute the above estimate back in (11) to obtain

$$\log x = \frac{y}{y^\beta - 1} = \frac{y}{(1 - \beta)(y^\alpha - 1)} \Big\{1 + O\Big(\frac{\log_2 y}{\log y}\Big)\Big\},$$

from which we deduce $y^\alpha = y^\beta\{1 + O(\beta \log_2 y)\}$, so that (12) also holds in this case.

The upper bound implied in Theorem 2 will follow by choosing $\sigma = \beta$ in (4). We note that

$$Z = \beta \log x + \frac{y}{\log y} \log \Big(\frac{1}{1 - y^{-\beta}}\Big). \tag{14}$$

By (4), we have

$$\Psi(x, y) \leq x^\beta \zeta(\beta, y) = \zeta(1, y) \exp\Big\{\beta \log x + \int_\beta^1 \varphi_y(\sigma)\, \mathrm{d}\sigma\Big\}. \tag{15}$$

When $y \leq (\log x)^2$, we certainly have $\beta \leq \frac{2}{3}$. Using (11), we can then write

$$\int_\beta^1 \varphi_y(\sigma)\, \mathrm{d}\sigma = \Big\{1 + O\Big(\frac{1}{\log y}\Big)\Big\} \frac{y}{\log y} \int_\beta^{2/3} \frac{\log y}{y^\sigma - 1} \frac{\mathrm{d}\sigma}{1 - \sigma} + O\Big(\int_{2/3}^1 y^{1-\sigma} \log y\, \mathrm{d}\sigma\Big).$$

The second error term is $\ll y^{1/3}$. The integral over σ equals

$$\begin{aligned}\int_\beta^{2/3} \frac{\log y}{y^\sigma - 1}\,\mathrm{d}\sigma + O\Big(\int_\beta^{2/3} \frac{\sigma \log y}{y^\sigma - 1}\,\mathrm{d}\sigma\Big) \\ &= \log\Big(\frac{1 - y^{-2/3}}{1 - y^{-\beta}}\Big) + O\Big(\frac{1}{\log y}\int_{\beta \log y}^{+\infty} \frac{t\,\mathrm{d}t}{e^t - 1}\Big) \\ &= \log\Big(\frac{1}{1 - y^{-\beta}}\Big) + O\Big(y^{-2/3} + \frac{1 + \beta \log y}{y^\beta \log y}\Big).\end{aligned}$$

It follows that

$$\int_\beta^1 \varphi_y(\sigma)\,\mathrm{d}\sigma = \Big\{1 + O\Big(\frac{1}{\log y}\Big)\Big\}\frac{y}{\log y}\log\Big(\frac{1}{1 - y^{-\beta}}\Big) + O\Big(\frac{y(1 + \beta \log y)}{y^\beta (\log y)^2}\Big).$$

We claim that the second error term is $\ll Z/\log y$. Indeed, when $y \le \log x$, we have $\beta \log y \ll 1$, from which $Z \gg y/\log y$, and when $\log x < y \le (\log x)^2$, we have $\beta \log y \gg 1$, from which

$$\frac{y(1 + \beta \log y)}{y^\beta (\log y)^2} \asymp \frac{\beta y^{1-\beta}}{\log y} \asymp \beta u \ll \frac{Z}{\log y}.$$

Substituting in (15), we obtain that the bound

$$\log \Psi(x, y) \le Z\Big\{1 + O\Big(\frac{1}{\log y}\Big)\Big\} \tag{16}$$

holds for $y \le (\log x)^2$.

If $(\log x)^2 < y \le x$, then $\beta \gg 1$. It hence follows from (11) that

$$\varphi_y(\sigma) = \frac{y^{1-\sigma} - 1}{1 - \sigma}\Big\{1 + O\Big(\frac{1}{\log y}\Big)\Big\} \qquad (\beta \le \sigma \le 1).$$

This immediately implies that

$$\int_\beta^1 \varphi_y(\sigma)\,\mathrm{d}\sigma = \Big\{1 + O\Big(\frac{1}{\log y}\Big)\Big\}\int_0^{(1-\beta)\log y} \frac{e^v - 1}{v}\,\mathrm{d}v \ll \frac{y^{1-\beta}}{(1 - \beta)\log y}.$$

Observing that $y^{1-\beta} = \Big(\dfrac{1}{y} + \dfrac{1}{\log x}\Big)^{-1} \asymp \log x$, we see that

$$\int_\beta^1 \varphi_y(\sigma)\,\mathrm{d}\sigma \ll \frac{\log x}{\log_2 x},$$

from which

$$\log \Psi(x, y) \le \beta \log x + O\big(\frac{\log x}{\log_2 x}\big) \le Z\Big\{1 + O\Big(\frac{1}{\log_2 x}\Big)\Big\}.$$

We have thus established, in all cases, the required upper bound implicit in (9). For the corresponding lower bound, we consider the smallest integer v such that $v \geq u$ and write $z := x^{1/v}$, $k := \pi(z)$. Then $\Psi(x,y) \geq \Psi(x,z)$ and it is clear that each k-tuple $(\nu_1, \ldots, \nu_k)$ of integers ≥ 0 such that $\sum_{i=1}^k \nu_i \leq v$ provides at least one integer counted in $\Psi(x,z)$, namely $\prod_{i=1}^k p_i^{\nu_i}$ (where p_i denotes the ith prime number). It follows that

$$\Psi(x,y) \geq \binom{k+v}{v},$$

from which, applying Stirling's formula,

$$\log \Psi(x,y) \geq (k+v)\log(k+v) - k\log k - v\log v + O\big(1 + \log\min(k,v)\big).$$

The main term of this lower bound can be rewritten as $\int_0^k \log(1+v/t)\,\mathrm{d}t$. We hence see that the error implied in replacing k by $z/\log z$ is

$$\ll \frac{z}{(\log z)^2}\log\left(1+\frac{v}{k}\right) \ll \frac{z}{(\log z)^2}\log\left(1+\frac{\log x}{z}\right).$$

We can thus write

$$\log \Psi(x,y) \geq W(z)\Big\{1 + O\Big(\frac{1}{\log z}\Big)\Big\} \tag{17}$$

with

$$W(z) := \frac{\log x}{\log z}\log\left(1+\frac{z}{\log x}\right) + \frac{z}{\log z}\log\left(1+\frac{\log x}{z}\right).$$

It may be easily checked that

$$W'(t) = -\frac{W(t)}{t\log t} + \frac{1}{\log t}\log\left(1+\frac{\log x}{t}\right).$$

This implies, after a small computation, that we have, for $\sqrt{y} \leq t \leq y^2$,

$$W'(t) \ll \begin{cases} \dfrac{\log x}{t\log t} & (y \leq (\log x)^4) \\[2mm] \dfrac{\log x \log_2 x}{t(\log t)^2} & ((\log x)^4 < y \leq x). \end{cases}$$

Bearing in mind that $(\log y)/\log z = 1 + O(1/u)$, it follows in a few lines that

$$Z - W(z) = W(y) - W(z) \ll \min(\log y, \log_2 x) \ll Z\Big(\frac{1}{\log y} + \frac{1}{\log_2 x}\Big).$$

Substituting in (17), we obtain

$$\log \Psi(x,y) \geq Z\Big\{1 + O\Big(\frac{1}{\log y} + \frac{1}{\log_2 x}\Big)\Big\}.$$

This completes the proof of Theorem 2.

Theorem 2 demonstrates a change in behaviour of $\Psi(x,y)$ when y passes through the threshold $\log x$. This phenomenon will be more fully described in the next sections. It may be explained, at least partially, by the fact that the relation

$$\prod_{p\le y} p = o(x),$$

valid for $y \le (1-\varepsilon)\log x$, imposes a particular structure on the integers counted by $\Psi(x,y)$ when y is "small", namely that certain exponents in the canonical prime decomposition *must* then be "large".

§ 5.2 The geometric method

We have seen in the course of the proof of the lower bound in Theorem 2 that the problem of $\Psi(x,y)$ can be interpreted as counting points with integral coordinates inside some polyhedron of $\mathbb{R}^k$, with $k := \pi(y)$. When k is not too large, a satisfactory estimate can quite easily be obtained.

Let $\{a_j\}_{j=1}^{\infty}$ be a sequence of positive real numbers. We set

$$N_k(z) := \Big|\Big\{(\nu_1,\ldots,\nu_k) : \nu_1 \ge 0,\ldots,\nu_k \ge 0,\ \sum_{i=1}^{k} \nu_i a_i \le z\Big\}\Big|.$$

Theorem 3. *For $k \ge 1$, $z \ge 0$, we have*

$$\frac{z^k}{k!}\prod_{i=1}^{k}\frac{1}{a_i} < N_k(z) \le \frac{\left(z+\sum_{i=1}^{k} a_i\right)^k}{k!}\prod_{i=1}^{k}\frac{1}{a_i}. \tag{18}$$

For the choice $k := \pi(y)$, $z := \log x$, $a_i := \log p_i$ (where p_i denotes the ith prime number), we have $N_k(z) = \Psi(x,y)$ and $\sum_{i=1}^{k} a_i = \theta(y) \ll y$. This hence immediately yields the following result.

Corollary 3.1 (Ennola, 1969). *Uniformly for $2 \le y \le \sqrt{(\log x \log_2 x)}$, we have that*

$$\Psi(x,y) = \frac{1}{\pi(y)!}\prod_{p\le y}\Big(\frac{\log x}{\log p}\Big)\Big\{1 + O\Big(\frac{y^2}{\log x \log y}\Big)\Big\}. \tag{19}$$

Proof of Theorem 3. We argue by induction on k. The result holds for $k=1$, since

$$\frac{z}{a_1} < N_1(z) = 1 + \Big[\frac{z}{a_1}\Big] \le \frac{z+a_1}{a_1}.$$

Suppose that the double inequality (18) is satisfied for $k-1$, with $k \ge 2$. We have

$$N_k(z) = \sum_{0\le\nu\le z/a_k} N_{k-1}(z-\nu a_k),$$

from which

$$\frac{1}{(k-1)!}\prod_{i=1}^{k-1}\frac{1}{a_i}\sum_{0\le\nu\le z/a_k}(z-\nu a_k)^{k-1}<N_k(z)\le\frac{1}{(k-1)!}\prod_{i=1}^{k-1}\frac{1}{a_i}\sum_{0\le\nu\le z/a_k}(w-\nu a_k)^{k-1},$$

with the notation $w := z + \sum_{i=1}^{k-1} a_i$.

Let $n := [z/a_k]$. The Euler–Maclaurin formula of order 0 (cf. Theorem I.0.4) implies that

$$\sum_{0\le\nu\le n}(z-\nu a_k)^{k-1} = \int_0^n (z-ta_k)^{k-1}\,\mathrm{d}t + \tfrac{1}{2}\left\{(z-na_k)^{k-1}+z^{k-1}\right\}$$
$$- (k-1)a_k\int_0^n B_1(t)(z-ta_k)^{k-2}\,\mathrm{d}t,$$

with $B_1(t) := \{t\} - \frac{1}{2}$. In absolute value, the last integral is bounded above by

$$-\tfrac{1}{2}\int_0^n \frac{\mathrm{d}}{\mathrm{d}t}\{(z-ta_k)^{k-1}\}\,\mathrm{d}t = \tfrac{1}{2}\left\{z^{k-1}-(z-na_k)^{k-1}\right\}.$$

Therefore

$$\sum_{0\le\nu\le n}(z-\nu a_k)^{k-1} \ge \frac{1}{a_k}\left\{\frac{z^k}{k}-\frac{(z-na_k)^k}{k}\right\}+(z-na_k)^{k-1} > \frac{z^k}{ka_k},$$

since $(z-na_k)/ka_k \le 1/k < 1$. This gives the desired lower bound.

In order to obtain the upper bound, we again apply the Euler–Maclaurin formula, but with z replaced by w. We obtain

$$\sum_{0\le\nu\le n}(w-\nu a_k)^{k-1} \le \int_0^n (w-ta_k)^{k-1}\,\mathrm{d}t + w^{k-1}$$
$$\le \frac{w^k+ka_kw^{k-1}}{ka_k} < \frac{(w+a_k)^{k-1}}{ka_k}.$$

This completes the proof.

In addition to the information it provides on the global behaviour of $\Psi(x,y)$, Ennola's theorem brings precision on the local nature of the variations of this function. When y is small, it can thus be seen that $\Psi(x,y)$ is very sensitive to irregularities in the distribution of primes. For example, it is easy to deduce from (19) that no continuous function of x and y is asymptotic to $\Psi(x,y)$ for $y \le \sqrt{(\log x)}$. Indeed, if y is a prime number, we then have

$$\frac{\Psi(x,y)}{\Psi(x,y-)} \sim \frac{\log x}{\pi(y)\log y} \sim \frac{\log x}{y} \to \infty.$$

We shall return to this point in Section 5.

§ 5.3 Functional equations

Like most quantities arising in sieve problems, $\Psi(x,y)$ satisfies a functional equation.

Theorem 4. *For $x \geq 1$, $y > 0$, we have*

$$\Psi(x,y) = 1 + \sum_{p \leq y} \Psi(x/p, p). \tag{20}$$

Proof. If $n > 1$ is counted in $\Psi(x,y)$, we can write $n = mp$ with $p = P^+(n)$. This last condition is equivalent to $P^+(m) \leq p$. Therefore

$$\Psi(x,y) = 1 + \sum_{p \leq y} \sum_{n \leq x,\, P^+(n)=p} 1 = 1 + \sum_{p \leq y} \sum_{m \leq x/p,\, P^+(m) \leq p} 1,$$

which is (20).

In practice, this equation is often used to investigate the difference $\Psi(x,z) - \Psi(x,y)$.

Corollary 4.1 (Buchstab's identity). *For $x \geq 1$, $z \geq y > 0$, we have*

$$\Psi(x,y) = \Psi(x,z) - \sum_{y < p \leq z} \Psi(x/p, p). \tag{21}$$

This identity allows one to approximate $\Psi(x,y)$ by induction on $[u] = [\log x/\log y]$. Indeed, we have $\Psi(x,y) = [x]$ if $y \geq x$, that is $u \leq 1$. Then, noting that $x/p \leq p$ in (21) whenever $y \geq \sqrt{x}$, we can write (with $z = x$) for $\sqrt{x} \leq y \leq x$

$$\Psi(x,y) = [x] - \sum_{y < p \leq x} [x/p] \sim x(1 - \log u),$$

by a straightforward application of the prime number theorem. Of course, it is then possible to feed this information back into (21). Choosing $z = \sqrt{x}$, we obtain for $x^{1/3} \leq y \leq x^{1/2}$

$$\Psi(x,y) \sim x(1 - \log 2) - \sum_{y < p \leq \sqrt{x}} \frac{x}{p}\left(1 - \log\left(\frac{\log(x/p)}{\log p}\right)\right) \sim x\rho(u),$$

with

$$\rho(u) := 1 - \log u + \int_2^u \log(v-1)\frac{\mathrm{d}v}{v} \qquad (2 \leq u \leq 3).$$

It is clear that this procedure yields, for any $\varepsilon > 0$, the asymptotic formula

$$\Psi(x,y) \sim x\rho(u) \qquad (x^\varepsilon < y \leq x), \tag{22}$$

where ρ is defined by the initial condition $\rho(u) = 1$ $(0 \leq u \leq 1)$ and the smooth version of equation (21), namely

$$\rho(u) = \rho(k) - \int_k^u \rho(v-1)\frac{\mathrm{d}v}{v} \qquad (k < u \leq k+1). \tag{23}$$

This function ρ was discovered by Dickman in 1930, and today bears his name. It is continuous at $u = 1$, differentiable for $u > 1$, and satisfies the *difference-differential equation*

$$up'(u) + \rho(u-1) = 0 \qquad (u > 1) \tag{24}$$

obtained by differentiating (23). The occurrence of an equation of this type, which reflects Buchstab's identity, is characteristic of sieve problems.

Theorem 5. *Dickman's function $\rho(u)$ satisfies the following properties:*

$$\begin{array}{lll} \text{(i)} & u\rho(u) = \int_{u-1}^{u} \rho(v)\,\mathrm{d}v & (u \geq 1) \\ \text{(ii)} & \rho(u) > 0 & (u > 0) \\ \text{(iii)} & \rho'(u) < 0 & (u > 1) \\ \text{(iv)} & \rho(u) \leq 1/\Gamma(u+1) & (u \geq 0). \end{array}$$

Proof. Part (i) follows immediately from (24) and the initial conditions for ρ. Indeed the two sides of (i) have the same derivative for $u > 1$ and the same value at $u = 1$.

Let us prove (ii). Let $\tau := \inf\{u : \rho(u) = 0\}$. If τ is finite, then $\tau > 1$ since ρ is continuous and satisfies $\rho(u) = 1$ for $0 \leq u \leq 1$. By (i), we can then write

$$0 = \tau\rho(\tau) = \int_{\tau-1}^{\tau} \rho(v)\,\mathrm{d}v.$$

The continuity of ρ then implies, by the very definition of τ, that the right-hand side is strictly positive. This shows that τ is not finite and (ii) follows.

Part (iii) follows immediately from (ii) and the functional equation (24).

We obtain (iv) by induction on $k := [u]$. The property holds for $k = 0$, since $\rho(u) = 1$ $(0 \leq u \leq 1)$. If $k \geq 1$, we deduce from (i), (ii), (iii) and the inductive assumption that

$$\rho(u) = \frac{1}{u}\int_{u-1}^{u} \rho(v)\,\mathrm{d}v \leq \frac{\rho(u-1)}{u} \leq \frac{1}{u\Gamma(u)} = \frac{1}{\Gamma(u+1)}.$$

We shall see (Corollary 9.3) that the asymptotic relation (22) remains true in a very large (x, y)-domain. A first attempt at an inductive use of Buchstab's identity is the object of the following theorem, which represents partial success in this direction.

Theorem 6. *Uniformly for $x \geq y \geq 2$, we have*

$$\Psi(x,y) = x\rho(u) + O\Big(\frac{x}{\log y}\Big). \tag{25}$$

Proof. Given the rapid decrease of $\rho(u)$ as $u \to \infty$, it is enough to give a proof for $u \leq 2\log_2 y$. Otherwise the error term in (25) is of larger order than the main term, and the result follows, for example, from Theorem 1.

So let $\Delta(x,y)$ be the quantity implicitly defined by

$$\Psi(x,y) = x\rho(u) + \frac{x\Delta(x,y)}{\log y}.$$

As we have previously noted, we have $\Delta(x,y) \ll 1$ for $y \geq 2$, $1 \leq u \leq 2$. Indeed relation (21) with $z = x$ gives in this case

$$\begin{aligned}\Psi(x,y) &= [x] - \sum_{y<p\leq x} [x/p] = x(1-\log u) + O\big(\pi(x)\big)\\ &= x\rho(u) + O\Big(\frac{x}{\log x}\Big).\end{aligned}$$

We use induction on $k \leq 1 + 2\log_2 y$ to show that the quantity

$$\Delta_k(y) := 1 + \sup\big\{|\Delta(X,Y)| : y \leq Y \leq X \leq Y^k\big\}$$

is finite and uniformly bounded. Let us hence assume that $\Delta_k(y) < \infty$, and consider X, Y such that $Y \geq y$, $Y^2 < X \leq Y^{k+1}$ $(k \geq 2)$. Applying (21) for (X,Y) with $z = \sqrt{X}$, and writing $U := (\log X)/\log Y$, we obtain

$$\begin{aligned}\Psi(X,Y) &= \Psi(X,\sqrt{X}) - \sum_{Y<p\leq\sqrt{X}} \Psi(X/p,p)\\ &= X\Big\{\rho(2) - \int_2^U \rho(v-1)\,\mathrm{d}L(v)\Big\} \qquad (26)\\ &\quad + \frac{X}{\log Y}\Big\{\frac{2}{U}\Delta(X,\sqrt{X}) - \sum_{Y<p\leq\sqrt{X}} \frac{\Delta(X/p,p)\log Y}{p\log p}\Big\},\end{aligned}$$

with

$$L(v) := -\sum_{Y<p\leq X^{1/v}} \frac{1}{p} = \log(v/U) + O\big(e^{-\sqrt{(\log Y)}}\big).$$

Property (iii) of Theorem 5 shows that ρ is non-decreasing. By partial summation, we hence immediately obtain that

$$\int_2^U \rho(v-1)\,\mathrm{d}L(v) = \int_2^U \rho(v-1)\frac{\mathrm{d}v}{v} + O\big(e^{-\sqrt{(\log Y)}}\big) = \rho(2) - \rho(U) + O\big(e^{-\sqrt{(\log Y)}}\big).$$

Moreover, for $Y < p \leq \sqrt{X}$, we have

$$1 \leq \frac{\log(X/p)}{\log p} \leq U - 1 \leq k.$$

For these values of p, we can thus bound $\Delta(X/p, p)$ from above by $\Delta_k(y) - 1$. Substituting in (26) it then follows that

$$\begin{aligned}1 + |\Delta(X,Y)| &\leq 1 + C_0\, e^{-\frac{1}{2}\sqrt{(\log Y)}} + (\Delta_k(y) - 1)\Big\{\frac{2}{U} + \sum_{Y<p\leq\sqrt{X}} \frac{\log Y}{p\log p}\Big\} \\ &\leq \Delta_k(y)\Big(1 + C_1\, e^{-\frac{1}{2}\sqrt{(\log Y)}}\Big),\end{aligned}$$

by the prime number theorem. Taking the supremum over X, Y such that $Y \geq y$, $k < U \leq k+1$ and iterating, it follows that

$$\Delta_{[2\log_2 y]}(y) \ll \Big(1 + C_1\, e^{-\frac{1}{2}\sqrt{(\log y)}}\Big)^{2\log_2 y} \ll 1,$$

which is the desired conclusion.

The proof of Theorem 6 can be refined by using a smooth main term whose behaviour mimics Buchstab's identity (satisfied by $\Psi(x,y)$) more closely. In this way, we improve the error term $O(x/\log y)$, but we cannot evade the inconvenience, inherent in the method, of a remainder term independent of u—and which is hence of little worth for large u. In this direction, de Bruijn (1951b) chose the main term

$$\Lambda(x,y) = \begin{cases} x\displaystyle\int_{-\infty}^{\infty} \rho(u-v)\,\mathrm{d}\big([y^v]/y^v\big) & (x \notin \mathbb{Z}^+) \\ \Lambda(x+,y) & (x \in \mathbb{Z}^+). \end{cases} \tag{27}$$

In Section 5 we shall meet a second heuristic justification of this definition, and in the Notes we give an asymptotic expansion which describes the behaviour of this function in a large (x,y)-domain. For the moment we content ourselves with the observation that

$$\Lambda(x,y) = [x] \qquad (x \leq y) \tag{28}$$

and

$$\Lambda(x,y) = \Lambda(x,z) - \int_y^z \Lambda\Big(\frac{x}{t},t\Big)\frac{\mathrm{d}t}{\log t} \qquad (y \leq z). \tag{29}$$

Thus $\Lambda(x,y)$ satisfies the same initial condition as $\Psi(x,y)$ and obeys a functional equation which is the smooth version of Buchstab's identity.

Employing a method analogous to that of Theorem 6, de Bruijn obtained the formula

$$\Psi(x,y) = \Lambda(x,y) + O\big(xu^2 L_\varepsilon(y)^{-1}\big) \qquad (x \geq y \geq 2), \tag{30}$$

with

$$L_\varepsilon(y) := \exp\big\{(\log y)^{(3/5)-\varepsilon}\big\}. \tag{31}$$

We shall see in Section 5 that the main term in (30) has order $x\rho(u)$ when $y \geq (\log x)^2$. From the upper bound (iv) of Theorem 5, we hence see that (30) can only yield an asymptotic formula for $\Psi(x,y)$ in the domain

$$\exp\big\{(\log x)^{(5/8)+\varepsilon}\big\} \leq y \leq x. \tag{32}$$

This region may be considered as the natural limit of Buchstab's iterative method.

De Bruijn's result has however been considerably improved by Hildebrand (1986a) using another functional equation, namely

$$\Psi(x,y)\log x = \int_1^x \Psi(t,y)\frac{\mathrm{d}t}{t} + \sum_{d\leq x,\, P^+(d)\leq y} \Lambda(d)\Psi\big(\frac{x}{d},y\big). \tag{33}$$

Equation (33) has two advantages over (21). On the one hand, for fixed y, it takes the form of an equation in one variable and is hence easier to iterate. On the other hand, by expressing $\Psi(x,y)$ as an average of itself, it also creates, by iteration, a powerful regulating effect. This latter feature was already present, but dampened, in the case of Buchstab's identity. The cause of this phenomenon can be traced back to the very source of the method, which consists in calculating $\Psi(x,y)$ starting with $\Psi(x,z)$ and deleting positive terms. As one might foresee, iterating Buchstab's identity results in alternating signs— and hence potential cancellations which have to be regarded as unexploited gains in the error terms.

Hildebrand established the validity of the formula

$$\Psi(x,y) = x\rho(u)\Big\{1 + O\Big(\frac{\log(u+1)}{\log y}\Big)\Big\} \tag{34}$$

uniformly, for each $\varepsilon > 0$, in the domain

$$\exp\big\{(\log_2 x)^{(5/3)+\varepsilon}\big\} \leq y \leq x. \tag{35}$$

We shall derive this result in Section 5 from a theorem which generalises both (30) and (34), and which depends on a purely analytic method. We therefore will not trouble ourselves here in deducing (34) from (33), but merely indicate how (33) can be established.

It actually suffices to calculate the quantity

$$S := \sum_{n \le x,\, P^+(n) \le y} \log n$$

in two ways. On the one hand, by partial summation,

$$S = \Psi(x,y)\log x - \int_1^x \Psi(t,y)\frac{\mathrm{d}t}{t},$$

and on the other, by the convolution identity for Λ,

$$S = \sum_{n \le x,\, P^+(n) \le y} \sum_{d|n} \Lambda(d) = \sum_{d \le x,\, P^+(d) \le y} \Lambda(d)\Psi(x/d, y).$$

For another application of (33), see Theorem 13 in the Notes.

§ 5.4 Dickman's function

Here we shall complete our analytic study of Dickman's function $\rho(u)$, introduced in the previous section, and, in particular, describe its asymptotic behaviour.

Recall that we have defined $\rho(u)$ as the solution, continuous at $u = 1$ and differentiable for $u > 1$, of the difference-differential equation

$$u\rho'(u) + \rho(u-1) = 0 \qquad (u > 1) \tag{36}$$

with the initial condition $\rho(u) = 1$ $(0 \le u \le 1)$. It is convenient to extend $\rho(u)$ to all of $\mathbb{R}$ in such a way that (36) is satisfied everywhere. We hence let

$$\rho(u) = 0 \qquad (u < 0). \tag{37}$$

Thus defined, $\rho(u)$ is right-continuous and has a single singularity, of the first kind, at $u = 0$. It is immediate that the derivative of order k, $\rho^{(k)}(u)$, is defined on $\mathbb{R} \smallsetminus \{0, 1, \ldots, k\}$ and also has discontinuities of the first kind at the exceptional points $u = j$ $(0 \le j \le k)$. We extend it to $\mathbb{R}$ by right-continuity.

Our method for studying the Dickman function rests on the computation of the Laplace transform

$$\widehat{\rho}(s) := \int_0^\infty e^{-st}\rho(t)\,\mathrm{d}t \tag{38}$$

and the evaluation of $\rho(u)$ by inverse Laplace transform. By Theorem 5(iv), the integral (38) is absolutely convergent for all complex s, and thus defines an entire function of s.

By change of variable $v = ts$, we can write for $s \in \mathbb{R}^+$

$$s\widehat{\rho}(s) = \int_0^\infty e^{-v}\rho(v/s)\,\mathrm{d}v, \tag{39}$$

from which we deduce, using (36), that

$$\frac{\mathrm{d}}{\mathrm{d}s}\{s\widehat{\rho}(s)\} = \frac{1}{s}\int_0^\infty e^{-v}\rho\Big(\frac{v}{s}-1\Big)\,\mathrm{d}v = \int_0^\infty e^{-ts}\rho(t-1)\,\mathrm{d}t = e^{-s}\widehat{\rho}(s).$$

Solving this first-order differential equation yields, for some suitable constant C, that

$$s\widehat{\rho}(s) = C\,e^{-J(s)}, \tag{40}$$

with

$$J(s) := \int_0^\infty \frac{e^{-s-t}}{s+t}\,\mathrm{d}t. \tag{41}$$

The integral $J(s)$ is evidently continuable as a holomorphic function on $\mathbb{C}\smallsetminus]-\infty,0]$. The following lemma, of crucial use here, gives supplementary information about this continuation. We set

$$I(s) := \int_0^s \frac{e^t-1}{t}\,\mathrm{d}t \qquad (s \in \mathbb{C}). \tag{42}$$

Lemma 7.1. *For* $s \in \mathbb{C}\smallsetminus]-\infty,0]$, *we have*

$$I(-s) + J(s) + \gamma + \log s = 0 \tag{43}$$

where γ *denotes Euler's constant.*

Proof. The classical formula $\gamma = -\Gamma'(1)$ immediately yields, after an integration by parts,

$$-\gamma = \int_0^{-1}\frac{e^t-1}{t}\,\mathrm{d}t + \int_{-1}^{-\infty} e^t\frac{\mathrm{d}t}{t}.$$

Without altering the value of this expression, we can replace the real path $[-1,-\infty[$ in the second integral by the polygonal line made up of the segment $[-1,-s]$ and the half-line $\{-s-t : t \geq 0\}$. It follows that

$$-\gamma = \int_0^{-1}\frac{e^t-1}{t}\,\mathrm{d}t + \int_{-1}^{-s}\frac{e^t-1}{t}\,\mathrm{d}t + \int_{-1}^{-s}\frac{\mathrm{d}t}{t} + \int_0^\infty \frac{e^{-s-t}}{s+t}\,\mathrm{d}t.$$

Rearranging the first two integrals, we obtain (43).

This result enables us to complete the calculation of $\widehat{\rho}(s)$.

Theorem 7. *For all complex numbers s, we have*

$$\widehat{\rho}(s) = e^{\gamma + I(-s)}. \tag{44}$$

Proof. By (40) and (43), we can write, when s is not real and negative,

$$s\widehat{\rho}(s) = C\,e^{-J(s)} = Cse^{\gamma+I(-s)}$$

from which

$$\widehat{\rho}(s) = C\,e^{\gamma+I(-s)}.$$

Now, on the one hand, it follows from (39) that

$$\lim_{s\to\infty} s\widehat{\rho}(s) = \rho(0+) = 1,$$

while on the other, (40) implies that

$$\lim_{s\to\infty} s\widehat{\rho}(s) = C.$$

Therefore, we must have $C = 1$, and the proof of (44) is complete.

In Exercise 2 we indicate an arithmetic proof, resting on Theorems 1 and 6, of the formula $\widehat{\rho}(0) = e^{\gamma}$. The value is then provided by Mertens' formula.

The regularity and the rapid decrease of $\rho(u)$ at infinity show that the inverse Laplace integral

$$\rho(u) = \frac{1}{2\pi i}\int_{\alpha-i\infty}^{\alpha+i\infty} \widehat{\rho}(s)e^{us}\,\mathrm{d}s \tag{45}$$

converges for all $u \neq 0$—cf. for example Widder (1946), theorem II.7.3 or II.7.5. In accordance with the principles of the saddle-point method—see de Bruijn (1970), chapter 5—it can be expected that, by choosing α to be a zero of the derivative of the integrand, the integral (45) as a whole will be dominated by the contribution from a small neighbourhood of the real point α. This will then allow the determination of an asymptotic formula for $\rho(u)$ by utilising the Taylor expansion of $\widehat{\rho}(s)e^{us}$ in the neighbourhood of $s = \alpha$.

The explicit formula (44) immediately yields the value of α. It is necessary to choose $\alpha = -\xi(u)$ where $\xi = \xi(u)$ is the unique real non-zero root of the equation

$$e^{\xi} = 1 + u\xi \qquad (u > 0,\ u \neq 1). \tag{46}$$

By convention, we set $\xi(1) = 0$.

The following two technical lemmas enable us to carry out the evaluation of the integral (45).

Lemma 8.1. *For $u \geq 3$, we have*

$$\xi(u) = \log(u \log u) + O\Big(\frac{\log_2 u}{\log u}\Big). \tag{47}$$

Proof. We trivially have $1 \ll \xi(u) \ll \log u$. This simple first estimate gives the result by iteration of (46). We thus have

$$\begin{aligned}\xi &= \log u + \log(\xi + 1/u) = \log u + \log\big(\log u + \log(\xi + 1/u) + 1/u\big) \\ &= \log(u \log u) + O\Big(\frac{\log(\xi + 1/u)}{\log u}\Big).\end{aligned}$$

Lemma 8.2. *For $u > 1$, $s = -\xi(u) + i\tau$, $\tau \in \mathbb{R}$, we have*

$$\widehat{\rho}(s) \ll \begin{cases} \exp\{I(\xi) - \tau^2 u/2\pi^2\} & (|\tau| \leq \pi) \\ \exp\{I(\xi) - u/(\pi^2 + \xi^2)\} & (|\tau| > \pi) \end{cases} \tag{48}$$

and

$$\widehat{\rho}(s) = \frac{1}{s}\Big\{1 + O\Big(\frac{1 + u\xi}{s}\Big)\Big\} \qquad (|\tau| > 1 + u\xi). \tag{49}$$

Proof. Note, first of all, that (49) follows immediately from (44) and (43) in the form

$$s\widehat{\rho}(s) = e^{-J(s)}. \tag{50}$$

Indeed it suffices to apply the trivial bound $J(s) \ll e^{-\sigma}|\tau|^{-1}$ with $\sigma = -\xi(u)$.

In order to establish (48), we introduce the quantity

$$H(\tau) := I(\xi) - \Re e\, I(-s) = \int_0^1 e^{h\xi}\, \frac{(1 - \cos(h\tau))}{h}\, \mathrm{d}h.$$

When $|\tau| \leq \pi$, we have $1 - \cos(h\tau) \geq 2\tau^2 h^2/\pi^2$, from which

$$H(\tau) \geq \frac{2\tau^2}{\pi^2} \int_0^1 h\, e^{\xi h}\, \mathrm{d}h.$$

The desired conclusion then follows from the lower bound

$$\int_0^1 h\, e^{h\xi}\, \mathrm{d}h \geq \tfrac{1}{2} \int_{1/2}^1 e^{h\xi}\, \mathrm{d}h \geq \tfrac{1}{4} \int_0^1 e^{h\xi}\, \mathrm{d}h = \tfrac{1}{4} u.$$

When $|\tau| > \pi$, we note that we can assume that u is sufficiently large since the conclusion is otherwise trivial. Taking (47) into account, we may hence suppose that

$$u > \pi^2 + \xi^2.$$

We can then write

$$\begin{aligned} H(\tau) &\geq \int_0^1 e^{h\xi}(1 - \cos(h\tau))\, \mathrm{d}h = u - \Re e\Big(\frac{e^{\xi+i\tau} - 1}{\xi + i\tau}\Big) \\ &= u - \Re e\Big(\frac{u\xi e^{i\tau} + e^{i\tau} - 1}{\xi + i\tau}\Big) \geq u\Big(1 - \frac{\xi\cos\theta}{\sqrt{(\pi^2+\xi^2)}}\Big) - \frac{2}{\sqrt{(\tau^2+\xi^2)}}, \end{aligned}$$

with $\theta := \tau - \arctan(\tau/\xi)$. The factor of u in the last expression is at least equal to $\pi^2/2(\pi^2 + \xi^2)$. It follows that

$$H(\tau) \geq \frac{u\pi^2}{2(\pi^2+\xi^2)} - \frac{2u}{\pi^2+\xi^2} > \frac{u}{\pi^2+\xi^2}.$$

We are now in a position to establish the following result which provides an asymptotic formula with remainder term for $\rho(u)$ as $u \to \infty$.

Theorem 8 (de Bruijn; Alladi). *For* $u \geq 1$, *we have*

$$\rho(u) = \sqrt{\frac{\xi'(u)}{2\pi}} \exp\{\gamma - u\xi + I(\xi)\}\Big\{1 + O\Big(\frac{1}{u}\Big)\Big\}. \tag{51}$$

Remarks. On differentiating (46) with respect to u, we immediately obtain $\xi'(u) = \xi/\big(1 + u(\xi - 1)\big)$. In particular, we have $\xi'(u) \sim 1/u$ as $u \to \infty$. It is also useful to note that the main term in (51) can be transformed by means of the identity

$$u\xi - I(\xi) = \int_1^u \xi(t)\, \mathrm{d}t \qquad (u > 0). \tag{52}$$

Proof. Let $\delta = \delta(u) := \pi\sqrt{(2\log(u+1)/u)}$ and

$$K(u) := \frac{1}{2\pi}\int_{-\delta}^{\delta} \widehat{\rho}(s)e^{us}\, \mathrm{d}\tau$$

where $s = -\xi(u) + i\tau$. As a first step, we shall show that the difference

$$\rho(u) - K(u) = \frac{1}{2\pi i}\int_{|\tau|>\delta} \widehat{\rho}(s)e^{us}\, \mathrm{d}s \tag{53}$$

is dominated by the error term in (51).

To this end, we use Lemma 8.2. The contribution to the integral (53) from the range $\delta < |\tau| \leq \pi$ is

$$\ll e^{-u\xi+I(\xi)} \int_\delta^\infty e^{-\tau^2 u/2\pi^2}\,\mathrm{d}\tau \ll \frac{e^{-u\xi+I(\xi)}}{\sqrt{u}} \int_{\log(u+1)}^\infty e^{-t}\frac{\mathrm{d}t}{\sqrt{t}} \ll \frac{e^{-u\xi+I(\xi)}}{u^{3/2}}.$$

Similarly, the contribution from the range $\pi < |\tau| \leq 1 + u\xi$ is

$$\ll (1+u\xi)\exp\{-u\xi + I(\xi) - u/(\pi^2+\xi^2)\}.$$

These bounds are clearly acceptable.

Finally, using (49), we can evaluate the contribution to the integral (53) from the range $|\tau| > 1 + u\xi$. We have

$$\int_{1+u\xi}^\infty \widehat{\rho}(s)e^{us}\,\mathrm{d}s = \int_{1+u\xi}^\infty \frac{e^{-u\xi+i\tau u}}{\tau}\Big\{1+O\Big(\frac{u\xi}{\tau}\Big)\Big\}\,\mathrm{d}\tau \ll e^{-u\xi}$$

where we have appealed to the second mean value formula in order to handle the term involving $1/\tau$. Since $I(\xi) \sim u$, this bound is also of the required order of magnitude.

It remains to evaluate $K(u)$. For this, we consider the Taylor expansion of $\widehat{\rho}(s)$ in the neighbourhood of $\tau = 0$. Noting that we have for $\Re e\, s = -\xi$, $k \geq 1$,

$$|I^{(k)}(s)| = \Big|\int_0^1 h^{k-1}e^{hs}\,\mathrm{d}h\Big| \leq I^{(k)}(\xi) \leq I'(\xi) = u,$$

it follows that

$$I(\xi - i\tau) = I(\xi) - i\tau u - \tfrac{1}{2}\tau^2 I''(\xi) - \tfrac{1}{6}i\tau^3 I'''(\xi) + O(u\tau^4).$$

We can then write for $\tau \ll u^{-1/4}$

$$\exp\{I(\xi - i\tau) + us\} = \exp\{I(\xi) - u\xi - \tfrac{1}{2}\tau^2 I''(\xi)\}(1 + h(u)),$$

with

$$h(u) = \exp\{-\tfrac{1}{6}i\tau^3 I'''(\xi) + O(u\tau^4)\} - 1 = -\tfrac{1}{6}i\tau^3 I'''(\xi) + O(u\tau^4 + u^2\tau^6).$$

Substituting in the integral defining $K(u)$ and noting that the contribution of the term in τ^3 vanishes by symmetry, we obtain

$$K(u) = \frac{1}{2\pi}e^{-u\xi+I(\xi)} \int_{-\delta}^{\delta} e^{-\tau^2 I''(\xi)/2}\{1 + O(u\tau^4 + u^2\tau^6)\}\,\mathrm{d}\tau. \tag{54}$$

The contribution of the error terms is

$$\ll \frac{e^{-u\xi+I(\xi)}}{u\sqrt{(I''(\xi))}}.$$

That of the main term is evaluated by extending the integral to infinity, and bounding above the contribution from the range $|\tau| > \delta$. Noting that

$$I''(\xi) = 1/\xi'(u) = u - (u-1)/\xi, \tag{55}$$

we obtain

$$\int_{-\delta}^{\delta} e^{-\tau^2 I''(\xi)/2}\,\mathrm{d}\tau = \sqrt{\frac{2\pi}{I''(\xi)}}\Big\{1+O\Big(\frac{1}{u}\Big)\Big\}.$$

Substituting this estimate in (54) and taking account of (55), we finally obtain the stated formula (51).

Corollary 8.3. *For any integer $k \geq 0$ and any real number $u_0 > 1$, we have*

$$\rho^{(k)}(u) = (-1)^k \xi(u)^k \rho(u)\{1+O(1/u)\} \qquad (u \geq u_0). \tag{56}$$

Proof. Differentiating the functional equation (36) k times, we obtain

$$u\rho^{(k+1)}(u) = -\rho^{(k)}(u-1) - k\rho^{(k)}(u) \qquad (u > 1). \tag{57}$$

By induction on k, this immediately implies that

$$(-1)^k \rho^{(k)}(u) > 0 \qquad (u > 1). \tag{58}$$

This gives (56) when u is bounded.

Relation (55) shows that

$$\xi'(u) \sim 1/u, \qquad \xi''(u) \sim -1/u^2 \qquad (u \to \infty). \tag{59}$$

We deduce from this that, for sufficiently large u,

$$\xi'(u-1) = \xi'(u)\{1+O(1/u)\}, \qquad \int_{u-1}^{u} \xi(t)\,\mathrm{d}t = \xi(u) + O(1/u). \tag{60}$$

Substituting in the asymptotic formula (51), written with (52) taken into account, we obtain

$$\rho(u-1) = \rho(u)e^{\xi(u)}\{1+O(1/u)\} = u\xi(u)\rho(u)\{1+O(1/u)\}. \tag{61}$$

This allows us to prove (56) by induction on k. Supposing that the formula is satisfied up to order k, we can write by (57), (59)—in the form $\xi(u-1) = \xi(u)\{1+O(1/u)\}$—and (61),

$$\begin{aligned} u\rho^{(k+1)}(u) &= (-1)^k \xi(u)^k\{-u\xi(u)\rho(u) + O(\xi(u)\rho(u))\}\{1+O(1/u)\} \\ &= (-1)^{k+1}\xi(u)^{k+1} u\rho(u)\{1+O(1/u)\}. \end{aligned}$$

The proof of (56) is thus complete.

Corollary 8.4. *Uniformly for $0 \le v \le u$, we have that*

$$\rho(u-v) \ll \rho(u)e^{v\xi(u)}. \tag{62}$$

Proof. When $u-1 < v \le u$, we have $\rho(u-v) = 1$. In this case the bound (62) follows from (51), since $I(\xi) \sim u$. When $0 \le v \le u-1$, we deduce from (51) and (52) that

$$\frac{\rho(u-v)}{\rho(u)} \ll \frac{u}{u-v} \exp\Big\{\int_0^v \xi(u-t)\,\mathrm{d}t\Big\}.$$

By (59), we have $\xi(u-t) \le \xi(u) - ct/u$ for some suitable absolute positive constant c. The required result then follows from the elementary inequality

$$\frac{cv^2}{2u} \ge \log\left(\frac{u}{u-v}\right) + O(1) \qquad (0 \le v \le u-1). \tag{63}$$

§ 5.5 Approximations to $\Psi(x,y)$ by the saddle-point method

For any $\alpha > 0$, Perron's formula yields the expression

$$\Psi(x,y) = \frac{1}{2\pi i}\int_{\alpha-i\infty}^{\alpha+i\infty} \zeta(s,y)x^s \frac{\mathrm{d}s}{s} \qquad (x \notin \mathbb{Z}^+). \tag{64}$$

We shall see that the saddle-point method, employed in the previous section to evaluate Dickman's function, also works for the integral (64).

The optimal choice for α is the (unique) solution $\alpha(x,y)$ to the transcendental equation

$$\frac{-\zeta'}{\zeta}(\alpha,y) = \sum_{p\le y} \frac{\log p}{p^\alpha - 1} = \log x. \tag{65}$$

As we have seen in the course of the proof of Theorem 2, $\alpha(x,y)$ can equally well be regarded as the optimal value of the parameter in Rankin's method—and formula (12) provides an asymptotic formula for $\alpha(x,y)$ as $y \to \infty$. However, the implicit nature of $\alpha(x,y)$ leads one to expect a certain lack of flexibility in the asymptotic formula resulting from such a treatment. The main aim of this section is to establish that, without loss of precision, $\alpha(x,y)$ can be replaced by an explicit approximation in a suitable (x,y)-range. The result is an extension of both the formulae (30) and (34) due to de Bruijn and Hildebrand respectively.

The explicit approximation for $\alpha(x,y)$ is suggested in a natural way by the following lemma. We recall the notation

$$u := \frac{\log x}{\log y}, \qquad L_\varepsilon(y) := \exp\big\{(\log y)^{(3/5)-\varepsilon}\big\},$$

which we use throughout this section.

Lemma 9.1. *Let $\varepsilon > 0$. There exists some $y_0 = y_0(\varepsilon)$ such that, under the conditions*

$$(66)\qquad y \ge y_0(\varepsilon), \qquad \sigma \ge 1 - (\log y)^{-(2/5)-\varepsilon}, \qquad |\tau| \le L_\varepsilon(y),$$

we have uniformly

$$(67)\qquad \zeta(s,y) = \zeta(s)(s-1)\log y\, \widehat{\rho}((s-1)\log y)\Big\{1 + O\Big(\frac{1}{L_\varepsilon(y)}\Big)\Big\}.$$

Proof. Observe first of all that, under the above hypotheses, we have

$$(68)\qquad -\frac{\zeta'}{\zeta}(s,y) = \sum_{n\le y} \frac{\Lambda(n)}{n^s} + O(y^{\frac12-\sigma}).$$

Indeed, the error term is in absolute value at most

$$\sum_{n>y,\, P^+(n)\le y} \frac{\Lambda(n)}{n^\sigma} = \sum_{p\le y} \sum_{\nu > \frac{\log y}{\log p}} \frac{\log p}{p^{\nu\sigma}} \ll \sum_{p\le\sqrt{y}} \frac{\log p}{y^\sigma} + \sum_{\sqrt{y}<p\le y} \frac{\log p}{p^{2\sigma}} \ll y^{\frac12-\sigma}.$$

For $\sigma \le 1$, the effective Perron formula (Corollary II.2.2.1) enables us to write the main term of (68) in the form

$$(69)\qquad \frac{-1}{2\pi i}\int_{\kappa-iT}^{\kappa+iT} \frac{\zeta'}{\zeta}(s+w)\frac{y^w}{w}\,dw + O\Big(\frac{y^{1-\sigma}\log^2 y}{T}\Big)$$

with $\kappa := 1-\sigma+1/\log y$, $T := L_\varepsilon(y)^3$. In order to estimate the integral in w, we move the segment of integration towards the left as far as $-\eta := 1-\sigma-\frac{\log T}{\log y}$. The real part of the argument of $-\zeta'/\zeta$ is still greater than or equal to

$$\sigma - \eta = 1 - \frac{\log T}{\log y} \ge 1 - 9(\log T)^{-(2+5\varepsilon)/(3-5\varepsilon)},$$

that is, for a suitable choice of $y_0(\varepsilon)$, the translated segment remains within Vinogradov's zero-free region for $\zeta(s)$—cf. Chapter II.3, Notes. Thus the segment of integration has crossed exactly two poles of the integrand, $w = 0$ and $w = 1-s$. By the residue theorem, the integral of (69) has the value

$$(70)\qquad -\frac{\zeta'}{\zeta}(s) + \frac{y^{1-s}}{1-s} - \frac{1}{2\pi i}\int_{\mathcal{W}} \frac{\zeta'}{\zeta}(s+w)\frac{y^w}{w}\,dw$$

where $\mathcal{W}$ denotes the polygonal line joining the points $\kappa \pm iT$, passing through $-\eta \pm iT$, and traced out clockwise. When $w \in \mathcal{W}$, we have—cf. Chapter II.3, Notes—that

$$-\frac{\zeta'}{\zeta}(s+w) \ll \log T \ll \log y.$$

Hence the contribution from the vertical segment of $\mathcal{W}$ to the integral is

$$\ll y^{-\eta} \int_0^T \frac{\log y}{\eta + t}\,\mathrm{d}t \ll y^{1-\sigma} L_\varepsilon(y)^{-2}.$$

That of the horizontal segments can be bounded by

$$\ll y^{1-\sigma} T^{-1} \log y \ll y^{1-\sigma} L_\varepsilon(y)^{-2}.$$

Taking account of (68), we finally obtain, for $\sigma \leq 1$,

(71) $$\frac{\zeta'}{\zeta}(s,y) = \frac{\zeta'}{\zeta}(s) + \frac{y^{1-s}}{s-1} + O\big(y^{1-\sigma} L_\varepsilon(y)^{-2}\big).$$

When $\sigma > 1$, an integration by parts shows, in view of (68), that this relation continues to be valid up to multiplying the error term by σ: we omit the details, which are standard. By integrating (71) thus modified on the half-line $\{s + t : t \geq 0\}$, it follows that

$$\zeta(s,y) = \zeta(s) e^{-J((s-1)\log y)} \big\{1 + O\big(L_\varepsilon(y)^{-1}\big)\big\}$$

with the notation (41) for $J(s)$. The stated result then follows from (50).

The fact that the inverse Laplace integral for $\widehat{\rho}(s)$ can be evaluated on the line $\sigma = -\xi(u)$ and the occurrence of this same function in the approximation of $\zeta(s,y)$ are two strong heuristic reasons for choosing the abscissa α in the Perron integral (64) in such a way that $(\alpha - 1)\log y = -\xi(u)$, that is $\alpha = \alpha_0$, with

(72) $$\alpha_0 := 1 - \frac{\xi(u)}{\log y}.$$

It is actually easy to verify, using estimates for

$$\varphi_y(\sigma) := \sum_{p \leq y} \frac{\log p}{p^\sigma - 1}$$

from the prime number theorem—cf. Exercise 1—, that α_0 is an excellent approximation of $\alpha(x,y)$ in a large (x,y)-range. We have

(73) $$\alpha(x,y) = \alpha_0 + O\Big(\frac{1}{L_\varepsilon(y)} + \frac{1}{\log x \log y}\Big)$$

uniformly for $x \geq x_0(\varepsilon)$, $(\log x)^{1+\varepsilon} \leq y \leq x$.

In order that Lemma 9.1 be applicable with $\sigma = \alpha_0$ as integration abscissa, it is necessary that $\xi(u) \le \log L_\varepsilon(y)$, that is, taking account of Lemma 8.1, that x and y belong to a region of the plane defined for some $\varepsilon > 0$ by

$$(H_\varepsilon) \qquad x > x_0(\varepsilon), \qquad \exp\left\{(\log_2 x)^{(5/3)+\varepsilon}\right\} \le y \le x.$$

Under this assumption, Lemma 9.1 hence suggests that the quantity

$$\frac{1}{2\pi i}\int_{\alpha_0-i\infty}^{\alpha_0+i\infty} \zeta(s)(s-1)\log y\,\widehat{\rho}\big((s-1)\log y\big)x^s\frac{\mathrm{d}s}{s} \tag{74}$$

is a good approximation to $\Psi(x,y)$. The change of variables $(s-1)\log y \mapsto s$ allows us to rewrite (74) in the form

$$\frac{x}{2\pi i}\int_{-\xi(u)-i\infty}^{-\xi(u)+i\infty} \widehat{\rho}(s)\zeta\Big(1+\frac{s}{\log y}\Big)\frac{s}{s+\log y}e^{us}\,\mathrm{d}s. \tag{75}$$

Now, it is easily verified that, for all complex s, we have

$$\frac{s}{s+\log y}\zeta\Big(1+\frac{s}{\log y}\Big) = \int_{-\infty}^{\infty} e^{-st}\,\mathrm{d}\Big(\frac{[y^t]}{y^t}\Big). \tag{76}$$

The convolution theorem then shows that the factor of x in (75) is the inverse Laplace integral for

$$\int_{-\infty}^{\infty} \rho(u-v)\,\mathrm{d}\big([y^v]/y^v\big) = \Lambda(x,y)/x.$$

Thus, in the context of an analysis by the saddle-point method, de Bruijn's function $\Lambda(x,y)$ appears as the natural approximation of $\Psi(x,y)$. In fact, the above heuristic considerations can actually be regarded as a genuine outline argument.

We are now in a position to state the main result of this section.

Theorem 9 (Saias, 1989). *Let $\varepsilon > 0$. For (x,y) in the domain (H_ε), we have that*

$$\Psi(x,y) = \Lambda(x,y)\Big\{1+O\Big(\frac{1}{L_\varepsilon(y)}\Big)\Big\}. \tag{77}$$

Prior to embarking on the proof, let us show how this theorem implies the asymptotic formula (34) of Hildebrand.

Lemma 9.2. *Let $\varepsilon > 0$. Under the condition*

$$x > x_0(\varepsilon), \qquad (\log x)^{1+\varepsilon} \le y \le x, \tag{78}$$

we have

$$\Lambda(x,y) = x\rho(u)\Big\{1 + O\Big(\frac{\log(u+1)}{\log y}\Big)\Big\}. \tag{79}$$

Proof. By partial summation, we can write

$$\Lambda(x,y) = x\rho(u) - \{x\} - x\int_0^u \rho'(u-v)\frac{\{y^v\}}{y^v}\,\mathrm{d}v. \tag{80}$$

By Corollaries 8.3 and 8.4, we have, for all v, $0 \le v \le u$,

$$\rho'(u-v) \ll \xi(u)\rho(u-v) \ll \log(u+1)\rho(u)e^{\xi(u)v}.$$

Now, in the domain (78),

$$\xi(u) \le \log_2 x + O(1) \le \big(1 - \tfrac{1}{2}\varepsilon\big)\log y.$$

This implies that

$$\begin{aligned}\int_0^u \rho'(u-v)y^{-v}\,\mathrm{d}v &\ll \log(u+1)\rho(u)\int_0^u \Big(\frac{e^{\xi(u)}}{y}\Big)^v \mathrm{d}v \\ &\ll \frac{\rho(u)\log(u+1)}{\log y},\end{aligned} \tag{81}$$

and yields the desired conclusion.

Corollary 9.3 (Hildebrand). *Let $\varepsilon > 0$. Uniformly under the condition*

$$x \ge 2, \qquad \exp\big\{(\log_2 x)^{(5/3)+\varepsilon}\big\} \le y \le x,$$

we have

$$\Psi(x,y) = x\rho(u)\Big\{1 + O\Big(\frac{\log(u+1)}{\log y}\Big)\Big\}. \tag{82}$$

Proof. When $x > x_0(\varepsilon)$, this follows immediately from (77) and (79). Since $\rho(u) > 0$ for $u \ge 1$, the conclusion is trivial for $2 \le x \le x_0(\varepsilon)$.

Of course, the information contained in (77) is much more precise than that in formula (82). In particular, an asymptotic expansion for $\Psi(x,y)$ in terms of powers of $\log(u+1)/\log y$ can be deduced from it—cf. the Notes.

The first step in the proof of Theorem 9 consists in truncating the Perron integral in such a way that Lemma 9.1 can be used for the integrand. We state the result below.

Lemma 9.4. *Let $\varepsilon > 0$. Set $T := L_{\varepsilon/2}(y)$. For x, y in (H_ε), we have*

$$\Psi(x,y) = \frac{1}{2\pi i}\int_{\alpha_0 - iT^2}^{\alpha_0 + iT^2} \zeta(s,y)\frac{x^s}{s}\,\mathrm{d}s + O\Big(\frac{x\rho(u)}{L_\varepsilon(y)}\Big). \tag{83}$$

Proof. We appeal to Theorem II.2.2. Let R denote the error term in (83); then we have

$$R \ll x^{\alpha_0} \sum_{P^+(n)\le y} \frac{n^{-\alpha_0}}{1 + T^2|\log(x/n)|} \ll \frac{x^{\alpha_0}\zeta(\alpha_0, y)}{T} + R_T,$$

with $R_T := \Psi(x + x/T, y) - \Psi(x - x/T, y)$. When u is not too large, we use the trivial bound

$$R_T \ll x/T.$$

Otherwise we proceed as in Exercise II.2.3, by introducing the weight function

$$w(t) := \frac{1}{2\pi}\Big(\frac{\sin(tT/2)}{tT/2}\Big)^2$$

with Fourier transform

$$\widehat{w}(\tau) := \int_{-\infty}^{\infty} w(t)e^{-it\tau}\,\mathrm{d}t = \frac{1}{T}\Big(1 - \Big|\frac{\tau}{T}\Big|\Big)^+.$$

We then have

$$R_T \ll \sum_{P^+(n)\le y}\Big(\frac{x}{n}\Big)^{\alpha_0} w\Big(\log\frac{x}{n}\Big) = \frac{1}{2\pi i}\int_{\alpha_0 - iT}^{\alpha_0 + iT}\zeta(s,y)x^s\widehat{w}(\tau)\,\mathrm{d}s.$$

Hence we can finally write

$$R \ll \frac{x^{\alpha_0}\zeta(\alpha_0, y)}{\sqrt{T}} + \min\Big\{\frac{x}{T},\, x^{\alpha_0}\max_{\sqrt{T}\le|\tau|\le T}\big|\zeta(\alpha_0 + i\tau, y)\big|\Big\}. \tag{84}$$

By Lemma 9.1 and Theorem 8, we have that

$$x^{\alpha_0}\zeta(\alpha_0, y) \sim x\,e^{-u\xi(u)}\zeta\Big(1 - \frac{\xi(u)}{\log y}\Big)(-\xi(u))\widehat{\rho}(-\xi(u)) \ll x\log y\sqrt{u}\,\rho(u). \tag{85}$$

The first term on the right-hand side of (84) is hence acceptable. The same holds for the second if $u \le 2\log L_\varepsilon(y)$. Indeed we then have

$$\frac{1}{T} \ll \frac{u^{-2u}}{\sqrt{T}} \ll \frac{\rho(u)}{L_\varepsilon(y)}.$$

When $u > 2\log L_\varepsilon(y)$, we can use Lemma 9.1 (with $\frac{1}{2}\varepsilon$ instead of ε) and Lemma 8.2. We have for $\sqrt{T} < |\tau| \le T$, $s = \alpha_0 + i\tau$,

$$\zeta(s,y) \ll \zeta(s)(s-1)\log y\,\widehat{\rho}\big((s-1)\log y\big) \ll \zeta(s) \ll \log T,$$

since $|\tau|\log y \ge \sqrt{T} \ge 1 + u\xi(u)$. By Theorem 8, we deduce from the above that the second term on the right-hand side of (84) is

$$\ll x^{\alpha_0}\log T \ll xe^{-u\xi(u)}\log y \ll x\rho(u)/L_\varepsilon(y)$$

where we have used the fact that $I(\xi) \sim u$ as $u \to \infty$. This completes the proof of Lemma 9.4.

We are now in a position to complete the proof of Theorem 9. The first step consists in replacing $\zeta(s,y)$ by its smooth approximation in the integral of (83). Applying Lemma 9.1 with $\frac{1}{2}\varepsilon$ instead of ε, we see that the error introduced by this process is

$$\ll \frac{x^{\alpha_0}\zeta(\alpha_0,y)}{T}\int_0^{T^2}\frac{\mathrm{d}\tau}{|\alpha_0+i\tau|}$$

with $T = L_{\varepsilon/2}(y)$. From (85), we get that this bound is of the required order of magnitude.

By carrying out the change of variables $(s-1)\log y \mapsto s$, we can rewrite the new main term in the form

$$\begin{aligned}P &:= \frac{x}{2\pi i}\int_{-\xi(u)-iT^2\log y}^{-\xi(u)+iT^2\log y}\widehat{\rho}(s)\zeta\Big(1+\frac{s}{\log y}\Big)\frac{s}{s+\log y}e^{us}\,\mathrm{d}s\\ &= \frac{x}{2\pi i}\int_{-\xi(u)-iT^2\log y}^{-\xi(u)+iT^2\log y}\widehat{\lambda}_y(s)e^{us}\,\mathrm{d}s\end{aligned}$$

where we have set

$$\text{(86)} \qquad \lambda_y(u) := \int_{-\infty}^{\infty}\rho(u-v)\,\mathrm{d}\big([y^v]/y^v\big) \qquad (y^u \notin \mathbb{Z}^+)$$

with the convention $\lambda_y(u) = \lambda_y(u+)$ if $y^u \in \mathbb{Z}^+$, so that $\Lambda(x,y) = x\lambda_y(u)$ identically.

By splitting the integral of (86) according to whether $v \le \frac{1}{2}u$ or $v > \frac{1}{2}u$, we immediately see that the bound

$$\text{(87)} \qquad \lambda_y(u) \ll \rho(u/2) + y^{-u/2}$$

holds uniformly for $u \geq 0$, $y \geq 2$. This implies that the inverse Laplace integral

$$\lambda_y(u) = \frac{1}{2\pi i}\int_{-\xi(u)-i\infty}^{-\xi(u)+i\infty} \widehat{\lambda}_y(s)e^{us}\,\mathrm{d}s \tag{88}$$

converges whenever $\xi(u) < \frac{1}{2}\log y$, $y^u \notin \mathbb{Z}^+$, and therefore certainly in the domain (H_ε) excluding pairs (x, y) such that $y^u \in \mathbb{Z}^+$. When $x = y^u \in \mathbb{Z}^+$, the integral (88) converges in principal value to

$$\tfrac{1}{2}\left(\lambda_y(u) + \lambda_y(u-)\right) = \lambda_y(u) + \frac{1}{2x}.$$

We can thus assume, from now on, that $x \in \frac{1}{2} + \mathbb{Z}$. It follows that

$$P = \Lambda(x, y) + O\Big(x\int_{\substack{\sigma=-\xi(u)\\ |\tau|>T^2\log y}} \widehat{\lambda}_y(s)e^{us}\,\mathrm{d}s\Big). \tag{89}$$

When $s = -\xi(u) + i\tau$, $|\tau| > T^2$, we have by Lemma 8.2 that

$$s\widehat{\rho}(s) = 1 + O\Big(\frac{1+u\xi}{s}\Big) \tag{90}$$

and it follows from the usual estimates for the zeta function that

$$\frac{1}{s+\log y}\zeta\Big(1+\frac{s}{\log y}\Big) \ll |s|^{-1/2}. \tag{91}$$

The error term in (89) is thus

$$\begin{aligned}
&\ll x\int_{\substack{\sigma=-\xi(u)\\ |\tau|>T^2\log y}} \zeta\Big(1+\frac{s}{\log y}\Big)\frac{e^{us}}{s+\log y}\,\mathrm{d}s + O\Big(x\frac{e^{-u\xi(u)}(1+u\xi(u))}{T\surd(\log y)}\Big)\\
&= \int_{\substack{\sigma=\alpha_0\\ |\tau|>T^2}} \zeta(s)\frac{x^s}{s}\,\mathrm{d}s + O\Big(\frac{x\rho(u)}{L_\varepsilon(y)}\Big).
\end{aligned}$$

In order to estimate the last integral, we use the approximation of $\zeta(s)$ by the partial sum of the series in the form of Corollary II.3.5.1, viz.

$$\zeta(s) = \sum_{n\leq|\tau|} n^{-s} - \frac{|\tau|^{1-s}}{1-s} + O(|\tau|^{-\sigma}) = \sum_{n\leq|\tau|} n^{-s} + O(|\tau|^{-\sigma}).$$

It follows that

$$\int_{\substack{\sigma=\alpha_0\\|\tau|>T^2}} \zeta(s)x^s\frac{\mathrm{d}s}{s} = \sum_{n=1}^{\infty}\int_{|\tau|\geq\max(n,T^2)}\Big(\frac{x}{n}\Big)^s\frac{\mathrm{d}s}{s} + O\Big(\frac{x^{\alpha_0}}{T}\Big).$$

The last term is clearly acceptable. Formula (II.2.7) allows us to estimate the general term of the sum over n by

$$\ll \Big(\frac{x}{n}\Big)^{\alpha_0}\frac{1}{1+(n+T^2)|\log(x/n)|}.$$

Considering separately the two cases obtained in comparing $|x-n|$ to $x^{3/4}$,we therefore obtain that the n-sum is

$$\ll \sum_{n=1}^{\infty}\frac{x^{\alpha_0}}{n^{3/2}+T^2} + x^{3/4} \ll \frac{x\rho(u)}{T}.$$

This shows that the error term in (89) is of the required order, and thus completes the proof of Theorem 9.

When condition (H_ε) is not fulfilled, the saddle-point method is still efficient, provided that one chooses the theoretical abscissa of integration $\sigma = \alpha(x,y)$ defined by (65). This route was followed by Hildebrand and the author (1986), who obtain the following result, valid for $x \geq y \geq 2$. We write

$$\varphi_y(\sigma) = \sum_{p\leq y}\frac{\log p}{p^\sigma - 1}, \qquad \varphi_y'(\sigma) = \frac{\mathrm{d}\varphi_y(\sigma)}{\mathrm{d}\sigma} = \sum_{p\leq y}\frac{p^\sigma(\log p)^2}{(p^\sigma-1)^2}.$$

Theorem 10 (Hildebrand–Tenenbaum). *Uniformly for $x \geq y \geq 2$, we have*

(92) $$\Psi(x,y) = \frac{x^\alpha\zeta(\alpha,y)}{\alpha\sqrt{(2\pi\varphi_y'(\alpha))}}\Big\{1+O\Big(\frac{1}{u}+\frac{\log y}{y}\Big)\Big\},$$

(93) $$\varphi_y'(\alpha) = \Big(1+\frac{\log x}{y}\Big)\log x\log y\Big\{1+O\Big(\frac{1}{\log(u+1)}+\frac{1}{\log y}\Big)\Big\}.$$

Moreover, we have for $0<\varepsilon<\frac{1}{2}$, $y \geq (\log x)^{1+\varepsilon}$,

(94) $$\Psi(x,y) = x\rho(u)\exp\Big\{O\Big(\frac{\log(u+1)}{\log y}+\frac{u}{L_\varepsilon(y)}+\frac{1}{u}\Big)\Big\}.$$

We shall not prove this result here, and restrict ourselves to some comments. First of all observe that estimate (12) for α and formula (93) imply that

$$\log y \ll \alpha\sqrt{(2\pi\varphi_y'(\alpha))} \sim \log\Big(1+\frac{y}{\log x}\Big)\sqrt{2\pi u\Big(1+\frac{\log x}{y}\Big)} \ll \sqrt{\frac{y}{\log y}}. \tag{95}$$

This allows us to compare Rankin's upper bound with the true order of magnitude of $\Psi(x,y)$. Although it never gives the exact order, Rankin's method is remarkably efficient: when, for instance, $y=\log x$, every estimate available in the literature imply an error factor $\gg \exp\{y^{1+o(1)}\}$.

Next, we note that (94) is almost equivalent to Corollary 9.3: it needs only a slight strengthening of Theorem 6 to obtain the result.

Finally, we draw the reader's attention to the interest of a formula like (92), depending on an implicit parameter such as $\alpha(x,y)$. As we have already seen, inserting an explicit approximation for α yields a genuine asymptotic formula. Moreover, there exists an application of another type, resting on the fact that small variations in $\alpha(x,y)$ are relatively easy to study. This allows an investigation of the *local behaviour* of $\Psi(x,y)$, even in regions where the global behaviour is not completely understood. The following result, characteristic of the method, can be obtained by an immediate extension of Theorem 3 of Hildebrand & Tenenbaum (1986).

Theorem 11. *Uniformly for $x \ge y \ge 2$, $c \ge 1$ and with $t := (\log c)/\log y$, we have*

$$\Psi(cx,y) = \Psi(x,y)c^{\alpha(x,y)}\Big\{1+O\Big((t^2+1)\Big(\frac{1}{u}+\frac{\log y}{y}\Big)\Big)\Big\}. \tag{96}$$

Thus, for example, we have

$$\Psi(2x,y)\sim\Psi(x,y) \Leftrightarrow y\le(\log x)^{1+o(1)}$$

and

$$\Psi(2x,y)\sim 2\Psi(x,y) \Leftrightarrow (\log y)/\log_2 x\to\infty.$$

For another application of Theorem 11, see Exercise 7.

Estimate (92) also provides information on the local behaviour with respect to the variable y. It is easy, for instance, to deduce from (92) and the estimate (12) for $\alpha(x,y)$ that

$$\Psi(x,y)/\Psi(x,y-)\sim(\log x)/y \tag{97}$$

provided that $y\le(\log x)^{1-\varepsilon}$. Hence in this region there does not exist a continuous function asymptotic to $\Psi(x,y)$. Hildebrand (1986f) showed that a continuous approximation cannot be "too precise" in the range $y\le(\log x)^{2-\varepsilon}$.

The following theorem gives a uniform one-sided estimate for the local behaviour of $\Psi(x,y)$. It is an simple consequence of (92).

Theorem 12. *We have uniformly for $x \geq y \geq 2$, $c \geq 1$,*

$$\Psi(cx, y) \leq c^{\alpha}\Psi(x, y)\Big\{1 + O\Big(\frac{1}{u} + \frac{\log y}{y}\Big)\Big\}. \tag{98}$$

Proof. Let $\alpha_1 := \alpha(cx, y)$. Then $\alpha_1 \leq \alpha$. By (92), we can write

$$\Psi(cx, y) = \frac{(cx)^{\alpha_1}\zeta(\alpha_1, y)}{\alpha_1\sqrt{(2\pi\varphi_y'(\alpha_1))}}\Big\{1 + O\Big(\frac{1}{u} + \frac{\log y}{y}\Big)\Big\}.$$

By definition of α_1, we have $(cx)^{\alpha_1}\zeta(\alpha_1, y) \leq (cx)^{\alpha}\zeta(\alpha, y)$. Furthermore, a routine calculation enables us to verify that $\sigma \mapsto \sigma\sqrt{(\varphi_y'(\sigma))}$ is a decreasing function of σ. Applying (92) a second time after replacing α_1 by α on the right-hand side, we obtain the stated inequality.

Notes

§ **5.1.** It is often of importance in number theory to decompose an integer into two or more factors determined by the size of their prime factors. This partially explains the frequent occurrrence of the function $\Psi(x, y)$ in the analytic theory of numbers. Daboussi (1984) showed that the prime number theorem can be proved by analysing the limiting case in the model consisting of integers n with $P^+(n) \leq y$. In fact, this result is only one example of application of a fertile general method, which rests on properties of integers without large prime factors—cf. Daboussi (1989).

Theorem 2 is slightly more precise than de Bruijn's original result, where the error term only tends to 0 if $u \to \infty$.

For historical surveys of the numerous works dedicated to the asymptotic behaviour of $\Psi(x, y)$, see Norton (1971), Hildebrand & Tenenbaum (1993b).

§ **5.2.** Ennola (1969) also gives a more complicated asymptotic formula when $y \leq (\log x)^{3/4}$.

§ **5.3.** The iterative method of Hildebrand is simply illustrated in the proof of the following result.

Theorem 13 (Hildebrand, 1986). *Let $\varepsilon > 0$. There exists a constant $C_3 > 0$ such that, uniformly for $x \geq y \geq 2$, we have*

$$\Psi(x,y) \geq x\rho(u)\exp\{-C_3 u/L_\varepsilon(u)\}. \tag{99}$$

Proof. We may assume that $u \leq y^2$. Otherwise Theorem 8 easily shows that the right-hand side of (99) does not exceed 1.

When $u \leq y^2$, we can suppose that y is sufficiently large, by modifying C_3 if need be. Then let us fix $y \geq y_0$ and set

$$\delta(u) := \inf_{0 \leq v \leq u} \Psi(y^v, y)/y^v \rho(v).$$

The functional equation (33) allows us to write

$$\Psi(x,y)\log x \geq \sum_{d \leq y} \Lambda(d)\Psi(x/d,y) \geq x\delta(u)S_{\frac{1}{2}} + x\delta\left(u - \tfrac{1}{2}\right)\left(S_1 - S_{\frac{1}{2}}\right), \tag{100}$$

where

$$S_\theta := \sum_{d \leq y^\theta} \frac{\Lambda(d)}{d}\rho\left(u - \frac{\log d}{\log y}\right) \qquad (0 \leq \theta \leq 1).$$

Assume for the time being the estimate

$$S_\theta = \log y \int_0^\theta \rho(u-v)\,\mathrm{d}v + O\left(\rho(u)\left\{1 + \frac{u\log(u+1)}{L_\varepsilon(y)}\right\}\right). \tag{101}$$

Since $\int_0^1 \rho(u-v)\,\mathrm{d}v = u\rho(u)$, by substituting in (100) and then dividing by $x\rho(u)\log x = xu\rho(u)\log y$, we obtain

$$\frac{\Psi(x,y)}{x\rho(u)} \geq \delta(u)r(u) + \delta\left(u - \tfrac{1}{2}\right)\left\{1 - r(u) + O\left(R_\varepsilon(u)\right)\right\}$$

with

$$r(u) := \frac{1}{u\rho(u)}\int_0^{\frac{1}{2}} \rho(u-v)\,\mathrm{d}v, \qquad R_\varepsilon(u) := \frac{1}{u\log y} + \frac{\log(u+1)}{L_\varepsilon(y)}.$$

Since ρ is decreasing, we have that $r(u) \leq \frac{1}{2}$. Similarly, since δ is also decreasing we have

$$\delta(u)r(u) + \delta\left(u - \tfrac{1}{2}\right)\left(1 - r(u)\right) \geq \tfrac{1}{2}\left\{\delta(u) + \delta\left(u - \tfrac{1}{2}\right)\right\}$$

and hence, for $y \geq y_0$, $u \leq y^2$,

$$\delta(u) \geq \delta\big(u - \tfrac{1}{2}\big)\Big\{1 + O\big(R_\varepsilon(u)\big)\Big\} \geq \delta\big(u - \tfrac{1}{2}\big)\exp\Big\{O\big(R_\varepsilon(u)\big)\Big\}.$$

Iterating, it follows that

$$\begin{aligned}\delta(u) &\geq \exp\Big\{O\Big(\sum_{k\leq u} R_\varepsilon\big(\tfrac{1}{2}\,k\big)\Big)\Big\} \geq \exp\Big\{O\Big(\frac{\log(u+1)}{\log y} + \frac{u\log(u+1)}{L_\varepsilon(y)}\Big)\Big\}\\ &\geq \exp\big\{O\big(u/L_{2\varepsilon}(u)\big)\big\}.\end{aligned}$$

It remains to show (101). The prime number theorem implies that

$$\sum_{d\leq y^v} \frac{\Lambda(d)}{d} = v\log y - \gamma + O\big(1/L_\varepsilon(y^v)\big).$$

Hence

$$S_\theta = \int_0^\theta \rho(u-v)\,\mathrm{d}\{v\log y\} + \Big[O\Big(\frac{\rho(u-v)}{L_\varepsilon(y^v)}\Big)\Big]_0^\theta + O\Big(\int_0^\theta \frac{\rho'(u-v)}{L_\varepsilon(y^v)}\,\mathrm{d}v\Big).$$

The first term is indeed the main term of (101). Set $\kappa := \frac{3}{5} - \varepsilon$. Corollary 8.4 allows us to bound the second term by

$$\ll \rho(u) + \rho(u)\exp\Big\{\theta\xi(u) - (\theta\log y)^\kappa\Big\} \ll \rho(u)\Big\{1 + \frac{u\log(u+1)}{L_\varepsilon(y)}\Big\}.$$

By Corollaries 8.3 and 8.4, we see that the third term is

$$\ll \rho(u)\log(u+1)\int_0^1 \exp\big\{v\xi(u) - (v\log y)^\kappa\big\}\,\mathrm{d}v.$$

If $\xi(u) \leq \frac{1}{2}\,(\log y)^\kappa$, the last integral is at most

$$\int_0^1 \exp\Big\{-\tfrac{1}{2}\,(v\log y)^\kappa\Big\}\,\mathrm{d}v \ll \frac{1}{\log y} \ll \frac{1}{\log(u+1)}.$$

If $\xi(u) > \frac{1}{2}\,(\log y)^\kappa$, it does not exceed

$$\begin{aligned}&\int_0^{1/2} e^{v\xi(u)}\,\mathrm{d}v + \int_{1/2}^1 \exp\big(v\xi(u) - (\tfrac{1}{2}\,\log y)^\kappa\big)\,\mathrm{d}v\\ &\qquad\qquad \ll \frac{e^{\xi(u)/2}}{\log(u+1)} + \frac{e^{\xi(u)}}{L_{2\varepsilon}(y)\log(u+1)} \ll \frac{u}{L_{2\varepsilon}(y)}.\end{aligned}$$

This establishes (101) and completes the proof.

§ **5.4.** In the form presented here, Theorem 8 was established by Alladi (1982), improving on an asymptotic formula of de Bruijn (1951a). The method employed by these authors is different from ours, and among other things depends on the following result.

Theorem 14 (de Bruijn; Alladi). *Let $f(u)$ be a continuous function for $u > 0$, satisfying*

$$uf(u) = \int_{u-1}^{u} f(t)\,\mathrm{d}t \qquad (u > 1). \tag{102}$$

Then, for some suitable constant C and all λ with $0 < \lambda < \frac{1}{4}$, we have

$$f(u) = \{C + O(e^{-u^\lambda})\}\rho(u) \qquad (u \to \infty).$$

The original proof uses the theory of Volterra equations. An alternative route consists in tackling $f(t)$ directly by inverse Laplace transform via the saddle-point method. Hildebrand and the author (1993a) obtain in this way the following result.

Theorem 15 (Hildebrand–Tenenbaum). *Let $0 < a < 2\pi^2$. In addition to the hypotheses of Theorem 14, suppose also that $f(0+) = 1$. We then have*

$$f(u) = \Big\{C + O\big(e^{-au/\log^2(u+1)}\big)\Big\}\rho(u) \qquad (u \to \infty)$$

with $C := 1 + \displaystyle\int_{-\infty}^{+\infty}\int_0^1 \frac{vf'(v)e^{-vt}}{t\widehat{\rho}(t)}\,\mathrm{d}v\,\mathrm{d}t.$

The method utilised here to estimate $\rho(u)$ easily yields an asymptotic expansion in powers of $1/u$. For a detailed proof of such a result and its extension to (positive real) convolution powers of $\rho(u)$, see Smida (1991), who extends a theorem of Hensley (1986). The case of complex convolution powers of $\rho(u)$ has been treated by Hildebrand (1990).

§ **5.5** The estimate (73) for $\alpha(x, y)$ is established in Hildebrand & Tenenbaum (1986), §7. See also Lemma 6.1.1.

As Saias (1989) showed, one can obtain an asymptotic expansion for $\Lambda(x, y)$. The most pleasing way to arrive at this result consists in writing down the Taylor–Lagrange formula for $\rho(u)$—cf. Fouvry & Tenenbaum (1991), lemma 4.2. If we set

$$r_{mj} := \rho^{(j)}(m) - \rho^{(j)}(m-) \qquad (0 \le m \le j),$$

for $u \ge 0$, $v \ge 0$, $k \ge 0$, we have

$$\begin{aligned}\rho(u - v) = {} & \sum_{j=0}^{k} \frac{(-1)^j}{j!}\rho^{(j)}(u)v^j \\ & + \sum_{j=0}^{k} \frac{(-1)^{j+1}}{j!} \sum_{\substack{u-v<m\le j\\ m\le u}} r_{mj}(v + m - u)^j \\ & + \frac{(-1)^{k+1}}{k!}\int_0^v (v - w)^k \rho^{(k+1)}(u - w)\,\mathrm{d}w.\end{aligned} \tag{103}$$

By integrating this formula with respect to the measure $d\big([y^v]/y^v\big)$ we obtain for all fixed $\varepsilon > 0$, $k \geq 0$,

$$\Lambda(x,y) = x\sum_{j=0}^{k} a_j \frac{\rho^{(j)}(u)}{(\log y)^j} + O\Big(x\frac{\rho^{(k+1)}(u)}{(\log y)^{k+1}}\Big) \tag{104}$$

uniformly under the conditions

$$x \geq 2, \qquad (\log x)^{1+\varepsilon} \leq y \leq x, \qquad \min_{0\leq j\leq k,\, j\leq u}\Big(\frac{u-j}{k+1-j}\Big) \geq \frac{\log_2 y}{\log y}, \tag{105}$$

where the a_j are the Taylor coefficients at $s = 0$ for $s\zeta(s+1)/(s+1)$. The last of the conditions (105) is indeed necessary—cf. Saias (1989).

Although Theorem 9 was obtained in a thesis supervised by the author, the proof given here is somewhat simpler than the original one.

Hildebrand (1984a) showed that extending the validity range of the formula

$$\Psi(x,y) = x\rho(u)\{1 + O\big(\log(u+1)/\log y\big)\}$$

to $y \geq (\log x)^{2+\varepsilon}$ is equivalent to the Riemann hypothesis.

The method of proof of Lemma 9.4 provides an estimate for the number of integers without large prime factor in short intervals. See Lemma 6.7.3, and Theorem 4 of Hildebrand & Tenenbaum (1986).

A more complete account on the saddle-point method and its arithmetic applications may be found in the author's survey (1988).

Exercises

1. For $\varepsilon > 0$, $y \geq 2$, set $L_\varepsilon(y) := \exp\{(\log y)^{(3/5)-\varepsilon}\}$.

(a) Using a strong form of the prime number theorem, show that, for any $\varepsilon > 0$, and uniformly for $y \geq 2$, $\sigma > 0$, one has

$$\sum_{p\leq y}\frac{\log p}{p^\sigma - 1} = \Big\{1 + O\Big(\frac{1}{L_\varepsilon(y)}\Big)\Big\}\int_2^y \frac{dt}{t^\sigma - 1} + O(1).$$

[*One can integrate by parts the contribution to the p-sum from the "remainder" $R(t) := \theta(t) - t$, and then consider separately the three cases obtained according to the position of σ relative to the points $\frac{1}{2}$ and $1 - 1/(2L_\varepsilon(y))$.*]

(b) Let $\delta > 0$. Show that, uniformly for $\sigma \geq \delta$, one has

$$\int_2^y \frac{\mathrm{d}t}{t^\sigma - 1} = \frac{1 + O(1/L_\varepsilon(y))}{1 - y^{-\sigma}} \int_1^y \frac{\mathrm{d}t}{t^\sigma} + O(1).$$

Show that the conclusion continues to hold for $\sigma \geq 3\log_2 2y/\log y$, replacing $L_\varepsilon(y)$ by $\log y$ if necessary.

[*One can use the relation* $(t^\sigma - 1)^{-1} = t^{-\sigma} + O\big(t^{-2\sigma}(1 - 2^{-\sigma})^{-1}\big)$ $(t \geq 2)$.]

(c) Suppose now that $0 < \sigma \leq \min\big(2/3, (3\log_2 2y)/\log y\big)$. Establish the relations

$$\int_1^y \frac{\mathrm{d}t}{t^\sigma} = \big\{1 + O(y^{-1/6})\big\} \int_{\sqrt{y}}^y \frac{\mathrm{d}t}{t^\sigma}, \quad \int_2^y \frac{\mathrm{d}t}{t^\sigma - 1} = \Big\{1 + O\Big(\frac{\log y}{y^{1/6}}\Big)\Big\} \int_{\sqrt{y}}^y \frac{\mathrm{d}t}{t^\sigma - 1}.$$

Show that

$$\int_{\sqrt{y}}^y \Big(\frac{1}{1 - t^{-\sigma}} - \frac{1}{1 - y^{-\sigma}}\Big) \frac{\mathrm{d}t}{t^\sigma} \ll \frac{y}{(y^\sigma - 1)\log y}.$$

[*One can use the inequality* $1 - (y/t)^{-\sigma} \leq \sigma \log(y/t)$.]

(d) Establish the following formula, uniformly for $y \geq 2$, $\sigma > 0$,

$$\sum_{p \leq y} \frac{\log p}{p^\sigma - 1} = \frac{1 + O(E)}{1 - y^{-\sigma}} \int_1^y \frac{\mathrm{d}t}{t^\sigma} + O(1)$$

with $E \ll_\delta 1/L_\varepsilon(y)$ if $\sigma \geq \delta$, and $E \ll 1/\log y$ for all $\sigma > 0$. Show that $(1 - y^{-\sigma})\log y \ll y^{1-\sigma} - 1$ for $0 < \sigma \leq 1$ and deduce formula (11) of § 5.1.

2. *An arithmetical proof of the formula* $\widehat{\rho}(0) = e^\gamma$.

(a) Deduce from Theorem 1 that, for $x \geq y \geq 2$, one has

$$\sum_{n > x,\, P^+(n) \leq y} \frac{1}{n} \ll e^{-u/2} \log y.$$

(b) Deduce from Theorem 6 that, for $x \geq y \geq 2$, one has

$$\sum_{n \leq x,\, P^+(n) \leq y} \frac{1}{n} = \log y \int_0^u \rho(v)\,\mathrm{d}v + O(u).$$

(c) Using Mertens' formula, show that we have for $u \geq 1$, $y \geq 2$,

$$\int_0^u \rho(v)\,\mathrm{d}v = e^\gamma + O\Big(\frac{u}{\log y} + e^{-u/2}\Big)$$

and deduce, with a suitable choice of $u = u(y)$, that $\widehat{\rho}(0) = e^\gamma$.

3. *Mean value of* $\log P^+(n)$. Let $S(x) := \sum_{n\le x} \log P^+(n)$.

(a) Show that $S(x) = x\log x - \int_1^x \big(\Psi(x,y)/y\big)\,\mathrm{d}y + O(\log x)$.

(b) Applying Theorem 6, show that

$$S(x) = \alpha x \log x + O(x \log_2 x)$$

with $\alpha := 1 - \int_1^{+\infty} \frac{\rho(v)}{v^2}\,\mathrm{d}v \approx 0.62433$.

(c) Deduce from (a) and from Theorem 9 that, for all $\varepsilon > 0$, one has

$$S(x) = x\log x - \int_1^x \Lambda(x,y)\frac{\mathrm{d}y}{y} + O\Big(x \exp\big\{-(\log x)^{(3/8)-\varepsilon}\big\}\Big).$$

(d) Show that, for $|s| < 1$, one has

$$\int_{1-}^{\infty} t^{-s}\,\mathrm{d}\Big(\frac{[t]}{t}\Big) = \frac{s\zeta(s+1)}{s+1}$$

where the integral is uniformly convergent on each compact subset. Deduce that

$$\int_1^x \log t\,\mathrm{d}\Big(\frac{[t]}{t}\Big) = 1 - \gamma + O\Big(\frac{\log x}{x}\Big),$$

where γ is Euler's constant.

(e) Show that

$$\begin{aligned}\int_1^x \Lambda(x,y)\frac{\mathrm{d}y}{y} &= x\int_{1-}^x \log\Big(\frac{x}{t}\Big)\int_{1-(\log t)/\log x}^{\infty} \frac{\rho(v)}{v^2}\,\mathrm{d}v\,\mathrm{d}\Big(\frac{[t]}{t}\Big)\\ &= (1-\alpha)x\log x + \alpha(1-\gamma)x + O(\log x)\end{aligned}$$

and deduce the asymptotic formula

$$S(x) = \alpha x\log x - \alpha(1-\gamma)x + O\Big(x\exp\big\{-(\log x)^{(3/8)-\varepsilon}\big\}\Big).$$

4. Let δ, $0 < \delta < 1$, be some fixed real number. Set $g_\delta(n) := \sum_{p|n} (\log p)^\delta$.

(a) Show that $\sum_{n\le x} g_\delta(n) \sim (x/\delta)(\log x)^\delta \quad (x \to \infty)$.

(b) Let $N(x;\lambda) := \big|\{n \le x : g_\delta(n) \le (\lambda/\delta)(\log x)^\delta\}\big| \quad (\lambda > 1)$. Show that $N(x;\lambda) \ge \{(\lambda - 1)/\lambda + o(1)\}x \quad (x \to \infty)$.

(c) For $n > 1$, write $\alpha_n := \big(\log P^+(n)\big)/\log n$. Show that α_n has a distribution function, and determine it.

(d) Show that, for all $n > 1$, one has $g_\delta(n) \ge (\alpha_n)^{\delta-1}(\log n)^{\delta-1} g_1(n)$. Using Exercise 3.3 and the previous question, show that, for each $\lambda > 0$, there exists a quantity $c(\lambda,\delta) > 0$ such that $N(x;\lambda) \le \{1 - c(\lambda,\delta) + o(1)\}x \quad (x \to \infty)$.

(e) Show that g_δ fails to have a non-decreasing normal order.

5. Let $N_k(z)$ be the quantity defined in §5.2.

(a) Calculate $F(\sigma) := \int_{0-}^{\infty} e^{-\sigma t}\, \mathrm{d}N_k(t)$.

(b) Show that $N_k(z) \le e^k F(k/z)$.

(c) Deduce the estimate

$$N_k(z) \le \left(\frac{ez}{k}\right)^k \exp\left\{\frac{k}{2z}\sum_{i=1}^{k} a_i\right\}\prod_{i=1}^{k}\frac{1}{a_i}.$$

For which values of z is this bound more precise than that of Theorem 3 ?

6. *Integers for which the product of the small prime factors is large.*

(a) Using the large sieve (Corollary I.4.6.1) show that, for $x \ge y \ge 2$, one has

$$\Phi(x,y) := \left|\{n \le x : P^-(n) > y\}\right| \ll \frac{x}{\log y}.$$

(b) Let $\Theta(x,y,z) := \left|\{n \le x : \prod_{p^\nu \| n,\, p \le y} p^\nu > z\}\right|$. Show that

$$\Theta(x,y,z) \le \sum_{z < a \le x/y,\ P^+(a) \le y} \Phi(x/a, y) + \Psi(x,y).$$

Deduce that, uniformly for $x \ge z \ge y \ge 2$, $v := (\log z)/\log y$, one has

$$\Theta(x,y,z) \ll x e^{-v/2}.$$

(c) Using Theorems 2 and 10, show that, for each $\varepsilon > 0$ and uniformly for $x \ge y \ge 2$, one has

$$\Psi(x,y) \ll x u^{-u} + x^{\varepsilon}.$$

Deduce that, for any $\varepsilon > 0$, and uniformly for $x \ge z \ge y \ge 2$ with $v := (\log z)/\log y$, one has

$$\Theta(x,y,z) \ll_\varepsilon x v^{-v} + x z^{-1+\varepsilon}.$$

7. *Squarefree integers without large prime factors.* Using Theorems 9 and 11, as well as the estimate (12), show that, uniformly for $x \ge y \ge 2$, one has

$$\sum_{n \le x,\ P^+(n) \le y} \mu(n)^2 = \{1/\zeta(\beta_1) + o(1)\}\Psi(x,y)$$

with $\beta_1 := \max(1, 2\beta(x,y))$ where $\beta(x,y)$ is defined by (13). [*For a more precise result, see Ivić & Tenenbaum (1986), Naïmi (1988).*]

III.6

Integers free of small prime factors

§ 6.1 Introduction

Here we undertake a study dual to that of the previous chapter, namely the evaluation of the quantity

$$\Phi(x,y) := \big|\{n \leq x : P^-(n) > y\}\big| \qquad (x \geq y \geq 2).$$

As is the case with $\Psi(x,y)$, this function occurs constantly in analytic and probabilistic number theory. In particular, it is of fundamental use in sieve problems.

We keep to the systematic notation

$$u := \frac{\log x}{\log y} \qquad (x \geq y \geq 2)$$

introduced in Chapter 5. We have seen (Theorem I.4.2) that Brun's pure sieve provides the formula

$$\text{(1)} \qquad \Phi(x,y) = \frac{x}{\zeta(1,y)}\Big\{1 + O\Big(\frac{1}{(\log y)^2}\Big)\Big\} \qquad (2 \leq y \leq x^{1/10\log_2 x})$$

while the fundamental lemma of the combinatorial sieve (Theorem I.4.3) implies the following estimate, valid uniformly for $x \geq y \geq 2$,

$$\text{(2)} \qquad \Phi(x,y) = \frac{x}{\zeta(1,y)}\{1 + O(u^{-u/2})\} + O(\Psi(x,y)).$$

The appearance of $\Psi(x,y)$ as error term in the evaluation of $\Phi(x,y)$ is not surprising. The characteristic function $\eta(n;y)$ of the set of integers n such that $P^-(n) > y$ is indeed multiplicative, and, by the Möbius inversion formula, it can be written as

$$\text{(3)} \qquad \eta(n;y) = \sum_{d|n,\, P^+(d)\leq y} \mu(d) \qquad (n \geq 1).$$

Summing over $n \le x$, we get

$$\Phi(x,y) = \sum_{d\le x,\, P^+(d)\le y} \mu(d)\big[x/d\big] \qquad (x \ge y \ge 2). \tag{4}$$

The main term of (2) corresponds to the quantity obtained from (4) by ignoring the square brackets and extending the summation to infinity. It is clear that the error implied in this process involves $\Psi(x,y)$.

In order to obtain, by this approach, our first result, we will use the following technical estimate concerning the saddle-point $\alpha(x,y)$, defined in (5.65) as the unique solution of the equation

$$\sum_{p\le y} \frac{\log p}{p^\alpha - 1} = \log x. \tag{5}$$

To this end, we introduce the further notation

$$\overline{u} := \min\big(u, y/\log y\big) = \min(\log x, y)/\log y. \tag{6}$$

Lemma 1.1. *Uniformly for $x \ge y \ge 2$, we have that*

$$\frac{y^{1-\alpha} - 1}{(1-\alpha)\log y} \asymp \overline{u}. \tag{7}$$

Proof. Formula (5.11) enables us to evaluate the left-hand side of (5). We obtain

$$u(1 - y^{-\alpha}) = \frac{y^{1-\alpha} - 1}{(1-\alpha)\log y}\Big(1 + O\Big(\frac{1}{\log y}\Big)\Big). \tag{8}$$

Now the estimate (5.12), namely

$$\alpha = \frac{\log(1 + y/\log x)}{\log y}\Big\{1 + O\Big(\frac{\log_2 y}{\log y}\Big)\Big\}, \tag{9}$$

implies that

$$1 - y^{-\alpha} \asymp \min\big(1, y/\log x\big)$$

provided y is sufficiently large. Actually, this estimate remains true when y is bounded: on the one hand, by (5), $\log 2/(2^\alpha - 1) \le \log x$, hence $1 - y^{-\alpha} \ge 1 - 2^{-\alpha} \gg 1/\log x$, and on the other hand, by (9), $1 - y^{-\alpha} \le \alpha \log y \ll 1/\log x$.

We immediately deduce the desired result by substituting in (8).

As in Chapter 5, let $\xi(u)$ denote the unique real non-zero root of the equation $e^{\xi} = 1 + u\xi$ for $u > 0$, $u \neq 1$, and define $\xi(1) = 0$. The estimates (5.47) and (5.59) for $\xi(u)$ and $\xi'(u)$ allow us to deduce immediately from Lemma 1.1 that

$$\alpha = 1 - \frac{\log\left(\overline{u}\log(\overline{u}+1)\right) + O(1)}{\log y} \qquad (x \geq y \geq 2). \tag{10}$$

Theorem 1. *Uniformly for $x \geq y \geq 2$, we have that*

$$\Phi(x,y) = \frac{x}{\zeta(1,y)} + O\big(\Psi(x,y)\big). \tag{11}$$

Remark. This estimate is only non-trivial for

$$y \leq x^{c \log_3 x/\log_2 x},$$

where $c = c(x) \to 1$. Outside such a range, formula (11) is weaker than the sieve upper bound (cf. for example Corollary I.4.6.1)

$$\Phi(x,y) \ll \frac{x}{\log y} \qquad (x \geq y \geq 2).$$

Proof. As we just observed, we can assume that u is sufficiently large, so that (9) and (10) imply

$$(1-\alpha)\log y \geq c_0 > 0. \tag{12}$$

By (4), the error term in (11) is

$$-\sum_{\substack{d \leq x \\ P^+(d) \leq y}} \mu(d)\Big\{\frac{x}{d}\Big\} - x \sum_{\substack{d > x \\ P^+(d) \leq y}} \frac{\mu(d)}{d} \ll \Psi(x,y) + x\int_x^{+\infty} |M(t,y)| \frac{\mathrm{d}t}{t^2},$$

with

$$M(x,y) := \sum_{d \leq x,\, P^+(d) \leq y} \mu(d).$$

Bounding trivially $|M(t,y)|$ by $\Psi(t,y)$, we obtain, by Theorem 5.12, a result slightly inferior to the required estimate. We have indeed

$$x\int_x^{\infty} \Psi(t,y)\frac{\mathrm{d}t}{t^2} \ll x^{1-\alpha}\Psi(x,y)\int_x^{\infty} t^{\alpha-2}\,\mathrm{d}t \ll \frac{\Psi(x,y)}{1-\alpha}.$$

However, a result of the author (1990), also proved by the saddle-point method, states that

$$M(x,y) \ll \Psi(x,y)\Big(\frac{\exp\{-c_1 u/\log^2(u+1)\}}{\log y} + \exp\big\{-(\log y)^{(3/2)-\varepsilon}\big\}\Big) \tag{13}$$

uniformly for $x \geq y \geq 2$. Thus we can certainly divide the above upper bound for the t-integral by $\log y$, and the required result then follows from (12).

§ 6.2 Functional equations

In common with $\Psi(x,y)$ and most functions arising in sieve problems for primes, $\Phi(x,y)$ too has a functional equation.

Theorem 2. *For $x \geq 1$, $y \geq 1$, we have*

$$\Phi(x,y) = 1 + \sum_{y<p\leq x} \sum_{\nu\geq 1} \Phi(x/p^\nu, p). \tag{14}$$

Remark. As in Chapter 5, this equation is normally used in the Buchstab form

$$\Phi(x,y) = \Phi(x,z) + \sum_{y<p\leq z} \sum_{\nu\geq 1} \Phi(x/p^\nu, p) \qquad (x \geq z \geq y \geq 1). \tag{15}$$

Proof. If $n > 1$ is counted by $\Phi(x,y)$, then n can be uniquely written in the form $n = p^\nu m$, with $p = P^-(n)$, $p \nmid m$. These two conditions are equivalent to $P^-(m) > p$. It follows that

$$\Phi(x,y) = 1 + \sum_{y<p\leq x} \sum_{n\leq x,\, P^-(n)=p} 1 = 1 + \sum_{y<p\leq x} \sum_{\nu\geq 1} \sum_{m\leq x/p^\nu,\, P^-(m)>p} 1,$$

which is (14).

By using the trivial upper bound x/p^ν of $\Phi(x/p^\nu, p)$ for $\nu \geq 2$, we deduce from (15) the approximate equation

$$\Phi(x,y) = \Phi(x,z) + \sum_{y<p\leq z} \Phi(x/p, p) + O(x/y) \qquad (x \geq z \geq y \geq 1). \tag{16}$$

If we had replaced the inequality $P^-(n) > y$ in the definition of $\Phi(x,y)$ by the weaker $P^-(n) \geq y$, then the approximate functional equation (16) would, in fact, hold without an error term. Here, however, we continue to adhere to the definition as previously stated: this is more convenient when decompositions of integers into products with terms defined according to the size of their prime factors are involved.

As in Section 5.3, the functional equation may be used in order to estimate the sieve function. One argues by induction on $[u]$ starting from the trivial case $u \leq 1$. A consideration of the first two steps indicates the general form of the approximation.

When $\sqrt{x} < y \leq x$, the inner sum of (14) is identically equal to 1 (since $P^-(1) = \infty$!) and we have $\Phi(x,y) = \pi(x) - \pi(y) + 1$. Substituting in (16) with $z = \sqrt{x}$ we deduce that, for $x^{1/3} < y \leq x^{1/2}$,

$$\begin{aligned}\Phi(x,y) &= \Phi(x,\sqrt{x}) + \sum_{y<p\leq\sqrt{x}} \Phi(x/p,p) + O(x^{2/3}) \\ &= \frac{x}{\log x} + \sum_{y<p\leq\sqrt{x}} \frac{x}{p\log(x/p)} + O\Big(\frac{x}{\log^2 x}\Big).\end{aligned}$$

For $2 \le v \le u$, let us set

$$H(v) := \sum_{x^{1/v} < p \le \sqrt{x}} \frac{1}{p} = \log(v/2) + O\big(e^{-\sqrt{(\log y)}}\big) \tag{17}$$

where the estimate follows, in a standard way, from the prime number theorem. The sum over p in the last evaluation of $\Phi(x, y)$ has value

$$\frac{x}{\log x} \int_{2-}^{x} \frac{v}{v-1} \, \mathrm{d}H(v) = \frac{x}{\log y} \Big\{ \frac{1 + \log(u-1)}{u} \Big\} + O\Big(\frac{x}{(\log x)^2} \Big),$$

by partial summation and an appeal to (17).

If, for $1 \le u \le 3$, we define a function $\omega(u)$ by $u\,\omega(u) := 1$ $(1 \le u \le 2)$, $u\,\omega(u) := 1 + \log(u-1)$ $(2 < u \le 3)$, we have then proved that

$$\Phi(x, y) = \frac{x\omega(u) - y}{\log y} + O\Big(\frac{x}{\log^2 y} \Big) \qquad (x^{1/3} \le y \le x). \tag{18}$$

When $y \le x/\log x$, the second main term $-y/\log y$ is absorbed by the error term. If $x^{1/4} < y \le x^{1/3}$, we can use (18) to evaluate $\Phi(x/p, p)$ in (16). For as long as (18) remains satisfied, we can iterate this procedure; the function $\omega(u)$ is then defined inductively on intervals of length 1 by the relation arising from (16) and (18)

$$\frac{x\omega(u)}{\log y} \sim \frac{x}{\log x} + \sum_{y < p \le \sqrt{x}} \frac{x}{p \log p} \omega\Big(\frac{\log x}{\log p} - 1 \Big) \sim \frac{x}{\log x} \Big\{ 1 + \int_2^u \omega(v-1) \, \mathrm{d}v \Big\},$$

that is

$$u\omega(u) = 1 + \int_1^{u-1} \omega(v) \, \mathrm{d}v \qquad (u > 2). \tag{19}$$

This function was discovered (precisely in this way) by Buchstab in 1937. Today it bears his name. For $u > 1$ it is the unique continuous solution to the difference-differential equation

$$\big(u\omega(u)\big)' = \omega(u-1) \qquad (u > 2) \tag{20}$$

with initial condition

$$u\omega(u) = 1 \qquad (1 \le u \le 2). \tag{21}$$

We extend the definition of $\omega(u)$ to 0 for $u < 1$, so that (20) is satisfied for $u \in \mathbb{R}\smallsetminus\{1,2\}$. For purposes of future reference, we now observe that, by induction on the integer $k := [u]$, (19) and (21) easily imply the bounds

$$\tag{22} \tfrac{1}{2} \le \omega(u) \le 1 \qquad (u \ge 1).$$

The iterative method for the evaluation of $\Phi(x,y)$ presented above leads in a natural way to the following result, which is analogous to Theorem 5.6 for $\Psi(x,y)$. However, in spite of the similarity of method, there is a remarkable discrepancy in the quality of the approximations obtained: while Theorem 5.6 is only of use in a restricted domain, here we obtain a genuine asymptotic formula with remainder, uniform for $y \to \infty$, $y \le x/2$.

Theorem 3. *Uniformly for $x \ge y \ge 2$, we have*

$$\tag{23} \Phi(x,y) = \frac{x\omega(u) - y}{\log y} + O\Big(\frac{x}{(\log y)^2}\Big).$$

Proof. The result being trivial for bounded y, we suppose that $y \ge y_0$, where y_0 is a sufficiently large constant. In addition, we may also assume that $u > 3$ since we have already established the result for $1 \le u \le 3$.

Let $\Delta(x,y)$ be the function implicitly defined by the formula

$$\tag{24} \Phi(x,y) = \frac{x}{\log y}\Big\{\omega(u) + \frac{\Delta(x,y)}{\log y}\Big\}.$$

We shall establish, by induction on the integer $k \ge 3$, that the quantity

$$\Delta_k := \sup\big\{|\Delta(x,y)| : y \ge y_0,\, 2 < u \le k\big\}$$

is finite and bounded independently of k.

Plainly, $\Delta_3 < \infty$. Let $k \ge 3$ be such that $\Delta_k < \infty$. If x, y satisfy $y \ge y_0$, $2 < u \le k+1$, we obtain by inserting (24) into (16) with $z = \sqrt{x}$ that

$$\tag{25} \Phi(x,y) = \Phi(x,\sqrt{x}) + \sum_{y<p\le\sqrt{x}} \frac{x}{p\log p}\Big\{\omega\Big(\frac{\log x}{\log p} - 1\Big) + \frac{\theta_p\Delta_k}{\log p}\Big\} + O\Big(\frac{x}{y}\Big)$$

with $\theta_p = \theta_p(x) \in [-1,1]$. We estimate $\Phi(x,\sqrt{x})$ by the prime number theorem. Furthermore, partial summation enables us to write, provided that y_0 is sufficiently large,

$$\tag{26} \sum_{p>y} \frac{1}{p(\log p)^2} = \frac{\frac{1}{2} + O\big(e^{-\sqrt{(\log y)}}\big)}{(\log y)^2} \le \frac{3}{4(\log y)^2}.$$

Finally, we note that

$$\sum_{y<p\leq\sqrt{x}} \frac{1}{p\log p}\,\omega\Big(\frac{\log x}{\log p}-1\Big) = \frac{1}{\log x}\int_{2-}^{u} v\,\omega(v-1)\,\mathrm{d}H(v)$$

where $H(v)$ is defined by (17). It follows from (20) and (22) that the function $v \mapsto v\,\omega(v-1)$ is continuous for $v \neq 2$, has a discontinuity of the first kind at $v = 2$, and is differentiable for $v \neq 2, 3$, with uniformly bounded derivative.

Using (17), we can then write

$$\begin{aligned}\int_{2-}^{u} v\,\omega(v-1)\,\mathrm{d}H(v) &= \int_{2}^{u} \omega(v-1)\,\mathrm{d}v + \int_{2-}^{u} v\omega(v-1)\,\mathrm{d}\{O\big(e^{-\sqrt{(\log y)}}\big)\}\\ &= u\,\omega(u) - 1 + \int_{2-}^{u} O\big(e^{-\sqrt{(\log y)}}\big)\,\mathrm{d}\{v\,\omega(v-1)\}\\ &= u\,\omega(u) - 1 + O\big(ue^{-\sqrt{(\log y)}}\big).\end{aligned}$$

Substituting in (25) and taking (26) into account, we obtain

$$\Phi(x,y) = \frac{x}{\log y}\Big\{\omega(u) + O\big(e^{-\sqrt{(\log y)}}\big)\Big\} + \frac{x\big(\theta\Delta_k + O(1)\big)}{(\log y)^2},$$

with $|\theta| \leq \frac{3}{4}$. We hence deduce the existence of some absolute constant C such that

$$\Delta_{k+1} \leq \tfrac{3}{4}\Delta_k + C.$$

Since $\Delta_k \leq \Delta_{k+1}$, it finally follows that $\Delta_{k+1} \leq 4C$. This completes the proof.

Corollary 3.1. *We have*

$$\omega(u) = e^{-\gamma} + O\big(u^{-u/2}\big) \qquad (u \geq 1). \tag{27}$$

Proof. The real number $u \geq 1$ being given, we set $y := \exp\{u^{u/2}\}$, $x := y^u$. By Theorem 5.8 and Corollary 5.9.3, we then have

$$\Psi(x,y) \ll xu^{-u}.$$

Comparing (11) and (23) and estimating $\zeta(1,y)$ by Mertens' formula, we readily obtain (27).

It is not hard to improve (27) by working directly with the difference-differential equation (20) provided the existence and value of the limit of $\omega(u)$ at infinity are known a priori. We make precise the result which can be obtained. Define $\omega'(u)$ by right-continuity at $u = 1$ and $u = 2$.

Theorem 4. *We have that*

$$|\omega'(u)| \leq \rho(u) \qquad (u \in \mathbb{R}), \tag{28}$$

$$\omega(u) = e^{-\gamma} + O\Big(\frac{\rho(u)}{\log(u+1)}\Big) \qquad (u \geq 1). \tag{29}$$

Proof. For $1 < u < 2$, we have $\omega'(u) = -1/u^2$ and $\rho(u) = 1 - \log u$. An elementary calculus computation then shows that $|\omega'(u)| < \rho(u)$. This inequality is also true for $u = 2$, since $\omega'(2) = \omega'(2+) = 1/4$. Writing

$$\tau := \inf\{u > 1 : |\omega'(u)| \geq \rho(u)\},$$

we hence have $\tau > 2$. Now, since $\omega(u)$ is continuous for $u > 1$, we deduce from (20) that, for $u > 2$,

$$u\,\omega'(u) = \omega(u-1) - \omega(u) = -\int_{u-1}^{u} \omega'(t)\,\mathrm{d}t. \tag{30}$$

If τ is finite, we obtain for $u = \tau$ that

$$\tau|\omega'(\tau)| \leq \int_{\tau-1}^{\tau} |\omega'(t)|\,\mathrm{d}t < \int_{\tau-1}^{\tau} \rho(t)\,\mathrm{d}t = \tau\rho(\tau)$$

and hence $|\omega'(\tau)| < \rho(\tau)$, which is a contradiction. Therefore τ is not finite, and (28) follows.

Since $\rho(u)$ is rapidly decreasing as $u \to \infty$, we may write for $u > 1$

$$\omega(u) - e^{-\gamma} = -\int_u^{\infty} \omega'(t)\,\mathrm{d}t \ll \int_u^{\infty} \rho(t)\,\mathrm{d}t.$$

We deduce (29) by using Corollary 5.8.3 in the form

$$\int_u^{\infty} \rho(t)\,\mathrm{d}t \ll \int_u^{\infty} \frac{-\rho'(t)}{\log(t+1)}\,\mathrm{d}t \ll \frac{\rho(u)}{\log(u+1)}. \tag{31}$$

§ 6.3 Buchstab's function

The object of this section is to undertake the asymptotic study of $\omega(u)$ by the saddle-point method. Apart from the explicit calculation of the Laplace transform

$$\widehat{\omega}(s) := \int_0^\infty e^{-su}\omega(u)\,\mathrm{d}u, \tag{32}$$

this also leads to a remarkable sharpening of Theorem 4.

By differentiation under the integral sign, we can write for $\sigma > 0$

$$\frac{\mathrm{d}}{\mathrm{d}s}\,\widehat{\omega}(s) = -\int_0^\infty e^{-su}u\omega(u)\,\mathrm{d}u = \left[\frac{e^{-su}u\omega(u)}{s}\right]_0^\infty - \frac{1}{s}\int_0^\infty e^{-su}\,\mathrm{d}\{u\omega(u)\}.$$

The term in the square brackets is zero. Denoting by δ_1 the Dirac measure at $u = 1$, we deduce from (20) and (21) the following equality of measures

$$\mathrm{d}\{u\omega(u)\} = \delta_1 + \big(u\omega(u)\big)'\,\mathrm{d}u = \delta_1 + \omega(u-1)\,\mathrm{d}u.$$

After the change of variable $(u-1) \mapsto u$, we obtain

$$\frac{\mathrm{d}}{\mathrm{d}s}\,\widehat{\omega}(s) = -\frac{e^{-s}}{s}\big(1 + \widehat{\omega}(s)\big)$$

from which it follows that $1 + \widehat{\omega}(s) = Ce^{J(s)}$ for some suitable constant C and

$$J(s) := \int_0^\infty \frac{e^{-s-t}}{s+t}\,\mathrm{d}t.$$

Lemma 5.7.1 then shows that

$$1 + \widehat{\omega}(s) = \frac{C}{s\widehat{\rho}(s)} \tag{33}$$

and in particular that

$$\lim_{s\to\infty} \widehat{\omega}(s) = C - 1,$$

since

$$s\widehat{\rho}(s) = \int_0^\infty \rho(u/s)e^{-u}\,\mathrm{d}u \to \rho(0) = 1 \qquad (s \to \infty).$$

Now, it clearly follows from (22) that $\widehat{\omega}(s) \to 0$ as $s \to \infty$. We hence have $C = 1$, and we can state the following result.

Theorem 5. *The function $\widehat{\omega}(s)$, defined by (32) for $\sigma > 0$, extends to a meromorphic function on $\mathbb{C}$, given explicitly by the formula*

$$1+\widehat{\omega}(s)=\frac{1}{s\widehat{\rho}(s)} \qquad (s\neq 0). \tag{34}$$

When s is not real and negative, we have

$$1+\widehat{\omega}(s)=e^{J(s)}. \tag{35}$$

Remarks. (i) One can establish (34) in a purely arithmetical way by using the elementary estimates for $\Phi(x,y)$ and $\Psi(x,y)$ deduced from the functional equations (Theorem 3 and Theorem 5.6) and the natural duality between these quantities—cf. Exercise 1.

(ii) Conversely, one can use (34) to recover analytically that

$$\lim_{u\to\infty}\omega(u)=e^{-\gamma},$$

which we proved in Corollary 3.1 by means of Mertens' formula. For this, observe that (34) implies that $\lim_{s\to 0+} s\widehat{\omega}(s)=1/\widehat{\rho}(0)=e^{-\gamma}$. Bearing in mind the positivity of ω, Karamata's theorem (II.7.5) implies that

$$\int_0^u \omega(t)\,\mathrm{d}t=\big(e^{-\gamma}+o(1)\big)u \qquad (u\to\infty).$$

Now it follows from (20) and (22) that $u\omega'(u)\ll 1$. For a suitable function $\varepsilon(u)\to 0$ and all h, $0<h\leq u$, we obtain, on the one hand, that

$$\frac{1}{h}\int_u^{u+h}\omega(t)\,\mathrm{d}t=e^{-\gamma}+O\Big(\varepsilon(u)\frac{u}{h}\Big)$$

and, on the other hand, that

$$\frac{1}{h}\int_u^{u+h}\omega(t)\,\mathrm{d}t=\omega(u)+\frac{1}{h}\int_u^{u+h}O\Big(\frac{h}{u}\Big)\,\mathrm{d}t=\omega(u)+O\Big(\frac{h}{u}\Big).$$

It thus suffices to choose $h=u\sqrt{(\varepsilon(u))}$.

It is easy to deduce from the difference-differential equation (20)–(21) satisfied by the Buchstab function that it is of class C^j on $\mathbb{R}\smallsetminus\{1,2,\ldots,j+1\}$ for each integer $j\geq 0$. At the exceptional points, successive derivatives have discontinuities of the first kind, and we extend the definition of $\omega^{(j)}(u)$ to the entire real line by right-continuity.

In particular, $\omega'(u)$ is of bounded variation on any bounded interval. When combined with the rapid decrease at infinity (Theorem 4), this property suffices to imply the convergence (to the value $\omega'(u)$) of the inverse Laplace integral for $u\neq 1$ or 2, for any abscissa of integration. We can therefore use the saddle-point method in order to evaluate $\omega'(u)$ at infinity. With the notation

$$H(u):=\exp\big\{u/\log^2(u+2)\big\} \qquad (u\geq 0),$$

we obtain the following result.

Theorem 6. *For a suitable absolute positive constant a, we have*

$$\begin{matrix} \omega(u) - e^{-\gamma} \\ \omega'(u) \end{matrix} \ll \rho(u)H(u)^{-a} \qquad (u \geq 0). \tag{36}$$

Proof. We can suppose that $u > 2$. First observe that the second stated upper bound easily implies the first, since

$$\omega(u) - e^{-\gamma} = -\int_u^\infty \omega'(t)\,\mathrm{d}t.$$

The conclusion then follows by using the fact that $H(u)^{-a}$ is decreasing and appealing to the estimate (31).

Let $h(s)$ be the Laplace transform of $\omega'(u)$. As we already noted, we have

$$\omega'(u) = \frac{1}{2\pi i}\int_{\kappa-i\infty}^{\kappa+i\infty} h(s)e^{us}\,\mathrm{d}s \qquad (u > 2) \tag{37}$$

for any real number κ. Now

$$h(s) = s\widehat{\omega}(s) - e^{-s} \tag{38}$$

where the second term arises from the discontinuity of $\omega(u)$ at $u = 1$. Taking account of (34), it follows that

$$h(s) = \widehat{\rho}(s)^{-1} - e^{-s} - s. \tag{39}$$

Recall that

$$\widehat{\rho}(s) = e^{\gamma + I(-s)} \qquad \Big(I(s) := \int_0^s \frac{e^v - 1}{v}\,\mathrm{d}v\Big).$$

In accordance with the underlying principles of the saddle-point method, and temporarily ignoring the influence of the term $-e^{-s} - s$ in this calculation, it is then necessary to choose the abscissa of integration κ in such a way that at least one of the complex solutions to the equation

$$e^{-s} = 1 + su \tag{40}$$

has abscissa κ. Contrary to the situation we met in the study of Dickman's function, the saddle-point equation (40) does not have a real solution. However, we shall see that, for sufficiently large u, it has a solution with real part close to $-\xi(u)$.

Let us indeed make the change of variable $s = -\xi(u) + i\pi - z$ in (40). We get

$$z\lambda(z) = w \tag{41}$$

with

$$\lambda(z) := (1 + u\xi)\Big(\frac{e^z - 1}{z}\Big) - u, \qquad w := -i\pi u - 2.$$

Since $\xi(u) \sim \log u$ $(u \to \infty)$—cf. Lemma 5.8.1—it follows that there exists some absolute positive constant u_0 such that

$$|z\lambda(z)| > |w| \qquad \big(|z| = 2\pi/\log(u+1),\ u > u_0\big).$$

Since $\lambda(z)$ does not vanish for $|z| < 2\pi/\log(u+1)$, $u > u_0$, this implies, by Rouché's theorem, that (41) has in this disc a unique solution. Lagrange's theorem—see for example Whittaker and Watson (1927), §7.32—actually states that this solution is an analytic function of w, and provides an explicit formula for the Taylor expansion. We make no use of this last piece of information in this proof.

We could evaluate the integral (37) by choosing precisely $\kappa = -\xi(u) - \Re e\, z$. This would lead to a slightly sharper result—see the Notes. However, here we content ourselves working with the simpler choice $\kappa = -\xi(u)$, which is heuristically justified by the fact that (41) has a solution approaching 0 as $u \to \infty$. This will enable us to use the estimates obtained in Chapter 5 for $\widehat{\rho}(s)$ on the line $\sigma = -\xi(u)$.

For the remainder of this proof, we hence let $\kappa = -\xi(u)$. We first consider the contribution to (37) from the domain $|\tau| > e^{\xi}$. From the estimate

$$J(s) = \frac{e^{-s}}{s}\Big(1 - \frac{1}{s}\Big) + O\Big(\frac{e^{-s}}{|\tau|^3}\Big) \qquad (\tau \neq 0),$$

it follows using (38) and (35) that, for $s = -\xi(u) + i\tau$, $|\tau| > e^{\xi}$, we have

$$\begin{aligned} h(s) &= s\Big\{1 + \frac{e^{-s}}{s}\Big(1 - \frac{1}{s}\Big) + \frac{e^{-2s}}{2s^2}\Big(1 - \frac{1}{s}\Big)^2 + O\Big(\frac{e^{3\xi}}{|\tau|^3}\Big)\Big\} - s - e^{-s} \\ &= -\frac{e^{-s}}{i\tau} + \frac{e^{-2s}}{2i\tau} + O\Big(\frac{e^{3\xi}}{\tau^2}\Big). \end{aligned}$$

We can estimate the contribution of the main terms to the integral by the second mean value theorem; estimating that of the remainder in a straightforward manner, we obtain that the contribution to (37) from the domain $|\tau| > e^{\xi}$ is

$$\ll e^{-u\xi + 2\xi} \ll \rho(u)e^{-u/2}$$

where the second estimate follows from Theorem 5.8.

In order to evaluate the contribution from the rest, we use $h(s)$ in the form (39). We shall show that, for some suitable positive constant a, we have

$$(42) \qquad \widehat{\rho}(s)^{-1} \ll \widehat{\rho}(-\xi)H(u)^{-2a} \qquad (\sigma = -\xi,\, |\tau| \le e^{\xi}).$$

From this upper bound, it follows that

$$\int_{|\tau|\le e^{\xi}} h(s)e^{su}\,\mathrm{d}s \ll e^{-u\xi+\xi}\{\widehat{\rho}(-\xi)H(u)^{-2a} + e^{\xi}\} \ll \rho(u)H(u)^{-a},$$

by Theorem 5.8.

It remains to establish (42), that is (setting $T := u/\log^2(u+1)$)

$$(43) \qquad \Re e\{I(\xi+i\tau) + I(\xi)\} \ge 2aT + O(1) \qquad (|\tau| \le e^{\xi}).$$

The left-hand side is

$$\begin{aligned}
&\int_0^1 \Big\{e^{v\xi}\big(1+\cos(\tau v)\big) - 2\Big\}\frac{\mathrm{d}v}{v}\\
&\quad\ge -\int_0^{\frac12} \big(1-\cos(\tau v)\big)\frac{\mathrm{d}v}{v} + \int_{\frac12}^1 \Big\{e^{v\xi}\big(1+\cos(\tau v)\big) - \frac{2}{v}\Big\}\,\mathrm{d}v\\
&\quad= -2\int_0^{\frac14\tau} \sin^2 v\,\frac{\mathrm{d}v}{v} + \left[\frac{e^{v\xi}}{\xi} + \Re e\Big(\frac{e^{v(\xi+i\tau)}}{\xi+i\tau}\Big)\right]_{\frac12}^{1} - \log 4\\
&\quad\ge O\big(\log(1+\tau^2)\big) + e^{\xi}\Big\{\frac{1}{\xi} + \frac{\cos\lambda}{\sqrt{(\xi^2+\tau^2)}}\Big\} + O\Big(\sqrt{\frac{u}{\xi}}\Big)
\end{aligned}$$

with $\lambda := \tau - \arctan(\tau/\xi)$. There exists some absolute constant τ_0 such that $\cos\lambda \ge 0$ for $|\tau| \le \tau_0$. For such τ, the above lower bound is $\ge u + O\big(\sqrt{(u/\xi)}\big)$. For $|\tau| > \tau_0$, it is

$$\ge e^{\xi}\Big(\frac{1}{\xi} - \frac{1}{\sqrt{(\xi^2+\tau_0^2)}}\Big) + O\Big(\sqrt{\frac{u}{\xi}}\Big) \gg \frac{e^{\xi}}{\xi^3} + O\Big(\sqrt{\frac{u}{\xi}}\Big) \gg T + O(1).$$

This establishes (43) and completes the proof of Theorem 6.

Corollary 6.1. *Let j be an integer ≥ 1. We have*

$$(44) \qquad \omega^{(j)}(u) \ll \rho^{(j)}(u)H(u)^{-a} \qquad (u \to \infty).$$

Proof. Equation (20) and our convention of extending ω and all of its derivatives by right-continuity enable us to write, after j differentiations,

$$(45) \qquad u\omega^{(j+1)}(u) = \omega^{(j)}(u-1) - (j+1)\omega^{(j)}(u) \qquad (u \in \mathbb{R}).$$

By induction on j and using Corollary 5.8.3 in the form

$$\begin{aligned}
\rho^{(j)}(u-1) &\ll \rho(u-1)\log^j(u+1) = -u\rho'(u)\log^j(u+1) \qquad (u \ge 1)\\
&\ll u\rho(u)\log^{j+1}(u+1) \ll u\rho^{(j+1)}(u),
\end{aligned}$$

this implies (44).

§ 6.4 Approximations to $\Phi(x, y)$ by the saddle-point method

As with $\Psi(x, y)$, the function $\Phi(x, y)$ can be estimated by the saddle-point method, starting from its Perron integral. Nevertheless, two differences stand out. On the one hand, the associated Dirichlet series $\zeta(s)/\zeta(s, y)$ now has a pole at $s = 1$; the possibility of making an optimal choice for the abscissa of integration will thus be dependent on a preliminary use of the residue theorem. On the other hand, it is here no longer a question of determining the asymptotic behaviour of the function under study, but of estimating the error term relative to an explicit main term—cf. Theorem 3.

Throughout this section we retain some of the notation from Chapter 5, such as the domain

$$(H_\varepsilon) \qquad x > x_0(\varepsilon), \quad \exp\{(\log_2 x)^{(5/3)+\varepsilon}\} \leq y \leq x,$$

and the function

$$(46) \qquad L_\varepsilon(y) := \exp\{(\log y)^{(3/5)-\varepsilon}\}.$$

In addition, we introduce the following notation:

$$(47) \qquad Y_\varepsilon := \exp\{(\log y)^{(3/2)-\varepsilon}\}, \quad E(x, y) := H(u)^{-c} L_\varepsilon(y)^{-1} + Y_\varepsilon^{-1},$$

$$(48) \qquad \mu_y(u) := \int_0^\infty \omega(u - v) y^{-v} \,\mathrm{d}v, \quad W(x, y) := x\mu_y(u) \frac{e^\gamma \log y}{\zeta(1, y)}.$$

The letter c, with or without index, stands for some absolute positive constant.

The main purpose of this section consists in establishing the following result.

Theorem 7. *Let* $\varepsilon > 0$. *For* $x \geq y \geq 2$, *we have*

$$(49) \qquad \Phi(x, y) - W(x, y) \ll \begin{cases} \Psi(x, y)\, E(x, y) & \text{(in the domain } H_\varepsilon) \\ \Psi(x, y) & \text{(elsewhere).} \end{cases}$$

We begin by proving the second upper bound. It is an easy consequence of the following lemma.

Lemma 7.1. *Let* $\delta > 1$. *Uniformly for* $x \geq y \geq 2$, *we have that*

$$(50) \qquad \left|\mu_y(u) e^\gamma \log y - 1\right| \ll \rho(u) H(u)^{-c_1} + x^{-1}(e^{\delta \overline{u}} + y).$$

Assuming this estimate is valid, we argue as follows. By Theorem 5.13, we have

$$x\rho(u) H(u)^{-c_1} \ll \Psi(x, y) H(u)^{-c_1/2}.$$

Furthermore, it is easy to see that the approximation Z for $\log\Psi(x,y)$ from Theorem 5.2 satisfies

$$Z := u\log\Big(1+\frac{y}{\log x}\Big)+\frac{y}{\log y}\log\Big(1+\frac{\log x}{y}\Big)\geq \overline{u}\log 4$$

and also that, when $u\geq 2$,

$$Z\geq \tfrac{1}{2}\overline{u}+\log y+O(1).$$

For $1<\delta<\log 4-\frac{1}{3}$, we therefore have

$$y+e^{\delta\,\overline{u}}\ll\Psi(x,y)e^{-\overline{u}/3}.$$

Inserting in (50), we thus obtain

$$W(x,y)-\frac{x}{\zeta(1,y)}\ll\frac{\Psi(x,y)}{\log y}H(\overline{u})^{-c_1/2}. \tag{51}$$

The second bound in (49) follows from this estimate and from Theorem 1.

Proof of Lemma 7.1. A simple integration by parts enables us to write

$$\mu_y(u)\log y=\omega(u)-\frac{y}{x}-\int_0^\infty\omega'(u-v)y^{-v}\,dv, \tag{52}$$

where the second term on the right-hand side arises from the discontinuity of $\omega(u)$ at $u=1$.

Taking into consideration the bound for $\omega(u)-e^{-\gamma}$ in Theorem 6, we see that the proof of (50) reduces to showing that, for $x\geq y\geq 2$, we have

$$\int_0^\infty|\omega'(u-v)|y^{-v}\,dv\ll\rho(u)H(u)^{-c_1}+e^{\delta\,\overline{u}}x^{-1}. \tag{53}$$

Observe first of all that, for any $\varepsilon>0$,

$$H(u-v)^{-a}\ll_\varepsilon H(u)^{-a}e^{\varepsilon v}\qquad(0\leq v\leq u)$$

where a is the absolute constant of Theorem 6. Put $y_\varepsilon := ye^{-\varepsilon}$. Using Theorem 6 we then obtain

$$\int_0^\infty|\omega'(u-v)|y^{-v}\,dv\ll_\varepsilon H(u)^{-a}\int_0^\infty\rho(u-v)y_\varepsilon^{-v}\,dv. \tag{54}$$

Let λ be a parameter ≥ 1. We write the last integral in the form

$$y_\varepsilon^{-u}\int_0^u \rho(t)y_\varepsilon^t\,\mathrm{d}t \leq y_\varepsilon^{-u}\int_0^\infty \rho(t)y_\varepsilon^t\lambda^{u-t}\,\mathrm{d}t = \left(y_\varepsilon/\lambda\right)^{-u}\widehat{\rho}\big(-\log(y_\varepsilon/\lambda)\big).$$

If $y_\varepsilon \geq e^{\xi(u)}$, we may choose $\lambda = y_\varepsilon e^{-\xi(u)}$; taking account of Theorem 5.8, we obtain that the left-hand side of (54) is

$$\ll_\varepsilon H(u)^{-a}e^{-u\xi(u)}\widehat{\rho}\big(-\xi(u)\big) \ll H(u)^{-a/2}\rho(u).$$

The bound (53) is therefore satisfied in this case.

When $y_\varepsilon < e^{\xi(u)}$, we appeal to the upper bound (28) for $|\omega'(u)|$ in Theorem 4. The left-hand side of (54) does not exceed

$$\int_0^\infty \rho(u-v)y^{-v}\,\mathrm{d}v = \int_0^u \rho(t)y^{t-u}\,\mathrm{d}t \leq y^{-u}\widehat{\rho}(-\log y) = y^{-u}e^{\gamma+I(\log y)}.$$

Since $I(\log y) \sim y/\log y$ as $y \to \infty$ and $I\big(\varepsilon+\xi(u)\big) \sim e^\varepsilon u$ as $u \to \infty$, we certainly have

$$e^{\gamma+I(\log y)} \ll e^{\delta\,\overline{u}}$$

in the region under consideration provided that $\varepsilon = \varepsilon(\delta)$ is sufficiently small.

This completes the proof of (53) and hence that of Lemma 7.1.

We now turn our attention to the first upper bound in Theorem 7, namely that concerning the domain (H_ε). By Perron's formula, we have for all real $\kappa > 1$

$$\text{(55)}\qquad \Phi(x,y) = \frac{1}{2\pi i}\int_{\kappa-i\infty}^{\kappa+i\infty}\frac{\zeta(s)x^s}{\zeta(s,y)s}\,\mathrm{d}s \qquad (x \notin \mathbb{Z}^+).$$

The residue at $s = 1$ has value $x/\zeta(1,y)$. The integrand is a holomorphic function of s for $s \neq 0$ or 1, and tends to 0 as $|\tau| \to \infty$ in every vertical strip $0 < \sigma_0 \leq \sigma \leq 1$. We can hence move the abscissa of integration to the left as far as

$$\sigma = \alpha_0 := 1 - \frac{\xi(u)}{\log y}.$$

We obtain

$$\text{(56)}\qquad \Phi(x,y) = \frac{x}{\zeta(1,y)} + \frac{1}{2\pi i}\int_{\alpha_0-i\infty}^{\alpha_0+i\infty}\frac{\zeta(s)x^s}{\zeta(s,y)s}\,\mathrm{d}s.$$

Lemma 5.9.1 and Theorem 5 enable us, for $|\tau| \leq L_\varepsilon(y)$, to approximate the integrand of (56) by

$$\frac{xe^{uz}\big(1+\widehat{\omega}(z)\big)}{1+z/\log y} \qquad \big(z := (s-1)\log y\big).$$

Ignoring for the time being the influence of the domain $|\tau| > L_\varepsilon(y)$ on the integral in (56), we obtain the heuristic approximation to $\Phi(x,y)$

$$W_1(x,y) := \frac{x}{\zeta(1,y)} + \frac{x}{2\pi i}\int_{-\xi(u)-i\infty}^{-\xi(u)+i\infty} \frac{1+\widehat{\omega}(s)}{s+\log y} e^{us}\,\mathrm{d}s. \tag{57}$$

We shall see below that this integral is indeed convergent. By Theorem 5, $\widehat{\omega}(s)$ has a simple pole at $s=0$, with residue $e^{-\gamma}$. Moving the line of integration of (57) to the right as far as $\sigma = \kappa > 0$, it follows that

$$W_1(x,y) = \frac{x}{\zeta(1,y)} - \frac{xe^{-\gamma}}{\log y} + \frac{x}{2\pi i}\int_{\kappa-i\infty}^{\kappa+i\infty} \frac{1+\widehat{\omega}(s)}{s+\log y}\, e^{us}\,\mathrm{d}s. \tag{58}$$

Now, $\big(1+\widehat{\omega}(s)\big)/(s+\log y)$ is the Laplace transform of the function

$$t \mapsto y^{-t} + \int_0^\infty \omega(t-v)y^{-v}\,\mathrm{d}v$$

which is continuous and of bounded variation on each bounded interval. The Laplace inversion theorem—cf. Widder (1946), theorem II.7.3—then yields the explicit formula

$$W_1(x,y) = 1 + x\Big\{\frac{1}{\zeta(1,y)} - \frac{e^{-\gamma}}{\log y} + \mu_y(u)\Big\}. \tag{59}$$

Now we observe that, in the domain (H_ε), Lemma 7.1 implies that

$$\begin{aligned} W_1(x,y) - W(x,y) &= 1 + x\Big(\frac{1}{\zeta(1,y)} - \frac{e^{-\gamma}}{\log y}\Big)\big(1-\mu_y(u)e^{\gamma}\log y\big) \\ &\ll x\rho(u)H(u)^{-c_1}L_\varepsilon(y)^{-1} \ll \Psi(x,y)E(x,y). \end{aligned} \tag{60}$$

This difference is of the same order as the error term stated in (49). We can therefore regard $W(x,y)$ as the natural approximation for $\Phi(x,y)$ arising from the saddle-point method.

It remains to make rigorous the argument sketched above. On several occasions we shall appeal to the following estimate, which is a consequence of Theorem 5.8, Lemma 5.9.1 and Corollary 5.9.3. We have, uniformly in (H_ε),

$$x^{\alpha_0}\zeta(\alpha_0,y) \asymp \Psi(x,y)\surd(u)\log y. \tag{61}$$

Set $L := L_{\varepsilon/2}(y)$. We proceed to establish the following three estimates

$$\int_{\substack{\sigma=\alpha_0\\|\tau|>L}} \frac{\zeta(s)x^s}{\zeta(s,y)s}\,\mathrm{d}s \ll \Psi(x,y)E(x,y), \tag{62}$$

$$\text{(63)} \qquad \frac{1}{L}\int_{\substack{\sigma=\alpha_0\\|\tau|\le L}} \left|\frac{\zeta(s)x^s}{\zeta(s,y)s}\right| ds \ll \Psi(x,y)E(x,y),$$

$$\text{(64)} \qquad x\int_{\substack{\sigma=-\xi\\|\tau|>L}} \frac{1+\widehat{\omega}(s)}{s+\log y} e^{us}\, ds \ll \Psi(x,y)E(x,y).$$

This clearly suffices to establish Theorem 7: the bound (62) enables us to truncate the integral (55), the estimate (63) permits us to neglect the error term created by employing Lemma 5.9.1 (smooth approximation to $\zeta(s,y)$), and finally (64) allows us to extend to infinity the integral obtained by replacing $\zeta(s,y)$ by its smooth approximation.

The most difficult of these is the estimate (62). We will need the following two auxiliary results.

Lemma 7.2. *For (x,y) in (H_ε), $s=\alpha_0+i\tau$, we have*

$$\text{(65)} \qquad \zeta(s,y) \ll \zeta(\alpha_0,y)\exp\Big\{-\frac{c_2\tau^2 u}{(1-\alpha_0)^2+\tau^2}\Big\} \qquad \Big(\frac{1}{\log y}\le|\tau|\le Y_\varepsilon\Big),$$

$$\text{(66)} \qquad \zeta(s,y)^{-1} \ll \zeta(\alpha_0,y)H(u)^{-c_2} \qquad (|\tau|\le Y_\varepsilon).$$

Proof. Let us first prove (65). An easy calculation shows that

$$\text{(67)} \quad \left|\frac{1-p^{-\alpha_0}}{1-p^{-s}}\right| = \Big(1+\frac{2(1-\cos(\tau\log p))}{p^{\alpha_0}(1-p^{-\alpha_0})^2}\Big)^{-1/2} \le \exp\Big\{-\frac{1-\cos(\tau\log p)}{p^{\alpha_0}}\Big\}$$

from which

$$|\zeta(s,y)|\zeta(\alpha_0,y)^{-1}\le e^{-X}, \quad \text{with} \quad X:=\sum_{p\le y}\frac{1-\cos(\tau\log p)}{p^{\alpha_0}}.$$

We need a lower bound for X. We have

$$\text{(68)} \qquad X\log y \ge A(\alpha_0) - \Re e\, A(\alpha_0+i\tau) + O\big(y^{\frac{1}{2}-\alpha_0}\big)$$

with

$$A(s) := \sum_{n\le y}\frac{\Lambda(n)}{n^s}.$$

The error term in (68) takes into account the contribution of the non-prime integers n in $A(s)$. The O-estimate is obtained as in Lemma 5.9.1.

The effective Perron formula (Corollary II.2.2.1) allows us to write

$$A(s) = \frac{-1}{2\pi i}\int_{\kappa-iT}^{\kappa+iT} \frac{\zeta'}{\zeta}(s+w)\frac{y^w}{w}\,dw + O\Big(\frac{\log y}{y^{\alpha_0}}\Big)$$

with $\kappa := 1-\alpha_0+1/\log y$, $T := Y_\varepsilon^2$. We move the segment of integration as far as $\Re e\, w = 1-\alpha_0-(\log T)^{-2/3-\varepsilon/10}$, with the result that the point $s+w$ remains constantly in Vinogradov's zero-free region. The only pole encountered is the point $w = 1-s$. The contribution from the vertical segment is

$$\ll (\log T)^2 y^{1-\alpha_0}\exp\Big\{-\frac{\log y}{(\log T)^{(2/3)+\varepsilon/10}}\Big\} \ll y^{1-\alpha_0}\exp\{-(\log y)^{\varepsilon/2}\}$$

and it is checked without difficulty that the same bound also holds for the contributions from the horizontal segments. We thus obtain

$$A(s) = \frac{y^{1-s}}{1-s} + O\Big(y^{1-\alpha_0}\exp\{-(\log y)^{\varepsilon/2}\}\Big).$$

Substituting in (68), we infer that

$$X\log y \geq \frac{y^{1-\alpha_0}}{1-\alpha_0}\,\Re e\Big(1-\frac{1-\alpha_0}{1-s}\Big) + O\Big(y^{1-\alpha_0}\exp\{-(\log y)^{\varepsilon/2}\}\Big)$$

and therefore

$$X \geq \frac{e^{\xi(u)}\tau^2}{\xi(u)\big((1-\alpha_0)^2+\tau^2\big)} + O\Big(\exp\{\xi(u)-(\log y)^{\varepsilon/2}\}\Big) \gg \frac{u\tau^2}{(1-\alpha_0)^2+\tau^2},$$

since for $|\tau| \geq 1/\log y$ we have that

(69) $$\frac{\tau^2}{(1-\alpha_0)^2+\tau^2} \geq \frac{1}{\xi(u)^2+1} \gg \frac{1}{\log^2 y}.$$

This establishes (65).

The proof of (66) is analogous. The starting point is

$$\big|(1-p^{-s})(1-p^{-\alpha_0})\big| = \big(1-p^{-2\alpha_0}\big)\bigg(1-\frac{2\big(1+\cos(\tau\log p)\big)}{p^{\alpha_0}\big(1+p^{-\alpha_0}\big)^2}\bigg)^{1/2}$$
$$\leq \exp\Big\{-\frac{1+\cos(\tau\log p)}{4p^{\alpha_0}}\Big\},$$

from which it follows that

$$\left|\zeta(s,y)\zeta(\alpha_0,y)\right|^{-1}\le e^{-V/4},\quad\text{with}\quad V:=\sum_{p\le y}\frac{1+\cos(\tau\log p)}{p^{\alpha_0}}.$$

If $|\tau|>1/\log y$, the lower bound established for X is also valid for V: it suffices to consider $A(\alpha_0)+\Re e\,A(\alpha_0+i\tau)$. Taking (69) into account, we certainly have (66) in this case. For $|\tau|\le 1/\log y$, we can write

$$V\ge\frac{1+\cos 1}{\log y}\sum_{p\le y}\frac{\log p}{p^{\alpha_0}}\gg\frac{y^{1-\alpha_0}-1}{(1-\alpha_0)\log y}=u.$$

This completes the proof of Lemma 7.2.

Lemma 7.3. *Let $\varepsilon>0$. Uniformly for (x,y) in (H_ε) and $1\le z\le Y_\varepsilon$,* we have

$$\Psi\big(x+x/z,y\big)-\Psi(x,y)\ll\Psi(x,y)\surd(u)\log y\big\{1/z+e^{-c_3u}\big\}.\tag{70}$$

Proof. We proceed as in Lemma 5.9.4. Setting

$$w(t):=\frac{1}{2\pi}\Big(\frac{\sin tz}{tz}\Big)^2,\quad \widehat{w}(\tau):=\int_{-\infty}^{\infty}w(t)e^{-i\tau t}\,\mathrm{d}t=\frac{1}{2z}\Big(1-\Big|\frac{\tau}{2z}\Big|\Big)^+,$$

we see that the left-hand side of (70) is

$$\begin{aligned}&\ll\sum_{P^+(n)\le y}\Big(\frac{x}{n}\Big)^{\alpha_0}w\Big(\log\frac{x}{n}\Big)=\frac{1}{2\pi i}\int_{\alpha_0-2iz}^{\alpha_0+2iz}\zeta(s,y)x^s\widehat{w}(\tau)\,\mathrm{d}s\\&\ll\frac{1}{z}\,x^{\alpha_0}\zeta(\alpha_0,y)+x^{\alpha_0}\max_{1\le|\tau|\le 2z}\big|\zeta(\alpha_0+i\tau,y)\big|.\end{aligned}$$

The stated result then follows from (65) and (61).

Proof of (62). Set

$$T:=\begin{cases}L & \text{if}\quad 1\le u\le(\log y)^{3(1-\varepsilon)/5},\\ Le^{c_4u} & \text{if}\quad (\log y)^{3(1-\varepsilon)/5}<u\le(\log y)^{(3-\varepsilon)/2},\\ Y_{\varepsilon/3} & \text{if}\quad (\log y)^{(3-\varepsilon)/2}<u\le L_\varepsilon(y),\end{cases}\tag{71}$$

where c_4 is a suitable absolute constant. When $T\ne L$ (that is when u exceeds $\big(\log y\big)^{3(1-\varepsilon)/5}$), we have by (66), estimating $\zeta(s)$ by Theorem II.3.7,

$$\int_{\substack{\sigma=\alpha_0\\ L<|\tau|\le T}}\frac{\zeta(s)x^s}{\zeta(s,y)s}\,\mathrm{d}s\ll x^{\alpha_0}\zeta(\alpha_0,y)H(u)^{-c_2}\frac{T^{1-\alpha_0}}{(1-\alpha_0)^2}\ll\Psi(x,y)E(x,y)$$

for some suitable choice of the constant c appearing in the definition of $E(x,y)$, formula (47). Thus it remains to show that the integral

$$R := \int_{\substack{\sigma=\alpha_0\\|\tau|>T}} \frac{\zeta(s)x^s}{\zeta(s,y)s}\,\mathrm{d}s$$

also has the same order of magnitude as that of the previous upper bound.

The crude form of the approximate functional equation of the zeta function (Corollary II.3.5.1) enables us to write, for $s=\alpha_0+i\tau$,

$$\zeta(s) = \sum_{n\le|\tau|} n^{-s} + O(|\tau|^{-\alpha_0}). \tag{72}$$

It follows that

$$\frac{\zeta(s)}{\zeta(s,y)} = \sum_{n\le|\tau|}\sum_{P(m)\le y} \frac{\mu(m)}{(mn)^s} + O\Big(\frac{\zeta(\alpha_0,y)}{|\tau|^{\alpha_0}}\Big). \tag{73}$$

The contribution to R from the error term in (73) is

$$\ll x^{\alpha_0}\zeta(\alpha_0,y)\int_T^\infty \frac{\mathrm{d}\tau}{\tau^{1+\alpha_0}} \ll \frac{x^{\alpha_0}\zeta(\alpha_0,y)}{\sqrt{T}} \ll \Psi(x,y)E(x,y),$$

where we have again appealed to (61).

The first effective Perron formula, in the form of the bound (II.2.7), enables us to write, uniformly for $z>0$,

$$\int_{\substack{\sigma=\alpha_0\\|\tau|>T}} \frac{z^s}{s}\,\mathrm{d}s \ll \frac{z^{\alpha_0}}{1+T|\log z|}.$$

The contribution to R from the main term in (73) can therefore be estimated as follows:

$$\begin{aligned}\sum_{n=1}^\infty \sum_{P^+(m)\le y} \mu(m)\int_{\substack{\sigma=\alpha_0\\|\tau|\ge\max(T,n)}} \Big(\frac{x}{mn}\Big)^s\frac{\mathrm{d}s}{s}\\ \ll \sum_{n=1}^\infty\sum_{P^+(m)\le y} \frac{(x/mn)^{\alpha_0}}{1+(T+n)|\log(x/mn)|}.\end{aligned} \tag{74}$$

The contribution to (74) arising from pairs (m,n) such that $|\log(x/mn)|>1$ is

$$\ll x^{\alpha_0}\zeta(\alpha_0,y)\sum_{n=1}^\infty \frac{1}{n^{\alpha_0}(T+n)} \ll \frac{x^{\alpha_0}\zeta(\alpha_0,y)}{\sqrt{T}} \ll \Psi(x,y)E(x,y).$$

For the remaining pairs (m, n), with $|\log(x/mn)| \le 1$, it is useful to bear in mind that $T \le x^{\varepsilon}$ whenever x and hence y is sufficiently large. Let S_1, S_2 denote the respective contributions corresponding to the cases $m \le x/T$, $m > x/T$. On the one hand

$$\begin{aligned} S_1 &\le \sum_{m \le x/T,\, P^+(m) \le y} \; \sum_{x/em < n \le ex/m} \frac{1}{1 + |n - x/m|} \\ &\ll \sum_{m \le x/T,\, P^+(m) \le y} \log(ex/m) \ll \Psi(x/T, y) \log x \le (x/T)^{\alpha_0} \zeta(\alpha_0, y) \log x \\ &\ll \Psi(x, y) E(x, y) \end{aligned}$$

where the penultimate inequality is just Rankin's upper bound for $\Psi(x/T, y)$.

When (m, n) is counted in S_2, we certainly have $n \le eT$. From this we obtain that

$$\begin{aligned} S_2 &\ll \sum_{n \le eT} \; \sum_{\substack{x/en < m \le ex/n \\ P^+(m) \le y}} \frac{1}{1 + T|x/mn - 1|} \\ &\ll \sum_{n \le eT} \left\{ \frac{\Psi(x/n, y)}{\sqrt{T}} + \sum_{\substack{|m - x/n| \le x/(n\sqrt{T}) \\ P^+(m) \le y}} 1 \right\}. \end{aligned}$$

The sum over m is, by Lemma 7.3,

$$\ll \Psi(x/n, y) E(x, y)^2$$

provided that $u > (\log y)^{3(1-\varepsilon)/5}$. In the contrary case, we appeal to the trivial bound

$$\ll x/(n\sqrt{T}) \ll (x/n) u^{-2u} T^{-1/3} \ll \Psi(x/n, y) E(x, y)^2.$$

Hence, in any case, we have that

$$\begin{aligned} S_2 &\ll \sum_{n \le eT} \Psi(x/n, y) E(x, y)^2 \le x^{\alpha_0} \zeta(\alpha_0, y) E(x, y)^2 \sum_{n \le eT} n^{-\alpha_0} \\ &\ll x^{\alpha_0} \zeta(\alpha_0, y) E(x, y)^2 T^{1-\alpha_0}/(1 - \alpha_0) \ll \Psi(x, y) E(x, y). \end{aligned}$$

This completes the proof of (62).

Proof of (63). Using (66) and then (61), we see that the left-hand side of (63) is bounded above by

$$\ll \frac{x^{\alpha_0} \zeta(\alpha_0, y)}{L} H(u)^{-c_2} \log L \ll \Psi(x, y) E(x, y).$$

Proof of (64). Formula (35) for $1+\widehat{\omega}(s)$ and the trivial bound $J(s) \ll e^{-\sigma}|\tau|^{-1}$ imply that

$$1+\widehat{\omega}(s) = 1 + O\Big(\frac{1+u\xi(u)}{s}\Big) \qquad (|\tau|>L).$$

This shows that the integral in (64) is conditionally convergent. The second mean value theorem allows us to write

$$\int_{\substack{\sigma=-\xi\\ |\tau|>L}} \frac{1+\widehat{\omega}(s)}{s+\log y}\, e^{us}\,\mathrm{d}s = \int_{\substack{\sigma=-\xi\\ |\tau|>L}} \frac{e^{us}}{s}\Big\{1+O\Big(\frac{u\xi(u)+\log y}{s}\Big)\Big\}\,\mathrm{d}s \ll e^{-u\xi}L^{-1}.$$

Since Theorem 5.8 and Corollary 5.9.3 imply that

$$\Psi(x,y) \gg xe^{-u\xi}e^{u/2}$$

in the domain (H_ε), we certainly obtain the required estimate (64).

This completes the proof of Theorem 7.

Corollary 7.4. *Let $\varepsilon>0$. For (x,y) in (H_ε), we have that*

$$\Phi(x,y) - \frac{x}{\zeta(1,y)} \ll \frac{x\rho(u)}{\log y}\Big(H(u)^{-c_5} + Y_\varepsilon^{-1}\Big). \tag{75}$$

Proof. This follows immediately from (49) and (51).

Corollary 7.5. *Let $\varepsilon>0$. For (x,y) in (H_ε), we have that*

$$\Phi(x,y) = \big(x\omega(u)-y\big)\frac{e^\gamma}{\zeta(1,y)} + O\bigg(\frac{x\rho(u)}{\log^2 y}\Big\{H(u)^{-c_6}+Y_\varepsilon^{-1}\Big\}\bigg). \tag{76}$$

Proof. The error term in (76) is compatible with the first upper bound of (49). It thus suffices to show that (76) is valid with $\Phi(x,y)$ replaced by $W(x,y)$. By (52), we have

$$W(x,y) = \big(x\omega(u)-y\big)\frac{e^\gamma}{\zeta(1,y)} - \frac{xe^\gamma}{\zeta(1,y)}\int_0^\infty \omega'(u-v)y^{-v}\,\mathrm{d}v.$$

By (54), with $\varepsilon=1$, and Corollary 5.8.4, the last integral is

$$\ll H(u)^{-a}\rho(u)\int_0^\infty \Big(\frac{e^{1+\xi(u)}}{y}\Big)^v \mathrm{d}v \ll \frac{H(u)^{-a}\rho(u)}{\log y}$$

since in (H_ε), we have, for example, that $e^{1+\xi(u)} \le \sqrt{y}$ whenever x and hence y is sufficiently large. This implies (76).

Corollary 7.6. *Uniformly for $x \geq 2y \geq 5$ we have that*

$$\Phi(x,y) = \frac{e^\gamma\big(x\omega(u) - y\big)}{\zeta(1,y)}\Big\{1 + O\Big(\frac{e^{-u/3}}{\log y}\Big)\Big\}. \tag{77}$$

Proof. The result is (amply) implied by (76) when (x,y) is in (H_ε). In the contrary case, we deduce from (52) and (53) that

$$W(x,y) - \big(x\omega(u) - y\big)\frac{e^\gamma}{\zeta(1,y)} \ll \Psi(x,y) \ll \frac{xe^{-u/3}}{(\log y)^2}$$

where the last estimate follows from Theorem 5.1. This yields the stated result.

Notes

§ **6.3.** A recent result of Hildebrand (1990) states that

$$\omega(u) - e^{-\gamma} = P(u)\{\cos\theta(u) + O(1/u)\} \qquad (u \geq 1) \tag{78}$$

where $P(u)$ is a positive, strictly decreasing function, satisfying

$$P(u) = \rho(u)H\big(u\big)^{-\pi^2/2+o(1)} \qquad (u \to \infty) \tag{79}$$

and $\theta(u)$ is a real function such that

$$\theta'(u) = 2\pi\Big(1 + \frac{1}{\log u}\Big) + O\Big(\frac{\log_2 u}{(\log u)^2}\Big) \qquad (u \to \infty). \tag{80}$$

In particular, this implies that $\omega(u) - e^{-\gamma}$ changes sign infinitely often. If λ_n denotes the nth zero of this quantity, Hildebrand deduces from (80) that

$$\lambda_{n+1} - \lambda_n = 1 - \frac{1}{\log n} + O\Big(\frac{\log_2 n}{(\log n)^2}\Big).$$

His proof rests essentially on the saddle-point method with a "good" choice of the abscissa of integration, such as was described in the proof of Theorem 6.

For a generalisation of these results to any solution of a difference-differential equation of the type

$$uf'(u) + af(u) + bf(u-1) = 0$$

with arbitrary complex numbers a, b, see Hildebrand & Tenenbaum (1993a).

§ **6.4.** By writing out the kth order Taylor–Lagrange formula for $\omega(u)$ (cf. Tenenbaum (1990), Lemma 6) we can generalise formula (52) in the following way.

Lemma. *Let* $\omega_{mj} := \omega^{(j)}(m) - \omega^{(j)}(m-)$ $(j \geq 0, 1 \leq m \leq j+1)$. *For any integer* $k \geq 0$, *we have*

$$\begin{aligned}\mu_y(u)\log y = \sum_{j=0}^{k} \frac{(-1)^j \omega^{(j)}(u)}{(\log y)^j} &+ \sum_{\substack{1\leq m\leq k+1\\ m\leq u}} y^{m-u} \sum_{m-1\leq j\leq k} \frac{(-1)^{j+1}\omega_{mj}}{(\log y)^j} \\ &+ \frac{(-1)^{k+1}}{(\log y)^k} \int_0^\infty \omega^{(k+1)}(u-v) y^{-v}\,\mathrm{d}v.\end{aligned} \tag{81}$$

This allows us to sharpen Corollary 7.5 by providing, in the domain (H_ε), an asymptotic expansion of $\Phi(x,y)$ in powers of $1/\log y$. It is indeed easy to show that, in this region, the integral in (81) is

$$\ll_k \frac{x\rho(u)}{(\log y)^k} H(u)^{-c}.$$

In the same way that (78) and (79) show that Theorem 6 is optimal up to the value of the exponent of $H(u)$, it can be established, in an analogous sense, that Corollary 7.4 is essentially optimal—see Friedlander, Granville, Hildebrand & Maier (1991). In this article the authors study sign changes of $\Phi(x,y)-x/\zeta(1,y)$ in order to deduce oscillation theorems concerning the distribution of prime numbers. The original idea underlying this somewhat surprising connection is due to Maier (1985). See also Hildebrand & Maier (1989) and Friedlander & Granville (1989).

Theorem 7 is far more precise than the classical result of de Bruijn (1950)

$$\Phi(x,y) = W(x,y) + O\Big(\frac{x}{L_\varepsilon(y)}\Big) \qquad (x \geq y \geq 2). \tag{82}$$

This is clear when (x, y) belongs to (H_ε). In the contrary case, the error term $O\big(\Psi(x,y)\big)$ of Theorem 7 is certainly

$$\ll x \exp\{-L_\varepsilon(y)\}.$$

It can be shown that Lemma 7.2 holds for $x \geq y \geq \log x$, subject to replacing α_0 by the actual solution $\alpha(x,y)$ to equation (5)—cf. Hildebrand & Tenenbaum (1986), lemma 8, and Tenenbaum (1990), corollary to lemma 1.

Exercises

1. (a) Establish the following identity for $x \geq y \geq 2$:

$$[x] = \sum_{n \leq x/y,\, P^+(n) \leq y} \Phi(x/n, y) + \Psi(x, y) - \Psi(x/y, y).$$

(b) Using Theorems 5.6 and 6.3, deduce from (a) the convolution identity

$$\rho(u) + \int_0^u \rho(v)\omega(u-v)\,\mathrm{d}v = 1 \qquad (u \geq 0).$$

Hence recover (34).

In the following exercises, we let $\Psi_k(x, y, z)$ *denote the number of integers* $n \leq x$ *satisfying*

$$\Omega(n; y, z) := \sum_{z < p \leq y,\, p^\nu \| n} \nu = k.$$

We shall also consistently set $x = y^u$, $z = y^\lambda$.

2. For $0 \leq \lambda \leq 1$, $u \geq 0$, set

$$\theta_0(\lambda, u) := \rho(u/\lambda) + \int_0^u \rho(v/\lambda)\,\omega(u-v)\,\mathrm{d}v$$

with the convention $\theta_0(0, u) = 0$.

(a) Show that, for $u \geq 1$, we have

$$\theta_0(\lambda, u) \leq \rho(u/\lambda) + \lambda \int_0^{(u-1)/\lambda} \rho(v)\,\mathrm{d}v$$

and deduce that $\theta_0(\lambda, u) \leq e^\gamma \lambda$ for $u \geq 1$.

(b) Establish the bounds

$$\tfrac{1}{2}\lambda \leq \theta_0(\lambda, u) \leq e^\gamma \lambda \qquad (0 \leq \lambda \leq 1,\, u \geq 1 + \lambda).$$

(c) Show that

$$\lim_{u \to \infty} \theta_0(\lambda, u) = \lambda.$$

3. (a) Establish Buchstab's identity

$$\Psi_0(x,y,z) = \Psi(x,z) + \sum_{y<p\le x} \Psi_0(x/p, p-1, z).$$

(b) By using, in addition to other results, that of the previous exercise, show that, uniformly for $y \ge z \ge 2$, $x \ge yz$,

$$\Psi_0(x,y,z) = x\theta_0(\lambda,u)\{1 + O(1/\log z)\}.$$

4. (a) Show that

$$\lambda\frac{\partial\theta_0}{\partial\lambda}(\lambda,u) = \theta_0(\lambda, u-\lambda) \qquad (0 \le \lambda \le 1,\, u \ge 1).$$

(b) By using the results of Exercise 3 (a) & (b), show that, for $0 \le \lambda \le 1$, $u \ge 1$,

$$u\frac{\partial\theta_0}{\partial u}(\lambda,u) = \theta_0(\lambda,u-1) - \theta_0(\lambda,u-\lambda).$$

(c) Show the existence of some absolute constant A such that

$$\left|\frac{\partial\theta_0}{\partial u}(\lambda,u)\right| \le A\lambda\rho(u) \quad (3 \le u \le 4)$$

and deduce that

$$\theta_0(\lambda,u) = \lambda\{1 + O(\rho(u))\} \qquad (u \ge 1).$$

5. Define inductively a sequence of functions $\theta_k(\lambda,u)$ by the formula

$$k\,\theta_k(\lambda,u) = \int_\lambda^1 \theta_{k-1}(\lambda,u-v)\frac{\mathrm{d}v}{v} \qquad (k \ge 1).$$

By using the results for Exercises 2 and 4, show that, for $k \ge 0$, $0 \le \lambda \le 1$, $u \ge k+1+\lambda$,

$$\text{(a)} \qquad \tfrac{1}{2}\frac{\lambda}{k!}\Big(\log\frac{1}{\lambda}\Big)^k \le \theta_k(\lambda,u) \le e^\gamma\frac{\lambda}{k!}\Big(\log\frac{1}{\lambda}\Big)^k,$$

$$\text{(b)} \qquad \theta_k(\lambda,u) = \frac{\lambda}{k!}\Big(\log\frac{1}{\lambda}\Big)^k\{1 + O(\rho(u-k))\}.$$

6. Show, uniformly under the conditions $y \ge z\big(1 + 1/L_\varepsilon(z)\big)$, $zy^{k+1} \le x$, $k \le \log(1/\lambda)L_\varepsilon(z)$, that

$$\Psi_k(x,y,z) = x\theta_k(\lambda,u)\{1 + O(1/\log z)\}.$$

7. (a) Show that $\theta_k(\lambda,u) = 0$ if $u < k\lambda$.

(b) Deduce, for $0 < \lambda \le 1$, $u > 0$, that $\sum_{k=0}^\infty \theta_k(\lambda,u) = 1$.

(c) Write $\kappa(\lambda,u) := 1/u \quad (\lambda \le u \le 1)$, $\kappa(\lambda,u) := 0$ otherwise. Show that

$$k!\theta_k(\lambda,u) = \kappa^{*k} * \theta_0(\lambda,u)$$

where the convolutions are carried out with respect to the variable u.

(d) Calculate the Laplace transform $\widehat{\kappa}_\lambda(s) := \int_0^\infty e^{-us}\kappa(\lambda, u)\,du$ and deduce, by part (b), that

$$\widehat{\theta}_0(\lambda, s) := \int_0^\infty e^{-us}\theta_0(\lambda, u)\,\mathrm{d}u = \frac{1}{s}\exp\Big\{-\int_{\lambda s}^{s} e^{-v}\frac{\mathrm{d}v}{v}\Big\}.$$

(e) Recover the above result by using the definition of $\theta_0(\lambda, u)$ given in Exercise 2.

8. Let $Y(u)$ denote the Heaviside function $Y(u) = \begin{cases} 1 & (u \geq 0) \\ 0 & (u < 0) \end{cases}$.

(a) Show, for $0 \leq \lambda \leq 1$, $u \geq 0$, that

$$\theta_0(\lambda, u) = \sum_{k=0}^{\infty} \frac{(-1)^k}{k!}(\kappa^{*k} * Y)(\lambda, u)$$

and deduce, for all $k \geq 0$, that

$$\theta_k(\lambda, u) = \frac{1}{k!}\sum_{j=0}^{\infty} \frac{(-1)^j}{j!} F_{k+j}(\lambda, u)$$

with

$$F_0(\lambda, u) = 1, \quad F_k(\lambda, u) = \int_{\substack{[\lambda,1]^k \\ \sum v_i \leq u}} \frac{\mathrm{d}v_1}{v_1}\cdots\frac{\mathrm{d}v_k}{v_k}.$$

(b) Recover the differential equation 4(a).

9. Let $\widehat{\rho}_\lambda(s) := \int_0^\infty e^{-us}\rho(u/\lambda)\,du$. Show, for $s \neq 0$, that

$$s\widehat{\rho}(s)\widehat{\theta}_0(\lambda, s) = \widehat{\rho}_\lambda(s) = \lambda\widehat{\rho}(\lambda s)$$

and deduce the validity of the convolution formula

$$\int_0^u \theta_0(\lambda, v)\rho(u - v)\,\mathrm{d}v = \lambda\int_0^{u/\lambda} \rho(v)\,\mathrm{d}v.$$

10. (a) Show, for all $\xi \in \mathbb{C}$, that

$$\sum_{k=0}^{\infty} \theta_k(\lambda, u)\xi^k = \sum_{k=0}^{\infty} \frac{1}{k!}F_k(\lambda, u)(\xi - 1)^k$$

where the functions $F_k(\lambda, u)$ are as defined in Exercise 8.

(b) Deduce, for all $\xi \geq 1$, that

$$\sum_{k \geq \xi\log(1/\lambda)} \theta_k(\lambda, u) \leq \lambda^{Q(\xi)} \qquad (Q(\xi) := \xi\log\xi - \xi + 1).$$

11. Let $\widehat{\theta}_k(\lambda, s) := \int_0^\infty e^{-us}\theta_k(\lambda, u)\,du$.

(a) —By using the formula established in Exercise 7, write down the expansion of $\widehat{\theta}_0(\lambda, s)$ at $s = 0$ of order 1.

(b) Show that

$$\int_0^\infty \{\theta_0(\lambda, u) - \lambda\}\,du = \lambda(1-\lambda).$$

(c) Calculate $\widehat{\theta}_k(\lambda, s)$, and write down the expansion at $s = 0$ of order 1.

(d) Show, for $k \geq 1$, that

$$\int_0^\infty \left\{\theta_k(\lambda, u) - \frac{\lambda}{k!}\Big(\log\frac{1}{\lambda}\Big)^k\right\}du = \frac{\lambda(1-\lambda)}{k!}\Big(\log\frac{1}{\lambda}\Big)^{k-1}\Big\{\log\frac{1}{\lambda} - k\Big\}.$$

Bibliography

K. Alladi,

1982. The Turán–Kubilius inequality for integers without large prime factor, *J. reine angew. Math.* **335**, 180–196.

1988. Probabilistic Number Theory and Brun's sieve, in: C. Goldstein (ed.), *Séminaire de Théorie des Nombres, Paris 1986-87*, Prog. Math. 75 (Birkhäuser), 1–26.

K. Alladi & P. Erdős,

1977. On an additive arithmetic function, *Pacific J. Math.* **71**, no. 2, 275–294.

1979. On the asymptotic behavior of large prime factors of integers, *Pacific J. Math.* **82**, no. 2, 295–315.

E. Aparicio Bernardo,

1981. Sobre unas sistemas de números algebraicos de D. S. Gorshkov y sus aplicaciones al cálculo, *Revista Matematica Hispano-Americana* **41**, 3–17.

R. Ayoub,

1963. *An introduction to the analytic theory of numbers*, AMS Math. Surveys 10 (Providence).

G.J. Babu,

1973. Some results on the distribution of additive arithmetic functions, II, *Acta Arith.* **23**, 315–328.

1992. Smoothness of the distributions of arithmetic functions, in: F. Schweiger & E. Manstavičius (eds.), *Analytic and probabilistic methods in number theory*, New Trends in Probab. and Statist. 2, 191-199, VSP/TEV.

C.G. Bachet, sieur de Meziriac,

1624. *Problèmes plaisans et délectables qui se font par les nombres*, second edition; first edition: 1612.

M. Balazard,

1987. *Sur la répartition des valeurs de certaines fonctions arithmétiques*, Thèse, Université de Limoges.

1990. Unimodalité de la distribution du nombre de diviseurs premiers d'un entier, *Ann. Inst. Fourier* (Grenoble), **40**, no. 2, 255–270.

M. Balazard, H. Delange & J.-L. Nicolas,

1988. Sur le nombre de facteurs premiers des entiers, *C. R. Acad. Sci. Paris,* **306**, Série I, 511–514.

M. Balazard & A. Smati,

1990. Elementary proof of a theorem of Bateman, in: B. Berndt, H. Diamond, H. Halberstam & A. Hildebrand (eds.), *Analytic Number Theory* (Urbana, 1989), Prog. Math. 85 (Birkhäuser), 41–46.

P. Bateman,

1972. The distribution of values of the Euler function, *Acta Arith.* **21**, 329–345.

F. Behrend,

1935. On sequences of numbers not divisible one by another, *J. London Math. Soc.* **10**, 42–44.

A.C. Berry,

1941. The accuracy of the Gaussian approximation to the sum of independent variates, *Trans. Amer. Math. Soc.* **49**, 122–136.

A.S. Besicovitch,

1934. On the density of certain sequences, *Math. Annalen* **110**, 336–341.

N.H. Bingham, C.M. Goldie & J.L. Teugels,

1987. *Regular variation*, Cambridge University Press.

A. Blanchard,

1969. *Initiation à la théorie analytique des nombres premiers*, Dunod, Paris.

H. Bohr,

1910. *Bidrag til de Dirichlet'ske Raekkers Theori*, Thesis, Copenhagen (= Collected Mathematical Works III, S1).

1935. Ein allgemeiner Satz über die Integration eines trigonometrischen Polynoms, *Prace Matem. Fiz.*, 273–288 (= Collected Mathematical Works II, C 36).

E. Bombieri,

1965. On the large sieve, *Mathematika* **12**, 201–225.

1974. Le grand crible dans la théorie analytique des nombres, *Astérisque* **18** (Société mathématique de France).

E. Bombieri & H. Davenport,

1966. Small differences between prime numbers, *Proc. Roy. Soc.* Ser. A **293**, 1–18.

E. Bombieri & H. Iwaniec,

1986. On the order of $\zeta(\frac{1}{2}+it)$, *Ann. Sc. Norm. Sup. Pisa Cl. Sc.* (4) **13**, 449–472.

J.D. Bovey,

1977. On the size of prime factors of integers, *Acta Arith.* **33**, 65–80.

N.G. de Bruijn,

1950. On the number of uncancelled elements in the sieve of Eratosthenes, *Proc. Kon. Ned. Akad. Wetensch.* **53**, 803–812.

1951a. The asymptotic behaviour of a function occurring in the theory of primes, *J. Indian Math. Soc.* (N.S.) **15**, 25–32.

1951b. On the number of positive integers $\leq x$ and free of prime factors $> y$, *Nederl. Akad. Wetensch. Proc. Ser.* A **54**, 50–60.

1966. On the number of positive integers $\leq x$ and free of prime factors $> y$, II, *Nederl. Akad. Wetensch. Proc. Ser.* A **69**, 239–247 = *Indag. Math.* **28**, 239–247.

1970. *Asymptotic methods in Analysis*, North Holland (Amsterdam), third edition; new printing: Dover (New York), 1981.

N.G. de Bruijn, C. van Ebbenhorst Tengbergen & D. Kruyswijk,

1949–51. On the set of divisors of a number, *Nieuw Arch. f. Wisk. Ser. II* **23**, 191–193.

V. Brun,

1917. Sur les nombres premiers de la forme $ap+b$, *Arkiv. for Math. og Naturvid.* B **34**, no. 14, 9 pp.

1919a. Le crible d'Eratosthène et le théorème de Goldbach, *C. R. Acad. Sci. Paris* **168**, 544–546.

1919b. La série $1/5+1/7+1/11+1/13+1/17+1/19+1/29+1/31+1/41+1/43+1/59+1/61+\ldots$ où les dénominateurs sont nombres premiers jumeaux est convergente ou finie, *Bull. Sci. Math.* (2) **43** 100–104; 124–128.

1922. Das Sieb des Eratosthenes, *5 Skand. Mat. Kongr., Helsingfors* 197–203.

1924. Untersuchungen über das Siebverfahren des Eratosthenes, *Jahresber. Deutsch. Math.-Verein.* **33**, 81–96.

1967. Reflections on the sieve of Eratosthenes, *Norske Vid. Selsk. Skr.* (Trondheim) no. 1, 9 pp.

A.A. Buchstab,

1937. An asymptotic estimation of a general number-theoretic function, *Mat. Sbornik* (2) **44**, 1239–1246.

D.A. Burgess,

1962. On character sums and $\mathcal{L}$-series, I, *Proc. London Math. Soc.* **12**, 193–206.

1963. On character sums and $\mathcal{L}$-series, II, *Proc. London Math. Soc.* **13**, 524-536.

E. Cahen,

1984. Sur la fonction $\zeta(s)$ de Riemann et sur des fonctions analogues, *Ann. Ec. Norm.* (Ser. 3) **11**, 75–164.

H. Cartan,

1961. *Théorie élémentaire des fonctions analytiques d'une ou plusieurs variables complexes*, Hermann (Paris).

E.D. Cashwell & C.J. Everett,

1959. The ring of number theoretic functions, *Pacific J. Math.* **9**, 975–985.

J.-R. Chen,

1983. The exceptional set of Goldbach numbers (II), *Sci. Sinica* **26**, 714–731.

E. Cohen,

1960. Arithmetical functions associated with the unitary divisors of an integer, *Math. Z.* **74**, 66–80.

1964. Some asymptotic formulas in the theory of numbers, *Trans. Amer. Math. Soc.* **112**, 214–227.

J.B. Conrey,

1989. More than two fifths of the zeros of the Riemann zeta function are on the critical line, *J. reine angew. Math.* **399**, 1–26.

J.G. van der Corput,

1922. Verschärferung der Abschätzung beim Teilerproblem, *Math. Annalen* **87**, 39–65.

1923. Neue zahlentheoretische Abschätzungen, erste Mitteilung, *Math. Annalen* **89**, 215–254.

1928a. Zum Teilerproblem, *Math. Annalen* **98**, 697–716.

1928b. Zahlentheoretische Abschätzungen, mit Anwendung auf Gitterpunktprobleme, *Math. Z.* **28**, 301–310.

1929. Neue Zahlentheoretische Abschätzungen, zweite Mitteilung, *Math. Z.* **29**, 397–426.

1936–37. Über Weylsche Summen, *Mathematica B*, 1–30.

H. Cramér,

1970. *Random variables and probability distributions*, third edition, Cambridge Tracts in Mathematics and Mathematical Physics no. 36, Cambridge.

H. Daboussi,

1974. See: Daboussi & Delange (1974).

1979. On the density of direct factors of the set of positive integers, *J. London Math. Soc.* (2) **19**, 21–24.

1981. On the limiting distribution of non-negative additive functions, *Compositio Math.* **43**, 101–105.

1982. La méthode de convolution, *Théorie Elémentaire et Analytique des Nombres* (J. Coquet ed.), Journées SMF–CNRS de Valenciennes, Dép. Math. Univ. Valenciennes, 26–29.

1984. Sur le théorème des nombres premiers, *C. R. Acad. Sc. Paris* **298**, Série I, no. 8, 161–164.

1989. On a convolution method, in: E. Aparicio, C. Calderón & J.C. Peral (eds.), *Congreso de Teoria de los Números (Universitad del País Vasco)*, 110–137.

H. Daboussi & H. Delange,

1974. Quelques propriétés des fonctions multiplicatives de module au plus égal à 1, *C. R. Acad. Sci. Paris Sér.* A **278**, 657–660.

1982. On multiplicative arithmetical functions whose modulus does not exceed one, *J. London Math. Soc.* (2) **26**, 245–264.

1985. On a class of multiplicative functions, *Acta Sci. Math.* **49**, 143–149.

H. Davenport,

1980. *Multiplicative Number Theory* (second edition), Springer, New York, Heidelberg, Berlin.

H. Davenport & P. Erdős,

1937. On sequences of positive integers, *Acta Arith.* **2**, 147–151.

1951. On sequences of positive integers, *J. Indian Math. Soc.* **15**, 19–24.

H. Davenport & H. Halberstam,

1966. The values of a trigonometric polynomial at well spaced points, *Mathematika* **13**, 91–96. Corrigendum and Addendum, *Mathematika* **14** (1967), 229–232.

H. Delange,

1954. Généralisation du Théorème de Ikehara, *Ann. Sci. Ec. Norm. Sup.* (3) **71**, fasc. 3, 213–242.

1959. Sur des formules dues à Atle Selberg, *Bull. Sc. Math.* 2° série **83**, 101–111.

1961. Sur les fonctions arithmétiques multiplicatives, *Ann. Scient. Ec. Norm. Sup.*, 3° série, **78**, 273–304.

1971. Sur des formules de Atle Selberg, *Acta Arith.* **19**, 105–146.

1982. Théorie probabiliste des nombres, in: *Journées Arithmétiques*, Metz 1981, *Astérisque* **94**, 31–42.

J.-M. Deshouillers, F. Dress & G. Tenenbaum,

1979. Lois de répartition des diviseurs, 1, *Acta Arith.* **23**, 273–285.

H.G. Diamond,

1982. Elementary methods in the study of the distribution of prime numbers, *Bulletin (N.S.) of the Amer. Math. Soc.* **7**, 553–589.

1984. A number theoretic series of I. Kasara, *Pacific J. Math.* **111**, no. 2, 283–285.

1988. Review no. 88a: 40006, *Mathematical Reviews.*

H.G. Diamond & H. Halberstam,

1985. The Combinatorial sieve, in: K. Alladi (ed.), *Number Theory*, Proc. 4th Matsci. Conf. Ootacamund, India, 1984, *Springer Lecture Notes* 1122, 63–73.

K. Dickman,

1930. On the frequency of numbers containing prime factors of a certain relative magnitude, *Ark. Math. Astr. Fys.* **22**, 1–14.

P.G.L. Dirichlet,

1837. Beweis des Satzes, daß jede unbegrenzte arithmetische Progression, deren erstes Glied und Differenz ganze Zahlen ohne gemeinschaftlichen Factor sind, unendlich viele Primzahlen enthält, *Abh. Akad. Berlin* (1837 math. Abh. 45–71).

F. Dress,

1983–84. Théorèmes d'oscillation et fonction de Möbius, *Séminaire de Théorie des nombres (Bordeaux),* exposé no. 33 (33 pp.).

F. Dress, H. Iwaniec & G. Tenenbaum,

1983. Sur une somme liée à la fonction de Möbius, *J. reine angew. Math.* **340**, 53–58.

Y. Dupain, R.R. Hall & G. Tenenbaum,

1982. Sur l'équirépartition modulo 1 de certaines fonctions de diviseurs, *J. London Math. Soc.* (2) **26**, 397–411.

P.D.T.A. Elliott,

1979. *Probabilistic number theory: mean value theorems*, Grundlehren der Math. Wiss. 239, Springer-Verlag, New York, Berlin, Heidelberg.

1980. *Probabilistic number theory: central limit theorems*, Grundlehren der Math. Wiss. 240, Springer-Verlag, New York, Berlin, Heidelberg.

1985. *Arithmetic functions and integer products*, Grundlehren der Math. Wiss. 272, Springer-Verlag, Berlin, New York, Tokyo.

P.D.T.A. Elliott & C. Ryavec,

1971. The distribution of values of additive arithmetical functions, *Acta Math.* **126**, 143–164.

V. Ennola,

1969. On numbers with small prime divisors, *Ann. Acad. Sci. Fenn. Ser.* AI **440**, 16 pp.

W.J. Ellison & M. Mendès France,

1975. *Les nombres premiers*, Hermann, Paris.

P. Erdős,

1935. On the normal number of prime factors of $p-1$ and some related problems concerning Euler's φ-function, *Quart. J. Math.* (Oxford) **6**, 205–213.

1935/37/38. On the density of some sequences of numbers I, *J. London Math. Soc.* **10**, 120–125; II, *ibid.* **12**, 7–11; III, *ibid.* **13**, 119–127.

1939. On the smoothness of the asymptotic distribution of additive arithmetical functions, *Amer. J. Math.* **61**, 722–725.

1946. On the distribution function of additive functions, *Ann. of Math.* **47**, 1–20.

1949. On a new method in elementary number theory which leads to an elementary proof of the prime number theorem, *Proc. Nat. Acad. Sci.* (Washington) **35**, 374–384.

1969. On the distribution of prime divisors, *Aequationes Math.* **2**, 177–183.

1979. Some unconventional problems in number theory, *Astérisque* **61**, 73–82.

P. Erdős & R.R. Hall,

1974. On the distribution of values of certain divisor functions, *J. Number Theory* **6**, 52–63.

P. Erdős, R.R. Hall & G. Tenenbaum,

1994. On the densities of sets of multiples, *J. reine angew. Math.* **454**, 119-141.

P. Erdős & M. Kac,

1939. On the Gaussian law of errors in the theory of additive functions, *Proc. Nat. Acad. Sci. U.S.A.* **25**, 206–207.

1940. The Gaussian law of errors in the theory of additive number theoretic functions, *Amer. J. Math.* **62**, 738–742.

P. Erdős & J.-L. Nicolas,

1981a. Sur la fonction : nombre de facteurs premiers de N, *Ens. Math.* **27**, 3–27.

1981b. Grandes valeurs d'une fonction liée aux produits d'entiers consécutifs, *Ann. Fa. Sci. Toulouse* **3**, 173–199.

1989. Grandes valeurs de fonctions liées aux diviseurs premiers consécutifs d'un entier, in: J.-M. De Koninck & C. Levesque (eds.), *Théorie des nombres/Number Theory*, W. de Gruyter, 169–200.

P. Erdős, B. Saffari & R.C. Vaughan,

1979. On the asymptotic density of sets of integers II, *J. London Math. Soc.* (2) **19**, 17–20.

P. Erdős & A. Sárközy,

1994. On isolated, respectively consecutive large values of arithmetic functions, *Acta Arith.* **66**, 269–295.

P. Erdős, A. Sárközy & E. Szemerédi,

1967. On an extremal problem concerning primitive sequences, *J. London Math. Soc.* **42**, 484–488.

P. Erdős & H.N. Shapiro,

1951. On the changes of signs of a certain error function, *Canad. J. Math.* **3**, 375–385.

P. Erdős & G. Tenenbaum,

1989a. Sur les densités de certaines suites d'entiers, *Proc. London Math. Soc.* (3) **59**, 417–438.

1989b. Sur les fonctions arithmétiques liées aux diviseurs consécutifs, *J. Number Theory* **31**, 285–311.

P. Erdős & A. Wintner,

1939. Additive arithmetical functions and statistical independence, *Amer. J. Math.* **61**, 713–721.

C.G. Esseen,

1942. On the Liapounoff limit of error in the theory of probability, *Ark. Mat. Astr. Fysik* **28** A, 1–19.

1945. Fourier analysis of distribution functions. A mathematical study of the Laplace–Gaussian law, *Acta Math.* **77**, nos. 1–2, 1–125.

1966. On the Kolmogorov–Rogozin inequality for the concentration function, *Z. Wahrscheinlichkeitstheorie verw. Geb.* **5**, 210–216.

W.J. Feller,

1970. *An introduction to probability theory and its applications*, vol. 1, John Wiley (1st ed. 1950).

1971. *An introduction to probability theory and its applications*, vol. 2, John Wiley (1st ed. 1966).

E. Fouvry & F. Grupp,

1986. On the switching principle in sieve theory, *J. reine angew. Math.* **370**, 101–125.

E. Fouvry & G. Tenenbaum,

1991. Entiers sans grand facteur premier en progressions arithmétiques, *Proc. London Math. Soc.* (3) **63**, 449–494.

G. Freud,

1952. Restglied eines Tauberschen Sätzes I, *Acta Math. Acad. Sci. Hung.* **2**, 299–308.

1953. Restglied eines Tauberschen Sätzes II, *Acta Math. Acad. Sci. Hung.* **3**, 299–307.

1954. Restglied eines Tauberschen Sätzes III, *Acta Math. Acad. Sci. Hung.* **5**, 275–289.

G. Freud & T. Ganelius,

1957. Some remarks on one sided approximation, *Math. Scand.* **5**, 276–284.

J.B. Friedlander & A. Granville,

1989. Limitations to the equidistribution of primes I, *Ann. Math.* **129**, 363–382.

J.B. Friedlander, A. Granville, A. Hildebrand & H. Maier,

1991. Oscillation theorems for primes in arithmetic progressions and for sifting functions, *J. Amer. Math. Soc.* **4**, 25–86.

J. Galambos,

1970. Distribution of arithmetical functions. A survey, *Ann. Inst. H. Poincaré Sect. B (N.S.)* **6**, 281–305.

1971. On the distribution of strongly multiplicative function, *Bull. London Math. Soc.* **3**, 307–312.

1976. The sequences of prime divisors of integers, *Acta Arith.* **31**, 213–218.

J. Galambos & P. Szüsz,

1986. On the distribution of multiplicative arithmetical functions, *Acta Arith.* **47**, 57–62.

P.X. Gallagher,

1967. The large sieve, *Mathematika* **14**, 14–20.

T. Ganelius,

1971. *Tauberian Remainder Theorems*, Springer Lecture Notes 232.

A. Gelfond,

1946. Comments to the papers 'On the determination of the number of prime numbers not exceeding a given quantity' and 'On prime numbers' (in Russian), in: *Collected works of P. L. Chebyshev*, vol. 1, Moscow-Leningrad, 285–288.

A. Gelfond & Y. Linnik,

1965. *Méthodes élémentaires dans la théorie analytique des nombres*, Gauthier-Villars, Paris.

L. S. Gorshkov,

1956. On the deviation of polynomials with rational integer coefficients from zero on the interval [0, 1] (in Russian), in: *Proceedings of the 3rd All-union congress of Soviet mathematicians*, vol. 3, 5–7, Moscow.

S.W. Graham,

1981a. The distribution of squarefree numbers, *J. London Math. Soc.* (2) **24**, 54–64.

1981b. On Linnik's constant, *Acta Arith.* **39**, 163–179.

S.W. Graham & G. Kolesnik,

1991. *Van der Corput's method of exponential sums*, London Math. Soc. Lecture Notes 126, Cambridge University Press.

S.W. Graham & J.D. Vaaler,

1981. A class of extremal functions for the Fourier transform, *Trans. Amer. Math. Soc.* **265**, 283–302.

1984. Extremal functions for the Fourier transform and the large sieve, in G. Halász (ed.) *Topics in classical number theory, Colloq. Budapest 1981*, vol. I, Colloq. Math. Soc. János Bolyai 34, 599–615.

E. Grosswald,

1982. Oscillation theorems, *Conference on the theory of arithmetic functions*, Springer Lecture Notes 251, 141–168.

G. Halász,

1968. Über die Mittelwerte multiplikativer zahlentheoretischer Funktionen, *Acta Math. Acad. Sci. Hung.* **19**, 365–403.

1971. On the distribution of additive and the mean values of multiplicative arithmetic functions, *Studia Scient. Math. Hung.* **6**, 211–233.

1972. Remarks to my paper 'On the distribution of additive and the mean values of multiplicative arithmetic functions', *Acta Math. Acad. Scient. Hung.* **23**, 425–432.

H. Halberstam & H.-E. Richert,

1974. *Sieve Methods*, Academic Press, London, New York, San Francisco.

1979. On a result of R.R. Hall, *J. Number Theory* (1) **11**, 76–89.

H. Halberstam & K.F. Roth,

1966. *Sequences*, Oxford; second edition, Springer, 1983.

R.R. Hall,

1974. Halving an estimate obtained from Selberg's upper bound method, *Acta Arith.* **25**, 347–351.

1978. A new definition of the density of an integer sequence, *J. Austral. Math. Soc.* Ser. A **26**, 487–500.

1981. The divisor density of integer sequences, *J. London Math. Soc.* (2) **24**, 41–53.

R.R. Hall & G. Tenenbaum,

1982. On the average and normal orders of Hooley's Δ-function, *J. London Math. Soc.* (2) **25**, 392-406.

1986. Les ensembles de multiples et la densité divisorielle, *J. Number Theory* **22**, 308–333.

1988. *Divisors*, Cambridge Tracts in Mathematics 90, Cambridge.

1991. Efffective mean value estimates for complex multiplicative functions, *Math. Proc. Camb. Phil. Soc.* **110**, 337–351.

D. Hanson,

1972. On the product of primes, *Canad. Math. Bull.* **15**, 33–37.

G.H. Hardy,

1914. Sur les zéros de la fonction $\zeta(s)$ de Riemann, *C. R. Acad. Sci. Paris* **158**, 1012–1014.

1916a. On Dirichlet's divisor problem, *Proc. London Math. Soc.* **15**, (2) 1–25.

1916b. The average orders of the functions $P(x)$ and $\Delta(x)$, *Proc. London Math. Soc.* **15** (2), 192–213.

1949. *Divergent Series*, Oxford at the Clarendon Press.

G.H. Hardy & J.E. Littlewood,

1913. Contributions to the arithmetic theory of series, *Proc. London Math. Soc.* (2) **11**, 411–478.

1921. The zeros of Riemann's zeta function on the critical line, *Math. Z.* **10**, 283–317.

1922. Some problems of partitio numerorum III: On the expression of a number as a sum of primes, *Acta Math.* **44**, 1–70.

1923. The approximate functional equation in the theory of the zeta function, with applications to the divisor problems of Dirichlet and Piltz, *Proc. London Math. Soc.* (2), **21**, 39–74.

1929. The approximate functional equations for $\zeta(s)$ and $\zeta(s)^2$, *Proc. London Math. Soc.* (2) **29**, 81–97.

G.H. Hardy & S. Ramanujan,

1917. The normal number of prime factors of a number n, *Quart. J. Math.* **48**, 76–92.

G.H. Hardy & M. Riesz,

1915. *The general theory of Dirichlet series*, Cambridge Tracts in Mathematics and Mathematical Physics no. 18, Cambridge University Press; 2nd impression: 1952.

G.H. Hardy & E.M. Wright,

1938. *An introduction to the theory of numbers*, Oxford (5th ed. 1979).

D.R. Heath-Brown,

1992. Zero-free regions for Dirichlet L-functions, and the least prime in an arithmetic progression, *Proc. London Math. Soc.* (3) **64**, no. 2, 265–338.

W. Hengartner & R. Theodorescu,

1973. *Concentration functions*, Academic Press, New York, London.

D. Hensley,

1986. The convolution powers of the Dickman function, *J. London Math. Soc.* (2) **33**, 395–406.

1987. The distribution of round numbers, *Proc. London Math. Soc.* (3) **54**, 412–444.

E. Heppner,

1974. Über die Iteration von Teilerfunktionen, *J. reine angew. Math.* **265**, 176–182.

A. Hildebrand,

1983. An asymptotic formula for the variance of an additive function, *Math. Z.* **183**, 145–170.

1984a. Integers free of large prime factors and the Riemann hypothesis, *Mathematika* **31**, 258–271.

1984b. Quantitative mean value theorems for non-negative multiplicative functions I, *J. London Math. Soc.* (2) **30**, 394–406.

1984c. Fonctions multiplicatives et équations intégrales, Séminaire de Théorie des Nombres, Paris 1982–83, *Prog. Math.* **51**, 115–124.

1986a. On the number of positive integers $\leq x$ and free of prime factors $> y$, *J. Number Theory* **22**, 289–307.

1986b. The prime number theorem via the large sieve, *Mathematika* **33**, 23–30.

1986c. On Wirsing's mean value theorem for multiplicative functions, *Bull. London Math. Soc.* **18**, 147–152.

1986d. A note on Burgess' character sum estimate, *C. R. Acad. Sci. Canada* **8**, 35–37.

1986e. Multiplicative functions at consecutive integers, *Math. Proc. Camb. Phil. Soc.* **100**, 229–236.

1986f. On the local behavior of $\Psi(x, y)$, *Trans. Amer. Math. Soc.* **296**, 265–290.

1987a. Multiplicative functions in short intervals, *Canad. J. Math.*

1987b. Quantitative mean value theorems for non negative mutliplicative functions II, *Acta Arith.* **48**, 209–260.

1990. The asymptotic behavior of the solutions of a class of differential-difference equations, *J. London Math. Soc.* (2) **42**, 11–31.

A. Hildebrand & H. Maier,

1989. Irregularities in the distribution of primes in short intervals, *J. reine angew. Math.* **397**, 162–193.

A. Hildebrand & G. Tenenbaum,

1986. On integers free of large prime factors, *Trans. Amer. Math. Soc.* **296**, 265–290.

1988. On the number of prime factors of an integer, *Duke Math. J.* **56**, no. 3 (1988), 471–501.

1993a. On a class of differential-difference equations arising in number theory, *J. d'Analyse* **61**, 145–179.

1993b. Integers without large prime factors, *J. Théorie des Nombres de Bordeaux* **5**, 411-484.

C. Hooley,

1979. A new technique and its applications to the theory of numbers, *Proc. London Math. Soc.* (3) **38**, 115–151.

L. Hörmander,

1954. A new proof and a generalization of an inequality of Bohr, *Math. Scand.* **2**, 33-45.

M.N. Huxley,

1972a. On the difference between consecutive primes, *Inventiones Math.* **15**, 164–170.

1972b. *The distribution of prime numbers*, Oxford.

1990. Exponential sums and lattice points, *Proc. London Math. Soc.* (3) **60**, 471–502; Corrigenda **66** (1993), 70.

1993a. Exponential sums and lattice points II, *Proc. London Math. Soc.* (3) **66**, 279–301.

1993b. Exponential sums and the Riemann zeta function IV, *Proc. London Math. Soc.* (3) **66**, 1–40.

M.N. Huxley & G. Kolesnik,

1991. Exponential sums and the Riemann zeta function III, *Proc. London Math. Soc.* (3) **62**, 449–468; Corrigenda **66** (1993), 302.

M.N. Huxley & N. Watt,

1988. Exponential sums and the Riemann zeta function, *Proc. London Math. Soc.* (3) **57**, 1–24.

S. Ikehara,

1931. An extension of Landau's theorem in the analytic theory of numbers, *J. Math. and Phys.* (Mass. Inst. of Techn.) **10**, 1–12.

A.E. Ingham,

1930. Notes on Riemann's ζ-function and Dirichlet's L-functions, *J. London Math. Soc.* **5**, 107–112.

1935. On Wiener's method in tauberian theorems, *Proc. London Math. Soc.* (2) **38**, 458–480.

1965. On Tauberian theorems, *Proc. London Math. Soc.* (3) **14A**, 157–173.

A. Ivić,

1985. *The Riemann zeta-function*, John Wiley, New York, Chichester, Brisbane, Toronto, Singapore.

A. Ivić & G. Tenenbaum,

1986. Local densities over integers free of large prime factors, *Quart. J. Math.* (Oxford), (2) **37**, 401–417.

H. Iwaniec,

1980. Rosser's sieve, *Acta Arith.* **36**, 171–202.

1981. Rosser's sieve—Bilinear Forms of the Remainder Terms—Some applications, in: H. Halberstam & C. Hooley (eds.), *Recent Progress in Analytic Number Theory*, Academic Press, London, New York, Toronto, Sydney, San Francisco, vol. 1, 203–230.

H. Iwaniec & C.J. Mozzochi,

1988. On the divisor and circle problems, *J. Number Theory* **29**, 1, 60–93.

B. Jessen & A. Wintner,

1935. Distribution functions and the Riemann Zeta function, *Trans. Amer. Math. Soc.* **38**, 48–88.

J. Kaczorowski & J. Pintz,

1986–7. Oscillatory properties of arithmetical functions, I, *Acta Math. Hung.* **48**, 173–185; II, *ibid.* **49**, 441–453.

J. Karamata,

1931. Neuer Beweis und Verallgemeinerung der Tauberschen Sätze, welche die Laplacesche und Stieltjesche Transformation betreffen, *J. reine angew. Math.* **164**, 27–39.

1952. Review of G. Freud's article (1952), *Zentralblatt für Math.* **44**, 324.

Y. Katznelson,

1968. *An introduction to harmonic analysis*, Dover, New York (second edition, 1976).

G. Kolesnik,

1981. On the estimation of multiple exponential sums, in: H. Halberstam & C. Hooley (eds.), *Recent Progress in Analytic Number Theory* Academic Press, London, New York, Toronto, Sydney, San Francisco, vol. 1, 231–246.

1985. The method of exponent pairs, *Acta Arith.* **45**, 115–143.

A. Kolmogorov,

1956. Two uniform limit theorems for sums of independent random variables, *Theor. Probab. Appl.* **1**, 384–394.

1958. Sur les propriétés des fonctions de concentration de M. P. Lévy, *Ann. Inst. Henri Poincaré* **16**, no. 1, 27–34.

J. Korevaar,

1954. A very general form of Littlewood's theorem, *Indag. Math.* **16**, 36–45.

N.M. Korobov,

1958a. Estimates of trigonometric sums and their applications (in Russian), *Usp. Mat. Nauk* **13**, 185–192.

1958b. Estimates of Weyl sums and the distribution of prime numbers (in Russian), *Dokl. Akad. Nauk SSSR* **123**, 28–31.

1958c. On zeros of $\zeta(s)$ (in Russian), *Dokl. Akad. Nauk SSSR* **118**, 231–232.

J. Kubilius,

1956. Probabilistic methods in the theory of numbers (in Russian), *Uspehi Mat. Nauk* **11** (no. 2), 31–66 = *Amer. Math. Soc. Transl.* **19** (1962), 47–85.

1964. *Probabilistic methods in the theory of numbers*, Amer. Math. Soc. Translations of Math. Monographs, no. 11, Providence.

1983a. Estimation of the central moment for strongly additive arithmetic functions (in Russian, abstracts in English and Lithuanian), *Litovsk. Mat. Sb.* **23** (no. 1), 122–133.

1983b. Estimate of the second central moment for any additive arithmetic functions (in Russian, abstracts in English and Lithuanian), *Litovsk. Mat. Sb.* **23** (no. 2), 110–117.

E. Landau,

1906. Über die Grundlagen der Theorie der Fakultätenreihen, *Sitzungsberichte der mathematische-physikalischen Klasse der Kgl. Bayerischen Akademie der Wissenschaften zu München,* **36**, 151–218.

1909. *Handbuch der Lehre von der Verteilung der Primzahlen* (2 vols.), Teubner, Leipzig; 3rd edition: Chelsea, New York (1974).

1912. Über die Anzahl der Gitterpunkte in gewissen Bereichen, *Göttingen Nachrichten*, 687–771.

1927. *Vorlesungen über Zahlentheorie* (3 vols; Hirzel, Leipzig). New printing by Chelsea, New York, 1969.

1933. Über den Wertervorrat von $\zeta(s)$ in den Halbebene $\sigma > 1$, *Göttingen Nachrichten*, 81–91.

1966. *Elementary Number Theory* (second edition) Chelsea; first edition: Hirzel, Leipzig, 1927.

J. Lee,

1989. *On the constant in the Turán–Kubilius inequality*, PhD Thesis, Univ. Michigan.

W.J. LeVeque,

1949. On the size of certain number theoretic functions, *Trans. Amer. Math. Soc.* **66**, 440–463.

B.V. Levin & N.M. Timofeev,

1971. An analytic method in probabilistic number theory (in Russian), *Vladimir. Gos. Ped. Inst. Učen. Zap.* **38**, 57–150.

N. Levinson,

1974. More than one third of the zeros of Riemann's zeta function are on $\sigma = \frac{1}{2}$, *Adv. Math.* **13**, 383–436.

1975. A simplification of the proof that $N_0(T) > \frac{1}{3}N(T)$ for Riemann's zeta-function, *Adv. Math.* **18**, 239–242.

P. Lévy,

1925. *Calcul des probabilités*, Gauthier-Villars, Paris.

1931. Sur les séries dont les termes sont des variables éventuelles indépendantes, *Studia Math.* **3**, 119–155.

1937. Théorie de l'addition des variables aléatoires, Gauthier-Villars, Paris (second edition, 1954).

Ju. V. Linnik,

1941. The large sieve, *Dokl. Akad. Nauk SSSR* **30**, 292–294 (in Russian).

J.E. Littlewood,

1971. The quickest proof of the prime number theorem, *Acta Arith.* **18**, 83–86.

M. Loève,

1963. *Probability Theory*, 2 vols. Springer; fourth edition 1977.

E. Lukacs,

1970. *Characteristic functions*, second edition, Griffin (London).

H. Maier,

1985. Primes in short intervals, *Michigan Math. J.* **32**, 221–225.

H. Maier & G. Tenenbaum,

1984. On the set of divisors of an integer, *Invent. Math.* **76**, 121–128.

H. von Mangoldt,

1895. Zu Riemann's Abhandlung: Über die Anzahl..., *J. reine angew. Math.* **114**, 255–305.

H.B. Mann,

1942. A proof of the fundamental theorem on the density of sums of sets of positive integers, *Ann. Math.* (2) **43**, 523–527.

M. Mendès France & G. Tenenbaum,

1993. Systèmes de points, diviseurs, et structure fractale, *Bull. Soc. Math. de France* **121**, 197–225.

R.J. Miech,

1969. A number theoretic constant, *Acta Arith.* **15**, 119–137.

H.L. Montgomery,

1968. A note on the large sieve, *J. London Math. Soc.* **43**, 93–98.

1971. *Topics in Multiplicative Number Theory*, Springer Lecture Notes 227, Springer, Berlin, Heidelberg.

1978a. The analytic principle of the large sieve, *Bulletin of the Amer. Math. Soc.* **84**, no. 4, 547–567.

1978b. A note on the mean values of multiplicative functions, *Inst. Mittag-Leffler*, Report no. 17.

1987. Fluctuations in the mean of Euler's phi function, *Proc. Indian Acad. Sci. Math. Sci.* **97**, nos. 1–3, 239–245.

H.L. Montgomery & R.C. Vaughan,

1973. On the large sieve, *Mathematika* **20**, 119–134.

1974. Hilbert's inequality, *J. London Math. Soc.* (2) **8**, 73–81.

1977. Exponential sums with multiplicative coefficients, *Invent. Math.* **43**, 69–82.

1979. Mean values of character sums, *Can. J. Math.* **31**, 476–487.

1981. The distribution of squarefree numbers, in: H. Halberstam & C. Hooley (eds.), *Recent Progress in Analytic Number Theory*, Academic Press; vol. 1, 247–256.

1994. Mean values of multiplicative functions, preprint.

L. Moser & J. Lambek,

1953. On monotone multiplicative functions, *Proc. Amer. Math. Soc.* **4**, 544–545.

M. Naïmi,

1988. Les entiers sans facteur carré $\leq x$ dont les facteurs premiers sont $\leq y$, *Groupe de travail en théorie analytique des nombres 1986-87*, 69-76, *Publ. Math. Orsay* 88-01, *Univ. Paris* XI, *Orsay.*

M. Nair,

1982a. On Chebyshev-type inequalities for primes, *Amer. Math. Monthly,* **89**, no. 2, 126–129.

1982b. A new method in elementary prime number theory, *J. London Math. Soc.* (2) **25**, 385–391.

P. Nanopoulos,

1975. Lois de Dirichlet sur $\mathbb{N}^*$ et pseudo-probabilités, *C. R. Acad. Sci. Paris* **280**, Série A, 1543–1546.

1977. Lois zêta et fonctions arithmétiques additives: loi faible des grand nombres, *C. R. Acad. Sci. Paris* **285**, Série A, 875–877.

1982. Lois zêta et fonctions arithmétiques additives: convergence vers une loi normale, *C. R. Acad. Sci. Paris,* Série I, **295**, 159–161.

D.J. Newman,

1980. Simple analytic proof of the prime number theorem, *Amer. Math. Monthly* **87**, no. 9, 693–696.

J.-L. Nicolas,

1974/75. Grandes valeurs des fonctions arithmétiques, *Sém. de Théorie des Nombres* (Delange-Pisot-Poitou) 16-ième année, no. G20.

1978. Sur les entiers N pour lesquels il y a beaucoup de groupes abéliens d'ordre N, *Ann. Inst. Fourier* **28**, 1–16.

1983a. Petites valeurs de la fonction d'Euler, *J. Number Theory* **17**, 375–388.

1983b. Autour de formules dues à A. Selberg, *Pub. Math. Orsay* 83-04, Colloque Hubert Delange, 122–134.

1984. Sur la distribution des nombres entiers ayant une quantité fixée de nombres premiers, *Acta Arith.* **44**, 191–200.

1988. On highly composite numbers, in: G. E. Andrews, R. A. Askey, B. C. Berndt, K. G. Ramanathan & R. A. Rankin (eds.), *Ramanujan Revisited*, Academic Press, 215–244.

K.K. Norton,

1971. *Numbers with small prime factors and the least kth power non-residue*, Mem. Amer. Math. Soc. no. 106.

1976. On the number of restricted prime factors of an integer I, *Illinois J. Math.* **20**, 681–705.

1978. Estimates for partial sums of the exponential series, *J. of Math. and Applications* **63**, 265–296.

1979. On the number of restricted prime factors of an integer II, *Acta Math.* **143**, 9–38.

1982. On the number of restricted prime factors of an integer III, *Enseign. Math.*, II Sér., **28**, 31–52.

E.V. Novoselov,

1964. A new method in probabilistic number theory (in Russian), *Izv. Akad. Nauk SSSR,* Ser. Mat., **28**, 307–364.

A. Oppenheim,

1926. On an arithmetic function, *J. London Math. Soc.* **1**, 205–211.

1927. On an arithmetic function, II, *J. London Math. Soc.* **2**, 123–130.

R.E.A.C. Paley,

1932. A theorem on characters, *J. London Math. Soc.* **7**, 28–32.

E. Phillips,

1933. The zeta function of Riemann: further developments of van der Corput's method, *Quart. J. Math.* (Oxford) **4**, 209–225.

J. Pintz,

1984. On the remainder term of the prime number formula and the zeros of Riemann's zeta function, in: H. Jager (ed.), *Number Theory, Noordwijkerhout 1983*, Springer Lecture Notes 1068, 186–197.

G. Pólya,

1919. Verschiedene Bemerkungen zur Zahlentheorie, *Jahresber. Deutsch. Math.-Ver.* **28**, 31–40.

C. Pomerance,

1984. On the distribution of round numbers, in: K. Alladi (ed.), *Number theory,* Proc. 4th Matsci. Conf. Ootacamund, India, 1984, Springer Lecture Notes 1122, 173–200.

K. Prachar,

1958. Über die kleinste quadratfrei Zahl einer arithmetischen Reihe, *Monatsch. Math.* **62**, 173–176.

S. Ramanujan,

1915. Highly composite numbers, *Proc. London Math. Soc.* (2) **14**, 347–409.

R. Rankin,

1938. The difference between consecutive prime numbers, *J. London Math. Soc.* **13**, 242–247.

A. Rényi,

1950. On the large sieve of Ju. V. Linnik, *Compositio Math.* **8**, 68–75.

1955. On the density of certain sequences of integers, *Publ. Inst. Math.* (Belgrade) **8**, 157–162.

1965. A new proof of a theorem of Delange, *Publ. Math. Debrecen* **12**, 323–329.

A. Rényi & P. Turán,

1958. On a theorem of Erdős–Kac, *Acta Arith.* **4**, 71–84.

G.J. Rieger,

1972. Über einige arithmetische Summen, *Manuscripta Math.* **7**, 23–34.

1983. On Wiener's method in prime number theory, *Abstracts Amer. Math. Soc.* **4**, no. 5, Abstract 802-10-86, p. 144; II, *ibid.*, no. 5, Abstract 83 T-10-345, p. 392.

B. Riemann,

1859. Über die Anzahl der Primzahlen unter einer gegebenen Grösse, Monatsberichte der Berliner Akademie, 671–680; see: *Œuvres de Riemann*, Albert Blanchard, Paris 1968, 165–176.

B.A. Rogozin,

1961. An estimate for concentration functions, *Theor. Probab. Appl.* **6**, 94–97.

J.B. Rosser & L. Shoenfeld,

1962. Approximate formulas for some functions of prime numbers, *Illinois J. Math.* **6**, 64–94.

1975. Sharper bounds for the Chebyshev functions $\theta(x)$ and $\psi(x)$, *Math. Comp.* **29**, 243–269.

K.F. Roth,

1965. On the large sieves of Linnik and Rényi, *Mathematika* **12**, 1–9.

W. Rudin,

1970. *Real and Complex Analysis*, McGraw-Hill.

I.Z. Ruzsa,

1982. Effective results in probabilistic number theory, in: J. Coquet (ed.), *Théorie élémentaire et analytique des nombres*, Dépt. Math. Univ. Valenciennes, 107–130.

1983. On the variance of additive functions, *Studies in Pure Mathematics*, Mem. of P. Turán, 577–586.

1984. Generalized moments of additive functions, *J. Number Theory* **18**, 27–33.

B. Saffari,

1976. On the asymptotic density of sets of integers, *J. London Math. Soc.* (2) **13**, 475–485.

1979. *Comportement à l'infini de la transformée de Fourier de μ_f pour f (fortement) additive telle que $f(p) = (\log p)^{-\alpha}$, $\alpha > 0$ fixé*, private communication.

E. Saias,

1989. Sur le nombre des entiers sans grand facteur premier, *J. Number Theory* **32**, no. 1, 78–99.

A. Sárközy,

1977a. Some remarks concerning irregularities of distribution of sequences of integers in arithmetic progressions, IV, *Acta Math. Acad. Sci. Hung.* **30**, 155–162.

1977b. Remarks on a paper of G. Halász, *Periodica Math. Hung.* **8**, 135–150.

L.G. Sathe,

1953. On a problem of Hardy and Ramanujan on the distribution of integers having a given number of prime factors, *J. Indian Math. Soc.* **17**, 63–141.

1954. (same title), *J. Indian Math. Soc.* **18**, 27–81.

L.G. Schnirelmann,

1930. Über additive Eigenschaften von Zahlen, *Annals Polyt. Inst. Novocherkassk.* **14**, 3–28; *Math. Annalen* **107** (1933), 649–690.

I.J. Schoenberg,

1928. Über die asymptotische Verteilung reeller Zahlen mod 1, *Math. Z.* **28**, 171–200.
1936. On asymptotic distributions of arithmetical functions, *Trans. Amer. Math. Soc.* **39**, 315–330.

L. Schoenfeld,

1976. Sharper bounds for the Chebyshev functions $\theta(x)$ and $\psi(x)$, II, *Math. Comp.* **30**, 337–360.

A. Selberg,

1942. On the zeros of Riemann's zeta-function on the critical line, *Skr. Norske Vid. Akad. Oslo*, no. 10.
1949. An elementary proof of the prime number theorem, *Ann. Math.* **50**, 305–313.
1954. Note on the paper by L.G. Sathe, *J. Indian Math. Soc.* **18**, 83–87.

H.N. Shapiro,

1950. On the number of primes less than or equal to x, *Proc. Amer. Math. Soc.* **1**, 346–348.
1959. Tauberian theorems and elementary prime number theory, *Comm. Pure Appl. Math.* **12**, 579–610.
1972. On the convolution ring of arithmetic functions, *Comm. Pure Appl. Math.* **25**, 287–336.

C.L. Siegel,

1936. Über die Classenzahl quadratischer Zahlkörper, *Acta Arith.* **1**, 83–86.
1966. Gesammelte Abhandlungen, Springer-Verlag, vol. 1, 406–409.

V. Sitaramaiah & M.V. Subbarao,

1993. The maximal order of certain arithmetic functions, *Indian J. Pure Appl. Math.* **24**(6), 347-355.

H. Smida,

1991. Sur les puissances de convolution de la fonction de Dickman, *Acta Arith.* **59**, 123–143.

A. Smith,

1980. On Shapiro's Tauberian theorem, *Carleton Math. Ser.* **170**, 4p.

A.V. Sokolovskii,

1979. Lower bounds in the "large sieve", *Zap. Naučn. Sem. Leningrad. Otdel. Mat. Inst. Steklov. (LOMI)* **91**, 125–133, 181–182; English translation: *J. Soviet. Math.* **17** (1981), 2166–2173.

E. Sperner,

1928. Ein Satz über Untermengen einer endlichen Menge, *Math. Z.* **27**, 544-548.

H. Squalli,

1985. *Sur la répartition du noyau d'un entier*, Thèse 3-ième cycle, Univ. Nancy I.

C.M. Stein,

1984. *On the Turán–Kubilius inequality*, Technical Report no. 220, Stanford University.

T.J. Stieltjes,

1887. Note sur la mutliplication de deux séries, *Nouvelles Annales de Mathématiques*, sér. 3, **6**, 210–215.

D. Suryanarayana & R.R. Sita Rama Chandra,

1973. The distribution of squarefull integers, *Arkiv. Math.* **11**, 195–201.

P. Szüsz,

1974. Remark to a theorem of P. Erdős, *Acta Arith.* **26**, 97–100.

A. Tauber,

1897. Ein Satz aus der Theorie der unendlichen Reihen, *Monatshefte für Mathematik und Physik,* **3**, 273–277.

G. Tenenbaum,

1982. Sur la densité divisorielle d'une suite d'entiers, *J. Number Theory* **15**, no. 3, 331-346.

1985. Sur la concentration moyenne des diviseurs, *Comment. Math. Helvetici* **60**, 411-428.

1987. Sur un problème extrémal en Arithmétique, *Ann. Inst. Fourier* (Grenoble) **37**, 2, 1–18.

1988. La méthode du col en théorie analytique des nombres, in: C. Goldstein (ed.), *Séminaire de Théorie des Nombres, Paris 1986–87*, Prog. Math. 75 (Birkhäuser), 411–441.

1990. Sur un problème d'Erdős et Alladi, in: C. Goldstein (ed.), *Séminaire de Théorie des Nombres, Paris 1988–89*, Prog. Math. 91 (Birkhäuser), 221–239.

E.C. Titchmarsh,

1939. *The theory of functions*, Oxford University Press (second edition, new printing in 1979).

1951. *The theory of the Riemann zeta-function*, Oxford University Press.

E.C. Titchmarsh & D.R. Heath-Brown,

1986. *The theory of the Riemann zeta-function*, Oxford.

K.C. Tong,

1956. On divisor problems II, III, *Acta Math. Sinica* **6**, 139–152; 515–541.

P. Turán,

1934. On a theorem of Hardy and Ramanujan, *J. London Math. Soc.* **9**, 274–276.

1936. Über einige Verallgemeinerungen eines Satzes von Hardy und Ramanujan, *J. London Math. Soc.* **11**, 125–133.

J.D. Vaaler,

1985. Some extremal functions in Fourier Analysis, *Bull. (N.S.) of the Amer. Math. Soc.* **12**, 183–216.

G. Valiron,

1955. *Théorie des fonctions* (second edition), Masson, Paris.

R.C. Vaughan,

1980. An elementary method in prime number theory, *Acta Arith.* **37**, 111–115.

A.I. Vinogradov,

1965. On the density hypothesis for Dirichlet *L*-functions, *Izv. Akad. Nauk SSSR,* Ser. Math. **29**, 903–934.

1966. Correction to the paper of A.I. Vinogradov 'On the density hypothesis for Dirichlet *L*-functions' , *Izv. Akad. Nauk SSSR,* Ser. Math. **30**, 719–720.

I.M. Vinogradov,

1954. The method of trigonometric sums in the theory of numbers, Interscience.
1958. A new estimate for $\zeta(1+it)$ (in Russian), *Izv. Akad. Nauk SSSR,* Ser. Math. **22**, 161–164.

G. Voronoï,

1903. Sur un problème de calcul des fonctions asymptotiques, *J. reine angew. Math.* **126**, 241–282.

M. Vose,

1984. Integers with consecutive divisors in small ratio, *J. Number Theory* **19**, 233–238.

A. Walfisz,

1963. *Weylsche Exponentialsummen in der Neueren Zahlentheorie*, VEB Deutscher Verlag, Berlin.

R. Warlimont,

1969. On squarefree numbers in arithmetic progressions, *Monatsh. Math.* **73**, 433–448.
1980. Squarefree numbers in arithmetic progressions, *J. London Math. Soc.* (2) **22**, no. 1, 21–24.

N. Watt

1989. Exponential sums and the Riemann zeta function II, *J. London Math. Soc.* (2) **39**, 385–404.

H. Weyl

1916. Über die Gleichverteilung von Zahlen mod. Eins, *Math. Ann.* **77**, 313–404.
1921. Zur Abschätzung von $\zeta(1+ti)$, *Math. Z.* **10**, 88–101.

E.T. Whittaker & G.N. Watson,

1927. *A course of modern analysis* (4th edition), Cambridge University Press (new printing: 1986).

D.V. Widder,

1946. *The Laplace transform*, Princeton University Press, Princeton, New Jersey.
1971. *An introduction to transform theory*, Academic Press, New York and London.

E. Wirsing,

1956. Über die Zahlen, deren Primteiler einer gegeben Menge angehören, *Arch. der Math.* **7**, no. 4, 263–272.
1967. Das asymptotische Verhalten von Summen über multiplikative Funktionen II, *Acta Math. Acad. Sci. Hung.* **18**, 411–467.

J. Wu,

1990. Sur la suite des nombres premiers jumeaux, *Acta Arith.* **55**, 365–394.

Index